core maths for the
biosciences

core maths for the
biosciences

Martin B Reed

Department of Mathematical Sciences
University of Bath

OXFORD
UNIVERSITY PRESS

OXFORD

UNIVERSITY PRESS

Great Clarendon Street, Oxford OX2 6DP

Oxford University Press is a department of the University of Oxford.
It furthers the University's objective of excellence in research, scholarship,
and education by publishing worldwide in

Oxford New York

Auckland Cape Town Dar es Salaam Hong Kong Karachi
Kuala Lumpur Madrid Melbourne Mexico City Nairobi
New Delhi Shanghai Taipei Toronto

With offices in

Argentina Austria Brazil Chile Czech Republic France Greece
Guatemala Hungary Italy Japan Poland Portugal Singapore
South Korea Switzerland Thailand Turkey Ukraine Vietnam

Oxford is a registered trade mark of Oxford University Press
in the UK and in certain other countries

Published in the United States
by Oxford University Press Inc., New York

British Library Cataloguing in Publication Data

Data available

Library of Congress Cataloguing in Publication Data

Data available

Typeset by Techset Composition Ltd, Salisbury, UK
Printed in Great Britain by Ashford Colour Press Ltd

ISBN 978-0-19-921634-5

10 9 8 7 6 5 4 3

For Helena and Marika

Overview of contents

The search for The Ultimate Answer · xxi

Introduction: why and how to use this book · xxii

Acknowledgements · xxv

Technical notes: using the electronic support that accompanies this book · xxvi

PART I **ARITHMETIC, ALGEBRA, AND FUNCTIONS** · 1

  1 Arithmetic and algebra · 3
  2 Units; precision and accuracy · 45
  3 Data tables, graphs, interpolation · 81
  4 Molarity and dilutions · 109
  5 Variables, functions, and equations · 125
  6 Linear functions and curve sketching · 151
  7 Quadratic and polynomial functions · 176
  8 Fitting curves; rational and inverse functions · 220
  9 Periodic functions · 259
  10 Exponential and logarithmic functions · 290
     Historical interlude: finding the roots of polynomials · 331

PART II **CALCULUS AND DIFFERENTIAL EQUATIONS** · 335

  11 Instantaneous rate of change: the derivative · 337
  12 Rules of differentiation · 366
  13 Applications of differentiation · 389
  14 Techniques of integration · 426
  15 The definite integral · 451
  16 Differential equations I · 481
  17 Differential equations II · 509
  18 Extension: dynamical systems · 537

Answers to odd-numbered problems · 563

Appendix: The Greek alphabet · 570

References · 571

Index · 573

Contents in full

The search for The Ultimate Answer   xxi

Introduction: why and how to use this book   xxii

Acknowledgements   xxv

Technical notes: using the electronic support that accompanies this book   xxvi

. . . .

PART I ARITHMETIC, ALGEBRA, AND FUNCTIONS   1

1 Arithmetic and algebra   3

  Case Study A1: Introduction to models of population growth   4

  Case Study B1: Introduction to models of cancer   4

  Case Study C1: Introduction to predator–prey relationships   5

  1.1 Numerical and algebraic expressions   5

  Case Study B2: Angiogenic cancer cells   6

  1.2 The real numbers   7

    1.2.1 Integers and reals   7

    1.2.2 The real line   8

  1.3 Arithmetic operations   9

    1.3.1 Negation   9

    1.3.2 Addition and subtraction   9

    1.3.3 Multiplication and division   10

    1.3.4 Absolute value   13

    1.3.5 Percentages   14

    1.3.6 Basic rules for manipulating equations   15

  1.4 Brackets and the distributive law   16

    1.4.1 How to use brackets   16

    1.4.2 Rule of precedence   18

    1.4.3 The distributive law   18

1.5 Exponents   19
 1.5.1 Definition of exponents   19
 1.5.2 Rules for exponents   20

Case study A2: Formula for geometric growth   22
 1.5.3 Products and factors   22

1.6 Roots   23
 1.6.1 Definition of roots   23
 1.6.2 Roots and exponents   24
 1.6.3 Irrational numbers   25
 1.6.4 Surds   25
 1.6.5 A third operation in manipulating equations   26

1.7 Evaluating expressions   26
 1.7.1 Order of operations   27
 1.7.2 Handling complex fractions   27
 1.7.3 Numerical expressions in Excel   28

Case Study A3: Birth and death rates   30

Case Study B3: Evaluating the angiogenic cancer cell density   31

1.8 Extension: intervals and inequalities   32
 1.8.1 Intervals on the real line   32
 1.8.2 Inequalities   33

Case Study A4: Birth rate, death rate and extinction   35

Case Study B4: Conditions for angiogenic cell line extinction   36

 Summary   37

 Problems   38

2 Units; precision and accuracy   45

2.1 Scientific notation   46
 2.1.1 Definition of scientific notation   46
 2.1.2 Converting numbers between decimal and scientific notation   47
 2.1.3 Performing addition and subtraction in scientific notation   48
 2.1.4 Performing multiplication and division in scientific notation   49
 2.1.5 An aside: floating point notation   49

2.2 SI units   50
 2.2.1 Base, supplementary, and derived SI units   50
 2.2.2 SI prefixes   53

Case Study C2: Velocity   54
 2.2.3 More problems with SI units; units of volume   55
 2.2.4 Non-SI units   56

2.3 Calculations using SI units   56

Case Study C3: Force and acceleration   57

2.4 Dimensional analysis   58

2.5 Rounding, precision, and accuracy   60
    2.5.1 Rounding numbers   60
    2.5.2 Significant figures   61
    2.5.3 Uncertainty intervals   62

2.6 Extension: accuracy and errors   64
    2.6.1 Errors in addition and subtraction   65
    2.6.2 Errors in multiplication and division   66
    2.6.3 Errors in exponentiation   68
    2.6.4 Delta notation   71

Case Study A5: Error analysis for geometric growth   72

    Summary   74

    Problems   75

3   Data tables, graphs, interpolation   81

3.1 Constructing a data table and a data plot   82
    3.1.1 Independent and dependent variables   82
    3.1.2 Data plots   82

3.2 Drawing graphs   85
    3.2.1 Three basic types of graph   85
    3.2.2 Drawing graphs in Excel   86

3.3 Straight-line graphs: finding the slope   88
    3.3.1 Direct proportion   89
    3.3.2 Linear relationship   91
    3.3.3 Calculating the slope   91

3.4 Inverse proportion   93

3.5 Application: allometry   94

3.6 Extension: interpolation   96
    3.6.1 Performing interpolation by hand   96
    3.6.2 Linear interpolation between two data values   97
    3.6.3 Piecewise linear interpolation   99
    3.6.4 Linear interpolation using Excel   101

Case Study A6: Cobwebbing   102

    Problems   104

4   Molarity and dilutions   109

4.1 Basic concepts   109
    4.1.1 Simple solutions   109

4.1.2 Atomic mass   110

4.1.3 The mole   110

4.1.4 The molar mass of a substance   111

4.1.5 The molarity of a solution   112

4.1.6 Application: measurements of cholesterol level   113

4.2 Calculations involving moles and molarity   **113**

4.2.1 Calculating the number of moles in a sample   113

4.2.2 Calculating the molar mass of a compound   114

4.2.3 Calculating the molarity of a solution   114

4.2.4 Calculating the moles present in a sample of solution   115

4.2.5 Calculating the moles to add in making a solution   115

4.2.6 Calculating the mass to add in making a solution   116

4.3 Calculations for dilutions of solutions   **116**

4.3.1 Calculating the new concentration after diluting   116

4.3.2 Calculating how much to dilute to obtain a specific concentration   117

4.3.3 Serial dilutions   119

4.3.4 Application: serial dilution in homeopathy   120

4.4 Excel spreadsheets   **121**

Problems   **122**

5   Variables, functions, and equations   **125**

5.1 What is a function?   **126**

5.2 Some simple functions   **127**

5.2.1 Functions on your calculator, and in Excel   127

5.2.2 Direct and inverse proportionality functions   128

5.2.3 Quadratic functions, parameters   129

5.3 Creating tables of values in Excel   **131**

Case Study B5: Graphing the relationship in Excel   **132**

5.4 Manipulating equations   **133**

5.4.1 What is an equation?   133

5.4.2 Linear equations and their graphs   134

5.4.3 Rearranging linear equations   135

5.4.4 Equation of a circle   135

5.4.5 Rearranging more complicated equations   138

5.5 Graphs of the direct and inverse proportion functions   **141**

5.6 Algebra of functions   **142**

5.6.1 Changing the argument   142

5.6.2 Arithmetic operations with functions   143

5.6.3 Composition of functions (or 'function of a function')   145

5.7 Extension: solving simultaneous linear equations   **145**

Problems   **149**

6 Linear functions and curve sketching   **151**

6.1 Graph sketching with FNGraph   **152**

6.2 Constant functions $y = c$ **155**

6.3 Linear functions   **156**
  6.3.1 The identity function $y = x$ 156
  6.3.2 Proportionality functions $y = mx$ 157

Case Study C4: Equations of motion 1   **159**
  6.3.3 General linear functions $y = mx + c$ **159**

Case Study C5: A constant-velocity chase   **164**
  6.3.4 Trend lines revisited: goodness of fit   165

6.4 Limits and asymptotes   **166**
  6.4.1 Limits of functions   166
  6.4.2 Horizontal asymptotes   167
  6.4.3 Vertical asymptotes   168
  6.4.4 Limits of sequences   168

6.5 Extension: linear transformations   **169**
  6.5.1 Vertical stretch/squash/flip: $y = a.f(x)$ 171
  6.5.2 Vertical shift: $y = f(x) + d$ 171
  6.5.3 Horizontal stretch/squash/flip: $y = f(bx)$ 171
  6.5.4 Horizontal shift: $y = f(x - c)$ 171

  Problems   173

7 Quadratic and polynomial functions   **176**

7.1 Simple quadratic functions   **176**
  7.1.1 The squaring function: $y = x^2$ 176
  7.1.2 The 'proportional to the square' function: $y = ax^2$ 178
  7.1.3 Adding a constant term: $y = ax^2 + d$ 179
  7.1.4 The 'shifted squaring' function: $y = a(x - \alpha)^2$ 180

7.2 General quadratic functions   **181**
  7.2.1 The 'completed square' form: $y = a(x - \alpha)^2 + d$ 181
  7.2.2 The polynomial form: $y = ax^2 + bx + c$ 184

Case Study C6: Equations of motion 2   **189**

Case Study A7: Fibonacci's Rabbits: geometric growth in a population
                structured by age   **191**

  7.2.3 Calculating the polynomial coefficients   192
  7.2.4 Application: Reduction of cholesterol level   193

Case Study B6: Derivation of the equilibrium density   **195**
  7.2.5 The factorized form: $y = a(x - \alpha)(x - \beta)$ 197

7.3 Cubic and higher-degree polynomials   198
  7.3.1 Power functions and geometric series   198
  7.3.2 General properties of polynomial functions   201
  7.3.3 Algebraic long division   201

7.4 Logistic growth   204
  7.4.1 The logistic function   204

Case Study C7: Fisheries management   205
  7.4.2 Logistic growth of populations   206

Case Study B7: Logistic growth of cancer cells   208

Case Study A8: 'Cobwebbing' of logistic growth model   208

7.5 Extension: quadratic interpolation   210

  Problems   213

8 Fitting curves; rational and inverse functions   220

8.1 Reciprocal functions   220
  8.1.1 Definition of the reciprocal of $f(x)$ 221
  8.1.2 Rational functions $y = \frac{1}{x}$, $y = \frac{1}{ax+b}$, $y = \frac{x}{ax+b}$ 221

Case Study C8: A hyperbolic model of animal speed   223

8.2 General rational functions $\frac{p(x)}{q(x)}$ 224
  8.2.1 Finding the x-intercepts   225
  8.2.2 Finding the y-intercept   225
  8.2.3 Finding the horizontal (and sloping) asymptotes   225
  8.2.4 Finding the vertical asymptotes   226
  8.2.5 Example of graph sketching   226

8.3 Fitting curves to data   227
  8.3.1 Inverse proportion   228
  8.3.2 Rational function $y = \frac{1}{ax+b}$ 230
  8.3.3 Quadratic functions   230
  8.3.4 Rational function $y = \frac{a}{x} + b$ 231

8.4 Application: enzyme kinetics   231
  8.4.1 The Michaelis–Menten equation   232
  8.4.2 The Lineweaver–Burk transformation   233
  8.4.3 Error analysis   234
  8.4.4 Allosteric regulation   235

8.5 Inverse functions   237
  8.5.1 Definition of the inverse of $f(x)$ 237
  8.5.2 The inverse of rational functions   238

8.6 Bracketing methods   239
  8.6.1 Root-finding algorithms   241
  8.6.2 Minimization algorithms   246

Case Study C9: Fisheries management: finding the Maximum Economic Yield 250

8.7 Extension: finding the equation of a trend line 251

Problems 255

9 Periodic functions 259

9.1 Sawtooth functions 259
9.1.1 Basic sawtooth function 259
9.1.2 Specifying the period and amplitude 261
9.1.3 Specifying the vertical shift and phase 262

9.2 Revision of school trigonometry 262

9.3 Measurement of angles in radians 265

9.4 The sine and cosine functions 268

9.5 Periodic functions of time 273
9.5.1 General sine and cosine functions 273

Case Study C10: A simple model of predator–prey population dynamics 274

9.5.2 Application: modelling tidal data 276
9.5.3 Application: modelling temperature variations 279

9.6 Reciprocal and inverse trigonometric functions 280
9.6.1 Reciprocal trigonometric functions 280
9.6.2 Inverse trigonometric functions 282

9.7 More trigonometric identities 283

9.8 The tangent function and the gradient of a curve 284
9.8.1 Definition of the tangent function 284
9.8.2 The tangent function and the slope of a line 284
9.8.3 The geometric tangent 285
9.8.4 An approximation to the gradient 287

Problems 289

10 Exponential and logarithmic functions 290

10.1 Exponential functions to the base a 290
10.1.1 Discrete and continuous models 290
10.1.2 Exponential function to the base a: $y = a^x$ 291

10.2 Exponential growth function $y = Ae^{kx}$ 294

Case study A9: Exponential growth of populations 296

10.3 Logarithms 296
10.3.1 Definition of logarithms to base a 296
10.3.2 Laws of logarithms 298
10.3.3 Logarithms to base 2 299
10.3.4 Logarithms to base 10 (common logarithms) 300
10.3.5 Logarithms to base e (natural logarithms) 301

10.4 Fitting exponential curves to data 302

 10.4.1 Fitting an exponential growth model 303

 10.4.2 Application: allometry 303

 10.4.3 Application: allosteric regulation 305

10.5 Exponential decay 306

 10.5.1 Exponential decay function: $y = Ae^{-kx}$ 307

Case Study C11: An exponential model of animal speed 309

 10.5.2 Application: sensitization and habituation 309

 10.5.3 Application: drug administration 310

 10.5.4 Example: radiocarbon dating 311

Case study A10: An equation for logistic growth 312

10.6 Example: reduction of cholesterol level 314

10.7 Extension: a stochastic model of exponential decay 317

Case study A11: Gompertz curve for population mortality 319

 Problems 320

 Revision Problems 328

Historical interlude: finding the roots of polynomials 331

PART II CALCULUS AND DIFFERENTIAL EQUATIONS 335

11 Instantaneous rate of change: the derivative 337

11.1 Introduction to the calculus 337

 11.1.1 Differential calculus 338

 11.1.2 Integral calculus 340

 11.1.3 Differential equations 341

Case Study B8: Constructing the angiogenic tumour model 342

11.2 Definition of the derivative 344

11.3 Differentiating polynomial functions 347

 11.3.1 The derivative of power functions $y = x^n$ 348

 11.3.2 Notation 351

 11.3.3 The derivative of linear functions 351

 11.3.4 The derivative of polynomial functions 352

Case Study C12: Differentiating the animal motion model 354

11.4 Differentiating roots and reciprocals 354

11.5 Differentiating functions of linear functions 355

11.6 Differentiating exponential functions 357

11.7 Extension: small changes and errors   360

Case Study C13: Deriving the exponential model of animal speed   360

Case Study A12: Differential equation for exponential growth   362

    Problems   363

12   Rules of differentiation   **366**

    12.1 Differentiable functions   366

    12.2 The chain rule   367

    12.3 The product and quotient rules   370
       12.3.1 The product rule   370
       12.3.2 The quotient rule   372

Case Study C14: Deriving the hyperbolic model of animal speed   376

    12.4 Differentiating trigonometric functions   376

    12.5 Implicit differentiation   379

    12.6 Differentiating logarithmic functions   381

    12.7 Differentiating inverse trigonometric functions   382

    12.8 Higher-order derivatives   383

    12.9 Summary of standard derivatives, and rules of differentiation   386

    Problems   387

13   Applications of differentiation   **389**

    13.1 Interpretation of graphs   389
       13.1.1 Gradients   391
       13.1.2 Roots   392
       13.1.3 Critical points   393
       13.1.4 Curvature   395

Case study A13: Analysing the Ricker update equation   397
       13.1.5 Summary   398

Case study A14: The point of inflection in the logistic growth curve   402

    13.2 Optimization   403
       13.2.1 Optimization in the biosciences   403
       13.2.2 One-dimensional unconstrained optimization   404

Case study C15: Fisheries management: using calculus to find the
               Maximum Economic Yield   406

       13.2.3 Application: tubular bones   407

    13.3 Related rates   411

    13.4 Polynomial approximation of functions   413

13.4.1 Linear approximation of $f(x)$ around $x = 0$ 414

13.4.2 Quadratic approximation of $f(x)$ around $x = 0$ 416

13.4.3 Maclaurin series expansions of functions   417

13.4.4 Taylor series expansions of functions   418

13.5 Extension: numerical methods for finding roots and critical points   **419**

13.5.1 Newton–Raphson method for finding roots   420

13.5.2 Newton's method for optimization   423

Problems   424

14   Techniques of integration   **426**

14.1 The integral as anti-derivative   **426**

14.1.1 Definition and notation   427

14.1.2 The integrals of power functions, and the coefficient rule   428

14.1.3 The sum rule, and the integrals of polynomial functions   430

14.1.4 Integrals of some standard functions   432

Case study C16: Integrating the hyperbolic and exponential models   **433**
of animal speed

14.2 Integration by substitution   **435**

14.3 Integration by parts   **438**

Case study A15: Solving the differential equation for exponential growth   **439**

14.4 Integration by partial fractions   **441**

14.5 Integrating trigonometric functions   **444**

14.5.1 The general sine and cosine functions   444

14.5.2 The tangent function   444

14.5.3 Powers of sines and cosines   445

14.5.4 Integrating $e^x \cos x$ 446

14.5.5 Integrating inverse trigonometric functions   447

14.6 Extension: integration using power series approximations   **448**

14.7 Summary of standard integrals   **449**

Problems   450

15   The definite integral   **451**

15.1 The integral as area under the curve   **451**

15.1.1 The link between the integral and area   451

15.1.2 Speed–time graphs   452

15.1.3 Definition of the definite integral   453

15.2 The integral as limit of a sum   **458**

15.2.1 The Riemann integral   458

15.2.2 Application: chemotherapy drug delivery   460

15.2.3 Application: laminar blood flow   462

15.3 Using techniques of integration with definite integrals 465
15.3.1 Integration by substitution 465
15.3.2 Integration by parts 466
15.3.3 Integration by partial fractions 467

15.4 Improper integrals 468

15.5 Extension: numerical integration 470
15.5.1 The trapezium rule 470
15.5.2 Simpson's rule 473
15.5.3 Using Simpson's rule with data-sets 476

Problems 479

16 Differential equations I 481

16.1 Overview of differential equations 481
16.1.1 Order of a differential equation 482
16.1.2 Boundary conditions 482
16.1.3 ODEs and PDEs 484

16.2 Solution by separation of variables 484
16.2.1 Right-hand side a function of *x* only 484
16.2.2 Right-hand side a function of *y* only 486
16.2.3 Variables separable 489

Case Study B9: The Gompertz model of tumour growth 490

Case Study A16: Solving the ODE for logistic growth 492

Case Study C17: A harvesting model for fish stocks 494
16.2.4 Change of variable 496

Case Study B10: The Gompertz model revisited 497

16.3 Linear first-order ODEs 499

16.4 Extension: partial differentiation 500
16.4.1 Reducing a PDE to an ODE 501
16.4.2 Error analysis in several variables 503
16.4.3 Minimization in two variables 504

Problems 505

17 Differential equations II 509

17.1 Numerical methods for first-order ODEs 509
17.1.1 Euler's method 510
17.1.2 Heun's method 515

Case Study C18: Numerical solution of fish harvesting model 518
17.1.3 Runge–Kutta method RK4 519

17.2 Systems of first-order ODEs — 519

    17.2.1 Lotka–Volterra models of predator–prey dynamics — 520

    17.2.2 Kermack–McKendrick model of epidemics — 526

Case Study A17: The peak of an epidemic — 527

17.3 Extension: analytic solutions — 529

    17.3.1 Solving second-order ODEs — 529

    17.3.2 Solving first-order systems — 531

    17.3.3 Solving partial differential equations — 532

    17.3.4 Further reading — 534

    Problems — 534

18 Extension: dynamical systems — 537

18.1 The butterfly effect — 537

    18.1.1 The birth of a new science — 538

    18.1.2 Numerical experiments — 538

18.2 Equilibria and stability — 542

    18.2.1 Points of equilibrium for differential equations — 542

    18.2.2 Stability of equilibria for differential equations — 543

Case Study C19: Analysing the equilibria of the harvesting model — 547

    18.2.3 Stability of equilibria for update equations — 549

    18.2.4 Numerical experiments with the update equation — 551

18.3 Bifurcations ... — 553

18.4 ... and Chaos — 557

18.5 Postscript — 559

    Problems — 561

Answers to odd-numbered problems — 563

Appendix: The Greek alphabet — 570

References — 571

Index — 573

The search for The Ultimate Answer

According to *The Hitchhiker's Guide to the Galaxy*, a race of vast hyper-intelligent, pan-dimensional beings constructed the second greatest computer in all of time and space, Deep Thought, to calculate The Ultimate Answer to The Great Question of Life, the Universe, and Everything. A 'simple answer' is requested. After seven and a half million years of computing cycles, Deep Thought's answer is 42.

'Forty two?!' yelled Loonquawl. 'Is that all you've got to show for seven and a half million years' work?'

'I checked it very thoroughly,' said the computer, 'and that quite definitely is the answer. I think the problem, to be quite honest with you, is that you've never actually known what the question is.'

from Wikipedia entry for 'Answer to Life, the Universe and Everything'[1]

---

[1] http://en.wikipedia.org/w/index.php?title=Answer_to_Life,_the_Universe,_and_Everything accessed July 2008. See Adams (2009), chapters 25 and 27.

Introduction: why and how to use this book

The study of biology in the nineteenth century was largely concerned with definition, description, and classification: collecting specimens, naming them, and arranging them in display cabinets, or drawing detailed anatomical diagrams and labelling them. Detailed description is still an important aspect of some fields of modern biosciences – cell biology, for example – but biology in the twenty-first century is all about **measuring and modelling change and development**, both of individual organisms and of populations. Because of this, mathematical and statistical techniques are now essential tools for the bioscientist.

There is a field of mathematics called the **calculus**, which is used to analyse rates of change. The second part of this book explains the calculus and shows how it is used in the biosciences. The first part of the book establishes the mathematical foundations needed for the calculus (what is referred to in the US as pre-calculus), though these foundations are also important for scientists in their own right. The principal foundation is the concept of a **function**; functions can be used to model the way a physical quantity changes depending on factors such as temperature and time (for example, the change in blood sugar levels as time passes after a meal).

We will use three modelling case studies, with instalments appearing in boxes throughout the book:

- Case Study A: Models of population growth
- Case Study B: Models of cancer
- Case Study C: Predator–prey relationships

These case studies will be used to illustrate the mathematical techniques of each chapter. They all involve real, practical biology, and though Case Study B initially looks more complex and difficult than Case Study A, and Case Study C is easier, they will all be using essentially the same mathematical concepts.

But this book is more than a summary of the mathematics taught at school or college. It is intended for students of the biosciences at both undergraduate and postgraduate degree level, and I will be presenting the mathematics from a university-level viewpoint also. My aim is not just to give you the rules and techniques you'll need to use, for you simply to memorize parrot-fashion, but to explain why they work, and how everything fits together consistently and elegantly. In this way, maths should not seem like a bewildering array of arbitrary rules. A wider aim is to convince you that mathematics can be fun, exciting, and beautiful, and is not just the domain of geeks and eggheads. I'm not referring here to mathematical puzzles and diversions, but the maths that can provide insight into the miracles and mysteries of

Nature. To this end, the final chapter of this book introduces concepts of dynamical systems and chaos theory, which is a field of much current research at the interface of mathematics, computing, science, and engineering.

The scope of this book covers 'only' mathematics, and not probability or statistics. There is one exception: we will give the mathematical derivation of the best-fit straight line through a set of data-points (Section 8.7), as this is a natural extension to finding the equation of a straight line through two points, and illustrates some important mathematical techniques. For a primer on the use of statistics in the biosciences, see the OUP companion volume *Biomeasurement: A Student's Guide to Biological Statistics* (Hawkins, 2005).

There are plenty of biological examples in the book, intended to be accessible to students of all fields in the life sciences. But I should emphasize that this is a mathematics textbook, not a bioscience book. Its structure is intended to build up all the mathematics systematically. Mathematics is a very linear subject: you need to know the rules of arithmetic in order to learn to use algebra; you can't handle functions without algebra; you need to know about functions in order to apply the ideas of calculus; and you can then use calculus to solve differential equations. By contrast, in biology you can study animal behaviour without knowing about the structure and function of the animal's internal organs, or the biology of the cells that make up those organs.

We will start with the basics of addition and subtraction, and present a self-contained body of mathematics which will lead to the solution of systems of differential equations (as used in epidemiology and predator–prey models). Important concepts, such as the roots of a function, may be defined more than once in different chapters, to cater for students who skip sections they believe they are familiar with. I would encourage you to at least skim through such sections, especially Chapter 1, where you will see the rules of arithmetic and algebra, which were shown to you piecemeal over a decade of schooling, used to construct what I would claim is humankind's greatest achievement – the Real Number system. Each chapter is intended to present something new and useful, whether you stopped studying maths at GCSE or went on to Maths or even Further Maths at A-Level.

As well as the algebra you will recall from school, the emphasis in this book (and particularly in the Extension sections) is on **computation**: techniques for finding numerical solutions to problems. As the quotation from *The Hitchhiker's Guide to the Galaxy* (Adams 2009) illustrates, even the most profound questions about the natural world can have a numerical solution! We'll see how numerical algorithms (sets of instructions, like computer programs) can by used to solve complex practical problems through a series of simple calculations. These numerical techniques are covered mainly in the **Extension sections** at the end of most chapters. These are not 'extensions' in the sense of being optional extras which only the mathematically gifted student would understand. On the contrary, I have found that even bioscience students with a positive aversion to maths will appreciate and enjoy applying these techniques, armed with a calculator and Excel. They are also techniques of increasing practical importance; in fact, reviewers of the book drafts have said that they are particularly important for today's bioscientists.

I won't be teaching you computer programming in this book. To perform calculations I'll assume that you have an electronic calculator, and you'll see that even a simple calculator can do more than you might expect. The other computing tool we'll use is **Microsoft Excel**®.[2]

---

[2] Excel is a trademark of the Microsoft group of companies. *Core Maths for the Biosciences* is an independent publication and is not affiliated with, nor has it been authorized, sponsored, or otherwise approved by Microsoft Corporation.

MS Excel (which we'll refer to henceforth simply as Excel) is a spreadsheet package. A spreadsheet is a grid of cells; each cell can be used either to store text or a number, or it can be programmed to make calculations using data from other cells. Other spreadsheet programs are available, which have very similar features, but Excel is by far the most widely used and best-known program (being part of the Microsoft Office suite). You have probably already learned how to use Excel at school or college. I'll introduce the features of Excel which we'll need as we go along, but it would be good if you have a basic prior knowledge. As well as showing you how to construct your own spreadsheets, I'll give you ready-made spreadsheets as 'numerical laboratories' to allow you to experiment with the mathematical concepts. (See the **Technical Notes** section below.)

Excel can be used to plot graphs from data, but for sketching the graph of a function we'll mainly use a Freeware package called **FNGraph**. I am very grateful to Dr Alexander Minza for permission to use the package, by including screenshots as diagrams in the book, and providing FNGraph files on the website. There is an introduction to FNGraph in Chapter 6.

The Online Resource Centre that accompanies this book (http://www.oxfordtextbooks. co.uk/orc/reed/) has downloads of all the Excel and FNGraph files, as well as the FNGraph program itself; see the **Technical Notes** section which follows. You'll have to obtain your own copy of Excel, I'm afraid. The Online Resource Centre also has solutions to problems and web pages with supplementary content (including external links). To get the most from this book, you should work with a calculator and computer, and open the Excel and FNGraph files when they are referred to in the text.

Mathematics is a way of thinking, and this can be practised in many different contexts. So the Problems include applications of maths in other areas, which I hope you'll find interesting. If you can work through all of the problems, I hope that you'll not only be able to apply maths in your degree course, but you'll be thinking like a mathematician. On that scary note, I'll leave you to get started with Chapter 1 ...

Send us your feedback

I would welcome any comments about the book, to help improve future editions, and would especially appreciate reports of any misprints or errors. Send your comments via http://www. oup.com/uk/orc/feedback/ or follow the 'Send us your feedback' link on the left of the main Online Resource Centre page.

Acknowledgements

I firstly want to thank my colleague at Bath Dr Chris Todd, Director of Teaching in the Department of Biology and Biochemistry, who collaborated on the first drafts of the early chapters, Chapter 4 in particular. Unfortunately he had to withdraw from the project at that point. I am next very grateful to the anonymous reviewers of the drafts; their constructive criticism, positive comments, and suggestions have hugely improved the final product. I should also repeat here my thanks to Dr Alexander Minza for his kind permission for FNGraph to be used freely in the book, and on the website.

Thanks are due to Jonathan Crowe, Commissioning Editor at OUP, for his support and advice throughout this mammoth project. And to my family, whose encouragement and exasperation drove me to get it finished.

Technical notes: using the electronic support that accompanies this book

Two software packages are used in this book: Microsoft Excel and FNGraph. You will need to have your own copy of Excel, but you can download a free copy of FNGraph as described below. Then you can download the files referred to in the book, from the CoreMaths website: http://www.oxfordtextbooks.co.uk/orc/reed/

Mac users: Excel is available for the Apple Macintosh. Files for the Mac graph-drawing package Grapher are also provided on the CoreMaths website.

. . . .

Excel

MS Excel is by far the most widely used spreadsheet package. Other packages such as Lotus 1-2-3 and Quattro Pro are very similar in their layout and features, and are becoming compatible with Excel files. There are self-teach manuals such as the 'Excel for Dummies' range, and you can find online tutorials by typing 'excel tutorial' into a search engine.[3] But you only need a very, very basic knowledge of typing formulas into cells to get started, and I'll give step-by-step instructions for constructing the first few spreadsheets.

The version I have used and described is Excel 2003. The toolbars etc. in later versions may differ slightly.

Filenames

You can download from the CoreMaths website an Excel workbook file for each of the chapters in this book. The workbook file for Chapter 5 is named cm-excel5.xls, for example. Within the workbook, individual spreadsheets are named 5.1, 5.2, etc., and referred to in the text as 'SPREADSHEET 5.1' etc. The heading of each spreadsheet cites the (sub)section of the text where it is used. Where an end-of-chapter problem asks you to produce a spreadsheet, the solution can be found in the files cm-excelsolutionsI.xls and cm-excelsolutionsII.xls for Part I and Part II chapters respectively.

[3] At time of writing, there are free online tutorials for MS Office packages at: http://www.baycongroup.com/tutorials.htm and http://phoenix.phys.clemson.edu/tutorials/excel.

Protection

Many of the spreadsheets have been protected, to prevent you from unintentionally deleting formulas. If you want to take the spreadsheet apart to see how it is programmed, click on `tools` in the main toolbar and choose `tools > protection > unprotect sheet ...`

There is no password.

Running macros

Some of the Excel spreadsheets in this book include a sliding scrollbar. This causes Excel to refuse to open the spreadsheet, claiming that it contains an unrecognized macro. To override this, click the 'enable macros' button on the warning window. If this doesn't make the scrollbar work, do the following: go to `tools > macro > security ...` and change the security level from High to Medium. This setting will be remembered, and in future when you open an .xls file containing macros you will be prompted to choose whether macros should be enabled; click on the 'Enable macros' button.

....

FNGraph

Installing FNGraph

FNGraph was created by Alexander Minza. The package version 2.61 can be downloaded from the Online Resource Centre for this book, or from many other download sites – just do a Google search for 'fngraph download'.[4]

The file `fngraph _ 2.61.exe` is 523KB in size, just run it to install. The installed package `fngraph.exe` is 340KB in size.

The package only runs on Windows, not Macs, unfortunately. It works with Windows Vista and Windows 7, but you may get an error message 'Failed to start because MSVBVM50. DLL was not found'. To install that file (part of Visual Basic 5.0 Virtual Machine) go to http://support.microsoft.com/kb/180071/EN-US/ and click on the link to download the file msvbvm50.exe, then run it to install the files.

The CoreMaths website also contains the files `fngraph_2.61.exe` and `msvbvm50.exe`.

FNGraph is freeware, and can be used without restriction. It does not require any plug-ins. The source code has been made available at http://fngraph.codeplex.com/

Filenames

Each FNGraph graph is in a separate file, which has a file extension of `.fng`. You need to have installed the FNGraph software, to read the file. The use of FNGraph begins in Chapter 6. The files are named `cm6-1.fng`, `cm6-2.fng`, etc., and are referred to in the text as 'FNGRAPH 6.1', etc. (So, the file that corresponds with FNGRAPH 6.1 is `cm6-1.fng`; the file that corresponds with FNGRAPH 6.2 is `cm6-2.fng`; etc.)

---

[4] The tucows site is http://www.tucows.com/preview/198040

Grapher

FNGraph does not work on Macs, but Mac users have a graph-drawing package called Grapher bundled with Mac OS X (version 10.4 onwards). It is found in /Applications/Utilities/Grapher.app. Grapher versions of the FNGraph files are also provided on the ORC website. The file-naming convention is the same as for the FNGraph files, but the filename extension is .gcx (files cm6-1.gcx and so on).

ARITHMETIC, ALGEBRA, AND FUNCTIONS

1 Arithmetic and algebra   3

2 Units; precision and accuracy   45

3 Data tables, graphs, interpolation   81

4 Molarity and dilutions   109

5 Variables, functions, and equations   125

6 Linear functions and curve sketching   151

7 Quadratic and polynomial functions   176

8 Fitting curves; rational and inverse functions   220

9 Periodic functions   259

10 Exponential and logarithmic functions   290

   Historical interlude: finding the roots of polynomials   331

Arithmetic and algebra

For much of its history, biology has been largely a descriptive science. Biologists made detailed observations of plants and animals, and used these to analyse, classify, and describe their anatomy, characteristics, and behaviour. In the middle of the nineteenth century, Darwin[1] and Mendel[2] – who have been described as the fathers of modern biology – also made careful observations, to answer the question 'What?', but then they started to ask 'How?' and 'Why?': 'How do different species evolve?'; 'Why does a plant have particular traits?' In effect, they started to ask about the **relationship** between one variable and another. Does the flower colour of a pea plant depend on the flower colour of the plants it was cross-bred from, or is the shape of a giant tortoise's shell related to the particular Galapagos island it came from? Once you start trying to define such relationships, and quantify them, you need the language of mathematics.

You probably remember maths from school as being a bunch of rules for handling numbers, called arithmetic, and another bunch of rules for handling xs and ys, called algebra. This chapter will revise this material, but not from the schoolboy/girl viewpoint. You can now appreciate a more abstract approach, which builds up the rules from basic concepts of real numbers and arithmetic operations. In this way we'll show how arithmetic and algebra form a coherent, consistent whole, and we'll hopefully answer not just questions like 'What is 12 divided by 3?' but also 'Why can't we divide 12 by zero?' So even if you remember all your school maths, we'd recommend you skim through this chapter to see the subject from a university-level viewpoint.

There will be three case studies appearing in instalments throughout this book:

Case Study A: Models of population growth

Case Study B: Models of cancer

Case Study C: Predator–prey relationships

As well as presenting the mathematics in a bioscience context, these case studies show how a range of different mathematical techniques, from basic to advanced, are needed in the development of a biological topic.

---

[1] Charles Robert Darwin FRS (1809–1882), English naturalist.

[2] Gregor Johann Mendel (1822–1884), Austro-Hungarian priest and scientist, lived in what is now the Czech Republic.

Here is the first instalment (labelled A1, B1, C1) of each Case Study:

CASE STUDY A1: Introduction to models of population growth

Population biology is the study of how populations of organisms (which might be bacteria or humans, or anything in between) grow and/or decay over time. Population growth depends on many factors, including: the reproductive cycle, availability of nutrients (food resources), presence of predators, occurrence of diseases, and genetic variation. You can see that population biology involves ecology, epidemiology, and genetics.

Let's consider the simplest case: a population of bacteria growing on nutrient in a Petri dish. We will denote the size of the population by N, and use subscripts to indicate the time at which the population is measured. So we say that there are N_0 cells in the initial population. After a certain time, each cell divides into two. So after 1 generation there are N_1 cells, where $N_1 = 2N_0$. In a further generation each of those N_1 cells divides, producing N_2 cells, where

$$N_2 = 2N_1 = 2(2N_0) = 4N_0.$$

This type of growth is called *geometric growth*. It doesn't just apply to bacteria. It can be used to model colonies of insects, rabbits, and even humans, in situations where the birth rate exceeds the mortality rate, and there are sufficient supplies of food, and an absence of predators or disease.

CASE STUDY B1: Introduction to models of cancer

Many cancer tumours which occur in the human body through cell mutations remain extremely small, because there is not enough blood supply available to feed them. However, a type of cancer cell can develop which is called *angiogenic*; this is able to recruit nearby blood vessels to supply it with nutrients, allowing it to grow and become malignant. Cancer researchers have developed a mathematical model that expresses the rates of change of the densities of the three types of cell (healthy, stage one cancer, and angiogenic cells) depending on the replication and death rates of each type, and other factors. Such equations involving rates of change are called **differential equations**, which we will study in Part II. Using these equations, we can predict the equilibrium levels of each type of cell in the tumour. We'll show one of these formulas in the next instalment.

The example is taken from Wodarz and Komarova (2005), p. 8.

These three case studies come from different areas of the biosciences – population biology, cancer biology, and ecology – but all involve the idea of growth, or more generally of **rates of change**. This is true of virtually all fields in the biosciences; even the study of fossils uses radiocarbon dating – a technique based on the decay of carbon atoms over time, which we'll look at in Chapter 10. An extremely powerful mathematical theory that analyses rates of change is called the **calculus**, and will be the subject of Part II of this book.

CASE STUDY C1: Introduction to predator–prey relationships

Predator–prey relationships form a branch of ethology, the study of animal behaviour. If you ask the public for examples of predator–prey relationships, they will think of lions and antelopes, cats and mice, spiders, and flies. But ladybirds and badgers are also predators. And we haven't mentioned the greatest predator of all: human beings.

Nature has evolved a wide array of mechanisms by which predators catch prey, and an even wider range of techniques by which the prey try to avoid capture. From Chapter 6 onwards we will consider the kinematics (the quantitative description of motion, i.e. velocity and acceleration) of the classical predator–prey scenario: the chase. But we will also look at an economic model of predation – how overfishing can cause the decline and eventual extinction of fish stocks. From these examples you can see that predator–prey relationships are linked to models of population growth, with two or more interacting populations. In Chapter 17 we will solve a pair of equations (the Lotka–Volterra equations) that model the variation of predator and prey populations over time.

After completing this chapter, you should be able to:

- evaluate numerical expressions involving fractions, exponents, and roots;
- simplify algebraic expressions involving fractions, exponents, and roots;
- manipulate simple equations.

1.1 Numerical and algebraic expressions

An **arithmetic expression** (or **numerical expression**) is a 'sum' involving numbers and **arithmetic operations** such as addition, division, or square roots. An example of an arithmetic expression is

$$\frac{6 \times 4^2 - 10\sqrt{3 \times 4.88 + 3}}{12 \times 4}. \tag{1.1}$$

Try evaluating this using your calculator. If you arrived at 1.125, then you still remember many of the basic rules of arithmetic. If you didn't then this chapter will explain them.

If an expression involves letters such as a, b, x, or y, as well as numbers, it is called an **algebraic expression**. For example,

$$\frac{6a^2 - 10\sqrt{3b + 3}}{12a} \tag{1.2}$$

is an algebraic expression. The letters a and b represent numbers, but their values are not fixed; they are therefore called **variables**. If we chose to set $a = 4$ and $b = 4.88$, we would obtain the numerical expression (1.1), which we can evaluate. But we could choose other values of a and b, which would give us different numerical expressions to evaluate. So it's important to know how to work with algebraic expressions, as these are much more general than a particular numerical 'sum'.

The good news is that the rules for handling algebraic expressions are just the same as the rules for handling numbers. This is because variables like a and b are simply numbers whose values we don't yet know. The bad news is that it's easy to make mistakes when dealing with algebraic expressions, which you (hopefully) wouldn't do when working with numbers. For example, it is tempting to 'simplify' expression (1.2) by dividing all the terms on top and bottom of the fraction by 3, to get

$$\frac{2a^2 - 10\sqrt{b+1}}{4a}$$

and then to cancel the as to get

$$\frac{1}{2}a - \frac{5}{2}\sqrt{b+1}.$$

Both of these steps are **wrong**. In this chapter we'll give you the rules for what you can do, and also point out some of the things you can't do with algebraic expressions. There will be separate sections dealing with arithmetic operations, exponents and roots, and fractions. By also explaining **why** the rules work, you'll understand what is wrong with the 'simplifying' steps above.

You can often see that a step is wrong, by trying it out on a simple example with numbers. For example, the step we made above in cancelling the as is like going from $(2+1)/2$ to $1+1$ by 'cancelling the 2s'. But $(2+1)/2 = 3/2 = 1.5$, and $1+1 = 2$, so this cancellation must be **wrong**. Some of the explanations in the following sections will look rather childish when demonstrated using numbers, but please be patient: the numerical examples will help you to get used to the algebraic rules.

Before discussing the rules for working with numbers, we should first be clear what a number is. Can we visualize the set of numbers? Is there some order to them? What do we mean by negative numbers (and is -3 'bigger' or 'smaller' than -2)? Are there different types of numbers? Can the same number be written in different ways (as a decimal or a fraction)? We'll answer these questions in the next section.

CASE STUDY B2: Angiogenic cancer cells

The rate at which angiogenic cancer cells in a tumour can multiply is controlled by the presence of chemicals that are angiogenesis inhibitors (such as endostatin and interferons) and others that are angiogenesis promoters (growth factors such as FGF and PDGF). If we use x_0, x_1 to represent respectively the densities of healthy cells and non-angiogenic cancer cells, and y to represent the density of angiogenic cancer cells in the tumour, then the rate of increase of the angiogenic cancer cells is given by

$$r.y.\left(1 - \frac{y}{k}\right) - d.y - \frac{y.p.(x_0 + x_1 + y)}{q.y + 1}.$$

Don't worry about where this expression comes from; we'll explore that in later instalments. The purpose of showing you this expression (and the one below) now is to demonstrate that in real, cutting-edge research, biologists still need to use the algebra in this chapter. Thus it would be wrong to cancel the ys in the final fraction.

From the above result, we can work out the equilibrium density y of angiogenic cells; again, we'll show how this is done later. We find that

$$y = \frac{-Q + \sqrt{Q^2 - 4rq[k(d-r) + pk(x_0 + x_1)]}}{2rq}$$

where

$$Q = kq(d-r) + r + kp.$$

The equilibrium density of angiogenic cells in the tumour is the density at which the tumour stabilizes. The equation is derived by Wodarz and Komarova (2005), p. 151.

Here, p and q are parameters measuring the amounts of angiogenesis inhibitors and promoters present, r and d are the angiogenic cell replication and death rate coefficients, and k is the maximum density of angiogenic cells under normal conditions. We'll give the mathematical definition of terms such as 'parameter' and 'coefficient' later.

We need to be able to evaluate this expression for different values of the variables. In particular, what is the effect of increasing p? If this would make the angiogenic cell line die out, then a new therapy involving administration of angiogenesis inhibitors to the patient might be successful in driving the cancer into remission. This is the subject of current research.

. . . .

1.2 The real numbers

1.2.1 Integers and reals

As a scientist, you will obtain your numerical data by taking measurements. These are usually made either by counting (e.g. the number of flowers in a field, or the number of cells in a Petri dish), or by reading from a calibrated scale (e.g. measuring length or temperature). In the first case, your data will be in 'whole numbers'. At primary school you learned to add, subtract, and multiply whole numbers without being aware that other types of 'non-whole' numbers could exist.

In mathematics, 'whole numbers' are called **integers**. As there can be negative as well as positive integers (together with zero, which is neither positive nor negative), we could show all the integers in a list, with the negative integers to the left of zero, and the positive integers to the right:

$$\ldots, -4, -3, -2, -1, 0, 1, 2, 3, 4, \ldots$$

For counting, we use just the **positive integers** 1, 2, 3, 4, ...; these are also called the **natural numbers**.

There is an informal naming convention for variables that represent integers: variables named i, j, k, l, m, n will usually be integer variables.

The integers are **discrete**, meaning they are like individual, isolated packets: there is a gap between one integer and the next. Physical quantities like length or temperature, on the other hand, vary continuously. Consider measuring the temperature of a flask of water using a thermometer. We can only read the thermometer's calibrated scale to a certain level of precision (we'll talk about precision in Chapter 2) — or if it's a digital thermometer we'll get a readout correct to 2 decimal places, say — but the temperature could be any value on a continuous scale from 0.0 to 100.0 degrees centigrade. In mathematics, these values are called **real numbers**.

1.2.2 The real line

We can't write down the real numbers in a list, as we did with the integers, but we can visualize them as points on a continuous line, marked like the scale on a thermometer. This line, which stretches to infinity in either direction, is called the **real number line** — or **the real line**, for short. It is shown in Figure 1.1. Each point on the line corresponds to a real number. The point that is one quarter of the way from 1 to 2 on the line represents the real number 1.25, for example, and the point one third of the way from 1 to 2 represents the real number 1.33333 ... (which we can also write as $1\frac{1}{3}$ or $\frac{4}{3}$).

Just as 'the real number line' is abbreviated to 'the real line', so we can speak about 'the reals', meaning the real numbers.

As well as giving us a picture of the reals, the real line also gives them an **order**:

> If we take any two real numbers a and b, we say that 'a **is less than** b'
> if a lies to the left of b on the real line. This is written as $a < b$.
>
> We could equivalently say that 'b **is greater than** a', written as $b > a$. (1.3)

Thus, $2 < 3$ since 2 lies to the left of 3 on the real line.

Real numbers lying to the right of zero are called **positive**, and those lying to the left of zero are called **negative**. Any positive number is greater than any negative number. So $4 > -2$, which seems obvious, but also $2 > -4$; don't be fooled by the fact that four is bigger than two.

What if we compare two negative numbers: which is greater, -3 or -2? Looking at the real line, we see that -3 is to the left of -2, so $-3 < -2$. In other words, -2 is greater than -3. Again, don't be fooled by the size of the actual digits.

If we are given two real numbers a, b then precisely one of the following statements must be true:

(i) $a < b$ (a lies to the left of b on the real line); or

(ii) $a > b$ (a lies to the right of b on the real line); or

(iii) $a = b$ (a and b lie at the same point on the real line).

In mathematical number theory, this fact is known as **the Trichotomy of Cardinals** (which sounds like a secret society in a Dan Brown novel, but isn't — yet).

Problems: Now try the end-of-chapter problems 1.1–1.2.

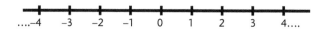

Figure 1.1 The real line

1.3 **Arithmetic operations**

Now that we know what numbers are, we can define the basic operations for combining numbers. At primary school you first learned to add numbers, then to subtract, and later learned about multiplying and dividing. You probably haven't seen them set out using the real line, though.

1.3.1 Negation

On your calculator there is a key marked ±. This produces the **negation** of the number you have entered.

Enter 7 and press ±, and you get −7.

> The **negation** of a is $-a$.   (1.4)

Now $-a$ is a number lying the same distance away from 0 as a on the real line, but on the other side. You can think of negating a number as flipping from the positive to the negative side of the real line, or vice versa. See Figure 1.2.

Now you have −7 displayed, press ± again; you return to 7. You have flipped from positive to negative and back again.

The negation of the negation of 7 is 7: $-(-7) = 7$.

This gives us our first useful result in algebra:

> $-(-a) = a.$ (1.5)

1.3.2 Addition and subtraction

To find $7 + 3$ using the real line, you start at 7 on the real line then move 3 units to the right (i.e. in the positive direction); you arrive at 10.

We can define subtraction to mean 'adding the negation': here's how it works. Start with 7. Add the negation of 3, i.e. add −3. Moving −3 units to the right means moving 3 units to the left: $7 + (-3) = 7 - 3 = 4$.

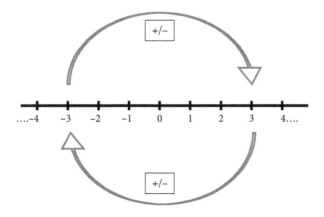

Figure 1.2 Negating numbers on the real line

This gives us our second useful fact of algebra:

$$a + (-b) = a - b. \tag{1.6}$$

1.3.3 Multiplication and division

We can define multiplication in terms of addition:

$$3 \times 7 = 7 + 7 + 7 = 21.$$

Mathematicians usually use a dot for multiplication, rather than ×, but not when multiplying two numbers in case it gets mistaken for a decimal point: we write $x.y$, $3.x$, but 3×7 rather than 3.7. Often we leave out a symbol for multiplication altogether: xy, $3x$.

When an expression is formed by multiplying numbers or variables together, it is called a **product**. Each term in the product is called a **factor**. So 7 is a factor of 56, since $56 = 8 \times 7$. A number that is a factor in an algebraic product is called a **coefficient**: in the expression $3x$, 3 is the coefficient.

Multiplying a positive number by a negative number produces a negative number:

$$3 \times (-7) = -7 - 7 - 7 = -21.$$

But multiplying two negative numbers together produces a positive.

$$a.(-b) = -a.b, \quad (-a).(-b) = a.b. \tag{1.7}$$

What about division: what do we mean by $\frac{3}{7}$? 'One seventh' of a cake involves dividing the cake into seven equal slices. The seven slices, added together, restore the whole cake. So

$$\text{If } x = \frac{1}{a}, \text{ it means that } a.x = 1. \tag{1.8}$$

$\frac{1}{a}$ is called the **reciprocal** of a. You have a reciprocal key on your calculator: ▣. If you enter 8 and press this key, you obtain 0.125, which is one eighth expressed as a decimal (you can also get this by the calculation '1 ÷ 8 ='). Press the reciprocal key again, and you get back to a. This demonstrates that

$$\frac{1}{1/a} = a. \tag{1.9}$$

Generalizing this to the idea of dividing one number by another:

$$\text{If } x = \frac{a}{b}, \text{ it means that } b.x = a. \tag{1.10}$$

So $x = \frac{3}{7}$ means that $7x = 3$.

'Three sevenths' is three slices of our cake, or three times one-seventh. In general,

$$\frac{a}{b} = a.\frac{1}{b}. \tag{1.11}$$

Now if $x = -\frac{3}{7}$ it means that $-7x = 3$, which is the same as $7x = -3$. So

$$\frac{a}{-b} = \frac{-a}{b} = -\frac{a}{b}. \tag{1.12}$$

The reciprocal of $\frac{a}{b}$ is $\frac{b}{a}$:

$$\frac{1}{a/b} = \frac{b}{a}. \tag{1.13}$$

The symbol we use most often for division in arithmetic is a forward-slash: a/b. We will still use ÷ in more complicated expressions, though.

At school you will have heard numbers such as $\frac{3}{7}$ called fractions. Mathematicians call them rational numbers (often abbreviated to 'rationals'). A **rational** is a real number that can be written as $\frac{a}{b}$, where a and b are integers. The name 'rational' comes from thinking of this as a ratio between the two numbers a and b (in our cake, $\frac{3}{7}$ means taking 3 slices out of 7). The numbers a and b are called the **numerator** and the **denominator** of the fraction $\frac{a}{b}$. We will discuss more complicated fractions in Section 1.7.

Rationals are real numbers, so they lie on the real line; the number $\frac{15}{7}$, for example, which is $2\frac{1}{7}$, is a point that is one seventh of the way from 2 to 3 on the real line. You can convert a rational number to its decimal representation using your calculator. For example, $\frac{3}{4} = 0.75$ and $\frac{15}{7} = 15 ÷ 7 = 2.1428571...$ You can see that some decimal representations **terminate**, and others don't. A terminating decimal is one in which the non-zero digits stop at a certain point; further digits would simply be 00000000 ...

A terminating decimal can be written as a fraction by moving the decimal point to the right to the end of the digits (to get the numerator) and then putting the corresponding power of 10 (10, 100, 1000, etc.) on the denominator. For example:

- 0.7 can be written as $\frac{7}{10}$;
- 0.73 can be written as $\frac{73}{100}$;
- $3.142 = \frac{3142}{1000}$;
- $3.14159 = \frac{314159}{100000}$.

Every rational is a real number, but there are real numbers that can't be written as a rational. They are called **irrational** numbers, and we'll mention them in Section 1.6.

To locate numbers on the real line, you usually need to look at their decimal representation, which you can find using your calculator. For example, the fraction $\frac{22}{7}$ and the terminating decimal 3.142 are often used as approximations to the irrational number π. As decimals, these are:

$$\frac{22}{7} = 3.142857142857142857142\ldots$$
$$3.142 = 3.1420000\ldots$$
$$\pi = 3.14159257\ldots$$

from which we can see that $\pi < 3.142 < \frac{22}{7}$.

Fractions may need to be simplified so that they are **in lowest terms**; that is, the numerator and denominator have no common factors. For example, 0.6 can be written as a rational as $\frac{6}{10}$, but there is a common factor of 2 in the numerator and denominator, which we can cancel:

$$0.6 = \frac{6}{10} = \frac{2 \times 3}{2 \times 5} = \frac{\cancel{2} \times 3}{\cancel{2} \times 5} = \frac{3}{5}.$$

We can now write down some further rules for handling fractions:

$$\frac{a.c}{b.c} = \frac{a}{b} \quad \text{Common factors can be cancelled from numerator}$$
$$\text{and denominator.}$$
$$\frac{a}{c} + \frac{b}{c} = \frac{a+b}{c} \quad \text{Fractions can be added or subtracted only if they have the}$$
$$\text{same denominator.} \qquad\qquad (1.14)$$
$$\frac{a}{b} \times \frac{c}{d} = \frac{a.c}{b.d} \quad \text{To multiply fractions, multiply the numerators, and}$$
$$\text{multiply the denominators.}$$
$$\frac{a}{b} \div \frac{c}{d} = \frac{a}{b} \times \frac{d}{c} = \frac{a.d}{b.c} \quad \text{To divide by a fraction, you multiply by its}$$
$$\text{reciprocal. See } (1.13).$$

If we need to add two fractions that have different denominators, we need to multiply each of them by a factor top and bottom, to make the denominators match. For example, to add $\frac{5}{8} + \frac{7}{6}$ we multiply the first fraction top and bottom by 3: $\frac{5 \times 3}{8 \times 3} = \frac{15}{24}$, and multiply the second fraction top and bottom by 4: $\frac{7 \times 4}{6 \times 4} = \frac{28}{24}$. Now we can add:

$$\frac{5}{8} + \frac{7}{6} = \frac{5 \times 3}{8 \times 3} + \frac{7 \times 4}{6 \times 4} = \frac{15}{24} + \frac{28}{24} = \frac{43}{24}.$$

You should remember this from school; 24 is the **lowest common denominator** of 8 and 6.

Here are some examples using numbers:

$$\frac{7}{12} - \frac{5}{12} = \frac{7-5}{12} = \frac{2}{12} = \frac{1}{6}$$

$$\frac{7}{12} - \frac{3}{20} = \frac{7 \times 5}{12 \times 5} - \frac{3 \times 3}{20 \times 3} = \frac{35}{60} - \frac{9}{60} = \frac{35-9}{60} = \frac{26}{60} = \frac{13}{30}$$

$$\frac{3}{7} \times \frac{5}{12} = \frac{3 \times 5}{7 \times 12} = \frac{15}{84}.$$

In the second example, we had to find the lowest common denominator. Here are some problems of the same type but involving variables:

$$\frac{2x+1}{3x} - \frac{x+1}{3x} = \frac{(2x+1)-(x+1)}{3x} = \frac{2x+1-x-1}{3x} = \frac{2x-x+1-1}{3x} = \frac{x}{3x} = \frac{1}{3}$$

$$\frac{2x+1}{3x} - \frac{x+1}{5x} = \frac{5(2x+1)}{15x} - \frac{3(x+1)}{15x} = \frac{(10x+5)-(3x+3)}{15x} = \frac{10x+5-3x-3}{15x} = \frac{7x+2}{15x}$$

$$\frac{3x^2}{7y} \times \frac{5y}{12xy} = \frac{3x^2 \times 5y}{7y \times 12xy} = \frac{15x^2 y}{84xy^2} = \frac{15.x.x.y}{84.x.y.y} = \frac{15x}{84y} = \frac{5x}{28y}.$$

Notes:

- We'll discuss brackets in Section 1.4, and exponents (such as x^2) in Section 1.5.
- There is a handy trick to spot whether a number has a factor of 3. Add up the digits, and if this sum is divisible by 3 then so is the original number. So in the last example above, the number 84 on the denominator has a factor of 3 because $8 + 4 = 12$, which is divisible by 3.

There are more examples to practise on in the problems at the end of this chapter.

These are the *only* rules you need for handling fractions. It's often tempting to invent your own 'mal-rules' to make things simple, for example to cancel the $3x$ terms in $\frac{3x}{3x+5}$ to get $\frac{1}{1+5} = \frac{1}{6}$, or to say $\frac{3}{4} - \frac{1}{2} = \frac{3-1}{4-2} = \frac{2}{2} = 1$. These are **wrong**.

Note: You cannot have 0 as the denominator of a fraction. $x = a/0$ would mean that $x \times 0 = a$. But any number multiplied by zero is zero, so this would only make sense if $a = 0$ too, in which case x could be anything. Thus, **division of a non-zero number by zero is impossible**, and **the fraction $\frac{0}{0}$ is undefined**. If you attempt to divide by zero on your calculator, or in Excel, you will get an error message.

Problems: Now try the end-of-chapter problems 1.3–1.14.

1.3.4 Absolute value

To find the **absolute value** of a, we throw away any negative sign. You can think of the absolute value as the 'size' of a, or the distance between 0 and a on the real line. The absolute value of a is written as $|a|$. So

$$|6| = 6, \quad |-7| = 7.$$

If a and b are two numbers on the real line, the **distance between them** is given by $|a - b|$.

$$|7 - 4| = |3| = 3, \quad |4 - 7| = |-3| = 3.$$

Problems: Now try the end-of-chapter problems 1.15–1.20.

1.3.5 Percentages

Percentages are simply fractions where the denominator is 100. Thus, as 0.25 can be written as a fraction as $\dfrac{25}{100}$, it is written as a percentage as 25%. The decimal 1.25 would become 125%. In general, to convert a decimal to a percentage, multiply it by 100 (by moving the decimal point two places to the right) and put on the % sign.

EXAMPLE

On average, the North of England experiences 10.8 days of rainfall in the month of June. That is, there is rain on 10.8 days out of 30. The proportion of rainy days is $\dfrac{10.8}{30} = 0.36$.

Expressed as a percentage, 0.36 is 36%, which is more immediately meaningful to people.
So we would report that in June, 36% of the days are rainy.
Another way of presenting the information would be to find the reciprocal:

$$\frac{10.8}{30} = \frac{1}{30/10.8} = \frac{1}{2.7777\ldots} \text{ using rule (1.13)},$$

which means that 1 day in every 2.8 days is rainy. Here we have rounded 2.7777 to 2.8, i.e. to a precision of 1 decimal place. We will discuss rounding, precision, and accuracy in Section 2.5.

We often talk about a **percentage increase** or percentage decrease. To find the percentage change in a population, the calculation is

$$\text{percentage change} = \frac{\text{new population} - \text{old population}}{\text{old population}} \times 100\% \tag{1.15}$$

EXAMPLE

The population of puffins on the Farne Islands, off the Northumberland coast, stood at 6,800 breeding pairs when first counted in 1969. By 2003 (34 years later) there were 55,674 breeding pairs. So the percentage increase over that period was

$$\frac{55674 - 6800}{6800} \times 100 = \frac{48874}{6800} \times 100 = 7.1873529 \times 100 = 718.74\% \quad \text{(correct to 2 decimal}$$

places). That is, there had been a more than sevenfold increase in the population. The average annual percentage increase had been $\dfrac{718.7}{34} = 21.14\%$ per year. If this annual percentage increase had occurred from 2003 to 2004, what population size would conservationists have expected to see?

An increase of 21.14% means that the population will increase from 100% to 121.14% of its 2003 level. So predicted 2004 population = 55674 × 121.14% = 55674 × 1.2114 = 67443 pairs. Here, we've converted a percentage to a decimal by dividing by 100 (i.e. moving the decimal point two places to the left).

In fact, the next census was taken in 2008, and the National Trust wardens found that the population had dropped to 36,500 pairs. The percentage change over this five-year period is

$$\frac{36500 - 55674}{55674} \times 100 = \frac{-19174}{55674} \times 100 = -0.3443977 \times 100 = -34.44\%$$

(correct to 2 decimal places).

A negative percentage change, here of −34.44%, means a **percentage decrease** of 34.44%.

Our decision to quote percentages correct to 2 decimal places is rather arbitrary, and such a degree of precision may not be justified. If we want to describe the situation in a way that the public will immediately understand, we could say that the population has dropped by more than a third over the past 5 years, or that there has been an annual drop of nearly 7% each year over the period.

Problems: Now try the end-of-chapter problems 1.21–1.27.

1.3.6 Basic rules for manipulating equations

When we have an equation describing a relation between numbers or variables, we can **rearrange** it by performing the same operation or sequence of operations on both sides, for example 'add 1 to both sides'. You can choose to do any operation except dividing by zero (you can multiply both sides by zero, but this is fairly pointless since it reduces any equation to 0 = 0). Provided you perform the same operation to both sides, the resulting equation remains valid.

For example, if I am given the equation $x + 3 = 5$, I can subtract 3 from both sides to solve for x:

$x + 3 = 5$

$(x + 3) - 3 = 5 - 3$

$x + 0 = 5 - 3$

$x = 2.$

The value of x we have found will satisfy the original equation; we have **solved** the equation.

You can think of this as 'moving the +3 from the l.h.s. to the r.h.s. and changing its sign: $x + 3 = 5 \rightarrow x = 5-3$' but the true process is that we are subtracting 3 from both sides.

Similarly, you learnt at school about 'cross-multiplying' to go from

$$\frac{y}{4} = \frac{3}{x}$$

to

$x.y = 12.$

In fact, what we are doing is multiplying both sides of the equation by $4.x$, the product of the denominators:

$$\cancel{4}\,x \times \frac{y}{\cancel{4}} = 4\,x \times \frac{3}{\cancel{x}}$$
$$x.y = 4 \times 3 = 12.$$

There is one further operation used in solving equations, which we'll see in Section 1.6.5.

. . . .

1.4 Brackets and the distributive law

There's a big problem with the arithmetic operations described in the previous section: the same expression can produce different answers, depending on the order in which you do the operations!

For example, the sum $3 + 4 \times 2$ can produce the answer 14 if you perform the addition first:

$$3 + 4 \times 2 = 7 \times 2 = 14,$$

or the answer 11 if you perform the multiplication first:

$$3 + 4 \times 2 = 3 + 8 = 11.$$

There are two ways we get round this. The first is by using brackets. The second is to set an order of precedence, which dictates which operation is performed first.

1.4.1 How to use brackets

With brackets, the rule is:

> **The calculation within the brackets must be done first.**

So if our expression is written as $(3 + 4) \times 2$ the calculation would be

$$(3 + 4) \times 2 = 7 \times 2 = 14,$$

whereas if it is written as $3 + (4 \times 2)$ the calculation would be

$$3 + (4 \times 2) = 3 + 8 = 11.$$

There are some expressions without brackets where the order of operations doesn't matter, for example expressions involving only additions and subtractions. The sum $3 + 4 - 2$ will produce the answer 5, no matter whether it's the addition of the subtraction which is done first. But beware: using brackets with subtraction can be tricky. You must calculate the quantity in brackets first:

$$12 - (3 + 5) = 12 - 8 = 4. \quad \text{This is different from} \quad 12 - 3 + 5 = 9 + 5 = 14.$$

In more complicated expressions there may be 'brackets within brackets'. Calculate the innermost bracket first, then work outwards:

$$1 + 4(3 - 2(4 + 1))$$
$$= 1 + 4(3 - 2 \times 5)$$
$$= 1 + 4(3 - 10)$$
$$= 1 + 4(-7)$$
$$= 1 - 28$$
$$= -27.$$

Practice:

(i) Evaluate $5(3(4 + 2) - 2(4 + 3)) + 2 \times 2$.

You should get the answer $5(18 - 14) + 4 = 24$.

(ii) Evaluate $2(7 + 6(3(2 + 1) - 2(5 - 1)))$.

You should get the answer 26.

To see which are the innermost pairs of brackets, try to visualize the brackets paired up, like this:

$$2 \, (7 + 6 \, (3 \, (2 + 1) - 2 \, (5 - 1) \,) \,).$$

To make this easier, we may use different styles of brackets (square, curly) in such expressions:

$$2\{7 + 6[3(2 + 1) - 2(5 - 1)]\}.$$

Problems: Now try the end-of-chapter problems 1.28–1.34.

EXAMPLE

The *Daily Telegraph* publishes a Mind Gym in its puzzle section; other UK newspapers have similar brain-training exercises. In these, you are given a number and instructions for performing a sequence of mental arithmetic steps on it. Here is an example, but with the initial number replaced by x:

x	TRIPLE IT	$- 14$	× BY ITSELF	$+ 10$	5/12 OF THIS	$- 1$	√	$+ 6$	75% OF THIS

Your task is to build up the algebraic expression, following these instructions and using brackets where necessary.

Solution: Here are the individual steps. You should be able to work it all out in one go, by performing 'mental algebra' instead of mental arithmetic!

x

$3x$

$3x - 14$

$(3x - 14)^2$

$(3x - 14)^2 + 10$

$\dfrac{5}{12}\big((3x - 14)^2 + 10\big)$

$\dfrac{5}{12}\big((3x - 14)^2 + 10\big) - 1$

$$\sqrt{\frac{5}{12}\left((3x-14)^2+10\right)-1}$$

$$\sqrt{\frac{5}{12}\left((3x-14)^2+10\right)-1}+6$$

$$\frac{3}{4}\left(\sqrt{\frac{5}{12}\left((3x-14)^2+10\right)-1}+6\right).$$

There are more examples to try in Problems 1.35–1.38.

1.4.2 Rule of precedence

The second way that we overcome the problem of ambiguity in evaluating an arithmetic expression, and avoid having brackets littering all our expressions, is to establish an **order of precedence for operations**. As we've just seen, it doesn't matter whether you do additions or subtractions first; the problem arises when additions (or subtractions) are combined with multiplications (or divisions). There is a convention which is universally recognized:

Perform multiplications (and divisions) before doing additions (and subtractions).

Using this convention, the expression $3 + 4 \times 2$, written without brackets, would be evaluated by performing the multiplication first:

$3 + 4 \times 2 = 3 + 8 = 11.$

This convention is part of the wider set of rules for operations described in Section 1.7.1.

1.4.3 The distributive law

It is sometimes possible to remove brackets by 'multiplying out'. This is done using the **distributive law**:

$$a(b + c) = a.b + a.c \tag{1.16}$$

For example

$3(4 + 2) + 2(4 + 3) = (3 \times 4 + 3 \times 2) + (2 \times 4 + 2 \times 3) = (12 + 6) + (8 + 6) = 18 + 14 = 32.$

But be careful when subtraction is involved:

$3(4 + 2) - 2(4 + 3) = (3 \times 4 + 3 \times 2) - (2 \times 4 + 2 \times 3)$
$= 12 + 6 - (8 + 6) = 18 - 14 = 4.$

The distributive law is more often used in algebraic expressions, e.g. $3(x + 4) = 3x + 12$.

It can also be needed if an expression in brackets is being subtracted, as in the example above:

$18 - (8 + 6) = 18 - 8 - 6 = 4.$

Here, we are using the distributive law, taking a as -1. Here's an algebraic example:

$x - (b + c) = x - 1 \times (b + c) = x - 1 \times b + -1 \times c = x - b - c.$

EXAMPLE

Simplify $3(x + 4y - 2) - 2(2x - y - 3)$

Solution:

$$\begin{aligned}
3(x + 4y - 2) - 2(2x - y - 3) &= (3x + 12y - 6) - (4x - 2y - 6) \\
&= 3x + 12y - 6 - 4x + 2y + 6 \\
&= 3x - 4x + 12y + 2y - 6 + 6 \\
&= -x + 14y.
\end{aligned}$$

. . . .

1.5 Exponents

We defined multiplication as meaning several copies of a number added together:

$$3 \times 7 = 7 + 7 + 7 = 21.$$

How about several copies of a number multiplied together, e.g. $7 \times 7 \times 7 = 343$? This will be written as 7^3, pronounced 'seven cubed'. The raised 3 is called an **exponent**. It is vital that you understand the rules for manipulating exponents.

1.5.1 Definition of exponents

Let's start with the simplest case.

The **square** of a number is that number multiplied by itself. So the square of 7 is 7×7, which is 49. We write the square of 7 as 7^2. Calculators have a button for finding the square of a number; it usually looks like ⬛.

If we square a negative number, this is the product of two negative numbers, which is positive.

$$(-7)^2 = (-7) \times (-7) = 49.$$

So

$$a^2 = a \times a, \quad (-a)^2 = a^2. \tag{1.17}$$

We have explained that the **cube** of a number is formed by multiplying three copies of the number together, e.g. $4^3 = 4 \times 4 \times 4 = 64$. This is also referred to as 'raising 4 to the power of three'.

In general, if we **raise a number a to the nth power**, its means we multiply n copies of a together. The result is denoted by a^n.

$$a^n = a \times a \times \cdots \times a \; (n \text{ times}). \tag{1.18}$$

In a^n, the number a is called the **base**, and n is called the **exponent** or **power** or **order**. We will refer to it as the exponent, as the other words are more vague, with additional meanings – and because 'exponent' is a more impressive-sounding word to non-mathematicians.

Here are some examples:

$$7^5 = 16807 \qquad 5^7 = 78125 \qquad (-3)^5 = -243 \qquad 3^2 = 9 \qquad 2^3 = 8.$$

Check these by hand, and also on your calculator. There is an exponent button marked
⬛. You enter the base x first, then press this button, then enter the exponent y, and
finally press =. To get −3 as the base, you'll need to enter 3 and then press the negation key
(Section 1.3.1).

1.5.2 Rules for exponents

If $4^3 = 4 \times 4 \times 4$ and $4^2 = 4 \times 4$, then what is $4^3 \times 4^2$?

$$4^3 \times 4^2 = (4 \times 4 \times 4) \times (4 \times 4) = 4 \times 4 \times 4 \times 4 \times 4 = 4^5.$$

In general, if you want to multiply a^m with a^n, you are multiplying m copies of a together with
n copies of a. That is, you are multiplying $m + n$ copies of a together. We write this as a^{m+n}.

This illustrates the first rule in the following rules for exponents:

$$
\begin{aligned}
a^m \times a^n &= a^{m+n} \\
a^m \div a^n &= a^{m-n} \\
(a^m)^n &= a^{m.n} \\
(a \times b)^n &= a^n \times b^n \\
\left(\frac{a}{b}\right)^n &= \frac{a^n}{b^n} \\
a^0 &= 1 \\
a^{-n} &= \frac{1}{a^n} \\
0^0 &= 1.
\end{aligned}
$$

(1.19)

Here are some examples, one for each of the first four rules:

(i) $6^7 \times 6^3 = 6^{7+3} = 6^{10}$ (iii) $(3^2)^4 = 3^{2\times4} = 3^8$

(ii) $6^7 \div 6^3 = 6^{7-3} = 6^4$ (iv) $(6 \times 3)^2 = 6^2 \times 3^2$.

We've already explored why the first rule works. To explore the second rule, look at the left-
hand side of example (ii). We can write this as a fraction, use the definition of a^n, and then
cancel common factors of 6 on numerator and denominator:

$$6^7 \div 6^3 = \frac{6^7}{6^3} = \frac{\cancel{6} \times \cancel{6} \times \cancel{6} \times 6 \times 6 \times 6 \times 6}{\cancel{6} \times \cancel{6} \times \cancel{6}} = 6 \times 6 \times 6 \times 6 = 6^4.$$

You should make similar explorations of the examples, to see why the third, fourth, and fifth
rules work.

Our definition of a^n, as meaning 'multiply a by itself n times', is fine if n is a positive integer.
But what if n is zero or negative? We can use the rules above to explore this, in the following
examples:

(i) $\dfrac{6^3}{6^2} = 6^{3-2} = 6^1$.   But $\dfrac{6^3}{6^2} = \dfrac{\cancel{6} \times \cancel{6} \times 6}{\cancel{6} \times \cancel{6}} = 6$.   So $6^1 = 6$.

(ii) $\dfrac{6^3}{6^3} = 6^{3-3} = 6^0$.   But $\dfrac{6^3}{6^3} = \dfrac{\cancel{6} \times \cancel{6} \times \cancel{6}}{\cancel{6} \times \cancel{6} \times \cancel{6}} = 1$.   So $6^0 = 1$.

(iii) $\dfrac{6^2}{6^4} = 6^{2-4} = 6^{-2}$.   But $\dfrac{6^2}{6^4} = \dfrac{\cancel{6} \times \cancel{6}}{\cancel{6} \times \cancel{6} \times 6 \times 6} = \dfrac{1}{6 \times 6} = \dfrac{1}{6^2}$.   So $6^{-2} = \dfrac{1}{6^2}$.

The first example tells us that, in general, $a^1 = a$. You might have expected this.
The second example shows that $a^0 = 1$. This is the sixth rule of exponents in (1.19).
The third example shows that $a^{-n} = \dfrac{1}{a^n}$. This is the last rule of exponents in (1.19).

   We also note that when 1 is multiplied by itself, any number of times, we just get 1. And when 0 is multiplied by itself any number of times we get 0. So $1^n = 1$, and $0^n = 0$ if n is positive.

   There is one more quantity we don't know: what would 0^0 be? Mathematicians have decided to take $0^0 = 1$.

Warning: The rules of exponents only apply when multiplication or division is occurring, and when the bases or the exponents are the same. Expressions such as $6^3 + 6^2$, or $6^3 \times 5^2$ **cannot** be simplified by these rules.

As we will see in the next chapter, there is a very important scientific use of exponents with a base of 10. Here we will observe that

$$10^2 = 10 \times 10 = 100 \qquad \text{(one hundred)}$$
$$10^3 = 10 \times 10 \times 10 = 1\,000 \qquad \text{(one thousand)}$$
$$10^4 = 10 \times 10 \times 10 \times 10 = 10\,000 \qquad \text{(ten thousand)}$$

and in general the number 10^n will be a 1 followed by n zeros. Thus,

   1 million = $1\,000\,000 = 10^6$,

   1 billion = 1 thousand million = $1\,000\,000\,000 = 10^9$, and

   1 trillion = 1 thousand billion = $1\,000\,000\,000\,000 = 10^{12}$.

The term 'a googol' to mean the number 10^{100} was coined in 1938 and has entered popular culture.[3]
   Also,

$$10^{-1} = \frac{1}{10} = 0.1$$
$$10^{-2} = \frac{1}{10^2} = \frac{1}{100} = 0.01$$
$$10^{-3} = \frac{1}{10^3} = \frac{1}{1000} = 0.001$$

so that the number 10^{-n} will be a decimal point followed by $n - 1$ zeros before the 1.
   Here are some physical illustrations, to give a feel for such large numbers:

- The age of the Universe is 1.37×10^{10} years (which is 13.7×10^9 or 13.7 billion years).
- A hydrogen bomb exploding releases 2×10^{17} joules of energy (which is $2 \times 10^8 \times 10^9$ or 200 million billion joules).

---

[3] Not least because Google is a mis-spelling of it.

CASE STUDY A2: Formula for geometric growth

In our geometric growth model, we supposed that at each generation the population size doubles. This can be written as

$$N_{p+1} = 2N_p$$

where N_p is the population size after p generations, and N_{p+1} is the population size at the next (the $(p + 1)$th) generation. For example, after 10 generations the population size is N_{10}. But $N_{10} = 2N_9$ and $N_9 = 2N_8$, so $N_{10} = 2N_9 = 2(2N_8) = 4N_8$ and so on:

$$N_{10} = 2N_9 = 4N_8 = 8N_7 = \ldots = 512N_1 = 1024N_0.$$

Using exponents: $N_{10} = 2N_9 = 2^2N_8 = 2^3N_7 = \ldots = 2^9N_1 = 2^{10}N_0$.

In general, the population size after p generations is $N_p = 2^pN_0$.

If each cell had divided into three at each generation, we would have $N_{p+1} = 3N_p$ and hence $N_p = 3^pN_0$. The most general model for geometric growth is where $N_{p+1} = \lambda N_p$, in which case $N_p = \lambda^pN_0$. λ (the Greek letter l, pronounced 'lambda') is called the *multiplication rate.*

- A typical atom has a diameter of 0.0000008 millimetres (which is $8 \times 10^2 \times 10^{-9}$ mm or 800 billionths of a millimetre.

- The US National Debt stands at \$11,046,247,657,049.48 (according to US Treasury Direct, 26 March 2009), i.e. 11×10^{12} or 11 trillion dollars.

1.5.3 Products and factors

When two or more numbers or expressions are multiplied together, the result is called a **product.** So the product of 3 and 8 is 24. Each number or expression involved in the multiplication is called a **factor.** So 3 is a factor of 24. Also, 2 and 4 and 6 and 8 and 12 are factors of 24. The expression $(x + 3)(x + 5)$ has two factors: $(x + 3)$ and $(x + 5)$.

But we can multiply out the product $(x + 3)(x + 5)$ using the distributive law (1.16):

$$(x + 3)(x + 5) = x(x + 5) + 3(x + 5) = x^2 + 5x + 3x + 15 = x^2 + 8x + 15,$$

so we can say that $(x + 3)$ and $(x + 5)$ are factors of $x^2 + 8x + 15$. In the problems you'll see more complicated examples, but they can all be multiplied out by using the distributive law repeatedly.

Similarly, $(x - 3)$ and $(x + 1)$ are factors of $x^2 - 2x - 3$, since this expression is what you get when you multiply out $(x - 3)(x + 1)$:

$$(x - 3)(x + 1) = x(x + 1) - 3(x + 1) = x^2 + x - 3x - 3 = x^2 - 2x - 3.$$

Notice that the middle terms partially cancel out: $+ x - 3x = -2x$. Similarly:

$$(x - 1)(x + 1) = x(x + 1) - 1(x + 1) = x^2 + x - x - 1 = x^2 - 1.$$

The process of converting such an expression (called a **polynomial** in x; we'll see them in Chapter 7) into factors is called **factorization**. You will have been taught this technique at school, and probably thought it was pointless. You were largely right; the expressions you were given to factorize had the coefficients carefully chosen to make the result come out in nice whole numbers, but the expressions you'll derive as a scientist from real-world experiment and observation are not so perfect, even if the coefficients look simple. Thus, $x^2 - 2x - 3 = (x - 3)(x + 1)$, but $x^2 - x - 3 = (x - 2.3027756)(x + 1.3027756)$, and $x^2 - 2x + 3$ can't be factorized at all! But there are some important factorizations which we'll use in this book:

$$
\begin{aligned}
(x-1)(x+1) &= x^2 - 1 \\
(x-1)(x^2 + x + 1) &= x^3 - 1 \\
(x-a)(x+a) &= x^2 - a^2 \\
(a+b)(c+d) &= a.c + b.c + a.d + b.d \\
(a+b)^2 &= a^2 + 2ab + b^2 \\
(a-b)^2 &= a^2 - 2ab + b^2 \\
(a+b)^3 &= a^3 + 3a^2b + 3ab^2 + b^3 \\
(a-b)^3 &= a^3 - 3a^2b + 3ab^2 - b^3.
\end{aligned}
\tag{1.20}
$$

You should check these results by multiplying out, using the distributive law, as we did above with the first result.

A useful fact is that the only way you can multiply two numbers together and get zero is if one or both of them are zero:

$$\text{If } a.b = 0, \text{ then either } a = 0 \text{ or } b = 0 \text{ or both.} \tag{1.21}$$

So if $(x - 3)(x + 1) = 0$ then either $x - 3 = 0$ or $x + 1 = 0$. That is, either $x = 3$ or $x = -1$.

Problems: Now try the end-of-chapter problems 1.39–1.50.

. . . .

1.6 **Roots**

1.6.1 Definition of roots

At the start of the last section, we defined the square of a number a as being that number multiplied by itself: $a^2 = a.a$. We now define the **square root** of a to be the positive number that produces a when you square it. That is,

$$x \text{ is the } \textbf{square root} \text{ of } a \left(\text{written } x = \sqrt{a} \right) \text{ if } x \text{ is positive } \quad \text{and} \quad x^2 = a. \tag{1.22}$$

For example, $\sqrt{49} = 7$ since $7^2 = 49$. You have a ▨ key on your calculator, for finding square roots. If we just talk about 'the root of a', we mean the square root. Note that $\sqrt{x^2} = \left(\sqrt{x} \right)^2 = x$.

We can generalze this idea to higher roots:

$$x \text{ is the } \textit{n}\textbf{th root} \text{ of } a \text{ if } \quad x^n = a. \tag{1.23}$$

The nth root of a is written as $\sqrt[n]{a}$. For example, 3 is the 4th root of 81, since $3^4 = 81$, and we could write $3 = \sqrt[4]{81}$.

1.6.2 Roots and exponents

Roots can be combined in a similar way to exponents, namely:

$$\sqrt{a.b} = \sqrt{a}.\sqrt{b} \quad \text{and} \quad \sqrt{\frac{a}{b}} = \frac{\sqrt{a}}{\sqrt{b}}. \tag{1.24}$$

As a consequence of these rules

$$c\sqrt{a.} = \sqrt{c^2}\sqrt{a} = \sqrt{c^2 a} \quad \text{and} \quad \frac{\sqrt{a}}{c} = \frac{\sqrt{a}}{\sqrt{c^2}} = \sqrt{\frac{a}{c^2}}. \tag{1.25}$$

So if you are bringing a coefficient inside the root sign, you have to square it. For example

$$\frac{\sqrt{17.64}}{3} = \sqrt{\frac{17.64}{9}} = \sqrt{1.96} = 1.4.$$

But there's a closer link between roots and exponents. The square root of a is actually the same as a raised to the power (or exponent) of 1/2. Why is this?

Suppose that $x = a^{1/2}$.

Then $x^2 = \left(a^{1/2}\right)^2 = a^{\frac{1}{2} \cdot 2} = a^1 = a$ using the rules for exponents from the previous section.

So $x^2 = a$, meaning that $x = \sqrt{a}$.

This can be generalized to nth roots. We have

$$\sqrt{a.} = a^{1/2} \qquad \sqrt[n]{a} = a^{1/n}. \tag{1.26}$$

So $3 = 81^{1/4}$, or $81^{0.25}$.

We have only considered the square roots of positive numbers (and should note that $\sqrt{0} = 0$). This is because **square roots of negative numbers don't exist**. You cannot find a number that produces a negative answer when multiplied with itself.[4]

You can find cube roots of negative numbers, though. For example, $\sqrt[3]{-8} = -2$, since $(-2) \times (-2) \times (-2) = -8$. Most calculators have a cube root key, perhaps as an inverse of the x^3 key. If not, or if you want to find other roots, you need to convert it using (1.26) and use the x^y key. For example, to find the fifth root $\sqrt[5]{7}$, write it as $7^{0.2}$. In this book we'll only need square roots (and the occasional cube root).

In some situations we need to consider negative square roots as well as positive ones. For example, $\sqrt{9} = 3$ (since $3^2 = 9$), but -3 is also a negative square root of 9, since $(-3)^2 = 9$. When we talk about '$\sqrt{a}$, the square root of a' we mean the positive root (which is what your calculator produces also), unless we specify '$-\sqrt{a}$, the negative root of a'. If we want to write a symbol to show we are intending to consider both roots of a, we can write $\pm\sqrt{a}$.

---

[4] Here's the proof:

Suppose that $\sqrt{-3}$ exists; call it z. So $z^2 = -3$. Then by the Trichotomy of Cardinals (Section 1.2.2), z would have to be either negative, zero or positive. If z were negative, the product z^2 would be positive, by (1.7). This is also the case if z is positive. And if z is zero then z^2 would be zero. So there is no real number whose square is a negative real number.

We could now define the **absolute value** of a, where a may be positive or negative, as

$$|a| = \sqrt{a^2}. \tag{1.27}$$

So $|8| = \sqrt{(8)^2} = \sqrt{64} = 8$, and $|-6| = \sqrt{(-6)^2} = \sqrt{36} = 6$.

Problems: Now try the end-of-chapter problems 1.51–1.58.

1.6.3 Irrational numbers

Some numbers have 'nice' square roots, e.g. $\sqrt{49} = 7$ and $\sqrt{13.69} = 3.7$. But what about $\sqrt{2}$? If I evaluate $\sqrt{2}$ on my calculator, I get 1.4142136. But this is giving the root correct to 8 decimal places (we'll discuss precision in the next chapter); if I use a more powerful calculator like the one provided with MS Windows, I get 1.4142135623730950488016887242097. And it doesn't stop there. The decimal value of $\sqrt{2}$ is a 'non-terminating decimal', similar to the decimals for 1/3 and 1/7:

$$\frac{1}{3} = 0.3333333333\ldots \text{ and } \frac{1}{7} = 0.142857142857142857\ldots$$

But with those fractions, there is a repeating pattern in the sequence of decimal digits. In the decimal expansion of $\sqrt{2}$ there is no pattern: to find the millionth digit I would need to evaluate all the previous ones. This is because $\sqrt{2}$ cannot be expressed as a rational number, i.e. you can't find integers m and n such that $\sqrt{2} = m/n$.

   We say that $\sqrt{2}$ is **irrational**. There are other real numbers that are irrational, for example $\sqrt{3}$ and π. There is more about rational and irrational numbers on our website, including a proof that $\sqrt{2}$ is irrational.

1.6.4 Surds

If you want to find the value of $z = \dfrac{2+\sqrt{3}}{5-\sqrt{3}}$, you could calculate $\sqrt{3}$ as 1.732, and insert this:

$$z = \frac{2+\sqrt{3}}{5-\sqrt{3}} = \frac{3.732}{3.268} = 1.14198 \quad \text{(correct to 5 decimal places)}.$$

But can we be sure that this value is really correct to 5 decimal places, when we only found $\sqrt{3}$ correct to 3 decimal places? We'll discuss this problem of accuracy in the next chapter. But it is actually possible to find z exactly. We use the factorization $(x-a)(x+a) = x^2 - a^2$ from (1.20). Putting $x = 5$ and $a = \sqrt{3}$, we have $\left(5-\sqrt{3}\right)\left(5+\sqrt{3}\right) = 5^2 - \left(\sqrt{3}\right)^2 = 25 - 3 = 22$. If we multiply both the numerator and the denominator of z by $\left(5+\sqrt{3}\right)$, we will get an integer on the denominator, and we can then divide out:

$$z = \frac{2+\sqrt{3}}{5-\sqrt{3}} = \frac{2+\sqrt{3}}{5-\sqrt{3}} \times \frac{5+\sqrt{3}}{5+\sqrt{3}} = \frac{\left(2+\sqrt{3}\right)\left(5+\sqrt{3}\right)}{\left(5-\sqrt{3}\right)\left(5+\sqrt{3}\right)} = \frac{2\times5+2\sqrt{3}+5\sqrt{3}+\sqrt{3}.\sqrt{3}}{22}$$

$$= \frac{10+7\sqrt{3}+3}{22} = \frac{13+7\sqrt{3}}{22} = \frac{1}{22}\left(13+7\sqrt{3}\right).$$

This final form of z is called a **surd,** which is a number made from a rational number plus a coefficient multiplying a root, i.e. of the form $a + b\sqrt{c}$.

1.6.5 A third operation in manipulating equations

In Section 1.3.6 we showed how equations can be manipulated by performing the same operation on both sides; we can:

- add/subtract the same nonzero quantity;
- multiply/divide by the same nonzero quantity.

A third operation is to raise both sides to the same nonzero exponent (i.e. power). This includes taking roots of both sides, since taking the nth root of x is equivalent to raising x to the power $\frac{1}{n}$:

$$x^3 = 5 \quad \Rightarrow \quad x = \sqrt[3]{5}$$

or, using exponents:

$$x^3 = 5 \quad \Rightarrow \quad (x^3)^{\frac{1}{3}} = (5)^{\frac{1}{3}} \quad \Rightarrow \quad x^{3 \cdot \frac{1}{3}} = 5^{\frac{1}{3}} \quad \Rightarrow \quad x^1 = 5^{\frac{1}{3}} \quad \Rightarrow \quad x = 5^{\frac{1}{3}}.$$

Remember that when taking square roots, there can be two possible solutions, coming from taking the positive or the negative root:

> **EXAMPLE**
>
> Solve $(5x - 2)^2 = 9$.
> **Solution:** $5x - 2 = \pm\sqrt{9} = \pm 3$ taking square roots of both sides
> $5x = 2 \pm 3 = 5$ or -1 adding 2 to both sides
> $x = \dfrac{5}{5}$ or $x = \dfrac{-1}{5} = -0.2$ dividing both sides by 5.
>
> Finally, you should check your answers by inserting them in the original equation.

Problems: Now try the end-of-chapter problems 1.59–1.64.

. . . .

1.7 Evaluating expressions

We have now established all the basic facts we need about numbers, and the rules for handling them (and for handling variables that represent numbers). Much of this will be vaguely familiar from maths lessons at school, but when it's all put together like this, you can hopefully see that mathematics is not just a jumble of arbitrary rules, but that it all fits together into a single, consistent, and rather elegant theory. Also, we can now talk like real mathematicians – about integers and rationals, for example, rather than 'whole numbers' and fractions. In this section we will use these principles to evaluate and manipulate complex numerical and algebraic expressions.

1.7.1 Order of operations

In Section 1.4 we said that you evaluate the expression in brackets first, and that you perform multiplications and divisions before doing additions and subtractions. If we put exponents and roots into this scheme, we arrive at BEDMAS. BEDMAS is a supposedly easy-to-remember acronym which tells the order of precedence for arithmetic operations:

> Brackets
>
> Exponents
>
> Division
>
> Multiplication
>
> Addition
>
> Subtraction

That is:

- Expressions within brackets must be evaluated first.
- Next, evaluate exponents and roots.
- Then do multiplications and divisions.
- Finally, do additions and subtractions.

Sometimes books give this as BODMAS, where the O stands for 'Orders', or as BIDMAS, where the I stands for 'Indices'. These are other words people use for Exponent, but they can have other meanings as well. Exponent is the clearest technical term for the operation a^n.
Roots are on a par with exponents, since, for example, $\sqrt{36+64}$ is the same as $(36+64)^{\frac{1}{2}}$. Multiplication and division are on equal footings, as are addition and subtraction. Sometimes the brackets are not written down, which makes life difficult:

- Complex fractions can be thought of as (numerator)/(denominator): $\dfrac{x+1}{y+1} = (x+1) \div (y+1)$. So you can't 'cancel the 1s' between numerator and denominator.
- The same pitfall occurs with exponents: a^{n+1} means $a^{(n+1)}$ so you must evaluate $(n+1)$ first. It's not the same as $a^n + 1$!

1.7.2 Handling complex fractions

At the start of this chapter, we asked you if you could evaluate the numerical expression

$$\frac{6 \times 4^2 - 10\sqrt{3 \times 4.88 + 3}}{12 \times 4}. \tag{1.1}$$

You should now be able to do this without problem. First you must evaluate the expression under the square root sign (doing the multiplication before the addition). Then take the square root, before multiplying the result by 10. Store this in the calculator. Evaluate the other term on the numerator (squaring the 4 before multiplying by 6). Perform the subtraction. Store the result. Now evaluate the denominator, and finally perform the division.

$$\frac{6\times 4^2 - 10\sqrt{3\times 4.88 + 3}}{12\times 4}$$

$$=\frac{6\times 4^2 - 10\sqrt{14.64 + 3}}{12\times 4}$$

$$=\frac{6\times 16 - 10\sqrt{17.64}}{12\times 4}$$

$$=\frac{96 - 10\times 4.2}{12\times 4} = \frac{96 - 42}{12\times 4} = \frac{54}{48} = \frac{\cancel{6}\times 9}{\cancel{6}\times 8} = \frac{9}{8} = 1.125.$$

A **complex fraction** properly refers to a fraction involving 'sub-fractions' on the numerator and/or the denominator. Here, the trick is to multiply top and bottom by the same quantity so that the sub-fractions disappear. Here is an example:

$$\frac{x^2 - \dfrac{1}{x-1}}{1 + \dfrac{1}{x-1}} = \frac{x^2(x-1) - 1}{(x-1) + 1} = \frac{(x^3 - x) - 1}{x - 1 + 1} = \frac{x^3 - x - 1}{x} = \frac{x^3}{x} - \frac{x}{x} - \frac{1}{x} = x^2 - 1 - \frac{1}{x}.$$

Here, we multiplied the numerator and denominator by $(x - 1)$ in order to remove the sub-fractions.

We could spend many more pages demonstrating ways of simplifying ever more complicated algebraic expressions. But in the Introduction we promised to keep the algebra to a minimum, so we shall leave these techniques until we actually need them. Instead, we shall say something about how numerical expressions can be programmed into Excel.

1.7.3 Numerical expressions in Excel

In this chapter we've indicated how to use a pocket calculator to perform the arithmetic we've described. A more powerful way of performing calculations is to program expressions into cells in Microsoft Excel. We are assuming now that you are sitting at a PC that has Excel installed. If you haven't used Excel before, see the Technical Notes for online tutorials.

To program a calculation into a cell, double-click the cell and start typing: the first character you type is =, to indicate that this is an expression, not a number or text. Then type the expression, using cell references instead of the variables. The expression is called a formula in Excel, and will appear in the formula bar above the spreadsheet grid. For example, if we want cell F9 to display the sum of the numbers contained in cells A5 and B5, you would double-click on the cell F9 and type

= a5 + b5

and press Enter. You don't have to type capital letters, Excel will capitalize them for you.

If you look at SPREADSHEET 1.1 (which you can download from the website for this book), you will see that this has been done. Click on cell F9 to see the formula in the formula bar.

In Excel, you use the * symbol for multiply, and the / symbol for divide. Otherwise, typing formulas in Excel is very similar to using a calculator. You must always type the multiplication: to calculate 8 times the value in cell A5, type '= 8 * A5', not '= 8 A5'. You also type brackets where they are sometimes not written in algebraic expressions, as we mentioned in Section 1.7.1.

Table 1.1 gives some algebraic expressions in the left-hand column, which are rewritten as Excel-type algebra in the middle column. The right-hand column contains the Excel formula;

Table 1.1 Some simple formulas in Excel

	Algebraic expression	Excel-type algebra	Excel formula
(i)	$a.c + b.c$	$a*c + b*c$	= A5*C5 + B5*C5
(ii)	$a(b + c)$	$a*(b + c)$	= A5*(B5 + C5)
(iii)	$\dfrac{a+c}{b-c}$	$(a + c)/(b - c)$	= (A5 + C5)/(B5-C5)
(iv)	$\dfrac{1}{a} - 1 - \dfrac{1}{a-1}$	$1/a - 1 - 1/(a - 1)$	= 1/A5-1-1/(A5-1)
(v)	$\dfrac{b - \dfrac{1}{a-1}}{1 + \dfrac{1}{a-1}}$	$(b - 1/(a - 1))/(1 + 1/(a - 1))$	= (B5-1/(A5-1))/(1 + 1/(A5-1))

here, we assume that the variables a, b, c are stored in the cells A5, B5, C5 respectively. SPREADSHEET 1.1 has these formulas (i) – (v) programmed into cells G7, H7, I7, J7, and K7.

Notice that Excel is clever enough to apply the BEDMAS rule that multiplications and divisions take precedence over additions and subtractions. Your calculator may be able to do this in simple calculations, but not in more complex ones.

You should try typing different numbers into cells A5, B5, C5 and check the values that appear in the cells in column F. In particular, when the values in B5 and C5 are equal, expression (iii) cannot be evaluated as we cannot divide by zero. What Excel displays is the error message #DIV/0!

If you want to calculate a^2, it's probably easiest to use A5*A5. But for higher exponents you can use the ∧ key to indicate an exponent: a^7 would be programmed as A5∧7. Remember you'll need to use brackets with exponents, as in Section 1.7.1: a^{b+1} would be written as A5∧(B5 + 1).

For a square root, you must use the **Excel function** SQRT(). For example, $\sqrt{a}$ would be programmed as SQRT(A5). With Excel functions, you must follow the function name with brackets enclosing the expression being acted on, even if it's just a single cell address; this is called the **argument** of the function. There is another Excel function for the absolute value: $|a|$ would be written as ABS(A5). Table 1.2 shows the translation of some more complex algebraic functions into Excel-type algebra, and then to Excel formulas.

Table 1.2 Some not-so-simple formulas in Excel

	Algebraic expression	Excel-type algebra	Excel formula
(vi)	$\sqrt{a^2 + b^2}$	$\sqrt{(a^2 + b^2)}$	= SQRT(A5*A5 + B5*B5)
(vii)	$\dfrac{6a^2 - 10\sqrt{3b+1}}{12a}$	$(6*a*a - 10*\sqrt{(3*b + 1)})/$ $(12*a)$	=(6*A5*A5-10*SQRT(3*B5 + 1))/ (12*A5)
(viii)	$\sqrt{|a - b|}$	$\sqrt{(|a - b|)}$	=SQRT(ABS(A5-B5))
(ix)	$\dfrac{\sqrt{a.b}}{\sqrt{a} - \sqrt{b}}$	$\sqrt{(a*b)}/(\sqrt{(a)} - \sqrt{(b)})$	=SQRT(A5*B5)/ (SQRT(A5)-SQRT(B5))

These are also programmed in SPREADSHEET 1.1.

The most common mistake people make in programming formulas into spreadsheets, is to forget the brackets around the denominator in example (vii). If you want to program $\frac{a-b}{3c}$ and you type (A5-B5) / 3*C5, Excel will calculate $\frac{a-b}{3}c$.

CASE STUDY A3: Birth and death rates

In geometric growth, the population size after $p+1$ generations is given by $N_{p+1} = \lambda N_p$. This is called an **update equation**; it allows us to update N from one generation to the next.

Using the multiplication rate λ makes sense when we are thinking about cells dividing, but when ecologists look at the growth of a colony of animals, they talk about the *birth rate B* and the *death* (or mortality) *rate D*. B is the number of live births per head of population, from one generation to the next. This can be expressed as a percentage, e.g. '$B = 2.5\%$' means that there are 2.5 live births per hundred, or 25 live births per thousand in the population. So a population of 8000 would see $\frac{2.5}{100} \times 8000$, or 25×8, i.e. 200 births in the next generation.

Of course, some individuals will also die; this is measured by the death rate D. If our population of 8000 had a death rate of 1.5%, there would be $\frac{1.5}{100} \times 8000 = 120$ deaths in the colony. Thus, in moving to the next generation we have 200 births but 120 deaths, meaning a net increase of $200 - 120 = 80$ in the population. So the new population is $8000 + 80 = 8080$.

If the population size is N_0, and the birth rate is B, then the number of live births in that generation is $B.N_0$. In our example, $N_0 = 8000$ and $B = 2.5\% = \frac{2.5}{100} = 0.025$. Similarly, if the death rate is D, the number of deaths in that generation is $D.N_0$.

The net growth rate per head of the population R is obtained by subtracting the death rate from the birth rate: $R = B - D$. Then if the initial population is N_0, the next-generation population will be given by

$$N_1 = N_0 + B.N_0 - D.N_0$$
$$= N_0 + (B - D).N_0$$
$$= N_0 + R.N_0.$$

We can see that this is still a model of geometric growth, since

$$N_1 = N_0 + R.N_0 = (1 + R)N_0.$$

This is identical to our original model, in which $N_1 = \lambda N_0$, and the relationship between the multiplication rate of the population λ, and the growth rate per head R, is

$$\lambda = 1 + R.$$

In general,

$$N_{p+1} = N_p + R.N_p. \tag{1.28}$$

The geometric growth model is programmed into EXCEL SPREADSHEET 1.2. You can enter the values of the initial population size, birth rate, and death rate per head into cells

B5, B6, and B7. The cells currently have the values used in the example above: 8000, 0.025, and 0.015 (note that the rates are written as decimals, not percentages). The worksheet calculates the growth rate and multiplication rate, and uses the latter to calculate the new population sizes for the first five generations (rounded to zero decimal places). Highlight the block of cells D10, E10, F10, G10, then left-click on the square button in the bottom right corner of the block, and drag it down through four or five rows below. The formulae in cells E10 and G10 are copied into the following rows, with the cell references changing so that each cell formula refers to the cell in the row above it. To ensure that the reference to the multiplication rate in cell B9 remains fixed, it is written with dollar signs $. So the formula in cell G10 is ' = $B$9*G9', and when you've dragged it down, the formula in cell G11 will be ' = $B$9*G10'. If you are not familiar with this feature of Excel, you should work through an introductory guide or online tutorial (see the Technical Notes).

CASE STUDY B3: Evaluating the angiogenic cancer cell density

Recall that we are using x_0, x_1, y to represent respectively the densities of healthy, non-angiogenic, and angiogenic cells in the tumour, and that the rate of increase of angiogenic cancer cells is

$$r.y.\left(1 - \frac{y}{k}\right) - d.y - \frac{y.p.(x_0 + x_1 + y)}{q.y + 1}$$

in which r is the replication rate, and d is the death rate of the angiogenic cells. Can you see that the first two terms in this expression look very similar to the expression $B.N - D.N$ used in Case Study A3, giving the growth of the population N, as the births minus the deaths? But why is there the extra factor $\left(1 - \frac{y}{k}\right)$ multiplying the replication (i.e. birth) rate? We'll find out in Chapter 7.

Also recall that at equilibrium the density of angiogenic cells y is given by the expression

$$y = \frac{-Q + \sqrt{Q^2 - 4rq[k(d - r) + pk(x_0 + x_1)]}}{2rq}$$

where

$$Q = kq(d - r) + r + kp.$$

We are given the following values for the different variables:

death rate coefficient $d = 0.1$
replication rate coefficient $r = 0.15$
maximum density of angiogenic cells $k = 100$
angiogenesis inhibitor value $p = 1.0$
angiogenesis promoter value $q = 10.0$.

Then

$$Q = 100 \times 10(0.1 - 0.15) + 0.15 + 100 \times 1 = 1000(-0.05) + 100.15$$
$$= -50 + 100.15 = 50.15.$$

So the expression for y becomes

$$y = \frac{-50.15 + \sqrt{(50.15)^2 - 4 \times 0.15 \times 10[100(0.1 - 0.15) + 1 \times 100(x_0 + x_1)]}}{2 \times 0.15 \times 10}$$

$$y = \frac{-50.15 + \sqrt{2515 - 6[-5 + 100(x_0 + x_1)]}}{3}$$

$$y = \frac{-50.15 + \sqrt{2545 - 600(x_0 + x_1)}}{3}$$

$$= -16.71 + \sqrt{282.78 - 66.67(x_0 + x_1)}.$$

Notice that the expression under the square root sign has been divided by 9, not 3. This is because $\frac{\sqrt{a}}{3} = \frac{\sqrt{a}}{\sqrt{9}} = \sqrt{\frac{a}{9}}$.

The formulae for Q and y have been programmed in SPREADSHEET 1.3. You should try entering different values for x_0 and x_1. Is it possible to reduce y to zero? Notice that if the number under the root sign becomes negative, Excel gives an error message: cell B18 shows #NUM! to indicate that the number cannot be calculated. In instalment B4 we'll find the conditions which can make y zero.

1.8 Extension: intervals and inequalities

In Section 1.2.2 we introduced the < symbol, meaning 'is less than', to produce an order for the set of real numbers. $a < b$ means that 'a is less than b', which is equivalent to saying that a lies to the left of b on the real line. We can also use the symbol $\leq$ to mean 'less than or equal to'. We can use these symbols to define parts of the real line, called intervals.

1.8.1 Intervals on the real line

If x represents any real number, I can write $x < 0$ to mean 'all real numbers x which are less than zero', i.e. the set of negative numbers. This would be the left-hand half of the real line. We could write $x > 0$ for the set of positive numbers. Mathematicians actually write sets using curly brackets, so we should write $\{x \mid x < 0\}$ to mean 'the set of all numbers x satisfying $x < 0$'. But in this book we'll just talk about the set of numbers $x < 0$.

Similarly, I could write $1 < x < 2$ to mean the set of all real numbers that are greater than 1 and less than 2. On the real line, this is the strip or interval between the points 1 and 2. The idea of intervals is very useful in numerical mathematics, as we'll see in the extensions of subsequent chapters.

What interval would be defined by $|x - 5| < 2$? Recall from Section 1.3.4 that $|a - b|$ means the distance between a and b on the real line. So $|x - 5| < 2$ means: 'the distance between x and 5 is less than 2 units', i.e. that x lies within 2 units of the point 5 on the real line. This would be the interval from 3 to 7 (since, starting from 5 and travelling 2 units to the left brings you to the point $5 - 2 = 3$, and, starting at 5 again and moving 2 units to the right brings you to 7).

Figure 1.3 The interval $|x - a| < d$

We say that the interval $|x - 5| < 2$ has 5 as its **midpoint**, and that its **semi-width** is 2 (its total width, from 3 to 7, is 4 units, twice the semi-width).

So an interval described by $|x - a| < d$ has a as its midpoint, and d is its semi-width (Figure 1.3).

The **endpoints** of the interval $|x - a| < d$ are $a - d$ on the left, and $a + d$ on the right. So we could also write the interval as $a - d < x < a + d$.

$$|x - a| < d \quad \text{means} \quad a - d < x < a + d. \tag{1.29}$$

A particular case is the interval $|x| < d$. This is the interval $-d < x < d$, with midpoint 0 and semi-width d.

Mathematicians distinguish between open intervals and closed intervals:

- An **open interval** is an interval which does not include its end-points, i.e. an interval of the form $a < x < b$. This is sometimes written as (a, b).
- A **closed interval** does include its end-points, i.e. it is of the form $a \leq x \leq b$. This is sometimes written using square brackets, as $[a, b]$.

We'll use intervals in Chapter 2 when we talk about measurements, precision, and uncertainty intervals. We will use closed intervals when we describe bracketing methods, in Section 8.6.

1.8.2 Inequalities

A statement with two expressions linked by a $<$ or $>$ sign, is called an **inequality**. If $a < b$ (so a lies to the left of b on the real line) then it will also be true that $a + 2 < b + 2$, since adding the same quantity to a and b means that the two points are both moved the same distance to the right on the real line; the order of the points doesn't change (see Figure 1.4a). The same is true if you subtract the same number from both sides of the inequality, and also if you multiply (or divide) both numbers by the same positive factor. We can thus write down some rules for manipulating inequalities:

Suppose that $a < b$, and that c is a positive number. Then

$$a + c < b + c$$
$$a - c < b - c$$
$$c.a < c.b \tag{1.30}$$
$$\frac{a}{c} < \frac{b}{c}.$$

If you multiply (or divide) both a and b by a negative number, the sign of the inequality changes: for example $2 < 4$ but $-2 > -4$ (check back to Section 1.2.2 if you aren't sure about this!). See Figure 1.4b.

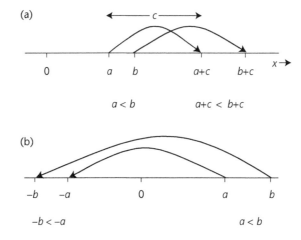

Figure 1.4 (a) and (b) Rules for inequalities

If a and b are both positive, then the following is true:

Suppose that $0 < a < b$. Then

$a^2 < b^2$

$\sqrt{a} < \sqrt{b}$.

But

$\dfrac{1}{a} > \dfrac{1}{b}$.

(1.31)

In the last inequality above, taking the reciprocal of a and b means that the sign of the inequality changes. For example, $2 < 3$ but $1/2 > 1/3$.

EXAMPLE

The American Council on Exercise states that the acceptable range of **Body Fat percentage** (BF) for adult males is 18–25%, and is 25–31% for women. It is not easy to measure a person's percentage of body fat, but it can be calculated from measuring their **body density** ρ (in kilograms per litre). The Siri formula for doing this is:

$$\mathrm{BF} = \left(\frac{4.95}{\rho} - 4.50 \right) \times 100\%.$$

(1.32)

What is the acceptable range of body density for adult women?

Inserting equation (1.32) into the acceptable body fat range $25 < \mathrm{BF} < 31$:

$$25 < \left(\frac{4.95}{\rho} - 4.50 \right) \times 100 < 31.$$

We want to solve this inequality to find an interval for ρ (this symbol is a Greek letter r and is pronounced 'rho').

We can divide through the inequality by 100:

$$0.25 < \left(\frac{4.95}{\rho} - 4.50 \right) < 0.31$$

and then add 4.50 to each expression:

$$0.25 + 4.50 < \frac{4.95}{\rho} < 0.31 + 4.50$$

$$4.75 < \frac{4.95}{\rho} < 4.81.$$

Now divide through by 4.95:

$$\frac{4.75}{4.95} < \frac{1}{\rho} < \frac{4.81}{4.95}.$$

If we finally take the reciprocal of each expression, the signs of the inequality change, and we get

$$\frac{4.95}{4.75} > \rho > \frac{4.95}{4.81}$$

$$1.042 > \rho > 1.029$$

which we can rewrite with < signs as:

$$1.029 < \rho < 1.042.$$

So an adult woman should have a body density of between 1.03 and 1.04 kilograms/litre.

Problems: Now try the end-of-chapter problems 1.63–1.64.

CASE STUDY A4: Birth rate, death rate and extinction

If N_0 is the initial size of a population of animals, then the population at the next generation is

$$N_1 = (1 + R)N_0,$$

where $R = B - D$ (the birth rate less the death rate).

We have assumed that the population is increasing in size, i.e. that $N_1 > N_0$. Is this always the case?

$N_1 > N_0$ means that

$$(1 + R)N_0 > N_0.$$

Divide both sides by N_0 (which must be positive; you can't have a population of negative size):

$$1 + R > 1.$$

Subtract 1 from both sides: $R > 0$.

That is, $B - D > 0$, or $B > D$.

So the population will increase provided the birth rate exceeds the death rate. If the birth rate falls below the death rate, the population will decrease and may become extinct.

CASE STUDY B4: Conditions for angiogenic cell line extinction

Recall that we are using x_0, x_1, y to represent respectively the densities of healthy, non-angiogenic, and angiogenic cells in the tumour, and that at equilibrium the density of angiogenic cells y is given by the expression

$$y = \frac{-Q + \sqrt{Q^2 - 4rq[k(d - r) + pk(x_0 + x_1)]}}{2rq}$$

where

$$Q = kq(d - r) + r + kp.$$

All the variables in this expression for y are positive quantities. For the angiogenic cell line to die out (i.e. $y < 0$), we would need the numerator to be negative. This will happen if the expression under the square root sign is less than Q^2 (so that the root will be less than Q):

We want

$$Q^2 - 4rq[k(d - r) + pk(x_0 + x_1)] < Q^2.$$

Subtract Q^2 from both sides:

$$-4rq[k(d - r) + pk(x_0 + x_1)] < 0.$$

Multiply through by –1; as this is multiplying by a negative number, the sign of the inequality changes:

$$4rq[k(d - r) + pk(x_0 + x_1)] > 0.$$

We can divide both sides by the factor of $4rq$ (and 0 divided by anything is still 0):

$$k(d - r) + pk(x_0 + x_1) > 0.$$

There is also a common factor of k which we can divide out:

$$(d - r) + p(x_0 + x_1) > 0.$$

Subtract $(d - r)$ from both sides:

$$p(x_0 + x_1) > -(d - r) = r - d.$$

So if there is sufficient angiogenesis inhibitor p and sufficient non-angiogenic cells $x_0 + x_1$, the angiogenic cell line will die out. The product $p(x_0 + x_1)$ should be greater than the difference between the angiogenic cell replication and death rates.

We'll see Case Study B again in Chapter 5, when we start to think about functions. But already in the next chapter we'll use Case Study A again to illustrate how numerical errors can grow when we use imprecise values of measured quantities in our calculations.

SUMMARY

Rules of algebra

$-(-a) = a$

$a + (-b) = a - b$

$a.(-b) = -a.b, \quad (-a).(-b) = a.b$

$a(b + c) = a.b + a.c$

If $a.b = 0$, then either $a = 0$ or $b = 0$ or both.

Rules for handling fractions

If $x = \dfrac{a}{b}$, it means that $b.x = a$

$$\frac{a}{b} = a.\frac{1}{b}, \quad \frac{a}{-b} = \frac{-a}{b} = -\frac{a}{b}$$

$$\frac{1}{a/b} = \frac{b}{a}$$

$$\frac{a.c}{b.c} = \frac{a}{b}, \quad \frac{a}{c} + \frac{b}{c} = \frac{a + b}{c}$$

$$\frac{a}{b} \times \frac{c}{d} = \frac{a.c}{b.d}, \quad \frac{a}{b} \div \frac{c}{d} = \frac{a}{b} \times \frac{d}{c} = \frac{a.d}{b.c}$$

Order of precedence for arithmetic operations (BEDMAS)

- Expressions within brackets must be evaluated first.
- Next, evaluate exponents and roots.
- Then do multiplications and divisions.
- Finally, do additions and subtractions.

Rules for handling exponents

Definition: $a^n = a \times a \times \ldots \times a$ (n times)

$$a^m \times a^n = a^{m+n}$$

$$a^m \div a^n = a^{m-n}$$

$$(a^m)^n = a^{m.n}$$

$$(a \times b)^n = a^n \times b^n$$

$$\left(\frac{a}{b}\right)^n = \frac{a^n}{b^n}$$

$$a^0 = 1$$

$$a^{-n} = \frac{1}{a^n}$$

Rules for handling roots

Definition: x is the n'th root of a if $x^n = a$

$$\sqrt{a.b} = \sqrt{a}.\sqrt{b} \text{ and } \sqrt{\frac{a}{b}} = \frac{\sqrt{a}}{\sqrt{b}}$$

$$\sqrt{a} = a^{1/2}, \text{ in general } \sqrt[n]{a} = a^{1/n}$$

Absolute value: $|a| = \sqrt{a^2}$

Important factorizations

$$(x-1)(x+1) = x^2 - 1$$
$$(x-1)(x^2 + x + 1) = x^3 - 1$$
$$(x-a)(x+a) = x^2 - a^2$$
$$(a+b)(c+d) = a.c + b.c + a.d + b.d$$
$$(a+b)^2 = a^2 + 2ab + b^2$$
$$(a-b)^2 = a^2 - 2ab + b^2$$
$$(a+b)^3 = a^3 + 3a^2b + 3ab^2 + b^3$$
$$(a-b)^3 = a^3 - 3a^2b + 3ab^2 - b^3$$

Rules for handling inequalities

Suppose that $a < b$, and that c is a positive number. Then

$$a + c < b + c$$
$$a - c < b - c$$
$$c.a < c.b$$
$$\frac{a}{c} < \frac{b}{c}$$

Suppose that $0 < a < b$. Then

$$a^2 < b^2$$
$$\sqrt{a} < \sqrt{b}$$

But

$$\frac{1}{a} > \frac{1}{b}$$

An open interval: $|x - a| < d$ means $a - d < x < a + d$.

? PROBLEMS

1.1–1.2: In each problem, six real numbers are given. Use your ruler to draw a real line, and mark the six numbers on it. List them in increasing order, e.g. $1 < 2 < 3 < 4 < 5 < 6$. Use your calculator to evaluate roots and π as decimals.

1.1 $-2, \sqrt{3}, -1.5, -\sqrt{2}, \dfrac{3}{4}, \dfrac{7}{8}$ 1.2 $-\dfrac{22}{7}, 3, \dfrac{10}{3}, \sqrt{10}, \pi, -3.15$.

1.3–1.10: Simplify the following fractions:

1.3 $\dfrac{1}{3} + \dfrac{1}{2}$ 1.4 $\dfrac{7}{8} - \dfrac{3}{4}$ 1.5 $\dfrac{5}{6} - \dfrac{11}{14}$

1.6 $\dfrac{3}{8} - \dfrac{7}{10}$ 1.7 $\dfrac{3-7}{8-10}$ 1.8 $\dfrac{12}{3/4}$

1.9 $\dfrac{84}{15} \times \dfrac{100}{21}$ 1.10 $\dfrac{40}{27} \div \dfrac{16}{21}$.

1.11–1.14: Simplify into a single fraction:

1.11 $\dfrac{1}{x} - \dfrac{1}{y}$

1.12 $\dfrac{x+y}{x} + \dfrac{x-y}{y}$

1.13 $\dfrac{x-y}{x} - \dfrac{x-y}{y}$

1.14 $\dfrac{4}{3x} - \dfrac{5}{2xy}$.

1.15–1.20: Evaluate:

1.15 $|\pi - 3.142|$

1.16 $|4-3| + |3-4|$

1.17 $|5-2| - |2-4|$

1.18 $\big| -4 - |4-7| \big|$

1.19 $\dfrac{|-4+9|}{|4-9|}$

1.20 $\Big\| \sqrt{2} - \sqrt{3} \big| - \big| \sqrt{2} + \sqrt{3} \Big\|$.

1.21: Find the percentage error (correct to four decimal places) in using 3.142 as an approximation to π.

1.22: Find the percentage error (correct to four decimal places) in using $\dfrac{22}{7}$ as an approximation to π.

1.23: Here is a table showing the resident population of the USA (in thousands) and the number of immigrants to the USA (in thousands), for the years 1930, 1935, 1940, 1945, 1950, and 1955. Calculate the percentage changes over each 5-year period (correct to two decimal places).

Year	Population	% change	Immigration	% change
1930	123 077	–	242	–
1935	127 250		35	
1940	131 954		71	
1945	132 481		38	
1950	151 868		249	
1955	165 069		238	

Type this table into an Excel spreadsheet, and program the cells in the 3rd and 5th columns to work out the percentage changes.

1.24–1.27: Calculate the percentage change over the time period, and the average annual percentage change (correct to 2 decimal places):

1.24 The normalized median monthly income in the UK was £377 in 2001/2, and £393 in 2007/8.

1.25 In 1993, the mean weight of men in the UK was 78.9 kg, and of women was 66.6 kg. In 2008 the figures were 83.6 kg and 70.2 kg respectively.

1.26 The average July temperature in London was 14.7 degrees centigrade in 1965, and 19.5 degrees centigrade in 1995.

1.27 The average amount of alcohol purchased from shops in the UK for home consumption was 763 ml per person per week in 2004, and 772 ml in 2007. The corresponding figures for alcohol purchased in pubs were 772 ml and 503 ml.

In one of the examples above, the percentage change you worked out is virtually meaningless! Which one, and why?

1.28–1.30: Evaluate:

    1.28 $6(5 - (4 - 3)/2)$

    1.29 $7 - (6 - (5 - 4(3 + 2)))$

    1.30 $7 - 6 - 5 - 4 \times 3 + 2.$

1.31–1.34: Simplify by multiplying out the brackets:

    1.31 $(a - b)(c - d)$ 1.32 $(a + b + c)^2$

    1.33 $a(b - (c - d)/e)$ 1.34 $a - (b - (c - d(e + f))).$

1.35–1.38: Here are some more algebraic MindGym puzzles, as in the example in Section 1.4.1. Build up the algebraic expressions they describe, using brackets where necessary:

| 1.35 | x | + 5 | ÷ 2 | − 3 | × BY ITSELF | − 25 | ABSOLUTE VALUE | √ | ÷ 2 | − 6 |

| 1.36 | x | × BY ITSELF | − 1 | × BY ITSELF | + 6 | √ | − 5 | × BY ITSELF | ÷ 3 | + 1 |

| 1.37 | x | + 1 | × 2 | + 1 | × 2 | + 1 | × BY ITSELF | × 2 | + 1 | √ |

| 1.38 | x | − 1 | ÷ 2 | − 1 | ÷ 2 | − 1 | √ | ÷ 2 | − 1 | ¾ OF THIS |

1.39–1.40: An algebraic expression that is a sum of terms involving powers of x, e.g. $x^4 + 3x^3 - 2x^2 + x - 5$, is called a **polynomial**. We will study polynomials in Chapter 7. To evaluate the polynomial at a particular value of x, say $x = 3$, you need to calculate $3^4 + 3 \times 3^3 - 2 \times 3^2 + 3 - 5.$

This is difficult to do in your head – especially if, like me, you have trouble remembering more than one number at a time. Working out each power of x separately, you will need to perform eight multiplications.

The calculation is much easier if you write the polynomial in nested form, as

$(((x + 3)x - 2)x + 1)x - 5.$

Check by multiplying out the brackets, that this is the same as our original polynomial. To evaluate it, work from left to right:

    $(((3 + 3)3 - 2)3 + 1)3 - 5$

$= ((6 \times 3 - 2)3 + 1)3 - 5$

$= (16 \times 3 + 1)3 - 5$

$= 49 \times 3 - 5 = 147 - 5 = 142.$

And we have only needed to perform three multiplications!

Now write the following polynomials in nested form, and evaluate them for the values of x given:

    1.39 $2x^4 - 3x^3 + 2x^2 - 5x - 3, \quad x = 2$

    1.40 $2x^5 - 5x^4 + 2x^2 - 3x + 1, \quad x = -2.$

1.41–1.44: Evaluate the following expressions without using a calculator:

1.41 $\dfrac{2^3 - 1}{3^2 + 1} \div \dfrac{(2-3)^5}{5^2 - 5^3}$

1.42 $\dfrac{2^3 . 3^2}{2^2 . 3^3} - \dfrac{3^2 - 1}{2^3 + 1}$

1.43 $\left(\dfrac{2^3 - 3^3 - 1}{3^2 - 2^2}\right)^3$

1.44 $\dfrac{5^{(3^2)} - (5^3)^2}{3^5 - 2^5}$.

1.45: When an ancient Indian mathematician invented the game of chess, the country's ruler was so pleased that he asked the man to name his price for the invention. Pointing at the chessboard, the mathematician asked to be given 1 grain of wheat for the first square, two grains for the second, four grains for the third, and so on, doubling the amount each time. The ruler thought that this paltry request proved the naïveté and unworldliness of the inventor (and of mathematicians in general). But how many grains would he need to pay for the final (64th) square on the board?

1.46: It is said that if sufficient monkeys were given typewriters and enough time, they would eventually produce the complete works of Shakespeare. Let's quantify this. Suppose that 18 monkeys were sat at a row of keyboards, and each monkey pressed a key. Assume that each keyboard has 27 keys (26 letters plus the space bar), so the probability of a particular monkey hitting a particular key is $\dfrac{1}{27}$. What is the probability that this will produce the string of characters 'to be or not to be'? If the monkeys repeated this process every second, how long do you think it might take before the monkeys come up with this string?[5] For comparison, the age of the Universe is about 13.7 billion years.

1.47–1.50: Simplify the following expressions involving exponents:

1.47 $\left(a^3 + ab^2\right)\left(\dfrac{1}{a^2} - \dfrac{1}{b^2}\right)$

1.48 $\dfrac{\dfrac{1}{a} - \dfrac{1}{b}}{1 + \dfrac{a}{b}}$

1.49 $\dfrac{a^2b + 3a^4b - a^3b^2}{ab^2 + 3a^3b^2 - a^2b^3}$

1.50 $(a^3b + a^{-2}b^2)(a^2 - a^{-3}b)$.

1.51–1.52: Without using a calculator, evaluate y from the formula in Case Study B, namely:

$$y = \dfrac{-Q + \sqrt{Q^2 - 4rq[k(d-r) + pk(x_0 + x_1)]}}{2rq}$$

where $Q = kq(d - r) + r + kp$, for the following sets of values of the variables:

[5] This example is sometimes used to 'prove' that evolution could never have produced a simple organism by blind chance mutations, even over millions of years. But if you model Natural Selection by keeping a correct letter once it has been produced, the odds on getting a line of Shakespeare are dramatically shortened. Using the 'Richard Dawkins Weasel demo' simulation at http://vlab.infotech.monash.edu.au/simulations/evolution/richard-dawkin-weasel/, I obtained 'to be or not to be' in just 74 attempts, starting with a random string of characters!

$$r = 0.1 \qquad\qquad r = 0.2$$
$$q = 2.0 \qquad\qquad q = 0.3$$
$$k = 1.0 \qquad\qquad k = -0.2$$
$$1.51 \quad d = 0.4 \qquad 1.52 \quad d = 0.2$$
$$p = 0.0 \qquad\qquad p = 1.0$$
$$x_0 = 0.3 \qquad\qquad x_0 = 0.15$$
$$x_1 = 0.2 \qquad\qquad x_1 = 0.15.$$

1.53: My car's fuel consumption is 20 miles per gallon (mpg) in the city, and 40 miles per gallon on the motorway. Convert these values to the European measure of fuel consumption, namely litres per 100 km (l/100 km). Note: 1 UK gallon is equivalent to 4.546 litres, and 1 UK mile is 1.609 kilometres.

If my fuel consumption is f mpg and this is equivalent to g l/100 km, find an equation relating f and g. Is there a fuel consumption at which $f = g$?

1.54: I drive my Ford Mondeo for d miles in London (consumption $f_1 = 20$ mpg) followed by a further d miles on the motorway (consumption $f_2 = 40$ mpg).

My sister Martine drives her Citroën C5 for d kilometres in Paris (consumption $g_1 = 7$ litres/100 km) followed by a further d kilometres on the autoroute (consumption $g_2 = 14$ litres/100 km).

Find expressions involving d for our average fuel consumptions on our two journeys, in the relevant units.

From your results, which do you think is a more sensible way of measuring fuel consumption: as 'volume per distance' or 'distance per volume'?

1.55–1.56: Evaluate the following expressions without using a calculator:

$$1.55 \quad \sqrt{\frac{1}{25 - \sqrt{25}} + \sqrt[3]{8 \times 10^{-3}}} \qquad 1.56 \quad \left(2\sqrt{3} - \frac{1}{\sqrt{3}}\right)^2.$$

1.57–1.58: Simplify the following expressions:

$$1.57 \quad \frac{2}{\sqrt{a} - 2\sqrt{b}} - \frac{1}{\sqrt{a} + 2\sqrt{b}} \qquad 1.58 \quad \frac{\dfrac{1}{1 + \sqrt{a}} - \dfrac{1}{1 - \sqrt{a}}}{1 + a}.$$

1.59–1.62: Convert the following expressions to surds (i.e. of the form $a + b\sqrt{c}$ or $\frac{1}{d}\left(a + b\sqrt{c}\right)$ where a, b, c are integers):

$$1.59 \quad \frac{5 - \sqrt{2}}{7 + \sqrt{2}} \qquad\qquad 1.60 \quad \frac{1}{2 - 3\sqrt{5}}$$

$$1.61 \quad \frac{2 + 3\sqrt{7}}{1 - 2\sqrt{7}} \qquad\qquad 1.62 \quad \frac{\sqrt{2} - 2\sqrt{3}}{\sqrt{2} + \sqrt{3}}.$$

1.63: In the wild, goldfish can live in water temperatures T in the range $41 < T < 104$ degrees Fahrenheit. Convert this interval into an interval in degrees Celsius, given the formula $C = \frac{5}{9}(F - 32)$ to convert from Fahrenheit to Celsius.

1.64: The new Audi A5 Sportback has a fuel consumption f in the range $30.4 < f < 48.7$ where f is in miles per gallon. Convert this into an interval which uses the European measure, namely litres per 100 kilometres, using the conversions given in problem 1.53. Give the consumptions in l / 100 km correct to 1 decimal place.

1.65: In Section 1.3.3 we stated a handy trick for telling if an integer is divisible by 3: add the digits together, and if the sum is divisible by 3 then so is the original number. So, the sum of the digits of 2385 is $2 + 3 + 8 + 5 = 18$ which is divisible by 3, so 2385 is divisible by 3 (actually $2385 = 3 \times 795$). Why does this trick work? Suppose you have a two-digit number n whose digits are a and b (e.g. if $n = 72$ then $a = 7$ and $b = 2$). Write down an equation linking n, a, and b. For the sum of the digits to be divisible by 3, it means that $a + b = 3k$, where k is another integer. Can you now deduce that n can be written as 3 multiplied by an integer? Do you think the trick will work for other divisors (e.g. if the sum of the digits is divisible by 7, will the original number be divisible by 7)?

1.66: We have two numbers c and d, and let $x = \sqrt[3]{c} + \sqrt[3]{d}$. Show that $x^3 = -3\sqrt[3]{cd}\,x - (c + d)$.

Now we let $c = \dfrac{b}{2} + \sqrt{\left(\dfrac{a}{3}\right)^3 + \left(\dfrac{b}{2}\right)^2}$, and $d = \dfrac{b}{2} - \sqrt{\left(\dfrac{a}{3}\right)^3 + \left(\dfrac{b}{2}\right)^2}$. Prove that $c.d = -\left(\dfrac{a}{3}\right)^3$,

and that $c + d = b$.

Hence deduce that $x^3 + ax - b = 0$.

We have found a value of x which makes the polynomial $x^3 + ax - b$ become zero, namely

$$x = \sqrt[3]{\dfrac{b}{2} + \sqrt{\left(\dfrac{a}{3}\right)^3 + \left(\dfrac{b}{2}\right)^2}} + \sqrt[3]{\dfrac{b}{2} - \sqrt{\left(\dfrac{a}{3}\right)^3 + \left(\dfrac{b}{2}\right)^2}}.$$

This is called a **root** of the polynomial. We will discuss roots of polynomials in Chapters 6 and 7, and you will see this result quoted in the Historical Interlude at the end of Part I. This root was found by Italian mathematicians in the fourteenth century. You will use this formula in problem 7.30.

And finally ...

1.67: You are working for MI5, and you are monitoring the email mailbox of a man who is spying for the government of Slaka. You know that the Slakans need to provide the man (who is using the name Ivo Panić) with a PIN code which he can use to draw out money from a bank account. You intercept and decrypt the following exchange of emails:

> From: olga@gov.slk
> To: I.Panic@gurglemail.com
> Comrade Ivo,
> Here is a number: 50841
> This is not the PIN number. But you should choose your own natural number — we'll call it b. Don't tell it to me, but multiply the number above by b, and tell me the result.
> Greetings, Olga

> From: I.Panic@gurglemail.com
> To: olga@gov.slk
> Here is the result: 864297. Ivo

> From: olga@gov.slk
> To: I.Panic@gurglemail.com
> Okay. Here is a new number: 123471. Divide this by your secret number b, and the result will be the PIN number you need. Olga.

Can you work out what is going on? If so, you should be able to work out the PIN number easily.

1.68: The following month, you intercept a new series of emails:

> From: olga@gov.slk
> To: I.Panic@gurglemail.com
> Comrade Ivo,
> You need a new PIN number. Our mathematicians have invented a more secure way of transmitting it to you.
> Choose a new natural number b. Use your calculator to raise the following number to the power b, and send me the result. The number is 3124943128.
> Comradely greetings, Olga

> From: I.Panic@gurglemail.com
> To: olga@gov.slk
> Here is the result: $2.979961062 \times 10^{47}$. Ivo

> From: olga@gov.slk
> To: I.Panic@gurglemail.com
> Okay. Here is a new number: $6.679390939 \times 10^{15}$. Take the bth root of this number, and the result will be the PIN number you need. Olga.

Now can you find out what the PIN number is? If not, you'll have to wait until the last question of Chapter 10 to see the answer!

This is an example of a type of **asymmetric key encryption**.[6] Here's the principle behind it: Suppose Olga wants to send Ivo a suitcase full of money, but she doesn't trust the courier who will take it to him. She puts a padlock on the case and gives it to the courier, keeping the key herself. When the courier gives the suitcase to Ivo, he puts a padlock of his own on the suitcase, and gives it back to the courier. The courier delivers the suitcase back to Olga; she unlocks and removes her padlock. The courier takes the suitcase – still locked with Ivo's padlock – back to Ivo, who unlocks it, opens it, and grabs the money.l

---

[6] I am grateful to Prof James Davenport for the idea for these questions.

Units; precision and accuracy

In the last chapter we treated numbers as abstract quantities that could be manipulated using the laws of arithmetic, and mathematicians are happiest when thinking of numbers in this way. They are also quite content to deal just with integers – whole numbers like 0, 1, −36, etc. But scientists encounter numbers as **data**: measurements of some physical characteristic such as length, speed, or population density. We call these characteristics **physical variables**. With an item of data,[1] we not only have to consider its numerical value, but also the **units** in which it is measured, and the **precision** to which it is measured. So if someone asks me how tall I am, I should not reply 'one eighty nine', but '189 centimetres, to the nearest centimetre', or '1.89 metres, correct to two decimal places' (whereas what I actually reply is 'six foot two and a half'). It is typical in conversation to leave out the units because we unconsciously expect what units are being used. In the sciences though, it is essential that we make explicit what units we are using. A nurse administering 50 milligrams of a drug instead of 50 micrograms, after the doctor told her to 'Give him a shot of fifty', would be a disaster.

Although I have quoted my height as 189 cm precise to the nearest centimetre, this may not be **accurate** (since I last measured my height 10 years ago), so there may be an **error** of several centimetres in using this value for my current height. Similarly, I may weigh a sample using high-quality electronic scales, and obtain a very precise reading correct to the nearest microgram – but this will not be very accurate if there are traces of a previously used substance stuck to the measuring tray!

> The **precision** of the measurement is how well I have made that measurement (which depends on the quality of the measuring device) whereas the **accuracy** of the measurement tells how close that measurement is to the true value (which may involve human error). (2.1)

In this chapter we will explain some things you need to know about units, precision, and accuracy.

After working through this chapter, you should be able to:

- convert decimal numbers to scientific notation, and vice versa;
- perform arithmetic with numbers in scientific notation;

---

[1] Pedantic note: 'data' is a plural noun. A single item of data is called a 'datum' or, more commonly, 'a single item of data'.

- use the system of SI units, including SI prefixes;
- perform calculations involving SI units, and find the units for the result;
- use non-SI units (gram, litre, and cubic decimetre) in calculations;
- use dimensional analysis to find the correct units for a quantity in an equation;
- round measurements to a required precision (in decimal places or significant figures).

2.1 Scientific notation

There are many different formats for expressing numbers. You may think that 3 and 3.0 mean the same thing, but they do not and a computer will store integers like 3 and reals like 3.0 in different ways. Similarly, a computer can store a decimal like 0.6, but not a fraction like 1/3.

The main number formats, or **notations**, you will be familiar with are:

- **Integer notation**, e.g. 3; only used for 'whole numbers'.
- **Fraction notation**, e.g. 3/5, 22/7.
- **Fixed-point notation** (the proper name for 'decimals'), e.g. 3.0, −0.6, 12.34.

2.1.1 Definition of scientific notation

Fixed-point notation is not efficient at handling numbers which are extravagantly large or small. For example, a typical atom has a diameter of 0.0000000008 metres, while the Earth's circumference is about 40,000,000 metres. (Is it a weird coincidence that this comes out to be such a nice round number? See Section 2.3 for the answer.) Scientists use **scientific notation** for such numbers: the atom has a diameter of 8.0×10^{-10} metres, and the Earth's circumference is 4.0×10^{7} metres.

> The general form of a number in scientific notation is $s \times 10^{n}$. Here, s is a fixed-point number which has one non-zero digit before the decimal point (i.e. it is a number between 1.0 and 9.99999 ...), and n is an integer (positive, negative, or zero).
>
> (2.2)

The fixed-point number s is called the **mantissa** (a word worth remembering, if only to use in Scrabble). If the number itself is negative, then s will be negative. The other part of the number consists of the **base** (which is 10 in scientific notation) raised to an **exponent** n.

Here are some more examples of numbers as decimals, and written in scientific notation:

Decimal form	Scientific notation
365.123	3.65123×10^{2}
−3.142	-3.142×10^{0}
0.00087	8.7×10^{-4}
−0.00111	-1.11×10^{-3}
0.999	9.99×10^{-1}

2.1.2 Converting numbers between decimal and scientific notation

We can use the properties of exponents discussed in Section 1.5 to explain the process of converting a decimal number to scientific notation, and *vice versa*. Recall that 10^n means 10 multiplied by itself n times, which is the number 1 followed by n zeroes: $10^3 = 1000$, for example.

Also, $10^1 = 10$ and $10^0 = 1$.

For negative exponents, $10^{-3} = 1/10^3 = 1/1000 = 0.001$.

In general, $10^{-n} = 1/10^n = 0.00 \ldots 01$ (with $n - 1$ zeroes between the decimal point and the 1).

Thus $1.234567 \times 10^3 = 1.234567 \times 1000 = 1234.567$ and $3.14159 \times 10^{-3} = 3.14159/1000 = 0.00314159$.

You can see that multiplying a decimal number by 10^n has the effect of moving the decimal point n places to the right, and multiplying by 10^{-n} (or dividing it by 10^n) moves the decimal point n places to the left.

To write a decimal number in scientific notation: Move the decimal point to left or right until there is just **one non-zero digit in front of the decimal point**.

- If you moved the point n places to the left, this is equivalent to dividing the number by 10^n, so you will need to multiply it by 10^n again, so that the result is equivalent to the number you started with. To write 123.4 in scientific notation, move the decimal point two places to the left, and multiply by 10^2, to get 1.234×10^2, which is equivalent to 123.4.

- If you moved the point n places to the right, this is equivalent to multiplying the number by 10^n, so you will need to divide it by 10^n to restore its original value; do this by multiplying by 10^{-n}. To write 0.000457 in scientific notation, move the decimal point four places to the left, and multiply by 10^{-4}, to get 4.57×10^{-4}, which is equivalent to 0.000457.

To convert $s \times 10^n$ to fixed-point (decimal) notation: Write down s and then move the decimal point n places to the right (when n is positive) or left (when n is negative).

Read this positive exponent to say 'Move the decimal point two places to the right'.

$$1.234 \times 10^2 = 123.4 \qquad 1.234 \times 10^{-2} = 0.01234$$

$$6.0 \times 10^5 = 600000 \qquad 6.0 \times 10^{-5} = 0.00006$$

Read this negative exponent to say 'Move the decimal point two places to the left'.

When my cheap calculator displays a number in scientific notation, it just displays s and n, that is, it displays 4.23×10^7 as 4.23 7, and 8.1×10^{-10} as 8.1 −10. More expensive calculators may follow the convention used in computers or Excel, indicating the exponent by E or e: 4.23 e7 or 8.1 E-10, for example.

In SPREADSHEET 2.1 you can test your understanding of scientific notation. Think of a decimal number and write down its equivalent in scientific notation. Then type your decimal in cell B9 and see if your answer corresponds to the scientific-notation equivalent which appears in cell E9. In row 19 you can do the reverse: type a scientific-notation number in cell B19 (e.g. type −2.3e − 3) and see it converted into a decimal in cell E19. Note that the cells are formatted to display 5 decimal places, so some trailing zeros may get displayed. You wouldn't write these in practice.

In Excel, you can format a cell (or a block of cells) to display numbers in scientific notation. Highlight the cells, and select FORMAT > CELLS . . . from the menu bar. In the pop-up window which appears, click on the Number tag and choose the option SCIENTIFIC. If you accept the default of displaying 2 decimal places in the mantissa, the number 4.23×10^7 will appear as 4.23E + 07, for example. Excel will display the sign of the exponent (even if it's positive) and use two digits, so it will show 4.23×10^7 as 4.23E + 07 rather than 4.23e7.

Problems: Now try the end-of-chapter problems 2.1–2.6

2.1.3 Performing addition and subtraction in scientific notation

Numbers in scientific notation **can only be added or subtracted if their exponents are the same:**

$$4.567 \times 10^{-5} - 2.14 \times 10^{-5} = (4.567 - 2.14) \times 10^{-5} = 2.427 \times 10^{-5}.$$

If the exponents are different, we have to adjust the decimal place on one mantissa. If we move the decimal point one place to the right (thus multiplying the mantissa by 10) but also subtract one from the exponent, we are left with the same value:

$$0.01234 \times 10^3 = 0.1234 \times 10^2 = 1.234 \times 10^1 = 12.34$$
$$= 123.4 \times 10^{-1} = 1234.0 \times 10^{-2}.$$

Conversely, if we increase the exponent by 1 (thus multiplying the number by 10) we'll need to move the decimal point in the mantissa one place to the left (thus dividing the mantissa by 10) to keep the overall value the same:

$$54.32 \times 10^3 = 5.432 \times 10^4.$$

We use this trick to convert numbers into a form suitable for adding or subtracting.

EXAMPLE

If $a = 1.234 \times 10^{-5}$, and $b = 3.764 \times 10^{-3}$, what is $a + b$?
Solution: Let's transform a into a number with the same exponent as that of b. We'll need to increase the exponent of a by 2, so it goes from −5 to −3. We then have to compensate by moving the decimal point in the mantissa two places to the left:

$$a = 1.234 \times 10^{-5} \quad \text{becomes} \quad a = 0.01234 \times 10^{-3}.$$

The sum then becomes

$$a + b = 0.01234 \times 10^{-3} + 3.764 \times 10^{-3} = (0.01234 + 3.764) \times 10^{-3}$$
$$= 3.77634 \times 10^{-3}.$$

Alternatively, we could have adjusted the exponent of b to match that of a (subtracting 2 from −3 to get −5):

$$b = 3.764 \times 10^{-3} \quad \text{becomes} \quad b = 376.4 \times 10^{-5}$$

so that the sum becomes

$$a + b = 1.234 \times 10^{-5} + 376.4 \times 10^{-5} = (1.234 + 376.4) \times 10^{-5}$$
$$= 377.634 \times 10^{-5}.$$

In this second case, the final step is to convert the result into proper scientific notation (where the mantissa has one non-zero digit before the decimal point):

$$377.634 \times 10^{-5} = (3.77634 \times 10^2) \times 10^{-5}$$
$$= 3.77634 \times 10^{-3}.$$

Note that both of these methods result in the same answer, and so either calculation can be used.

2.1.4 Performing multiplication and division in scientific notation

When multiplying two numbers in scientific notation, there is no need to adjust the exponents, we simply use the rule from (1.19) which defines how to multiply numbers a^m and a^n:

$$a^m \times a^n = a^{m+n}.$$

Using this rule,

$$(1.2 \times 10^{-5}) \times (3.7 \times 10^{-3}) = 1.2 \times 3.7 \times 10^{-5} \times 10^{-3}$$
$$= 4.44 \times 10^{(-5-3)} = 4.44 \times 10^{-8}.$$

So to multiply two numbers, we multiply the mantissas and add the exponents:

$$(s \times 10^m) \times (t \times 10^n) = (s \times t) \times 10^{m+n}. \qquad (2.3)$$

It may still be necessary to adjust the result so that the mantissa has one non-zero digit before the decimal point.

The rule for division using scientific notation is

$$(s \times 10^m) \div (t \times 10^n) = (s \div t) \times 10^{m-n}. \qquad (2.4)$$

Thus,

$$(1.2 \times 10^{-5}) \div (3.7 \times 10^{-3}) = (1.2 \div 3.7) \times 10^{-5-(-3)}$$
$$= 0.3243 \times 10^{(-5+3)} = 0.3243 \times 10^{-2}$$
$$= 3.243 \times 10^{-3} \text{ in scientific notation.}$$

Problems: Now try the end-of-chapter problems 2.7–2.12

2.1.5 An aside: floating-point notation

Computers use a slightly different format for storing numbers involving decimal points. Here, the first non-zero digit in the mantissa comes after the decimal point (i.e. the mantissa is a number between 0.1 and 0.99999 …). So 1234.56 would be written as 0.123456×10^4, and the computer would store the digits 123456 and the exponent 4. This is called **floating-point notation**. Typically a decimal number stored in a computer will consist of up to six digits for the mantissa, plus a sign (positive or negative) and two digits for the exponent (the actual situation is more complicated because computers use a base of 2 or 16 rather than 10!). When computers output the number, as with Excel they use the E notation and print the sign of the exponent even if it is positive. Thus $-64{,}123$ would be printed as $-0.64123\mathrm{E}+05$.

Decimal numbers can be stored in a computer in either single precision or double precision. A **single precision** number occupies 4 bytes (32 bits) of memory, which is enough for 5 or 6 digits in the floating-point mantissa. A **double precision** number occupies 8 bytes of memory, enough for 12 digits.

2.2 SI units

2.2.1 Base, supplementary, and derived SI units

The *Système Internationale d'unités*, or 'SI units', was adopted in Europe in 1960, and is now used (almost) universally for scientific measurement and calculation. It grew out of the metric system, which recognized the metre, kilogram, and second as the basic units of length, mass, and time. This was unveiled in 1799, as part of the French Revolution's sweeping away of old feudal traditions and practices, replacing them with rational, efficient new ones (e.g. inventing the guillotine as the modern, scientific way of executing people). The **metre** was declared to be the new unit of length 'for all people, for all time'; it was defined as one ten millionth of the distance along the meridian drawn from the North Pole to the Equator – and passing through Paris, naturally. This is why the Earth's circumference through the North and South Poles is close to 40 million metres. The **gram** was defined as the weight of one cubic centimetre of distilled water, but it was soon realized that this was too imprecise, and it was replaced as the base unit of mass by the **kilogram** (the official prototype kilogram was a platinum cylinder manufactured by the former royal jeweller); a gram was then one thousandth of the mass of this kilogram weight. From these foundations the system of SI units was developed, so that there is an SI unit for every measurable physical characteristic.

The SI system consists of:

- seven **principal** or **base units**, and two supplementary units (Table 2.1);
- a large number of **derived units**, some of which have their own names and symbols (the most important are in Tables 2.2 and 2.3);
- a set of **prefixes** used to create multiples or fractions of the above units (Table 2.4).

Table 2.1: Base and supplementary SI units (base units in bold)

Dimension	SI unit	Symbol
Length	**metre**	m
Mass	**kilogram**	kg
Amount of substance	**mole**	mol
Time	**second**	s
Electric current	**ampere**	A
Temperature	**kelvin**	K
Luminous intensity	**candela**	cd
Plane angle	radian	rad
Solid angle	steradian	sr

Table 2.2: Some SI derived units with their own names and symbols

Measurement	SI unit	Symbol	Definition	Form in base units
Force	newton	N	kg m s^{-2}	
Energy	joule	J	N m	kg m^2 s^{-2}
Pressure	pascal	Pa	N m^{-2}	kg m^{-1} s^{-2}
Power	watt	W	J s^{-1} or V A	kg m^2 s^{-3}
Electric charge	coulomb	C	A s	
Potential difference	volt	V	J C^{-1} or W A^{-1}	kg m^2 s^{-3} A^{-1}
Resistance	ohm	Ω	V A^{-1}	kg m^2 s^{-3} A^{-2}
Capacitance	farad	F	C V^{-1}	kg^{-1} m^{-2} s^4 A^2
Frequency	hertz	Hz	s^{-1}	
Radioactivity	becquerel	Bq	s^{-1}	

In Table 2.1, we have referred to the basic characteristics of length, mass, time, etc. as **dimensions**. Other measured quantities are combinations of these dimensions. For example, the average speed of a moving object is obtained as the distance (length) travelled, divided by the elapsed time. In terms of dimensions,

speed = length/time.

So the derived unit of speed is metres/second (pronounced 'metres per second'), as shown in Table 2.3. Using the symbols in Table 2.1, we could abbreviate metres/second to m/s.

It is an SI convention to use the exponent notation we saw in Chapter 1 for units, rather than division signs. Remember that

$$1/2 = 2^{-1} \quad \text{and} \quad \frac{a}{b} = a.b^{-1},$$

so we usually write 'metres per second' as m s^{-1} rather than m/s.

Table 2.3: Other important SI derived units

Measurement	SI unit	Form in base units
Area	square metre	m^2
Volume	cubic metre	m^3
Speed, velocity	metre per second	m s^{-1}
Acceleration	metre per second squared	m s^{-2}
Density	kilogram per cubic metre	kg m^{-3}
Specific weight	newton per cubic metre	kg m^{-2} s^{-2}
Luminance	candela per square metre	cd m^{-2}
Mass concentration	kilogram per cubic metre	kg m^{-3}
Molar concentration	mole per cubic metre	mol m^{-3}
Magnetic field	ampere per metre	A m^{-1}

Table 2.4: Prefixes used with SI units. (Prefixes in italics are not part of the official SI system)

Prefix	Symbol	Multiple/ fraction		Some typical biological examples
Multiples of 10:				
yotta	Y	10^{24}		
zetta	Z	10^{21}		
exa	E	10^{18}		
peta	P	10^{15}		
tera	T	10^{12}	1 000 000 000 000	
giga	G	10^{9}	1 000 000 000	
mega	M	10^{6}	1 000 000	Mbases
kilo	k *	10^{3}	1 000	kg, kbases
hecto	h	100		
deca	da	10		
Fractions of 10:				
deci	d	0.1		dm^3
centi	c	0.01		cm, cm^3 **
milli	m	10^{-3}	0.001	mg, ml **
micro	μ	10^{-6}	0.000 001	μg, μl
nano	n	10^{-9}	0.000 000 001	ng
pico	p	10^{-12}	0.000 000 000 001	pg
femto	f	10^{-15}		
atto	a	10^{-18}		
zepto	z	10^{-21}		
yocto	y	10^{-24}		

* kilo is the only standard multiple that has a non-capital symbol (see 2.2.3 below).

** cm^3 and ml are the same volume but have quite different prefixes.

Thus, if a car travels 100 metres in 5 seconds, its speed is $\frac{100}{5} = 20$ metres per second, which we could calculate using exponent notation as

$$100 \times 5^{-1} = 20 \text{ m s}^{-1}.$$

Acceleration is the change of speed over time, measured in metres per second per second, or m/s^2, which we will write as m s^{-2} using exponent notation again. Another SI convention is to write the trailing zeros in a large number in groups of three, separated by spaces, not commas. So we should write the Earth's circumference as 40 000 000 metres, not 40,000,000 metres as we did in Section 2.1.1. This will avoid any confusion arising from the Continental use of a comma instead of a decimal point (e.g. writing 17/18 as 2,125).

Notice that the names of SI units are not written with a capital letter, even if they are the name of a famous scientist. Thus the unit of force is the newton (not the Newton), but the symbol for the newton is N. Important quantities named after scientists include:

- **Force: 1 newton**, with symbol N, is the force required to give a mass of 1 kilogram an acceleration of 1 metre per second per second; you probably remember Newton's law:

force = mass × acceleration.

Using SI base units (Table 2.1), we could write 1 newton as 1 kg m s^{-2}.

- **Pressure** is a force per unit area,

  pressure = force ÷ area

  so would have units of N m^{-2} (newtons per square metre). The SI unit of pressure is the **pascal**, with symbol Pa. Using SI base units, 1 pascal would be 1 kg m s^{-2} m^{-2}, and we can combine the two references to metres:

  m m^{-2} = m^1 m^{-2} = m$^{1-2}$ = m^{-1}

  so that 1 pascal is one kg m^{-1} s^{-2}.

- **Energy** is a force acting over a unit distance,

  energy = force × distance

  so has units N m (newton metres), which in base units reduces to kg m^2 s^{-2}. The derived unit of energy is the **joule**, with symbol J. One joule[2] is the energy exerted by a force of 1 newton, in moving an object a distance of 1 metre.

  A pre-SI unit of energy was the **calorie**, still used by food nutritionists. One calorie (abbreviated c or cal) is the energy used in increasing the temperature of 1 gram of water by 1 degree centigrade. One calorie is equivalent to 4.2 J, approximately. One thousand calories is one kilocalorie (1 kcal).

- **Temperature** is measured in **kelvin** (symbol K). We don't say 'degree kelvin' or write °K, though we do say 'degree centigrade' and write °C. But 1 kelvin is the same increment of temperature as 1 °C. Celsius is now the correct name for centigrade, to avoid confusion with other scales based on 100 gradations.

- The level of **radioactivity** in a sample of radioactive material is measured by the number of nuclear decays which occur over a fixed period of time. An activity of N **becquerels** (symbol Bq) means that N nuclei decay per second. The units are thus s^{-1}. For a fixed sample, the level of radioactivity decays over time; this underlies the technique of radiocarbon dating (Section 10.5.4). So when scientists quote a value in becquerels for the radioactivity of a sample, they need to also state the time at which it was measured. Like the joule, 1 becquerel is a very small amount.

  A pre-SI unit of radioactivity was the **curie**. One curie (symbol Ci) is equivalent to 3.7×10^{10} Bq.

2.2.2 SI prefixes

The SI prefixes – shown in Table 2.4 – are used to make very large or small numbers manageable without having to resort to scientific notation.[3] SI prefixes increase or decrease by factors

---

[2] One joule is roughly the increase in potential energy of a basketball being shot up into the hoop by a basketball player.

[3] In May 2010, UK newspapers reported the prediction that the total amount of data stored on electronic devices around the world, would exceed one zettabyte by the end of the year. Earlier, a Facebook petition proposed that a new prefix, the hella, be recognized for 10^{27}: 'The analysis of many physical phenomena reveals natural quantities in excess of 27 orders of magnitude', it was claimed.

of 10^3. So if the metre is the SI base unit of length, the kilometre is 10^3 metres (1000 metres) and the millimetre is 10^{-3} metres (one thousandth of a metre, or 0.001 metres). By using the appropriate prefix, we can express any measurement as a decimal with only one, two, or three digits before the decimal place. Thus, a particle physicist can talk about a distance of 256 nanometres rather than 2.56×10^{-7} metres. A distance of 3570 nanometres can be expressed as 3.57 micrometres.

This is all very well in theory, but in real life we like to have units we can actually visualize, measure, and use. A metre is too long to draw with a ruler, while a millimetre is too short, so the **centimetre** (cm) is commonly used, though centi- is not an official SI prefix. The bioscientist will also come across the decimetre (dm) in its cube form to give a unit of volume, the **cubic decimetre** or dm^3. This volume is equal to one litre (see Section 2.2.3) and some bioscientists prefer to use this symbol in place of the litre, although like the litre the cubic decimetre is not an SI unit.[4]

The symbols for prefixes from 10^6 upwards are written with capital letters. In writing the symbols for an SI unit with a prefix, the prefix symbol is attached to the unit symbol without

---

CASE STUDY C2: Velocity

A gazelle runs 1.7 km in 1 minute. A cheetah runs 5.3 m in 180 ms. Usain Bolt set a world record of 19.19 seconds for the 200 m in 2009. Which is fastest?

We use the schoolboy formula: $\text{speed} = \dfrac{\text{distance}}{\text{time}}$. We need to evaluate each speed in the same units of m s^{-1} (metres per second).

The gazelle's average speed: $\dfrac{1700}{60} = 28.33$ m s^{-1}

The cheetah's average speed: $\dfrac{5.3}{0.180} = 29.44$ m s^{-1}

Usain Bolt's average speed: $\dfrac{200}{19.19} = 10.42$ m s^{-1}

The cheetah is fastest by this simple calculation, but there are many other factors to consider. The cheetah can only maintain this speed for a few seconds, while the gazelle can run for long distances. Bolt's time is measured from a standing start, so includes the initial acceleration. The gazelle may be able to produce a higher speed when measured over a shorter distance, but the cheetah has a greater power of initial acceleration. We will explore these factors in further instalments of this case study.

Also, there is no indication of the precision of the measurements quoted. We know that both the distance and the time are very precise in Bolt's race, but the measurements for the gazelle look very imprecise. We will discuss this at the end of this chapter.

'Velocity' is a more impressive word than 'speed', and implies movement in a particular direction. We will use it from now on.

---

[4] The non-SI units of decilitre and decagram are also in use in some parts of Europe. You could go into a wine/snack bar in Prague and order '3 deci vina a 5 deka salámi'.

a space, but there is a space between two units. Thus 1 kilojoule (1 kJ) is equivalent to 1 kilonewton metre, written as 1 kN m.

We have already remarked that the SI units of energy and radioactivity – the joule and the becquerel – are very small quantities, so SI prefixes are used in discussions. For example, the atomic bomb exploded over Hiroshima in 1945 was equivalent to 14–15 kilotons of TNT, and released about 60 terajoules of energy (1 ton of TNT is equivalent to 4.184 gigajoules). The initial radioactivity level has been estimated at 8 YBq (8×10^{24} becquerel).

Problems: Now try the end-of-chapter problems 2.13–2.17.

2.2.3 More problems with SI units; units of volume

As with all perfectly rational systems designed by a committee, the SI system soon started to unravel at the edges:

- The symbols for prefixes of multiples of 10 (kilo-, mega-, . . .) were all supposed to be written with capital letters, to distinguish them from the fractional prefixes (milli-, micro-, . . .). But kilo- is denoted by k, because K is used for the base unit of temperature, the kelvin.

- The kelvin is also a rather silly unit for practical use (imagine the weather forecast: 'Tonight there will be a light ground frost, with temperatures dropping as low as 269 kelvin, while tomorrow will be warm with a maximum of 293 kelvin'), and in practice we use the **degree Celsius**: 0 °C, the temperature at which water freezes, is 273.15 K. We have noted that we don't use a ° sign with kelvin, and shouldn't talk about degrees; water freezes at '273.15 kelvin', not '273.15 degrees kelvin'.

- The symbol μ (a Greek letter m, pronounced *mu*) for the prefix micro- is not available on the standard typing keyboard, so in the biosciences, and pharmacy in particular, a **microgram** is often abbreviated as ug or mcg (the latter does not mean millicentigram; one millicentigram would be 10 μg!).

- m is the symbol for the prefix milli- as well as for the base unit of length, the metre. So 'ms' (without a space) means millisecond, while 'm s' (with a space) would mean 'metre second'!

The biggest practical problem in using SI units is with units of volume. In measuring liquids, a cubic metre (m^3) is a ridiculously large unit for everyday use, while a cubic millimetre (mm^3) is too small to be a useful standard measure due to the difficulty in measuring it with sufficient precision. A useful small volume one could measure with reasonable precision and ease would be a cubic centimetre (cm^3), and a useful large unit would be 1000 of these, i.e. a volume of 10 cm × 10 cm × 10 cm. This volume (a cube of side length 100 mm) is called a **litre** (symbol l), and a *millilitre* (ml) is one thousandth of a litre. (In practice, when biological samples can be difficult to obtain in large enough quantities, the unit of volume used is the *microlitre* (μl), which is actually a cubic millimetre.)

So 1 litre = 10 cm × 10 cm × 10 cm = 10^3 cm^3 = 1000 cm^3, and as 1 litre = 1000 ml:

1 millilitre is the same volume as 1 cubic centimetre.   (2.5)

Further useful relationships are:

$$1 \text{ ml} = 1 \text{ cm}^3 = 1 \text{ cm} \times 1 \text{ cm} \times 1 \text{ cm} = 10^{-2} \text{ m} \times 10^{-2} \text{ m} \times 10^{-2} \text{ m} = 10^{-6} \text{ m}^3.$$

Then 1 litre $= 10^3 \text{ ml} = 10^3 \times 10^{-6} \text{ m}^3 == 10^{-3} \text{ m}^3$, or $1 \text{ m}^3 = 10^3 \text{ l}$ (1 cubic metre is equivalent to 1000 litres).

A litre can also be thought of as $1 \text{ dm} \times 1 \text{ dm} \times 1 \text{ dm}$, i.e. a cubic decimetre (dm^3).

One of the commonest tasks for the bioscientist is to make up a solution of a solute (typically a powder to be dissolved). We'll look in detail at the calculations involved in this in Chapter 4. But put simply, the working bioscientist might weigh out 30 grams of salt, and add to 4 litres of water, to obtain a solution with a mass concentration of $30/4 = 7.5$ grams/litre. But his *alter ego*, the SI bioscientist, would weigh the salt in kilograms (30×10^{-3} kg), and measure out the water in cubic metres (4×10^{-3} m^3). Then the mass concentration of the solution, in kilograms per cubic metre, will be

$$(30 \times 10^{-3}) \div (4 \times 10^{-3}) = 30 \div 4 \times 10^{-3-(-3)} = 7.5 \times 10^0 = 7.5 \text{ kg m}^{-3}.$$

So 1 gram per litre (1 g/l or 1 g l^{-1} or 1 g dm^{-3}) is the same concentration as 1 kilogram per cubic metre (1 kg/m^3 or 1 kg m^{-3}); a very useful relationship.

It's also important to realize that 1 gram per litre is also the same concentration as 1 milligram per millilitre (1 mg/ml or 1 mg ml^{-1}) and 1 microgram per microlitre (1 µg/µl or 1 µg µl^{-1}), etc. This works with other dimensions, e.g. a speed of 1 metre/second is the same as 1 millimetre per millisecond.

2.2.4 Non-SI units

We have already mentioned the decilitre and the decagram. Other non-SI units you may meet are:

- The **hectare:** 1 hectare (1 ha) is an area equivalent to 10^4 m^2, i.e. it is the area of a square of size 100 metres × 100 metres, or 1 square hectometre.

- The **angstrom unit:**[5] This is a distance of 0.1 nanometres, so $1 \text{ Å} = 10^{-10}$ m.

. . . .

2.3 Calculations using SI units

By using the prefixes, we avoid the need to continually use scientific notation when dealing with very large or very small quantities; a chromosome may have 30 megabases of DNA, while the atom's diameter is 0.8 nanometres. To convert between units with different prefixes, we multiply by the appropriate power of ten:

As 1 metre $= 10^9$ nanometres,

$$8.0 \times 10^{-10} \text{ metres} = (8.0 \times 10^{-10}) \times 10^9 \text{ nanometres}$$
$$= 8.0 \times 10^{-10+9} = 8.0 \times 10^{-1} = 0.8 \text{ nanometres}.$$

In calculations involving units with different prefixes, we first express all the quantities using base units:

---

[5] Named after Anders Ångström (1814–1874), Swedish physicist and father of the science of spectroscopy.

EXAMPLE

What is the force needed to give a mass of 0.3 milligrams an acceleration of 4.2 kilo-metre/second/second?

Force = mass × acceleration

$= 0.3 \text{ mg} \times 4.2 \text{ km s}^{-2}$ [Convert to SI base units]

$= (0.3 \times 10^{-6} \text{ kg}) \times (4.2 \times 10^{3} \text{ m s}^{-2})$

$= 0.3 \times 10^{-6} \times 4.2 \times 10^{3} \text{ kg m s}^{-2}$

$= 0.3 \times 4.2 \times 10^{-6+3} \text{ kg m s}^{-2}$ using the rule in (2.3)

$= 1.26 \times 10^{-3} \text{ N}$ since 1 newton is 1 kg m s^{-2}

$= 1.26 \text{ mN}$ using millinewtons to avoid scientific notation.

In practice you would not perform this calculation by keying the numbers into your calculator in scientific notation. Instead, you would work out 0.3×4.2 on the calculator, and work out the exponent (the power of ten) yourself. This is where errors can occur. There is only one thing worse than getting the power of 10 wrong, and that is not realizing that your answer is wrong by a factor of 10 or 100 or 1000! Before performing a calculation, try to form a rough idea of the result.[6] If you are finding the volume of a liquid to dispense with a micropipette, is it likely to be 0.5 ml, 5 ml, or 50 ml?

Finally, avoid giving answers in units where more than one prefix is used. For example, it would be true that dissolving 250 mg of solute in 50 ml of water produces a solution with a concentration of $250/50 = 5 \text{ mg ml}^{-1}$, but we should express this as 5 g l^{-1} (grams per litre). Similarly, a snake may be able to strike 30 centimetres towards its prey in 200 milliseconds,

CASE STUDY C3: Force and acceleration

A male lion weighs 200 kg, and can accelerate from rest with an acceleration of 6.25 m s^{-2}. A Thomson's gazelle weighs 25 kg and has an initial acceleration of 3.75 m s^{-2}. Usain Bolt weighs 86 kg, and can accelerate from a standing start at 2.13 m s^{-2}. Calculate the force required in each case.

We use: Force (newtons) = mass (kg) × acceleration (m s^{-2})

The lion: Force = $200 \times 6.25 = 1250 \text{ N} = 1.25 \text{ kN}$

The gazelle: Force = $25 \times 3.75 = 93.75 \text{ N}$

The athlete: Force = $86 \times 2.13 = 183 \text{ N}$

The lion's muscles must exert more than ten times as much force as the gazelle's. The athlete's muscles must exert twice as much force as the gazelle's (and using half the number of legs!).

[6] It is widely believed that spinach has an incredibly high iron content. This arose because the scientist who first measured it, in 1870, misplaced the decimal point. It was not corrected in the scientific literature until 1937, by which time the *Popeye* cartoons had picked up on this 'scientific fact' and imprinted the strength-giving properties of the vegetable in the public imagination.

but its average speed would be written as 1.5 m s^{-1} and not 0.15 centimetres per millisecond, since

$$0.15 \text{ cm ms}^{-1} = 0.15 \times 10^{-2} \text{ m ms}^{-1} = 0.15 \times 10^{-2} \times 10^{3} \text{ m s}^{-1}$$
$$= 0.15 \times 10 \text{ m s}^{-1} = 1.5 \text{ m s}^{-1}.$$

Problems: Now try the end-of-chapter problems 2.21–2.33.

2.4 Dimensional analysis

Dimensional analysis is a way of checking whether complicated equations involving physical variables make sense, and of determining what are the correct units for physical constants and coefficients.

The basic rule of dimensional analysis is that the dimensions of the expressions on left and right side of an equation must agree. So from the example in Section 2.3 (a force that gives a mass of 0.3 milligrams an acceleration of 4.2 kilometre/second/second):

Force = 0.3 mg × 4.2 km s^{-2} = 1.26 mN.

Even without checking the arithmetic, we can see that the units make sense. Replacing the units by the corresponding base units or dimensions, we have kg m s^{-2} on the left, and N on the right. Checking the dimensional form for newton in Table 2.2, we see that this agrees with the left-hand side dimension.

It follows from this that we cannot add or subtract quantities with different dimensions: the question 'What is 0.2 kg + 5.2 m?' is meaningless. But you can multiply or divide such quantities: 0.2 kg × 5.2 m = 1.04 kg m.

Dimensional analysis can also be used to find the dimensional form for a coefficient in an equation:

EXAMPLE

The average rate of diffusion R of a substance in a gradient along a pipe can be written as

$$R = -DA\,(c_{\text{right}} - c_{\text{left}})/L \tag{2.6}$$

where:

- A is the cross-sectional area of the pipe;
- c_{right} and c_{left} are the mass concentrations at the two ends of the pipe;
- L is the length of the pipe;
- D is a constant called the diffusion coefficient.

What units will the diffusion coefficient D be in?

Solution: Rate of diffusion is measured by the mass of substance transferred per second, so the left-hand side will have dimensional form kg s^{-1}. On the right-hand side, A is an area (dimension of m^2), the concentrations have dimensions of mass per unit volume, i.e. kg m^{-3} and L is a length. So, leaving D as unknown and writing the other quantities in dimensional form, on the right-hand side of (2.6) we have

r.h.s. $= D$ m^2 (kg m^{-3})/m $= D$ kg m^2 m^{-3} m^{-1} $= D$ kg m^{-2}.

So the dimensions on each side of the equation are

kg s^{-1} $= D$ kg m^{-2}.

We can cancel the kg dimension on both sides:

s^{-1} $= D$ m^{-2}.

To make the two sides of the equation agree, we need D to be of dimension m^2 s^{-1}. Then

r.h.s. $= D$ m^{-2} $=$ (m^2 s^{-1}) m^{-2} $=$ (m^2 m^{-2}) s^{-1} $=$ s^{-1} $=$ l.h.s.

So the diffusion coefficient would be given in SI units of square metres per second.

Notice that you can manipulate dimensions in just the same way as numbers, using the rules for exponents. Compare the line above with

$(7^2 \times 3^{-1})\ 7^{-2} = (7^2 \times 7^{-2})\ 3^{-1} = (7^{2-2})\ 3^{-1} = 7^0 \times 3^{-1} = 1 \times 3^{-1} = 3^{-1}.$

It may still seem strange that the dimension (m^2 m^{-2}) should simply disappear. But this dimension could be written as (m^2/m^2), i.e. 'square metres per square metre', which is one area divided by another area – the result is just a number, with no dimension.

Thus, there are some quantities that have no units, and are therefore **dimensionless**. As we have seen, this happens when we have a ratio of two quantities with the same dimension. Here are some examples:

- The equilibrium constant for a chemical reaction (an important characteristic useful in understanding the formation of products from starting materials) is the concentration of the products divided by the concentration of the starting materials; this is a ratio of concentrations and so is dimensionless.
- When an elastic material (e.g. animal skin) is stretched, the amount of strain it undergoes is defined as the extension divided by the original length. This is a ratio of two lengths, and so is dimensionless.
- More generally, percentages are dimensionless.
- Mathematical constants such as π are dimensionless.
- If an equation involves a base quantity raised to an exponent, the exponent should be dimensionless. The same applies to logarithms, which we'll see in Chapter 10.

A more general way of performing dimensional analysis, without referring to particular units, is to use generic dimensions of length L, mass M, time T, and temperature K. Thus, areas will have dimensions L^2, volumes will have dimensions L^3, velocities will have generic dimension L T^{-1}, and accelerations will have generic dimension L T^{-2}. Force (which is mass times acceleration) will have generic dimension M L T^{-2}, and pressure (which is force per unit area) will have generic dimension (M L T^{-2}) L^{-2}, i.e. M L^{-1} T^{-2}.

In the example above, the generic dimensions of each quantity in equation (2.6) are:

Quantity	Physical meaning	Dimensions
R	diffusion rate	$M\,T^{-1}$
A	area	L^2
c_{right} and c_{left}	mass concentrations	$M\,L^{-3}$
L	length	L
D	diffusion coefficient	?

Inserting the dimensions in (2.6), we have

$$M\,T^{-1} = ?\,(L^2)\,(M\,L^{-3})/L = ?\,(L^2)\,(M\,L^{-3})\,(L^{-1})$$

and solving for ?:

$$? = M\,T^{-1}\,(L^2\,M\,L^{-3}\,L^{-1})^{-1}$$
$$= M\,T^{-1}\,L^{-2}\,M^{-1}\,L^3\,L$$
$$= M\,M^{-1}\,L^{-2}\,L^3\,L\,T^{-1}$$
$$= M^{1-1}\,L^{-2+3+1}\,T^{-1}$$
$$= M^0\,L^2\,T^{-1} = L^2\,T^{-1}.$$

Thus, the dimensions of D are (length)^2 per unit of time, which in SI units is $m^2\,s^{-1}$.

Problems: Now try Problems 2.34–2.38. Note: the purpose of dimensional analysis is to provide you with information about new equations involving unfamiliar quantities. So these problems are deliberately not about biology, but involve things like the Hubble constant. It's all part of Life, the Universe and Everything.

. . . .

2.5 Rounding, precision, and accuracy

2.5.1 Rounding numbers

Data values can be measurements of either a discrete variable or a continuous variable. A **discrete variable** is one that can only take certain distinct values, usually integers e.g. the number of cells on a Petri dish. **Continuous variables** can take any value along the real line we saw in Chapter 1 (or an interval of it). We can only make an experimental measurement of a continuous variable to a certain degree or level of **precision**: we read the scale on the measuring cylinder to the nearest millilitre, or the stopwatch gives us the elapsed time to the nearest hundredth of a second. So we are **rounding** the data as we write it down, and the number of decimal places we use indicates the precision; the readings mentioned above might be 0.127 litres, or 38.26 seconds. If the time elapsed had been shown as exactly 38 seconds on the stopwatch, we would record this as 38.00 seconds; writing '38 seconds' would imply that this is only correct to the nearest second, rather than the nearest 1/100 of a second. We consider further down in this section what this level of precision means.

We apply another form of rounding after making a calculation from data:

EXAMPLE

In 2002, Cornelius Chirchir set a World Junior athletics record for the men's 1500 metres, with a time of 3 minutes 30.24 seconds. What was his average speed?

Solution: Using my calculator, the average speed works out at:

$$1500/210.24 = 7.1347032 \text{ metres per second.}$$

Of course, we cannot really know his speed correct to the nearest 100 nanometres per second! Only the first few of these decimal digits are meaningful, and we might give our answer as 7.135 m s^{-1} (correct to 3 decimal places). The **rule for rounding**, is that if the first digit being discarded (in this case 7) is 0, 1, 2, 3, or 4 we round down, if it is 5, 6, 7, 8, or 9 we round up. So to round to 3 decimal places we round up to 7.135, but to round to 4 decimal places we round down to 7.1347. To round 7.1347032 to 2 decimal places, the first digit being discarded is 4, so we round down to 7.13. *Do not look at the number rounded to 3 decimal places (7.135) and round that up to 7.14; you must start from the original number.* In a moment we will investigate how many decimal places are meaningful in such a calculation.

The rule for rounding also applies for rounding negative numbers. Thus, −7.1347 rounds to −7.135 to 3 decimal places, and to −7.13 to 2 decimal places. If you imagine the unrounded number x located on the real line, then the rule for rounding to n decimal places locates the nearest n-decimal-place number to x on the line. When the unrounded number lies exactly halfway between two n-decimal-place numbers, there is a convention called the 'banker's rule' for deciding which to use: round up or down so that the final digit in the rounded number is even. Thus, 7.135 or 7.1350000 would round to 7.14 to two decimal places, and 7.165 or 7.1650000 would round to 7.16.

Problems: Check that you have rounded correctly in problems 1.24–1.27.

2.5.2 Significant figures

Quoting Cornelius's average speed as 7.1347 metres per second (to 4 decimal places) would be equivalent to giving it as 7134.7 millimetres per second (to 1 decimal place); both of these are giving the same level of precision (to the nearest 0.1 millimetre per second). So the number of decimal places quoted for the same precision differs according to the units used. A way of defining the precision that is independent of the units used is to round to a certain number of **significant figures**. The speeds quoted above, of 7.1347 m s^{-1} and 7134.7 mm s^{-1}, are both rounded to 5 significant figures.

In this case, each digit in the number is a 'significant figure'. But not all digits in a decimal number will necessarily be significant.

The rules for identifying significant digits in a decimal number, are:

- All non-zero digits are significant, e.g. 12.3456 has six significant digits.

- Zeros that appear between non-zero digits are significant, e.g. 12.0406 has six significant digits.

- Zeros that are included at the end of a decimal number are significant. So 38.00 and 380.0 and 3.800 all have four significant digits.

- But zeros at the start of a decimal number are not significant: 0.003800 still only has four significant digits, as does 001234.

So a number given in scientific notation, with one non-zero digit before the decimal point and n digits after it, has $n + 1$ significant figures, e.g. 1.234×10^{-7} has 4 significant figures.

Problems: Now try the end-of-chapter problems 2.39–2.44.

A computer or calculator that stores decimals in floating-point notation with six digits in the mantissa (see Section 2.1.5) is using a form of rounding to 6 significant figures. In practice most modern calculators and computers display many more significant figures than are justified by the data used in the calculation. It is vital that after working out a calculation you write down your answer correct to an appropriate number of significant figures, and not unthinkingly copy down all the digits shown on your calculator screen. For example, the winner of the Guinness Gastropod Championship in 2001 was a snail called Archie, who completed the 330 mm course in 2 minutes 20 seconds (*New Scientist*, October 2001). Calculating his average speed on my calculator, I obtain 2.357142857 mm s^{-1}. How many of these digits are meaningful? We'll discuss this in the next section.

On the other hand, when performing a complicated series of calculations it is essential that you work very accurately throughout the calculation, and only round the final answer. If you know that you want the final answer to be correct to 3 significant figures, this does not mean that you can round to this precision during the calculation.

EXAMPLE

Suppose I want to perform the MindGym exercise in Section 1.4.1, with a starting value of $x = 1.25$, and that I want the final answer correct to 3 significant figures. Here is the exercise again:

x	TRIPLE IT	−14	× BY ITSELF	+10	5/12 OF THIS	−1	√	+6	75% OF THIS

Starting with 1.25 on my calculator and performing the operations without any rounding (except for the calculator's accuracy), I arrive at 9.638606177, which I would round to 9.64 (correct to 3 sig figs).

Now suppose I round to 3 significant figures after each operation. The numbers I have after each operation, to use in the next operation, are:

x	TRIPLE IT	−14	× BY ITSELF	+10	5/12 OF THIS	−1	√	+6	75% OF THIS
1.25	3.75	−10.3	106	116	48.3	47.3	6.88	12.9	9.68

The final answer is incorrect, even if rounded to 2 significant figures.

2.5.3 Uncertainty intervals

Returning to Cornelius's average speed, which we worked out as 7.1347032 metres per second: what would be the most appropriate level of precision to use in quoting his speed, given the precision of the measurements of distance and time?

Let us assume that the Olympics-grade track he was running on in Monaco was laid out very precisely, so that he ran exactly 1500 metres.

The stopwatch, giving a time of 210.24 seconds, was precise to the nearest hundredth of a second. This means that his actual time was somewhere between 210.235 and 210.245 seconds – all numbers between 210.235 and 210.245 would round to 210.24 to two decimal places. We can express this uncertainty in the time measurement, by writing it as 210.24 ± 0.005 seconds. This is called an **uncertainty interval**.

If we quote a measurement as $x = A \pm \varepsilon$, we mean that x lies between $A - \varepsilon$ and $A + \varepsilon$.

We can write this algebraically as $A - \varepsilon < x < A + \varepsilon$.

This range of values defines an **uncertainty interval** on the real line, centred at A and of semi-width ε: $\qquad(2.7)$

The symbol ε is a Greek letter e, called epsilon. It is used by mathematicians to represent a small positive quantity.

Does this mean that we can only quote two decimal places in our answer, i.e. 7.13 metres per second? Not necessarily. We need to work out an uncertainty interval for the speed.

We can see the uncertainty in the speed v by using those two extreme values of the time uncertainty interval, namely 210.235 and 210.245, in our calculation:

$1500 \div 210.235 = 7.1348729$ m s^{-1}

$1500 \div 210.245 = 7.1345335$ m s^{-1}

So $7.1345335 < v < 7.1348729$.

Both of these extreme values round to 7.135 to 4 significant figures. But if we quote an average speed of $v = 7.135$ m/s, this implies that we have a precision of the nearest mm/second, i.e. that the average speed was between 7.1345 and 7.1355; in fact, we can be a lot more precise than that. Suppose we work to one extra significant figure, and that we round the lower value down and round the higher number up. That is, 7.1345335 is rounded down to 7.1345 (to 5 sig figs), and 7.1348729 is rounded up to 7.1349. We can now say that the average speed v is somewhere between 7.1345 and 7.1349. This is an interval on the real line, centred around 7.1347, with a semi-width of 0.0002; we find the semi-width ε of the interval by $\varepsilon = \frac{1}{2}(7.1349 - 7.1345)$. This uncertainty interval would be more informative and justified to use: thus we should quote the average speed as $v = 7.1347 \pm 0.0002$ metres per second.

We rounded the lower number down and the higher number up, so that the original interval is enclosed within the new one:

$7.1345 < 7.1345335 < v < 7.1348729 < 7.1349.$

EXAMPLE

A gazelle's speed v is somewhere between 6.12278 and 6.12487 metres per second. We can write this as an interval:

$$6.12278 < v < 6.12487.$$

Both extreme values round to 6.12 m s^{-1}, to 3 significant figures, but to 4 sig figs they round to 6.123 and 6.125 respectively.

Rather than quoting $v = 6.12$ m s^{-1} (to 3 sig figs), we work to 4 significant figures and round the lower number down to 6.122, and round the higher number up to 6.125. So v is in the interval $6.122 < v < 6.125$.

The midpoint of the interval is $\frac{1}{2}(6.122 + 6.125) = 6.1235$, and its semi-width is $\varepsilon = \frac{1}{2}(6.125 - 6.122) = 0.0015$, so we can report the gazelle's speed as $v = 6.1235 \pm 0.0015$.

Problems: Now try the end-of-chapter problems 2.45–2.46.

We said at the start of this chapter, that there is a difference between precision and accuracy. In the final section, we analyse how inaccuracies in measurements can build up when we use those measurements in calculations.

. . . .

2.6 Extension: accuracy and errors

Measurements may be very precise, but they could still be inaccurate. Accuracy is expressed in the same way as precision, but it refers to the overall closeness of the numerical value to the true value. The difference between the two could be caused partly by a lack of precision, but there can be more significant factors. In the example of Chichir's race, the race official may have had slow reactions, so that he pressed the stopwatch about half a second after Chichir crossed the line. Or the stopwatch may not have been properly zeroed before the race began.

In mathematical parlance, the difference between an approximate value and the true value is called the **error**. The error may come from a lack of precision, a lack of accuracy, or both. The word 'error' does not imply that a mistake has been made; indeed it is an indication of the skill with which the measurement has been made. It is the mark of a true scientist, and therefore a bioscientist, that they recognize the limitations of the data they have collected.

Errors also arise every time the computer rounds a number from a calculation in order to store it in its memory; this is called **rounding error**. If a tiny rounding error occurs every time an addition or multiplication is performed and the result stored (this is called a **floating-point operation** or 'flop'), do these errors build up until the final result is meaningless? After all, modern computer chips can perform at a speed of 30 gigaflops (30 thousand million floating-point operations per second)! The investigation of this is called **error analysis**. You may not be called upon to perform error analyses, but it is important to gain a feel for concepts such as absolute and relative error.

2.6.1 Errors in addition and subtraction

Errors can be expressed in two ways – as absolute error or relative error:

> **absolute error = true value – approximate value**
>
> **relative error = absolute error/approximate value**   (2.8)

So an absolute error of 100 metres in a measured length sounds a lot, but if that length is the distance from the Earth to the Sun (about 150 million kilometres), it is a relative error of only 0.67×10^{-9}, or 0.000000067%. Of course, the true and approximate values must be quoted in the same units.

An absolute error has the same units as the quantity being measured, but the relative error is dimensionless, as it is a ratio of two quantities with the same units; see Section 2.4.

If we add two approximate values together, what is the error in the result? Let A and B be the (known) approximate values of the quantities being added, and a and b be the absolute errors. The true value is given by rearranging the equation above:

true value = approximate value + absolute error

so the true values of the quantities being added are $(A + a)$ and $(B + b)$.

We perform the addition of A and B, and obtain a result R :

$$R = A + B.$$

What is the error in R? The true value for the result would be given by adding the true values $(A + a)$ and $(B + b)$, so going back to the original expression:

$$\text{absolute error in result} = \text{true value of result} - \text{approximate value of result}$$
$$= [(A + a) + (B + b)] - [A + B]$$
$$= A + a + B + b - A - B = a + b.$$

For addition of two values, the absolute error in the result is the sum of the absolute errors in the two values.

What about subtraction? If we followed the algebra we have just done, for the calculation $R = A + B$, we would find that the error in R would be $a - b$, i.e. if both errors were positive they would partially cancel out. However, we don't know if the errors are positive or negative, we usually just know uncertainty intervals such as $A \pm \varepsilon$, which we introduced in Section 2.5.3. We can now think of the semi-width ε as the magnitude of the potential error. We must assume a worst-case scenario in which the actual errors have different signs, and build up. So we have the same potential error as with addition. The rule for addition and subtraction is:

> In addition or subtraction of two numbers A and B, the absolute error
> in $A + B$ or $A - B$ is the sum of the absolute errors in A and B.   (2.9)

From now on we will talk about an error ε meaning a potential error of $\pm \varepsilon$.

EXAMPLE

Suppose $A = 253.12$ (correct to 2 decimal places) and $B = 252.2$ (correct to 1 decimal place). What is the accuracy in the calculation of $A + B$ and $A - B$?

Solution: The uncertainty intervals for A and B are:

$$A = 253.12 \pm 0.005 \quad \text{and} \quad B = 252.2 \pm 0.05.$$

So the absolute error in $A + B$ and $A - B$ is $0.005 + 0.05 = 0.055$ in both cases.

Then $A + B = 505.32 \pm 0.055$, and $A - B = 0.92 \pm 0.055$.

The uncertainty intervals are

$$506.265 < A + B < 505.375, \quad \text{and} \quad 0.865 < A - B < 0.975.$$

This illustrates that **your result is only as accurate as the least accurate quantity in the calculation.** There is no point in measuring A even more accurately, in the hope that this will somehow make up for the lack of accuracy in B.

You can maybe see that there is a big difference in the relative errors of the two results:

$$\text{Relative error in } A = \frac{0.005}{253.12} = 0.0000197\ldots = 0.002\%$$

$$\text{Relative error in } B = \frac{0.05}{252.2} = 0.000198\ldots = 0.02\%$$

We only need to work out one or two significant figures in the errors, to get an idea of their size. To work out the errors accurately, is a contradiction in terms!

$$\text{Relative error in } A + B = \frac{0.055}{505.32} = 0.00011 = 0.011\%$$

$$\text{Relative error in } A - B = \frac{0.055}{0.92} = 0.060 = 6\%$$

The important lesson here is that **subtracting one large number from another can create a very big relative error.**

If we start with A and add or subtract an exact number (i.e. whose error is zero), then the absolute error in the result is just the absolute error from A.

2.6.2 Errors in multiplication and division

We follow the algebra of the previous section, to find the error in performing the multiplication

$$R = A.B.$$

The true value of the multiplication is $R + r$, where r is the absolute error in R, and is found by multiplying the true values:

$$R + r = (A + a).(B + b)$$

So $r = (A + a).(B + b) - R.$

Expanding the brackets, and replacing R by $A.B$, we get

$$r = (A.B + a.B + A.b + a.b) - A.B = a.B + A.b + a.b.$$

The final term on the right-hand side is the product of two small errors. This will be even smaller (e.g. if $a = b = 10^{-3}$, then $a.b = 10^{-6}$), and can be ignored in comparison with the other terms. Mathematicians call this **a second-order term**. The equation then becomes

$$r = a.B + A.b.$$

This becomes more meaningful if we express it in terms of relative errors. Divide both sides by R (which is $A . B$) :

$$\frac{r}{R} = \frac{a.B}{A.B} + \frac{A.b}{A.B} = \frac{a}{A} + \frac{b}{B}.$$

So, for multiplication of two values, the relative error in the result is the sum of the relative errors in the two values.

By some slightly trickier algebra (see Problem 2.52) we can obtain the same result for division. The result is thus:

> In multiplication or division of two numbers A and B, the relative error in $A.B$ or A/B is the sum of the relative errors in A and B. (2.10)

EXAMPLE

Let $A = 1.56$ and $B = 75.1$, both correct to 3 significant figures. What are the potential errors in (i) $A . B$, (ii) A/B and (iii) B/A?

Solution: Absolute errors: $A = 1.56 \pm 0.005$ and $B = 75.1 \pm 0.05$

$$\text{Relative error in } A = \frac{0.005}{1.56} = 0.0032 = 0.32\%$$

$$\text{Relative error in } B = \frac{0.05}{75.1} = 0.00067 = 0.067\%$$

So the relative errors in $A.B$, A/B and B/A will be $0.0032 + 0.00067$, which is 0.0039 or 0.39%.

The calculations are:

(i) $A.B = 1.56 \times 75.1 = 117.156$

(ii) $A/B = 1.56 \div 75.1 = 0.0207723$

(iii) $B/A = 75.1 \div 1.56 = 48.1410256$

We can find the absolute errors in the calculations, by

$$\text{absolute error} = \text{relative error} \times \text{approximate value}.$$

The absolute errors in the calculations are thus:

(i) $0.0039 \times 117.156 = 0.457$

(ii) $0.0039 \times 0.0207723 = 0.000081$

(iii) $0.0039 \times 48.1410256 = 0.19$

Although the relative errors are the same, the absolute errors are very different. It is therefore wrong to quote each calculation 'correct' to the same number of decimal places.

If we start with A and multiply or divide it by an exact number (i.e. whose error is zero), then the relative error in the result is just the relative error from A. The absolute error is the absolute error from A, multiplied by the exact factor. That is, if A has absolute error a, then $\frac{1}{3}A$ will have absolute error $\frac{1}{3}a$.

2.6.3 Errors in exponentiation

'Exponentiation' is a fancy term for raising a number to a power (or exponent). If we work out A^2, we are multiplying A by itself, so – by the previous section – the relative error in the result will be the sum of the relative error in A plus the relative error in A, i.e. twice the relative error in A. Raising A to the power n, i.e. working out A^n, means multiplying A by itself n times, so for the relative error we add up n copies of the relative error in A.

> **EXAMPLE**
>
> A roundworm infection called loiasis is caused in humans by the nematode worm *Loa loa*. The worms are 40–70 mm in length and about 0.5 mm diameter. Calculate the volume of a worm which is measured as 56.30 mm long and 0.58 mm diameter (correct to nearest 0.01 mm).
>
> **Solution:** We use the formula for the volume of a cylinder: $V = \pi r^2 h$. We will use the approximation 3.142 for π.
>
> $$\text{Relative error in our value of } \pi \text{ is } \frac{0.0005}{3.142} = 0.00016.$$
>
> The body diameter $d = 0.58 \pm 0.005$ mm, so the radius $r = 0.29 \pm 0.0025$ mm.
>
> $$\text{Relative error in } r \text{ is } \frac{0.0025}{0.29} = 0.0086.$$
>
> So relative error in r^2 is $0.0086 \times 2 = 0.0172$.
> The body length is $h = 56.30 \pm 0.005$ mm, so relative error in h is $\frac{0.005}{56.30} = 0.000089$.
> Calculation: $V = \pi r^2 h = 3.142 \times 0.58^2 \times 56.30 = 59.50734$
>
> Relative error in result $= 0.00016 + 0.0172 + 0.000089 = 0.01745$
>
> Absolute error in result $= 0.01745 \times 59.507 = 1.038$.
>
> The volume in mm^3 is then in the interval $58.47 < V < 60.55$
> We could quote the volume as $V = 59.5 \pm 1.04$ mm^3. Notice that it is the error in d which is the main contributor to the final error, rather than the approximation of π.

We could also obtain the result for the error in A^n in the same way as the previous sections, and this will allow us to generalize it. For this we need the **binomial theorem**.

In (1.20) we quoted the expansions of $(a + b)^2$ and $(a + b)^3$. The binomial theorem tells us that the first two terms of the expansion of $(a + b)^n$ are:

$$(a + b)^n = a^n + na^{n-1}b + \cdots$$

You'll be relieved to hear that we're not going to concern ourselves with the other terms in this expansion. Applying this to $(A + a)^n$, where a is the error in A:

$$(A + a)^n = A^n + nA^{n-1}a + \cdots \tag{2.11}$$

The remaining terms are comparatively small, involving a^2 and higher powers of a, so we can ignore them. The relative error in the calculation of A^n is then:

$$\frac{(A+a)^n - A^n}{A^n} = \frac{A^n + nA^{n-1}a - A^n}{A^n} = \frac{nA^{n-1}a}{A^n} = n\frac{a}{A}$$

i.e. n times the relative error in A, as we deduced for $n = 2, 3, 4, \ldots$ But if a is much smaller than A, equation (2.11) also applies when n is a fraction, or negative. In particular:

- If $n = -1$, we are calculating A^{-1}, which is $\frac{1}{A}$, the reciprocal of A. Our result tells us that the relative error in $\frac{1}{A}$ will be $\frac{a}{A}$. The absolute error in $\frac{1}{A}$ will be $\frac{a}{A} \times \frac{1}{A} = \frac{a}{A^2}$.

- If $n = \frac{1}{2}$, we are calculating $A^{\frac{1}{2}}$, which is $\sqrt{A}$, the square root of A. Our result tells us that the relative error in $\sqrt{A}$ will be $\frac{1}{2}\frac{a}{A}$. The absolute error in $\sqrt{A}$ will be $\frac{1}{2}\frac{a}{A} \times \sqrt{A} = \frac{a}{2\sqrt{A}}$.

> If n is a positive or negative integer or fraction, the relative error in calculating A^n will be n times the relative error in A. (2.12)

This result is sometimes used to find approximate values of roots and reciprocals of $1 + x$, where x is a small positive or negative number. By (2.11):

$$(1+x)^n \approx 1 + nx \quad \text{if } x \text{ is small.} \qquad (2.13)$$

The $\approx$ sign means 'is approximately equal to'. Thus

- $\frac{1}{1+x} = (1+x)^{-1} \approx 1-x$

- $\sqrt{1+x} = (1+x)^{\frac{1}{2}} \approx 1 + \frac{1}{2}x$

For example, $\sqrt{1.05} = \sqrt{1+0.05} \approx 1.025$. (Actually, $\sqrt{1.05} = 1.0246951$.)
For more complicated calculations, we need to use the three rules in combination:

EXAMPLE

We want to calculate $z = \frac{\sqrt{10}+3}{\sqrt{10}-3}$, and we are told that $\sqrt{10} = 3.162$ (correct to 3 d.p.).

We work out $z = \frac{6.162}{0.162}$. What precision can we use?

Solution: The absolute error in $\sqrt{10} = 3.162$ is $e = 0.0005$.

The absolute error in $\sqrt{10}+3$ will also be e (since we are adding an exact number), and the relative error will be $\frac{e}{\sqrt{10}+3} = \frac{0.0005}{6.162} = 0.00008$.

Similarly, the absolute error in $\sqrt{10}-3$ will also be e, and the relative error will be $\frac{e}{\sqrt{10}-3} = \frac{0.0005}{0.162} = 0.0031$.

Using the rule (2.10) for division, the relative error in our calculation of z will be $0.00008 + 0.0031 = 0.00318$. The absolute error in our value for z will be $0.00318 \times 38.037 = 0.121$. Thus, $z = 38.037 \pm 0.121$, or $37.916 < z < 38.158$.

But we can obtain a much more precise value for z if we use the trick we saw in Section 1.6.4, of converting z into a surd, using the identity

$$(a - b)(a + b) \equiv a^2 - b^2:$$

$$z = \frac{\sqrt{10}+3}{\sqrt{10}-3} \times \frac{\sqrt{10}+3}{\sqrt{10}+3} = \frac{\left(\sqrt{10}+3\right)^2}{\left(\sqrt{10}\right)^2 - 3^2} = \frac{\left(\sqrt{10}\right)^2 + 6\sqrt{10} + 9}{10 - 9} = 19 + 6\sqrt{10}.$$

Evaluating this, $z = 19 + 6\sqrt{10} = 19 + (6 \times 3.162) = 37.972$. What is the error in this value?

Absolute error in the calculation of $6\sqrt{10}$ is $6e = 0.0030$. Adding the exact number 19 does not affect the absolute error. So $z = 37.972 \pm 0.003$, or $37.969 < z < 37.975$.

The source of error in the first calculation is that the denominator is a small number. The lesson is: **avoid calculations which involve dividing by a small and imprecise number, if possible.**

Let us finally apply error analysis to the average speed of Cornelius Chirchir (see Section 2.5.1). Suppose the athletics track was laid out with a precision of ± 10 mm over 1500 metres.

The relative error in the distance would then be ± 0.01 m/1500 m $= 6.7 \times 10^{-6}$ (We only want to see the size of the error, so we only quote it to two significant figures.)

The relative error in the time measurement is ± 0.005 s/210.24 s $= 2.38 \times 10^{-5}$.

Allowing for a worst-case scenario where the distance error is positive and the time error is negative, so that the two magnitudes add together, the relative error in the speed would be

$$6.7 \times 10^{-6} + 2.38 \times 10^{-5} = 0.67 \times 10^{-5} + 2.38 \times 10^{-5} = 3.05 \times 10^{-5}$$

(remembering the trick in Section 2.2.2 for adding numbers in scientific notation).

Then

$$\text{absolute error in speed} = \text{relative error} \times \text{approximate value}$$
$$= (3.05 \times 10^{-5}) \times 7.135 = 0.00022 \text{ m s}^{-1}.$$

Compare this with the uncertainty interval of ± 0.0002 obtained in Section 2.5.1 when we ignored any error in the track length.

If we have a large number of data values to add up, it does look as if the errors will build up as we perform more and more floating-point operations. However, that is to assume a worst-case scenario where all the errors are of the same sign. In practice, the errors will be randomly distributed in size around zero, with just as many negative as positive errors, so they will largely cancel each other out. This is why it is common practice to quote the mean of a **large** set of values $x_1 x_2, \ldots, x_n$ to one extra significant figure compared to the accuracy of the data values themselves. Further discussion of means of data-sets is entering the realm of statistics, which is outside the scope of this book.

Problems: Now try the end-of-chapter problems 2.47–2.52.

2.6.4 Delta notation

Unless we are being really incompetent, the absolute error in a measurement will be much smaller than the measurement itself, though it has the same dimensions. Mathematicians have a notation for representing this situation: if x denotes the measured quantity, we use δx to denote a much smaller quantity with the same dimensions as x. The symbol δ is a Greek letter d, and is pronounced 'delta'. Note that δx does not mean 'delta times x'; it is really a single symbol for the error. We'll use this notation later in this book, so it's worth introducing it here, in a summary of our error analysis.

We could express our result (2.9) by saying that if $R = A + B$, then the error in R is given by $\delta R = \delta A + \delta B$. If δA was positive and δB was negative, the errors would partially cancel themselves out when they are added together. This would be lucky, but if we are trying to work out the possible error in our calculation of $A + B$ we can't rely on this happening; we should imagine a 'worst-case scenario' where the sizes of the errors add together. We can do this by considering the absolute values of the errors: $|\delta A|$ and $|\delta B|$. Look back at Section 1.3.4 for the definition of $|a|$, the absolute value of a. The use of the word 'absolute' here indicates that we are throwing away any minus sign (e.g. $|-3| = 3$), whereas the 'absolute' in 'absolute error' doesn't mean this, it is used to distinguish the two types of error: absolute and relative.

In real life, we don't know the actual error – if we did, we could work out the true value. But we may have an idea of the size of the error. So, using the 'worst-case scenario' approach, we can say that if $R = A + B$, and if the possible errors in A and B are $|\delta A|$ and $|\delta B|$, then $|\delta R|$, the size of the error in R, could in the worst case be $|\delta A| + |\delta B|$. We can express this using an inequality:

$$\text{If } R = A + B \text{ or } R = A - B, \quad \text{then} \quad |\delta R| \le |\delta A| + |\delta B|. \tag{2.14}$$

We can do the same thing in expressing the worst-case error when multiplying two quantities:

$$\text{If } R = A.B \text{ or } R = \frac{A}{B}, \quad \text{then} \quad \left|\frac{\delta R}{R}\right| \le \left|\frac{\delta A}{A}\right| + \left|\frac{\delta B}{B}\right|. \tag{2.15}$$

Mathematicians say that we have found an **upper bound** on the error.

Let's use this notation to produce a mathematical model of the race in the previous section. Let x be the distance run (which was 1500 m), and we'll denote the error in that measurement by δx. Similarly, let t denote the time taken, and denote the error in that measurement by δt.

The athlete's average speed v would be calculated as $v = \frac{x}{t}$.

Now the relative error in the distance is $\frac{\delta x}{x}$, and the relative error in the time is $\frac{\delta t}{t}$.

Our error analysis showed that in the worst-case scenario, the relative error in the speed is the sum of the relative errors in the distance and time:

$$\left|\frac{\delta v}{v}\right| \le \left|\frac{\delta x}{x}\right| + \left|\frac{\delta t}{t}\right|.$$

So the absolute error in the speed is given by

$$|\delta v| \le \left(\left|\frac{\delta x}{x}\right| + \left|\frac{\delta t}{t}\right|\right).v \tag{2.16}$$

We know that the speed is a positive quantity, so we don't need to write $|v|$.

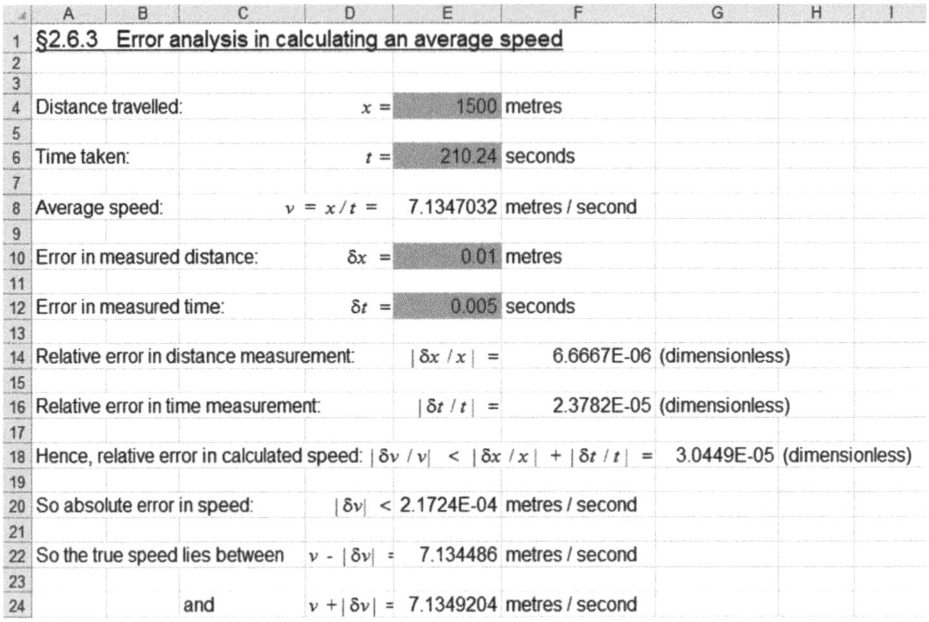

	A	B	C	D	E	F	G	H	I
1	§2.6.3 Error analysis in calculating an average speed								
2									
3									
4	Distance travelled:			$x =$	1500	metres			
5									
6	Time taken:			$t =$	210.24	seconds			
7									
8	Average speed:		$v = x/t =$		7.1347032	metres / second			
9									
10	Error in measured distance:			$\delta x =$	0.01	metres			
11									
12	Error in measured time:			$\delta t =$	0.005	seconds			
13									
14	Relative error in distance measurement:			$\lvert \delta x / x \rvert =$		6.6667E-06	(dimensionless)		
15									
16	Relative error in time measurement:			$\lvert \delta t / t \rvert =$		2.3782E-05	(dimensionless)		
17									
18	Hence, relative error in calculated speed: $\lvert \delta v / v \rvert < \lvert \delta x / x \rvert + \lvert \delta t / t \rvert =$						3.0449E-05	(dimensionless)	
19									
20	So absolute error in speed:			$\lvert \delta v \rvert <$	2.1724E-04	metres / second			
21									
22	So the true speed lies between		$v - \lvert \delta v \rvert =$		7.134486	metres / second			
23									
24		and		$v + \lvert \delta v \rvert =$	7.1349204	metres / second			

Figure 2.1 Error analysis in measuring a required concentration (SPREADSHEET 2.3)

SPREADSHEET 2.2 performs this analysis. Type the distance and time, and the associated errors, into the yellow cells as indicated. To work out the absolute value, for example in cell F14, Excel has the ABS() function.

You will need to round the answers to the appropriate number of significant figures!

Exactly the same analysis can be used in any calculation involving multiplication or division with measured quantities. Look back at Section 2.2.3, where our working bioscientist weighed out 30 grams of salt, and added it to 4 litres of water, to produce a solution with a mass concentration of 7.5 grams/litre. If the volume of solvent V was measured correct to the nearest millilitre (i.e. an error of ± 0.0005 litres), and the mass of solute M was weighed correct to the nearest milligram, what is the possible error in the concentration c? The calculation is programmed into SPREADSHEET 2.3; see Figure 2.1 for a screenshot.

CASE STUDY A5: Error analysis for geometric growth

In the geometric growth model, a population of initial size N_0 grows over succeeding generations according to the update equation

$$N_{p+1} = \lambda\, N_p.$$

In applying this model to a real insect colony, it is not possible to measure the multiplication rate λ precisely by observation. Suppose we introduce an ant colony with an initial population of 100 insects, and try to count the number after one week: because the little devils move around so much, we can only estimate the new population to be somewhere between 2000 and 2400 insects. So we can take the multiplication rate to be between 20 and 24. We use $\lambda = 22$ with an error of $\lvert \delta \lambda \rvert = 2$. (Notice that λ is a dimensionless variable.) Estimate the colony's size after a further five weeks.

The update equation involves multiplication, so the relative error in N_{p+1} is the sum of the relative errors in λ and N_p (see equation 2.8):

$$\frac{\delta N_{p+1}}{N_{p+1}} = \frac{\delta \lambda}{\lambda} + \frac{\delta N_p}{N_p}.$$

When we apply this update for different values of p, the quantities λ and $\delta \lambda$ remain constant:

$$\frac{\delta \lambda}{\lambda} = \frac{2}{22} = \frac{1}{11} = 0.091 \text{ (to 2 significant figures)}.$$

So

$$\frac{\delta N_{p+1}}{N_{p+1}} = 0.091 + \frac{\delta N_p}{N_p}.$$

We have counted the initial population precisely, so $\delta N_0 = 0$. Using the update equation for the relative error above, we see that:

- the relative error after 1 week is $\dfrac{\delta N_1}{N_1} = 0.091 + 0 = 0.091$

- the relative error after 2 weeks is $\dfrac{\delta N_2}{N_2} = 0.091 + 0.091 = 0.182$

and so on. The relative error after 6 weeks will be $6 \times 0.091 = 0.546$, or 54.6%.

Using $\lambda = 22$ we calculate the population after six weeks as $N_6 = \lambda^6 N_0 = 22^6 \times 100 = 11337990400$, or 1.1338×10^{10}. But this estimate may have an error of $0.546 N_6$, or 6×10^9. In SPREADSHEET 2.4 we have copied the geometric growth SPREADSHEET 1.2, but adapted the data to illustrate this example. We have used a birth rate of 21 and zero death rate, to produce a multiplication rate of 22. Try changing this to a multiplication rate of 20 or 24, to see the difference made to the population.

This example illustrates two points:

- If there is an error in a measured value which is then used repeatedly in a calculation, the error can build up and may soon swamp the actual data.

- The geometric growth model is unsatisfactory if it predicts that we will return to the forest in a few weeks to find 11 billion ants in the colony. If bacteria and insects really multiplied according to this model, the Earth would have been submerged in them a long time ago. In reality, there is a limited amount of nutrient (food source) available to support the colony, and this limits its growth. In Chapter 7 we will see how the geometric growth model can be modified to take this into account, producing a much more useful theory: the logistic growth model.

As well as the Greek small letter δx, mathematicians also use Δ (the Greek capital letter D) and write Δx to indicate a small or not-so-small distance in the x-direction. We will do this in the next chapter when we define the slope of a straight-line graph.

✳ SUMMARY

Scientific notation

The general form of a number in scientific notation is $s \times 10^n$. Here, s is a fixed-point number which has one non-zero digit before the decimal point (i.e. it is a number between 1.0 and 9.99999 …), and n is a positive or negative integer.

Numbers in scientific notation can only be added or subtracted if their exponents are the same. For multiplication and division:

$$(s \times 10^m) \times (t \times 10^n) = (s \times t) \times 10^{m+n}$$

$$(s \times 10^m) \div (t \times 10^n) = (s \div t) \times 10^{m-n}$$

SI units

Dimension	SI unit	Symbol
Length	**metre**	m
Mass	**kilogram**	kg
Amount of substance	**mole**	mol
Time	**second**	s
Electric current	**ampere**	A
Temperature	**kelvin**	K
Luminous intensity	**candela**	cd

Measurement	SI unit	Symbol	Definition	Form in base units
Force	newton	N	kg m s^{-2}	
Energy	joule	J	N m	kg m^2 s^{-2}
Pressure	pascal	Pa	N m^{-2}	kg m^{-1} s^{-2}
Power	watt	W	J s^{-1}	kg m^2 s^{-3}

Measurement	SI unit	Form in base units
Area	square metre	m^2
Volume	cubic metre	m^3
Speed, velocity	metre per second	m s^{-1}
Acceleration	metre per second squared	m s^{-2}
Density	kilogram per cubic metre	kg m^{-3}
Specific weight	newton per cubic metre	kg m^{-2}s^{-2}
Mass concentration	kilogram per cubic metre	kg m^{-3}
Molar concentration	mole per cubic metre	mol m^{-3}

Prefix	Symbol	Multiple/ Fraction	
Multiples of 10:			
giga	G	10^9	1 000 000 000
mega	M	10^6	1 000 000
kilo	k	10^3	1 000
hecto	h	100	
deca	da	10	
Fractions of 10:			
deci	d	0.1	
centi	c	0.01	
milli	m	10^{-3}	0.001
micro	μ	10^{-6}	0.000 001
nano	n	10^{-9}	0.000 000 001

1 millilitre is the same volume as 1 cubic centimetre.
A litre can be thought of as 10 cm × 10 cm × 10 cm, i.e. a cubic decimetre.

Significant figures

The rules for identifying significant digits in a decimal number, are:

- All non-zero digits are significant, e.g. 12.3456 has six significant digits.
- Zeros that appear between non-zero digits are significant, e.g. 12.0406 has six significant digits.
- Zeros that are included at the end of a decimal number are significant. So 38.00 and 380.0 and 3.800 all have four significant digits.
- But zeros at the start of a decimal number are not significant: 0.003800 still only has four significant digits, as does 001234.

So a number given in scientific notation, with one non-zero digit before the decimal point and n digits after it, has $n + 1$ significant figures, e.g. 1.234×10^{-7} has 4 significant figures.

? PROBLEMS

2.1–2.6: Write the following numbers in scientific notation:

2.1 31 200
2.2 0.00014
2.3 −1.025
2.4 −0.00000002
2.5 1 375 000
2.6 32.06.

2.7–2.14: Perform the following calculations in scientific notation:

2.7 $x + y$, where $x = 3.06 \times 10^6$ and $y = 8.30 \times 10^8$
2.8 $2a - 3b$, where $a = 2.1 \times 10^5$ and $b = 7.5 \times 10^7$

2.9 $c - d$, where $c = 4.2 \times 10^3$ and $d = 4.2 \times 10^{-1}$

2.10 $c - d \times 10^3$, where $c = 4.2 \times 10^3$ and $d = 4.2 \times 10^{-1}$

2.11 $c.d$, where $c = 4.2 \times 10^3$ and $d = 4.2 \times 10^{-1}$

2.12 $\dfrac{a}{b}$, where $a = 2.1 \times 10^5$ and $b = 7.5 \times 10^7$

2.13 $\dfrac{3.2 \times 10^5 - 4.7 \times 10^3}{4.7 \times 10^3 - 1}$

2.14 $\dfrac{3.2 \times 10^{-5}}{4.7 \times 10^{-3}} - 4.7 \times 10^{-3}$.

2.15–2.17: Write the following quantities in (combinations of) SI base units, using scientific notation:

2.15 The force required to give a car weighing 3.5 tonnes (1 tonne = 10^3 kg) an acceleration of 12 m s^{-1}.

2.16 The energy expended by a 50 gram weight dropping vertically downwards through a distance of 35 cm (acceleration due to gravity = 9.81 m s^{-2}).

2.17 The pressure caused by a force of 25 kilonewtons distributed uniformly over a circular disk of radius 7.5 cm.

2.18: Referring to problem 1.45: if each grain of wheat had a volume of 2 mm^3, what would be the volume of wheat (in m^3) on the final square of the chessboard?

2.19: Calculate the number of atoms in the Universe! Use scientific notation where appropriate. Here is some information to help you:

- Current estimate for the radius of the observable Universe is about 46.5 thousand million light-years.
- The speed of light is $c = 299\ 792\ 458$ metres per second.
- The average density of visible matter in the Universe is 3×10^{-31} grams per cm^3.
- The main elements in the Universe are hydrogen (740 000 parts per million), helium (240 000 ppm), oxygen (10 000 ppm) and carbon (10 000 ppm).
- The weight of 1 hydrogen atom is $m_H = 1.66 \times 10^{-27}$ kg.
- The atomic weights of helium, oxygen and carbon are 4, 16 and 12, meaning that 1 helium atom weighs the same as 4 hydrogen atoms, etc.

Method:

- Work out the distance travelled by light in one year. Hence find the radius of the Universe.
- Find the volume of the Universe using the formula for the volume of a sphere of radius r: $V = \frac{4}{3}\pi r^3$.
- Using the average density, find the mass M of the Universe.
- Let the total number of atoms in the Universe be N. Then how many hydrogen atoms are there, and what will be their mass? Repeat for the other three elements, and combine to find an equation linking N, m_H and M. Solve for N.

Express the result of each stage in appropriate SI units.

2.20: The WMAP space probe,[7] launched by NASA in 2001, has been used to make accurate measurements of the background microwave radiation from the Big Bang. Based on these, cosmologists say the Universe should have a critical mass density of 9.9×10^{-30} grams per cm^3 (to a precision of ±1%). They conclude that the mass of all the atoms in the visible Universe, as calculated in the previous question, comprises less than 5% of the total mass; the remainder is Dark Matter (23%) and Dark Energy (72%). Calculate the mass of the Dark Matter in the Universe.

2.21–2.24: Express the following volumes in litres:

  2.21 350 mm^3

  2.22 2.5 m^3

  2.23 A swimming pool which is 12 m long, 4 m wide, and with a sloping bottom from 1.5 m to 3.0 m deep.

  2.24 A 12 oz US Coke can which is 12.1 cm tall and has a diameter of 6.6 cm.

2.25: Sanjeev and Karpawich (2006) measured the length l and diameter d (in millimetres) of the superior vena cava (SVC) in a sample of growing children. They plotted these dimensions against the child's height h (in cm), and used linear regression to find least-squares equations relating l and d to h. The best-fit linear relations they obtained were:

  • $l = 14.3 + 0.25\,h$
  • $d = 3.2 + 0.093\,h$

Show that the predicted volume V of the vena cava for a child of height 1 metre is 115.1×10^{-6} litres.

Find a cubic equation in h (of the form $V = ah^3 + bh^2 + ch + d$), giving the SVC volume V in terms of the child's height h.

2.26: In the UK, the alcohol content of drinks is measured in units of alcohol. One unit is equivalent to 1 centilitre of pure alcohol. Calculate (to 3 decimal places) the units of alcohol for the following drinks. Check their prices, and hence work out the cost per unit of alcohol in each case:

  (i)   a four-pack of Bacardi Breezers from a UK supermarket;
  (ii) a Bacardi and Coke in a British pub;
  (iii) a pint of draught Guinness in a UK pub;
  (iv) a large plastic bottle of White Diamond cider from a supermarket;
  (v)   the *White Gold* Jeroboam of Dom Perignon.[8]

You will have to conduct your own experimental research to find the alcoholic strengths, volumes, and prices. Note: this does not have to involve purchasing or consuming the items mentioned.[9]

[7] See http://map.gsfc.nasa.gov/universe/uni_matter.html
[8] This also sounds like a secret society from a Dan Brown novel, but is in fact a bottle of champagne on sale at www.harrods.com
[9] Or at least, not all in the same evening.

2.27–2.29: Express the following rates in SI base units:

2.27 A rattlesnake strikes 5 cm forward in 80 milliseconds (velocity).

2.28 A litre of oil weighs 900 grams (density).

2.29 A cheetah can go from zero to 96 km/hr in 3 seconds (acceleration).

2.30: When a spring is stretched by a force F, it extends by a distance x. In Hooke's law, the relation between F and x is $F = -kx$, where k is the spring constant. What are the dimensions of k? Express the dimensions in SI units.

2.31: Three equations for types of energy are:

- Potential energy: $E = mgh$
- Kinetic energy: $E = \frac{1}{2}mv^2$
- Mass-energy equivalence: $E = mc^2$

Here, m is a mass, g is the acceleration due to gravity, h is a height above the ground, v is a velocity and c is the speed of light.

Check that energy has the same dimension, of joules, in each equation.

2.32: In the general law of gravity, the gravitational force F between two bodies with masses m_1 and m_2, which are a distance r apart, is

$$F = \frac{Gm_1m_2}{r^2}$$

where G is the gravitational constant. What are the dimensions of G?

In the 1960s, there were two alternative theories of the Universe: the Big Bang theory and the steady-state theory (now rejected). In the steady-state theory, the mass M of the Universe was given by:

$$M = \frac{c^3}{2GH}$$

where c is the velocity of light, G is the gravitational constant, and H is the Hubble constant. Using dimensional analysis, work out the SI units of H.

2.33: The **ideal gas law** links the pressure p, the volume V and the temperature T (in kelvin) of a body of gas:

$$p = \frac{RT}{V}$$

The other quantity in the equation is the **gas constant** R. Use dimensional analysis to show that the generic dimensions of R are M L^2 T^{-2} K^{-1}. Write the units of R in terms of SI base units.

2.34: A more sophisticated equation describing the state of a body of gas, is **van der Waal's equation:**

$$\left(p + \frac{a}{V^2} \right)(V - b) = kT$$

where k is the Boltzmann constant. Find the units for a, b, and k.

2.35–2.40: Round the following numbers to

(a) two decimal places

(b) three significant figures

2.35 38.2351

2.36 −0.0148

2.37 30.262

2.38 −0.0030262

2.39 −0.1030262

2.40 48.007.

2.41–2.44: Convert the following intervals to uncertainty intervals of the form $A \pm e$, where e has one or two significant figures:

2.41 $3.14159257 < p < 3.1428714$

2.42 $0.0123475 < x < 0.0123521$

2.43 $0.999789 < y < 1.000023$

2.44 $5.6868 < z < 5.7474$.

2.45–2.46: Usain Bolt set world records for both the 100 m and the 200 m sprint at the 2008 Beijing Olympic Games. He had a recorded time of 9.69 seconds in the 100 metres, and 19.30 seconds in the 200 metres. For each race, work out his average speed, and express it as precisely as possible using an uncertainty interval. Assume the track measurements are exact, and that the time is correct to the nearest 0.01 seconds.

2.45 100 metres

2.46 200 metres.

2.47–2.49: Two students, Chardonnay and Jedward, perform a simple experiment . . .

2.47 Jedward weighs out 5 grams of salt, using scales that are accurate to the nearest milligram. Chardonnay weighs out 25 mg of salt using the same scales. Write each mass using an uncertainty interval, and give the absolute error and relative error. If the two amounts were to be poured together, what would be the absolute and relative errors in the resulting mass of salt?

2.48 Jedward measures out 20 millilitres of water using a pipette that has a scale in millilitres. Chardonnay also measures out 20 ml of water, but her pipette is calibrated in 2 ml gradations. Both students read off 20 ml to the nearest gradation on their pipette's scale. Write each volume using an uncertainty interval, and give the absolute error and relative error. If the two amounts were to be poured together, what would be the absolute and relative errors in the resulting volume of water?

2.49 Jedward adds his 5 g of salt to his 20 ml of water, and stirs to dissolve it. Chardonnay does the same with her 25 mg of salt and 20 ml of water. State the salt concentration (in grams per litre) for each solution, and the absolute and relative errors in each case.

The two students now pour their salt solutions together. What is the concentration of the final solution, and what are the absolute and relative errors?

2.50: British newspapers such as the *Daily Mail* use their own system of measurement, which I'll call SJ (Système Journalistique). In this system, the base units are as follows;

- unit of mass: the double-decker bus;
- unit of height: Nelson's Column in London;
- unit of length: distance from London to Paris;
- unit of area: the football pitch;
- unit of volume: the Olympic swimming pool.

Use the World Wide Web to find the SI equivalents of each of these units. Hence translate the following piece of journalese into SI units:

'The Great Pyramid of Giza is three times as tall as Nelson's Column, and weighs as much as 450,000 buses. Its base covers an area equivalent to $7\frac{1}{2}$ football pitches. It could hold over 1,000 Olympic-size swimming pools inside it, and if it were dismantled and all its blocks laid end-to-end, they would stretch from London to Paris $13\frac{1}{2}$ times'.

2.51: Our decimal numeral system uses base 10. The decimal number 368 means $300 + 60 + 8$, i.e. $(3 \times 10^2) + (2 \times 10^1) + (8 \times 10^0)$. The binary numeral system used in computers works with base 2, so that the binary number 11010 means

$$(1 \times 2^4) + (1 \times 2^3) + (0 \times 2^2) + (1 \times 2^1) + (0 \times 2^0)$$

$$= (1 \times 16) + (1 \times 8) + (0 \times 4) + (1 \times 2) + (0 \times 1) = 16 + 8 + 2 = 26.$$

(a) Convert the binary numbers 10011 and 111011 to decimal numbers.

(b) What is the binary number for 42 (the Ultimate Answer to the Great Question of Life, the Universe and Everything)?

2.52: Using the notation of Section 2.6, we want to find the error r in the calculation of $R = \dfrac{A}{B}$, where a, b are the errors in A, B. Proceeding as in Section 2.6.2, we start with the equation $r = \dfrac{A + a}{B + b} - \dfrac{A}{B}$. Put the right-hand side over a common denominator, and manipulate the equation to obtain $\dfrac{r}{R} = \dfrac{\dfrac{a}{A} - \dfrac{b}{B}}{1 + \dfrac{b}{B}}$. We write this as $\dfrac{r}{R} = \left(\dfrac{a}{A} - \dfrac{b}{B} \right)\left(1 + \dfrac{b}{B} \right)^{-1}$, and expand the second bracket using (2.13). By discarding second-order terms, deduce the rule in (2.10) for the error in division of A by B.

3

Data tables, graphs, interpolation

A table of physical data is not just 'a set of numbers', and in the last chapter we considered issues such as units, dimension, and precision. We now move on to discuss how to analyse the data we collect by observation or experiment.

In the simplest case, we could collect the values of a single physical variable. For example, we could go out on a hillside to find ants' nests, and estimate the number of insects N in each one. We return with a set of values of just one variable: population size: N_1, N_2, N_3, and so on. To analyse it we use **statistics**: as a start, we would calculate the mean and standard deviation. If we can split up the data into two descriptive categories, e.g. whether the nest was in woodland or on open ground, we could find the mean and standard deviation of each category, and then use a t-test to see if there is a significant difference between the average sizes of woodland and heath-land nests. To learn about this you would need to find a statistics textbook, for example Hawkins (2005). If you find the use of Excel spreadsheets in this book useful, you may like Eddison (1999), which introduces statistics for bioscientists with extensive use of Minitab statistical analysis software.

The essence of mathematics is in finding patterns. We can do this if our data involves measurement of two or more physical variables. In our example, we could record the altitude h of each nest's location, as well as its size N, giving us a set of data-points (h_1, N_1), (h_2, N_2), (h_3, N_3), etc. Or, we could look at a single nest and measure its size at intervals over a period of time, as in Case Study A, producing a set of data-points (t_1, N_1), (t_2, N_2), (t_3, N_3), ...

Such a set of data-points can be plotted as points on a two-dimensional graph. We then need to detect a pattern in those points. Humans are naturally inclined to do this; the earliest civilizations were able to stare at a random scattering of pinpricks of light in the heavens, and 'join up the dots' to see a huntsman, a plough, and so on.

In this chapter we'll describe how to plot graphs of data-points on paper and in Excel. We'll also describe the basic patterns to look for: direct and inverse proportion, and linear relations. We can 'join up the dots', like our star-gazing ancestors, to see these relations as lines or curves, but we will still be dealing with just the set of discrete data-points. In Chapter 5 we will move on to analyse the relations themselves, using the concept (central to mathematics) of a function of a continuous variable.

3.1 Constructing a data table and a data plot

Once you have collected your set of data for two physical variables, the first step in analysing it is to tabulate it, then to plot the data values as points on a graph. But even before doing that: are you sure which variable is the independent variable, and which is the dependent variable? In the two examples given in the Introduction, the nest size N is the dependent variable. In the first experiment the nest size depends on the altitude at which the nest is constructed, and in the second experiment the nest size varies with time.

3.1.1 Independent and dependent variables

Here are some examples of experiments:

(a) The elasticity of a skin sample is tested in an extensometer. A rectangular sample 20 mm long of stratum corneum (prepared from the rear feet of guinea pigs) is connected to a load cell above, and to a motor-driven clamp below. The clamp stretches the skin sample at a rate of 0.01 mm s^{-1}. The resulting load in kilograms is recorded at extensions of 0.1, 0.2, up to 0.5 mm. This experiment is taken from a 1995 US patent for a cosmetic cream to enhance human skin elasticity.[1]

(b) The numbers of harebell plants in five neighbouring fields were counted. The area of each field was also calculated.

(c) The height of a sapling is measured when it is planted, and then at monthly intervals for the following four months.

Each of these experiments involves two variables that are being measured. One of these variables takes values chosen by the experimenter; this is the extension in experiment (a), the field area in (b), and the elapsed time in (c). This is called the **independent variable**. The other variable is called the **dependent variable**, because its value depends on the choice for the independent variable: in (a) the load depends on the extension applied; in (b) there will be more plants in the larger fields; in (c) the sapling height increases as time goes on. It is very important that in a description of an experiment, you can identify which is the independent and which is the dependent variable. Experiment (a) could have been performed by applying increasing load to the skin sample, and recording the resulting extensions; in that case, the load would be the independent variable, and the extension becomes the dependent variable. In experiment (b) the field size decides the number of plants, and not the other way round. If time is involved, it is sure to be the independent variable (since 'time and tide wait for no man').

3.1.2 Data plots

In a typical experiment, the experimenter allows the independent variable to increase (or decrease) and – usually at regular intervals – she measures the value of the dependent variable. The result is a **data table** with two columns: independent variable on the left, dependent variable on the right. On each row is a pair of values which, taken together, are called a **data-point**. Table 3.1 gives an example of such a data table, for each of our experiments from

[1] http://www.patentstorm.us/patents/5427772-description.html (Hagan 1995).

Table 3.1 Data tables

Experiment (a)	
Extension δ (mm)	Load W (kg)
0.1	0.021
0.2	0.043
0.3	0.064
0.4	0.085
0.5	0.106
Experiment (b)	
Area A (hectares)	No. of plants, N
0.27	13
0.35	23
0.56	35
0.78	45
0.95	66
Experiment (c)	
Time t (months)	Height H (m)
0	0.14
1	0.46
2	0.90
3	1.44
4	2.10
5	2.86
6	3.75
Experiment (d)	
Temperature T (°C)	Length l (m)
0	1.00000
10	1.00009
20	1.00018
30	1.00027

the previous section. The first row of the Table 3.1 (a) tells us that extending the skin sample by 0.1 mm caused a load of 0.021 kg on the skin; this is written as the data-point (0.1, 0.021).

In the above tables, we have chosen a letter to denote each variable: A for field area, W for load, etc. To describe a general situation, mathematicians use x for the independent variable and y for the dependent variable (that is, y depends on x). We could call the first data-point (x_1, y_1), the second data-point (x_2, y_2), and so on. It is difficult to see what sort of relationship is occurring just by looking at the data table, but we can interpret the data pictorially by drawing a **data plot**; the steps are shown in Figure 3.1:

- Draw a horizontal axis (a real line, see Section 1.2.2) for values of the independent variable x.

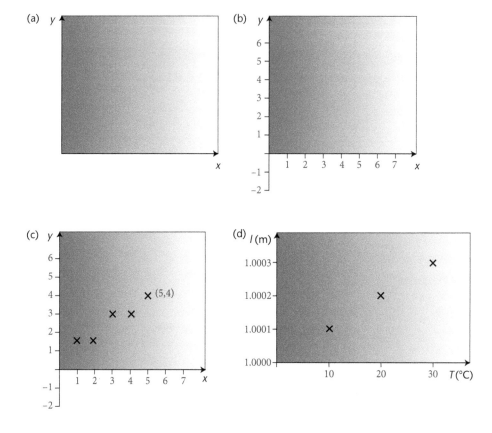

Figure 3.1 Stages in producing a data plot

- Draw a vertical axis (a second real line) for values of the dependent variable y; the axes intersect at (0,0), called the **origin**, often denoted by O. See Figure 3.1(a).

- Put suitable scales on each axis, and label them with the variable (and its units); see Figure 3.1(b). You must work out what scales you are going to use, and what the end-values of the horizontal and vertical axes will be, before you start to draw the graph. These must be chosen so that the graph will contain all the data-points, as well as the origin, which may not be in the centre of the graph. If all the x and y values are positive, for example, the origin will be near the bottom left corner of the graph.

- Plot each data-point as a cross: see Figure 3.1(c), where the data-points for the data table in Table 3.1(c) are plotted. For example, to plot the point (3, 1.44) from Table 3.1(c), find 3 on the horizontal axis, then move up until level with 1.44 on the vertical axis.

Sometimes it is not possible to include the origin (0,0) in your graph:

EXAMPLE

Until 1960, the metre was defined as the distance between two points marked on a proto-type bar of platinum/iridium alloy, kept in a laboratory in Sèvres, France. It was necessary to further specify that the bar should be kept at 0 °C, because metals expand slightly with temperature. If a metre-long bar was measured at higher temperatures, we might obtain a data table of bar length l against temperature T as in Table 3.1(d). Drawing a reasonable-sized graph by hand, with the data-points spaced out vertically, we might choose a scale of 2 cm representing 0.0001 m on the vertical l-axis. But then the graph would need to be 200 metres long, to reach down to the origin (0,0)! Instead, the horizontal and vertical axes will have to cross at the point (0, 1.00000), as in Figure 3.1(d).

If possible, the axes' crossing-point should be at 0 on the horizontal axis. If there are still problems, consider changing the variables. In the example above, we could have plotted l' against T, where $l' = l - 1$.

It is important to label each axis with the variable and its units, and to give a name to the graph. Data plots using axes without labels or scales are virtually useless (unless you are a politician or advertiser seeking to mislead the public).

The values x_1 and y_1 are called the **coordinates** of the data-point (x_1, y_1). This system for identifying points in a two-dimensional space using a horizontal x-axis and a vertical y-axis was proposed in 1637 by the French mathematician and philosopher René Descartes. It is known as the **Cartesian coordinate** system.

. . . .

3.2 Drawing graphs

The data-point plots help us to see the relationship between the variables, e.g. whether all the points lie along a straight line. To describe this relationship mathematically, we need to 'join up the dots' in some way, by putting a line or curve through them; this is called a **graph**. As we'll see in this section, it is possible to use a package such as Excel to produce data plots (called charts in Excel) and to fit lines or curves through the points. We'll see how to do this in Section 3.2.2. Figure 3.2 shows data plots from the data tables in Table 3.1(a)–(c). Figure 3.3 shows the corresponding Excel charts, with a line or curve fitted through the points. From these we can identify three basic types of graph.

3.2.1 Three basic types of graph

In Figure 3.3(a) the points all lie on a straight line, which also passes through the origin (since with a zero extension there will be no load). We say that the load is **directly proportional** to the extension applied; we will look at such a relationship in Section 3.3.

For experiment (b), in Figure 3.3(b) we have also put a straight line through the points, although they do not lie exactly on it – some are slightly above and others below. Such small discrepancies may be caused by experimental error, or by natural variation in the data. In our experiment, there is randomness in where plants seed and whether they survive, but in general there is a straight-line or **linear** relationship between plant population and field area. We therefore draw a **line of best fit** passing through the points. Not all the points will lie exactly on the line, but they will be close to it. This is sometimes called a **trend line**, because it shows the general trend of the relationship.

It would also be possible to draw a straight line through the data plot for experiment (c), but this is probably not correct. Unless the experimental errors or natural variation have occurred in a special non-random way, the relationship is actually nonlinear, and we can sketch a smooth curve through the points; see Figure. 3.2(c). Most of this book explores the mathematics needed to describe and analyse these curves.

The first three data tables in Table 3.1 thus illustrate three different categories of relationship:

3.1 (a) linear;

3.1 (b) approximately linear;

3.1 (c) nonlinear.

Table 3.1(d) also exhibits a linear relationship.

Strictly speaking, a graph is a set of points, a line or a curve in Cartesian coordinate space, but we often use the word 'graph' more generally to mean the Cartesian coordinate space itself, i.e. the plotting area with the x- and y-axes, before we've plotted any points or drawn any curves.

3.2.2 Drawing graphs in Excel

As a scientist you need to be able to construct plots and graphs on paper, but there is an easier way. Plots and graphs can be produced in Excel using the chart feature. Data plots in Excel for our three experiments are given in SPREADSHEET 3.1, which is reproduced in Figure 3.2. This and all other spreadsheets in this book are available on the textbook website. Here is how they were constructed:

SPREADSHEET 3.1

- Type the data table into two consecutive columns on the spreadsheet, with the independent variable in the left-hand column.
- Now highlight the data values by holding the left mouse button down and dragging over the table (excluding the headings).
- With the data values highlighted, click on Insert on the top toolbar, and choose chart . . . from the dropdown menu.
- Select the chart type 'XY (scatter)' and sub-type 'scatter'. You will now see a graph. Click the Series tab and type in a better name for the points than Series1 (or click the Legend tab at the next step and uncheck the Show Legend box to remove it).
- Then click the Next button. By clicking on the various tabs (Titles, Axes, Gridlines, Legend) you are now able to give the chart a title, label the axes, and remove the horizontal gridlines and legend. When finished, click Next, then Finish.
- The resulting plot can be enlarged by dragging one corner of the chart. If you hold the Shift key down while dragging, the chart dimensions will be preserved.

Finally we want to put a line or curve through the points – either (a) a straight line going exactly through the points, (b) a line of best fit, or (c) a smooth curve. This is done in SPREAD-SHEET 3.2, reproduced in Figure 3.3.

Nowadays with user-friendly packages such as Excel and Minitab widely available, which can take a table of data and produce a graph showing the points and the linear regression line

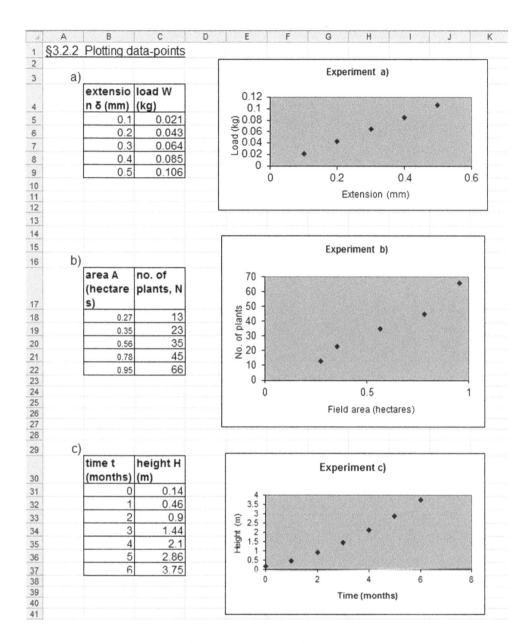

Figure 3.2 Plotting data-points (SPREADSHEET 3.1)

together with its equation, it is very tempting to use these, even if the data do not fully justify the assumption that a linear relationship exists. This is discussed more fully in statistics textbooks, but in Problem 3.21 I show some graphs from published research papers that illustrate what I mean.

A reminder that many of the spreadsheets you download from the book website have been protected, so that you can't mess them up by mistake by typing over cell contents, or deleting graphs. You can unprotect them via the PROTECTION ... option from the TOOLS menu. There is no password.

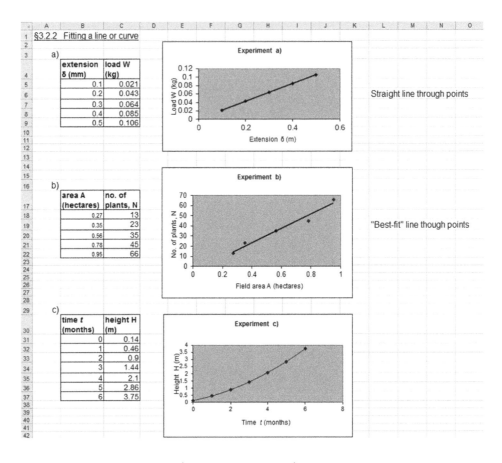

Figure 3.3 Fitting a line or curve (SPREADSHEET 3.2)

SPREADSHEET 3.2

To put a line or curve exactly through the points, highlight the chart by clicking on it, then right-click on it. Select `Chart Type...` from the menu that appears. Change the Chart Sub-type to 'Scatter with data-points connected by smoothed lines'. This has been done in (a) and (c) of SPREADSHEET 3.2.

Alternatively, to draw a 'best-fit' straight line through the points, as we did with the data for experiment (b), highlight the `Scatter` chart. Now on the top toolbar there should appear a new option called `Chart`. Click on this and select `Add Trendline ...` Choose the `Linear` option. This has been done on chart (b) in SPREADSHEET 3.2. In statistics, this line of best fit is called a regression line or trend line.

3.3 **Straight-line graphs: finding the slope**

When we put a straight line through a set of data-points, we are deciding that there is a **linear relationship** between the variables. The concept of linearity is central in science, and we'll study it mathematically in Chapter 6. Here is an introduction to it from the graphical point of view.

3.3.1 Direct proportion

The simplest relationship between two variables x and y is direct proportion, that is, the value of y is always a fixed multiple (or fraction) of the value of x. So if x doubles in size, y will double in size too. There is a mathematical symbol $\propto$ to indicate a proportionality relation. A more formal definition is:

> We say that the variable y is **directly proportional** to x (written as $y \propto x$) if the ratio of y to x $\left(\text{that is, } \dfrac{y}{x} \right)$ remains constant: $\dfrac{y}{x} = m$, say.
>
> The constant m is called the **constant of proportionality**.

We saw an example of direct proportionality in experiment (a) above; this was an illustration of Hooke's law, which says that for an elastic material, stress is proportional to strain. The strain is the relative extension, i.e. the extension divided by the original length. This is a dimensionless quantity, usually expressed as a percentage. The stress is the load per unit area of cross section. Hooke's law says that applying twice the extension produces twice as much load on the sample.

As we see if we extend the line in Figure 3.3(a), **the graph of a direct proportion relationship is a straight line passing through the origin.**

By multiplying both sides of the equation $\dfrac{y}{x} = m$ by x, we can write y in terms of x. A direct proportion relationship is then defined by:

$$y = m.x \tag{3.1}$$

where m is the constant of proportionality.

The graph of this relationship – a straight line passing through the origin – is shown in Figure 3.4(a). We will see this equation again in Section 6.3.2, when we study the graphs of functions. The value of m determines how steep the line is; this is called the **slope**. We'll explain in Section 3.3.3 how to calculate the slope from the graph.

Suppose we have two data-points (x_1, y_1) and (x_2, y_2) from a direct proportion relationship described by (3.1). That is,

$$y_1 = m.x_1 \quad \text{and} \quad y_2 = m.x_2.$$

We could rearrange these equations to

$$\frac{y_1}{x_1} = m \quad \text{and} \quad \frac{y_2}{x_2} = m.$$

The two left-hand sides must be equal, as they are both equal to m. So we can write

$$\frac{y_1}{x_1} = \frac{y_2}{x_2},$$

and cross-multiplying to remove the fractions:

$$x_2 y_1 = x_1 y_2. \tag{3.2}$$

We can use (3.2) to solve simple proportion problems.

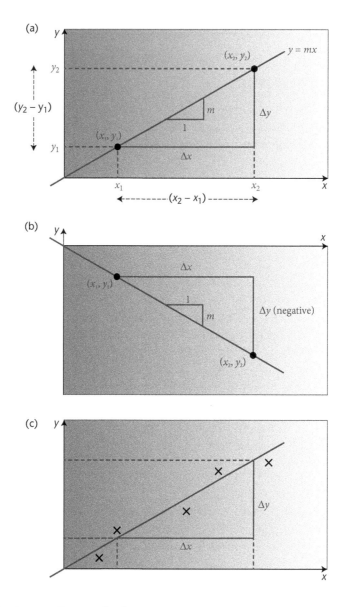

Figure 3.4 Direct proportion: graph of $y = m.x$

EXAMPLE

A hillside of 55 hectares is reforested; a total of 35,500 seedlings are used.

(a) The adjoining hillside has an area of 75 hectares; how many seedlings will be needed to reforest it?

(b) There are then 20,000 seedlings remaining in stock; how large an area could be reforested with these?

Solutions: Let x be the number of seedlings, and y be the area. Then $x_1 = 35,500$ and $y_1 = 55$.

(a) For the second hillside, $y_2 = 75$, and we want to find x_2. Inserting in (3.2),

$$x_2 \times 55 = 35,500 \times 75, \quad \text{so} \quad x_2 = \frac{35,500 \times 75}{55} = 48,409 \text{ seedlings.}$$

(b) We now have $x_2 = 20,000$, and we want to find y_2. Inserting in (3.2),
$$20,000 \times 55 = 35,500 \times y_2, \text{ so } y_2 = \frac{20,000 \times 55}{35,500} = 31.0 \text{ hectares.}$$

For these calculations we could just as happily choose to use x for the area, and y for the number of seedlings. In fact, if we were going to plot a graph of area against number of seedlings, that's what we should do: the number of seedlings depends on the area, not the other way round.

3.3.2 Linear relationship

In Experiment (b) our data had a straight-line or linear relationship, but the line of best fit did not pass through the origin. The equation of such a line (which we'll study in more detail in Section 6.3.3) is
$$y = m.x + c. \tag{3.3}$$

Here, m is again the slope of the line.

The other parameter, the constant c, is called **the y-intercept**, because it is the place at which the line crosses the y-axis, see Figure 3.5. On the y-axis the x-coordinate is zero, and when $x = 0$, $y = m.0 + c = c$.

Once you have found the value of m, you can find c by substituting into equation (3.3) the coordinates of a data-point that lies on the line. If the point (x_1, y_1) lies on the line, then $y_1 = m.x_1 + c$, so solving for c:
$$c = y_1 - m.x_1. \tag{3.4}$$

Alternatively, if the you have drawn the line on graph paper, simply extend the line until it crosses the y-axis; c is the value of y that you read off at that point. Notice that the direct proportion equation (3.1) is just a special case of the linear relationship (3.3) – the case where $c = 0$.

3.3.3 Calculating the slope

Suppose we have plotted our data-points on a graph, and drawn a straight line through them. How do we calculate its slope?

Pick two data-points; let's say they have coordinates (x_1, y_1) and (x_2, y_2), with (x_1, y_1) on the left. Draw a right-angled triangle under the line, with horizontal and vertical sides, and the sloping line between (x_1, y_1) and (x_2, y_2) forming the hypotenuse. Then the slope of the line is defined as:
$$\text{slope} = m = \frac{\text{length of vertical side}}{\text{length of horizontal side}}. \tag{3.5}$$

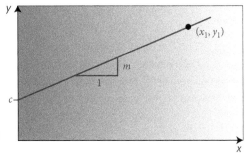

Figure 3.5 Linear relationship: graph of $y = m.x + c$

That is, if you imagine moving along the line from (x_1, y_1) to (x_2, y_2), the slope is the vertical distance by which you rise, divided by the distance you travel horizontally. This is sometimes expressed as 'rise over run'. The situation is shown in Figure 3.4(a). The proportion m can be thought of as the vertical distance you rise for each unit of horizontal distance.

As we are travelling from a height of y_1 to a height of y_2, the vertical distance will be the difference between the new height and the old height, or $(y_2 - y_1)$. Similarly, the horizontal distance travelled will be the difference between the new and old x-coordinates, i.e. $(x_2 - x_1)$. These differences give you the lengths of the vertical and horizontal sides of the triangle.

Mathematicians have a special notation for differences or distances in a particular direction. The symbol Δ is a capital D in the Greek alphabet. We will use it to indicate the distance between two values. We already (in Section 2.6.4) used δx to mean 'a small difference between two values of x' or 'a small distance in the x-direction'. We now use Δx to mean a not-so-small, finite-size distance in the x-direction. This does **not** mean Δ multiplied by x; it's all one symbol Δx. Similarly, Δy means 'a finite distance in the y-direction'. We will use this notation many times in the rest of this book. In particular, we use it in the final section of this chapter, on linear interpolation.

We can now write the slope as

$$\text{slope} = m = \frac{\text{change in } y}{\text{change in } x} = \frac{\Delta y}{\Delta x} = \frac{y_2 - y_1}{x_2 - x_1}. \tag{3.6}$$

If the line slopes downwards, we decrease in height as we travel from (x_1, y_1) to (x_2, y_2), so Δy will be negative. We are still travelling in the positive x direction, so Δx is still positive. Thus, dividing a negative number by a positive one, m will be negative; see Figure 3.4(b).

In example (a), we know that the line passes through the data-points (0.2, 0.043) and (0.3, 0.064) so the slope of this line is

$$m = \frac{y_1 - y_0}{x_1 - x_0} = \frac{0.064 - 0.043}{0.3 - 0.2} = \frac{0.021}{0.1} = 0.21.$$

Therefore, the equation of the line is $y = 0.21x$.

As we are dealing with a straight line, which has constant slope, we will get the same answer for m no matter which points we pick to construct our triangle. You can explore this in SPREADSHEET 3.3. Choose the values of x_1 and x_2 in cells D5 and D7. The triangle is drawn in the graph, and the calculation of equation (3.5) will always produce a value for the slope that equals the coefficient in cell E3. Try putting different coefficients (positive and negative) in cells E3 and G3 to see the effect on the line. Even if you make (x_1, y_1) lie to the right of (x_2, y_2), the slope is still calculated correctly by (3.6).

But what if you have produced a data plot by hand on graph paper, and drawn a straight line passing approximately through the data-points, i.e. a line of best fit, as in our example (b)? To calculate the slope of the line you **draw any right-angled triangle** under the line, using part of the line as the hypotenuse, measure the length Δx of the horizontal side and the length Δy of the vertical side, then calculate the slope as $\Delta y / \Delta x$, as in Figure 3.4(c).

In the case where the slope is negative, the value of Δy is negative because y decreases as x increases. So if you measure Δy as 3.8 cm in Figure 3.4(b), you use -3.8 in the calculation of m. It is important to make your triangle big, so that the errors involved in measuring the sides (± 0.5 mm if you can measure to the nearest millimetre) are small compared to the side lengths themselves. In the end-of-chapter questions, you will use the error analysis of

Section 2.6.2 to work out how these errors in measurement affect the accuracy of the gradient you calculate.

If you are working in Excel and your straight line is a trend line, you can obtain the line's equation, with the values of the parameters m and c, by right-clicking on the line. Choose the menu item 'Format Trendline . . .' In the window that appears, click the Options tab, and check the box 'Display equation on chart'. The equation now appears on the chart; in the case of example (b) in SPREADSHEET 3.2, it is $y = 70.602x - 4.6903$. This is illustrated in SPREADSHEET 3.4. We will look at trend lines again in Section 6.3.4, and in Section 8.7 we'll see how the equation of the trend line is calculated.

Problems: Now try the end-of-chapter problems 3.1–3.6.

3.4 Inverse proportion

Another type of proportionality is inverse proportion: when x doubles in size, the value of y halves. Mathematically, we say that y is **inversely proportional** to x if the product of x and y is constant: $x.y = k$. To write y in terms of x, divide both sides of the equation by x:

$$y = \frac{k}{x}. \tag{3.7}$$

We could rewrite (3.7) as $y = k.\frac{1}{x}$, so we could use the proportionality symbol from Section 3.3.1 and write $y \propto \frac{1}{x}$ to indicate that y is inversely proportional to x. The classic example of such a relationship in physics is Boyle's law, which states that for a fixed mass of gas at constant temperature, the product of the pressure P times the volume V is constant: $P.V = k$. If the gas in a balloon is compressed, decreasing its volume, then its pressure increases.

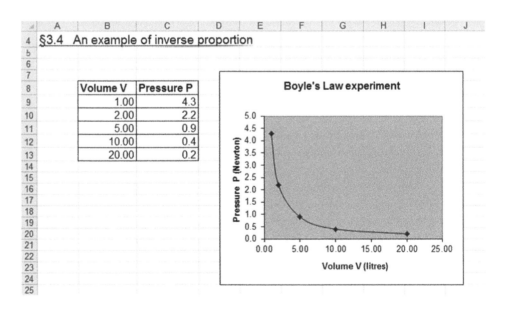

Figure 3.6 An example of inverse proportion (SPREADSHEET 3.5)

SPREADSHEET 3.5 shows some typical data from an experiment where a gas was allowed to expand, and its pressure was measured. The graph looks very different from those we have seen before. Notice that it never crosses either of the axes; this is because it is impossible for x or y in equation (3.7) to be zero (which would mean we have to divide by zero). We will study inverse proportion in more detail in Section 8.1.2.

EXAMPLE

Returning to the example at the end of Section 3.3.1, when we reforested the 55 hectare hillside with 35,500 seedlings, suppose that the average distance r between each tree was 30 cm. We hypothesize that the average distance between trees on a given hillside is inversely proportional to the number of seedlings x used:

$$r = \frac{k}{x}.$$

If we have two data-points (x_1, r_1) and (x_2, r_2), we can write

$$r_1 = \frac{k}{x_1} \text{ and } r_2 = \frac{k}{x_2}.$$

Solving each of these equations for k, and equating the right-hand sides, we find that

$$x_1 r_1 = x_2 r_2.$$

(a) Had we planted 50,000 seedlings on the hillside, what would the average distance between trees be?

(b) If we required a distance of 50 cm between trees, how many seedlings should we have used?

Solutions: From the information in the question, we have $x_1 = 35,500$ and $r_1 = 30$.

(a) Let $x_2 = 50,000$. So $35,500 \times 30 = 50,000 r_2$. Then $r_2 = \dfrac{35,500 \times 30}{50,000} = 21.3$ cm.

(b) Let $r_2 = 50$ cm. So $35,500 \times 30 = x_2 \times 50$. Then $x_2 = \dfrac{35,500 \times 30}{50} = 21,300$ seedlings.

Discussion question: do you think this hypothesized relationship is correct?

Problems: Now try the end-of-chapter problems 3.7–3.8.

. . . .

3.5 Application: allometry

If we look at a broad class of animals, e.g. fish or birds or herbivorous mammals, we find that species of roughly similar size tend to have similar values of physiological and ecological variables such as heart rate, gestation period, lifespan, territorial range, or population density. Moreover, such variables scale up or down in proportion to body size: the larger the species, the slower its heart rate, the longer its lifespan, the larger its territorial range, etc. The dependence is not usually linear, though: if we take the average body mass M (in kilograms) as the basic measure, then such variables may be directly or inversely proportional to M raised to some power (i.e. exponent). The study of these scalings is called allometry (Calder 2001).

Sometimes the scaling can be deduced from physical considerations; here is an example. The smaller the animal, the more of a problem it has with heat loss in cold weather. Its rate of heat loss is directly proportional to its exposed body surface area A, while the amount of body heat it contains will be directly proportional to its volume V. Assuming that animal bodies are of the same average density, volume will be proportional to mass: $V \propto M$. Volume has dimensions of (length)^3, while surface area has dimensions of (length)^2, so $V \propto L^3$ and $A \propto L^2$. By taking cube roots of both sides, the second relationship can be rewritten as $L \propto V^{\frac{1}{3}}$, and so $L \propto M^{\frac{1}{3}}$. Now use this in the third relationship: $A \propto \left(M^{\frac{1}{3}}\right)^2$, or $A \propto M^{\frac{2}{3}}$. The rate of heat loss will be proportional to the animal's mass raised to the power $\frac{2}{3}$. Thus, a deer may weigh 1,000 times more than a mouse, but its heat loss will only be 100 times greater. This $M^{\frac{2}{3}}$ scaling has been confirmed experimentally: measurements of skin surface area against body weight have produced relationships of $A = 1110M^{0.65}$ for mammals and $A = 813M^{0.67}$ for birds (Calder 2001).

Although some scalings can be justified by such arguments, allometry is largely an empirical science. One empirical conclusion is the 'quarter-power law': that many allometric scalings involve (roughly) $M^{\frac{1}{4}}$ or $M^{\frac{3}{4}}$. It has been argued that blood flow rate and respiratory volume scale as $M^{\frac{3}{4}}$. Other variables seem to match this quarter-power law by coincidence. Experimenters found that the stride length of mammals scales as $M^{0.38}$, while stride frequency scales as $M^{-0.14}$. Then walking speed, which is stride frequency multiplied by stride length (strides/hour × metres = metres/hour) will scale as

$$M^{-0.14} \times M^{0.38} = M^{0.38-0.14} = M^{0.24} \text{ (Calder 2001)}.$$

We can perform calculations using allometric scalings in the same way as the examples in the earlier direct and inverse proportion sections:

EXAMPLE

For eutherian mammals, the base metabolic rate P (the rate of energy consumption required for breathing, heartbeat, etc. while the animal is resting) scales roughly as $M^{0.74}$. A 12 kg deer has a base metabolic rate of 100 kJ/hour.

 (i) Estimate the base metabolic rate for a 2 kg rabbit.
(ii) What would be the body mass of a mammal with a metabolic rate of 50 kJ/hour?

Solution: (i) We are given that $P \propto M^{0.74}$, which we can write as an equation involving the constant of proportionality k: $P = kM^{0.74}$. For the deer we are given that $P_1 = 100$, $M_1 = 12$. Inserting in the equation:

$$100 = k\, 12^{0.74}$$
$$100 = 6.29\, k$$
$$k = \frac{100}{6.29} = 15.9.$$

For the rabbit we know $M_2 = 2$, so using this and the value of k in the equation:

$$P_2 = kM_2^{0.74} = 15.9 \times 2^{0.74} = 15.9 \times 1.67 = 26.56 \text{ kJ/hour}.$$

Alternatively, we could have written an equation involving P_1, M_1, P_2, M_2 as we did in the direct proportion example (Section 3.3.1) to obtain equation (3.2):

$$\frac{P_1}{M_1^{0.74}} = \frac{P_2}{M_2^{0.74}}.$$

(ii) In this case, we know that $50 = kM_2^{0.74} = 15.9 \times M_2^{0.74}$, so $M_2^{0.74} = \dfrac{50}{15.9} = 3.145$. To solve

for M_2, raise both sides to the power of $\dfrac{1}{0.74}$ (see Section 1.6.5):

$$\left(M_2^{0.74}\right)^{\frac{1}{0.74}} = (3.145)^{\frac{1}{0.74}}$$

use the $\boxed{x^y}$ key on your calculator

$$M_2 = (3.145)^{1.351} = 4.702 \text{ kg}.$$

You may wonder how the coefficients in an empirical allometric law such as $A = 1110M^{0.65}$ were deduced from a table of data-points (M_1, A_1), (M_2, A_2), …. If the relationship were linear, we could plot the points on a graph, fit a line of best fit though them, and then find its equation. This is in fact what we do: but we first have to transform the relationship to a linear one. This involves using logarithms; we'll see how in Section 10.4.2.

Problems: Now try the end-of-chapter problems 3.9–3.12.

. . . .

3.6 Extension: interpolation

One purpose of drawing a line or curve through our data-points, is so that we can estimate the value that the dependent variable would take, for some chosen value of the independent variable. For example, in experiment 3.1(a) we might like to know what load would be caused by an extension of 0.37 mm. This process is called **interpolation**.

3.6.1 Performing interpolation by hand

If we have drawn a graph with a line through the points, the job is easy: look on the x-axis for 0.37, then move up until you reach the line, then move across to the y-axis and read off the y-value, in this case 0.0814 kg. The 0.37 is the **interpolating x-value**, and the 0.0814 is the **interpolated y-value**. This is illustrated in Figure 3.7, taken from SPREADSHEET 3.6. The

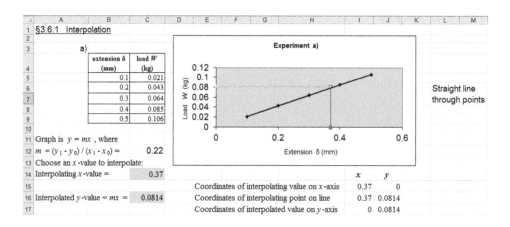

Figure 3.7 Interpolation (SPREADSHEET 3.6)

SPREADSHEET 3.6

Type an interpolating *x*-value into cell C14. This value is marked by the purple square on the *x*-axis of the chart. The vertical purple line projects up to the graph, then the horizontal yellow line moves across to the *y*-axis. The interpolated value of *y* appears in cell C16.

The purple line was added to the graph by highlighting the chart, and choosing `Add data` `...` from the `Chart` drop-down menu. Then the two data-points in the cells I15, I16, J15, J16 were highlighted and added as a new Series; check the box for *x*-values in the first column. The yellow line was created by adding the Series in cells I16, I17, J16, J17 in the same way.

value of *m* is calculated from the first two data-points, using equation (3.2), in cell C12. This is all we need to perform interpolation.

The same interpolation process could be used to interpolate from a curve drawn on graph paper.

If you have a curve on graph paper, you can also perform **inverse interpolation**. Interpolation asks the question: 'Here is a value of *x*, what is the corresponding value of *y*?' Inverse interpolation asks 'Here is a value of *y*. What value of *x* would produce this value of *y*?' You can answer this by finding the given value of *y* on the *y*-axis, moving across to the curve, then down to the *x*-axis. This is the reverse of the interpolation process. The processes of interpolation and inverse interpolation are shown in Figure 3.8. There are some numerical examples in the end-of-chapter problems.

Interpolating by drawing lines on graphs is not very precise. We need an algorithm for calculating interpolated values from the data-points.

3.6.2 Linear interpolation between two data values

Consider the data from experiment (c) in Table 3.1, and suppose we want to estimate the height of the sapling after $3\frac{1}{2}$ months. After 3 months it was 1.44 m tall, and after 4 months it was 2.10 m tall. Midway through the month, we might expect that, as a first approximation, the height was midway between 1.44 and 2.10, i.e. (1.44 + 2.10)/2 = 1.77 m tall. What we have performed is **linear interpolation**, where we assume that the dependent variable changes linearly, i.e. along a straight line, from one data-point to the next.

But what if we want to estimate the sapling's height after 3 months and 9 days? What formula could we use?

Suppose we know two adjacent data-points, which we'll call (x_0, y_0) and (x_1, y_1), and we want to estimate *y* for some value of *x* between x_0 and x_1. If we use linear interpolation between (x_0, y_0) and (x_1, y_1), the interpolated value of *y* will lie between y_0 and y_1. We express this algebraically as $x_0 < x < x_1$ and $y_0 < y < y_1$. But how do we estimate *y*, given *x*, (x_0, y_0) and (x_1, y_1)?

Let the distance from x_0 to x_1 be *h*. We can then write

$$x_1 = x_0 + h$$

or

$$h = x_1 - x_0. \tag{3.8}$$

We call *h* the **interval width**.

(a)   Direct interpolation

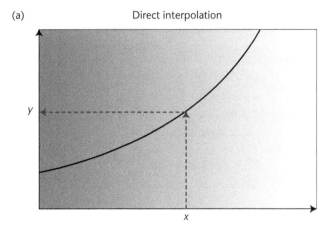

(b)   Inverse interpolation

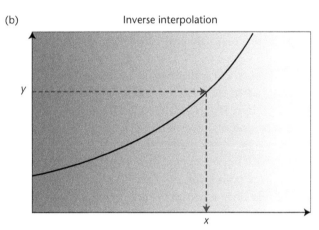

Figure 3.8 Direct and inverse interpolation

The distance from x_0 to our interpolating value x will be some fraction of h. For example, if x is half-way between x_0 and x_1 then $x = x_0 + \frac{1}{2}h$. We call this fraction the **interpolating factor** and write it as s, where $0 < s < 1$. That is,

$$x = x_0 + s.h. \tag{3.9}$$

This is illustrated on the x-axis of Figure 3.9. To find s for a given value of x, rearrange equation (3.9) to solve for s:

$$s = \frac{(x - x_0)}{h}. \tag{3.10}$$

Now in linear interpolation the value of y corresponding to x should be the same fraction s of the distance along the y-axis from y_0 to y_1.

Let Δy_0 be the distance from y_0 to y_1:

$$y_1 = y_0 + \Delta y_0$$

or

$$\Delta y_0 = y_1 - y_0. \tag{3.11}$$

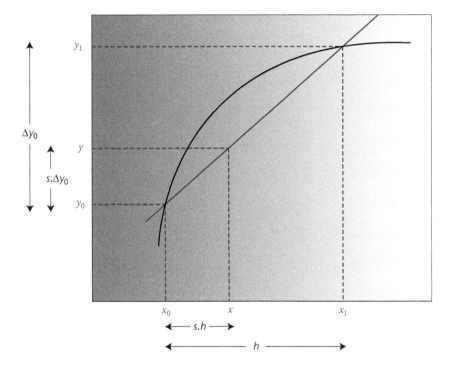

Figure 3.9 Linear interpolation

Δy_0 is called the **first difference** of y at y_0.

By the argument above, the interpolated value of y corresponding to x should be:

$$y = y_0 + s.\Delta y_0. \tag{3.12}$$

This algorithm is illustrated in SPREADSHEET 3.7. You can choose the x- and y-values of the two data-points, in the yellow cells B5, B6, C5, C6, and see the calculations of h and Δy_0. Then choose an interpolating x-value and enter it in cell C15. The interpolating factor appears in cell C17, and the interpolated y-value appears in cell I16. The chart shows the interpolation process.

Check the formulae in the various cells, to see how they match the equations above.

3.6.3 Piecewise linear interpolation

When we have a table of data-points and wish to interpolate, we first look to see which two data-points lie immediately to the left and right respectively of the interpolating x-value, then we use the algorithm of the previous section to perform linear interpolation between those two points. This is equivalent to joining up the data-points in the graph by a set of straight-line segments; it is called **piecewise linear interpolation**. In SPREADSHEET 3.8 (see Figure 3.10) you can see a table of data, a chart with a smooth curve drawn through the points, then another chart with these straight-line segments (obtained by using another of the Scatter sub-types in Chart Type). This looks like a good approximation to the curve. But note that you should never actually join up points by straight-line

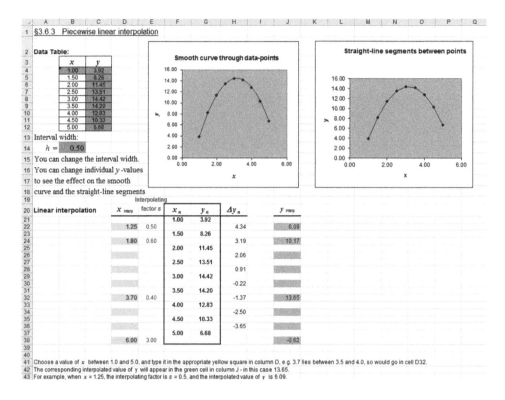

Figure 3.10 Piecewise linear interpolation (SPREADSHEET 3.8)

segments – always draw a smooth curve! We only do this to illustrate how piecewise linear interpolation works.

So here is the method of linear interpolation, in general (in the left column) and for a particular example (in the right column):

Here is the data table in SPREADSHEET 3.8.

x	y
1.00	3.92
1.50	8.26
2.00	11.45
2.50	13.51
3.00	14.42
3.50	14.20
4.00	12.83
4.50	10.33
5.00	6.68

Estimate the value of y when $x = 1.8$.

General method	Particular example
x is the interpolating x-value. The data x-values on the left and right of x are x_0 and x_1.	We want to interpolate at $x = 1.8$. This lies between the data x-values $x_0 = 1.5$ and $x_1 = 2.0$.
The interval width is $h = x_1 - x_0$.	The interval width is $h = 2.0 - 1.5 = 0.5$.
The interpolating factor is $$s = \frac{(x - x_0)}{h}.$$	The interpolating factor is $$s = \frac{(1.8 - 1.5)}{0.5} = \frac{0.3}{0.5} = 0.6.$$
The data y-values corresponding to x_0 and x_1 are y_0 and y_1.	The data y-values corresponding to $x_0 = 1.5$ and $x_1 = 2.0$, are $y_0 = 8.26$ and $y_1 = 11.45$.
The distance from y_0 to y_1 is $\Delta y_0 = y_1 - y_0$.	The distance from 8.26 to 11.45 is $\Delta y_0 = 11.45 - 8.26 = 3.19$.
The interpolated value is then $y = y_0 + s.\Delta y_0$.	The interpolated value is then $$\begin{aligned} y &= 8.26 + 0.6 \times 3.19 \\ &= 8.26 + 1.914 \\ &= 10.174. \end{aligned}$$

Here's an example where the y-values are decreasing. Estimate y when $x = 3.7$. We see that 3.7 lies between the x-values of 3.5 and 4.0. We first work out the interpolating factor s:

$$S = \frac{3.7 - 3.5}{4.0 - 3.5} = \frac{0.2}{0.5} = 0.4.$$

What this tells us is that 3.7 is 0.4, or 40%, of the distance from 3.5 to 4.0.

By linear interpolation, the y-value at 3.7 will be 0.4, or 40%, of the distance from 14.20 (the y-value at $x = 3.5$) to 12.83 (the y-value at $x = 4.0$):

$$y = 14.20 + 0.4 (12.83 - 14.20) = 14.20 + 0.4 \times (-1.37)$$
$$= 14.20 - 0.548 = 13.65 \text{ (to 2 decimal places)}.$$

3.6.4 Linear interpolation using Excel

Here is how SPREADSHEET 3.8 performs piecewise linear interpolation. The coordinates of the data-points are in columns F and G. The first differences of the y-values are in column H, programmed using equation (3.11), and written in the rows between the data. The interval width h is stored in cell B14. Look at the cells to see the formulae used.

In row 38 the linear interpolation formula is used with the last two data-points, to predict y-values for values of x which are greater than 5.0. In this case the interpolating factor s is greater than 1. This process of predicting y-values beyond the end of the data table is called **extrapolation**.

Problems: Now try the end-of-chapter problems 3.13–3.19.

SPREADSHEET 3.8

To use the spreadsheet, you type your *x*-value into the appropriate yellow cell in column D, for example 1.8 goes into cell D24. The value of the interpolating factor *s* appears on the same row in column E, using equation (3.10), and the interpolated *y*-value appears in the green square in column J, using equation (3.12). You should download this spreadsheet from the book website, and experiment with it. Try some of the end-of-chapter problems, and then check your answers on the spreadsheet. You can also adapt the spreadsheet to new problems by putting new data values in columns F and G. Have a look at the formulae in the cells. The formula in cell E22, to calculate *s* using equation (3.10), is:

    = (D22-F21)/$B$14

We used the $ signs for B14 so that this cell location didn't change as we copied the formula from E22 down the column to E24, E26, etc.

The formula in cell E22 is actually

    = IF(D22 = ' ', ' ', (D22-F21)/$B$14)

This is using the IF function, which has the format

    IF(logical test, value if true, value if false)

and in this case it means 'If cell D22 is blank, leave this cell blank, otherwise display the value of (D22-F23)/$B$14'.

The formula in cell H22, to form the first difference, is

    = G23-G21

and the formula in cell J22, to calculate *y* using equation (3.12), is:

    = G21 + E22*H22

CASE STUDY A6: Cobwebbing

If we are given an initial population size N_0 and a multiplication rate λ, we could use the update equation

$$N_{p+1} = \lambda N_p$$

to produce a table of data-points $(0, N_0)$, $(1, N_1)$, $(2, N_2)$, $(3, N_3)$, etc., and we could thus produce a graph of N_p against p. Such a graph is shown in Figure 3.11(a).

But there is another type of graph which biologists and mathematicians use to analyse the behaviour of a growth model. From Section 3.3 you can recognize the update equation as a relation of direct proportion, between the old population N_p and the new population N_{p+1}. The constant of proportionality is λ, and (assuming a positive growth rate R), we know that

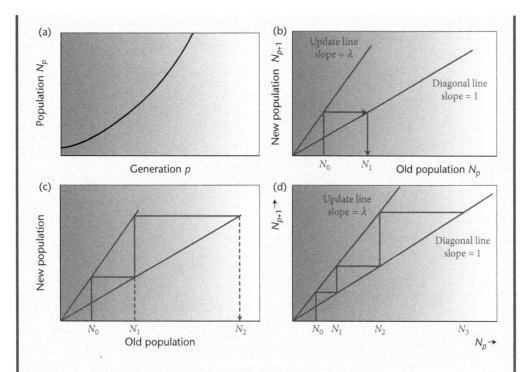

Figure 3.11 Cobwebbing

$\lambda > 1$. If we plotted a graph with the old population N_p along the horizontal axis, and the new population N_{p+1} along the vertical axis, then the points (N_0, N_1), (N_1, N_2), (N_2, N_3), etc. will lie on the line $N_{p+1} = \lambda N_p$, which passes through the origin and has slope $\lambda > 1$. This line is plotted in Figure 3.11(b) together with the diagonal line $N_{p+1} = N_p$, which has slope 1.

We can use this graph to visualize how the population grows from initial size N_0:

Locate N_0 on the horizontal axis, and move up from $(N_0, 0)$ to (N_0, N_1) on the update line. To locate N_1 on the horizontal axis, move across to the diagonal which will be met at the point (N_1, N_1), then down to find $(N_1, 0)$. This process is shown in Figure 3.11(b). To find N_2, repeat the process: move up to the update line, across to the diagonal line, and down; see Figure 3.11(c).

To visualize how the population is growing, we don't need to keep moving down to the horizontal axis and back up; we draw a sequence of horizontal and vertical lines between the update line and the diagonal, as shown in Figure 3.11(d). This plot is called a 'cobweb line', for reasons that will become clear when we apply it to the logistic growth model, in Chapter 7.

In SPREADSHEET 3.9 (see Figure 3.12) you can see a data table, a graph of N_p against p, and a cobweb graph. In the cobweb graph, the yellow line is the update line, and the purple line is the diagonal.

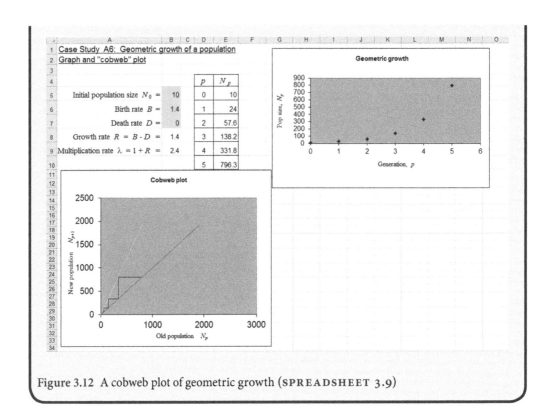

Figure 3.12 A cobweb plot of geometric growth (SPREADSHEET 3.9)

PROBLEMS

3.1–3.6: For the following data-sets, draw graphs on paper and in Excel. Decide if the relationship is linear, approximately linear, or nonlinear. If it is linear, find the y-intercept c and calculate the slope m. If it is approximately linear, draw your own line of best fit on the paper graph, and find m and c; then use Excel to draw the trend line and display its equation. Then draw Excel's trend line on your paper graph, by using its equation to plot two points, and joining them up. How closely do the two lines match?

3.1

x	y
2	7
4	11
7	17
9	21

3.2

x	y
−3	0.5
1	4.5
2	6.0
4	10.0

3.3

x	y
1	2.0
2	6.5
3	12.0
4	16.0
5	23.0

3.4	x	y
	0.5	26.50
	0.6	26.55
	0.9	26.70
	1.5	27.00
	2.0	27.25

3.5	x	y
	3.1	4.1314
	3.2	4.1632
	3.3	4.1939
	3.5	4.2528
	3.8	4.3350

3.6	x	y
	3.1	4.1253
	3.2	4.1571
	3.3	4.1756
	3.5	4.2242
	3.8	4.3101

3.7: Among females of the snake *Lampropeltis polyzona*, the tail length y is directly proportional to the total body length x. We find a specimen with a tail length of 70 mm and a total body length of 850 mm. On the basis of this single specimen:

(a) What would you expect the body length to be, for a specimen with a tail length of 130 mm?

(b) If we caught a specimen with a body length of 1 metre, what would you expect its tail length to be?

(c) Assuming that the measurements above are correct to the nearest millimetre, use error analysis (Section 2.6) to find uncertainty intervals for your calculated lengths in parts (a) and (b).

3.8: Seawater has a salinity of 35.0 grams per kilogram; that is, 1 kilogram of seawater contains 35.0 grams of salt. The density of seawater is 1.025 g ml^{-1}.

(a) What is the mass concentration of salt in seawater (in grams per litre)?

(b) How much salt is in 4 litres of seawater?

(c) I take 4 litres of seawater and dissolve 20 g of salt in it. What is the new concentration?

(d) I take 1 litre of seawater and add 200 ml of fresh water to it. What is the new concentration?

(e) I mix the water from (c) and (d) together. What is the salt concentration in the mixture?

3.9: The heart rate of mammals scales as $M^{-1/4}$. An 81 kg man has an average heart rate of 200 beats per minute (BPM). What would you predict to be the heart rate of (a) a gorilla weighing 160 kg, (b) a chimpanzee weighing 50 kg?

Assuming that the weight measurements are correct to the nearest kilogram, perform an error analysis (Section 2.6) to find uncertainty intervals for your answers.

3.10: Suppose that, from physiological considerations, we deduce that it would be physically impossible for a mammal's heart to beat faster than 100 times per second. What lower limit on mammal body weight does this impose?

3.11: The Ngorongoro Conservation Area in Tanzania, which includes the crater itself as well as surrounding farmland and villages, has a total area of 8,300 km^2. It is home to 7,000 wildebeest (body mass approximately 200 kg), 4,000 zebra (weighing approximately 270 kg), and 3,000 gazelles (which have an average weight of approximately 20 kg). Assuming that population density D of a herbivorous mammal of mass M in the wild is $D = 91.2M^{-0.73}$ individuals per km^2, estimate the possible size of these herds if the whole Conservation Area were given over to wildlife plains.

The feeding rate (grams per day) of these animals scales as $M^{0.73}$. What can you conclude from this?

3.12: The Field Metabolic Rate (energy consumption when they are active) for herbivorous wild animals in Ngorongoro is estimated as $5.95M^{0.81}$ kJ/day. The feeding rate of mammals eating only grass is $0.577M^{0.73}$ g/day. The extractable energy from ryegrass is 16.5 kJ/g. Use this information to predict the maximum size of grass-eating mammal which could survive on the plains of Ngorongoro.

3.13–3.20: The following table shows the CO_2 concentration in the atmosphere (annual average, in parts per million) at 10-year intervals:

Year	Atmospheric CO_2 (ppm)
1959	315.98
1969	324.62
1979	336.78
1989	352.90
1999	368.14
2009	387.35

Source: National Oceanic and Atmospheric Administration, quoted on CO2now.org website.

Use linear interpolation to estimate the CO_2 concentration in the following years:

3.13 1972   3.14 1987   3.15 2007.

It is also possible to use the linear interpolation formulae to estimate values beyond the end of a table: this is called **extrapolation**. In this case the value of s in equation (3.9) will be greater than 1. Use extrapolation with the data from 1999 and 2009, to predict the CO_2 levels in the following years:

3.16 2012   3.17 2020.

3.18–3.19: In linear interpolation, we are given two data-points (x_0, y_0) and (x_1, y_1), and an interpolating value of x, and asked to estimate the corresponding value of y. We use the equations:

$$h = x_1 - x_0 \tag{3.8}$$

$$s = \frac{(x - x_0)}{h} \tag{3.10}$$

$$\Delta y_0 = y_1 - y_0 \tag{3.11}$$

$$y = y_0 + s.\Delta y_0. \tag{3.12}$$

We can ask the reverse question: given a value of y, what is the corresponding value of x? This is called **inverse interpolation**. In this case, we know y_0, y_1 and y, so we can solve (3.12) for s:

$$s = \frac{y - y_0}{\Delta y_0} \qquad\qquad (3.13)$$

Then we can solve (3.10) to find x:

$$x = x_0 + sh.$$

Use inverse interpolation with appropriate pairs of data-points to estimate the years when CO_2 levels reached the following concentrations:

3.18 330.00 ppm   **3.19** 380.00 ppm.

3.20: We can also perform inverse extrapolation. Use the data for 1999 and 2009 to predict the year in which CO_2 levels will reach 450.00 ppm.

3.21: Whenever you have obtained a set of data-points from experiment or observation, you can enter them into a spreadsheet or a statistical package such as Minitab, and instantly produce a graph with a trend line, as we have done with experiment (b) in Section 3.1.2. We could have done the same with the data from experiment (c), but this would have been wrong: a look at the plot of the points on the graph suggests that there is a nonlinear relationship, which is best fitted by a curve. Statisticians ask for certain conditions to be met if linear regression is to be used validly:

- the data-points should be distributed uniformly along the line;
- the distances of the points from the line should all be of the same magnitude as you look along the line;
- the points should (nearly) all lie reasonably close to the line. There are special ways of dealing with one or two 'outliers' which may be due to experimental error.

In practice, it may not be possible to obtain sufficient data for these conditions, but this is the ideal.

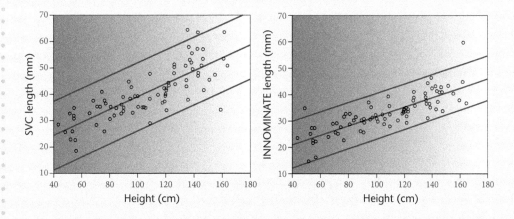

Figure 3.13 Best-fit lines (from Sanjeev and Karpowich 2006)

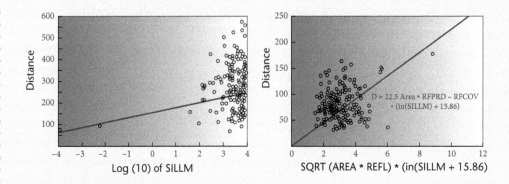

Figure 3.14 Best-fit lines (from Elliott et al. 1977)

In Figure 3.13 there are some published graphs;[2] do you think that the above conditions have been satisfied? What about the graphs[3] in Figure 3.14? This question will be discussed again in Section 6.3.4, where you'll find a way of measuring how good a fit to the data your trend line is.

[2] From Sanjeev and Karpowich (2006).
[3] From Elliott et al. (1977).

4

Molarity and dilutions

As all the living things on this planet require water in order to grow, it is crucial that bioscientists have a clear understanding of the nature of materials in solution. It is also important that they are able to work with solutions within a laboratory or field setting in order to carry out the appropriate experiments.

In this chapter we set out the calculations required to work effectively with materials in solutions.

4.1 Basic concepts

A **solution** consists of a certain amount of solid or liquid chemical (called the **solute**) dissolved in a volume of liquid (the **solvent**, which for biological purposes is usually water). The **concentration** of a solution can be expressed in a number of different ways.

4.1.1 Simple solutions

Using SI base units, the SI derived unit of concentration would be kilograms per cubic metre, or $kg\ m^{-3}$. In order to make such a solution one simply measures the required mass, dissolves it in water, and then makes it up to the required volume with more water. However, your lab will soon run out of stocks if you use chemicals by the kilogram, and you will struggle to find flasks that hold several cubic metres. As we will see, the standard measures of mass and volume are instead the gram and the litre. Other simple systems are used, such as **percentage by weight** (w/v) or **percentage by volume** (v/v). For example, a 10% v/v solution of ethanol would be made by adding 10% of the volume as pure, undiluted ethanol and 90% of the volume as water.

There is a significant drawback to such simple systems. When looking at biological systems we look at the interactions between individuals, which in this context are individual molecules. As molecules of different substances have different masses, placing the same mass, or the same liquid volume, of different substances together does not mean that you are combining equal numbers of molecules.

4.1.2 Atomic mass

As an example, consider the simple reaction in which one molecule of oxygen (O_2) combines with two molecules of hydrogen (H_2) to produce two molecules of water (H_2O). If, however, we added 10 grams of oxygen to 20 grams of hydrogen, we would obtain only 11.25 grams of water, and most of the hydrogen would remain. To combine different quantities of hydrogen and oxygen together to make water, so that both of the gases are completely used up in the reaction, requires using 160 grams of oxygen for every 20 grams of hydrogen. As the reaction involves twice as many hydrogen molecules as oxygen molecules, we conclude that one molecule of oxygen has 16 times the mass of one molecule of hydrogen.

This was the reaction considered by Dalton,[1] one of the founders of modern atomic theory. Using such experiments, nineteenth century chemists established values of atomic mass for each of the known elements. The atomic mass of hydrogen was set as 1, while oxygen was set as an atomic mass of 16, carbon was set as an atomic mass of 12, and so on. (**Atomic mass** is a relative measure, so it is dimensionless, like a percentage.) The term 'amu' is sometimes used for the unit of atomic mass; biologists and biochemists prefer to use the term 'dalton', with the symbol Da. The latter is particularly used for proteins, which being large molecules are typically referred to in kilodaltons, or kDa, with 1 kilodalton being 1000 daltons. An atomic mass of 1 dalton was defined as one twelfth of the mass of an atom of the standard isotope of carbon, written as ^{12}C, and containing 6 neutrons, 6 protons, and 6 electrons. Protons and neutrons have very nearly the same mass (about 1.67×10^{-27} kg), and the mass of an electron is about 0.05% that of a proton, so the atomic mass is essentially the total number of neutrons and protons in the atom. Some older textbooks may still refer, incorrectly, to atomic mass as relative atomic mass, or (relative) atomic weight.

The atomic mass of an element may not be given as a whole number because the naturally occurring substance may contain two or more isotopes with differing atomic masses. For example, there are two common isotopes of chlorine, ^{35}Cl and ^{37}Cl, with atomic masses of 35 and 37 daltons respectively. Naturally occurring chlorine contains 76% of ^{35}Cl and 24% of ^{37}Cl, so the atomic mass of chlorine is found by taking the average of the masses, weighted according to the percentages:

$$0.76 \times 35 + 0.24 \times 37 = 35.48.$$

Because of small differences in the masses of protons and neutrons, the value commonly used for the atomic mass of chlorine is actually 35.453.

4.1.3 The mole

Using the concept of atomic mass, we can say that 1 gram of hydrogen, 12 grams of carbon, and 16 grams of oxygen, will each contain the same number of atoms. If we want to measure out a sample of a substance in order to use it in a chemical reaction, it is the number of molecules in the sample which is important, rather than its weight or volume. For this purpose, the concept of the **mole** was developed.

A mole is a specific number of molecules. One mole of the amino acid glycine, for example, will contain the same number of molecules as one mole of the amino acid tryptophan.

---

[1] John Dalton (1766–1844), English chemist, meteorologist, and physicist. Dalton also did some work on the causes of colour-blindness, from which he himself suffered, and was thus important in the development of bioscience.

One mole of a substance is defined as 'that amount of the substance which contains as many molecules of the substance as there are atoms of carbon in 12 grams of carbon-12'.[2]

The mole is the SI unit for 'amount of substance' with the symbol mol. It is a measure of number, like 'a dozen', so it has no units. The SI prefixes can be used, i.e. 1 **millimole** (abbreviated as 1 mmol) is one thousandth of a mole.

The actual number of molecules in a mole is 6.02×10^{23} (or more accurately $6.02214179 \times 10^{23}$ with an uncertainty of $\pm 3.0 \times 10^{16}$), which is a very large number – to quote Bill Bryson (2004), that number of soft drink cans would cover the Earth's surface to a depth of 200 miles! This number is called the **Avogadro number** or Avogadro constant, written as L or N_A. In 1811 Avogadro[3] proposed that equal volumes of gases at equal pressures contained equal numbers of molecules but, like some other important breakthroughs (in chemistry as well as in mathematics), the importance of this was not recognized until much later. Avogadro did not even try to determine the number of molecules in a mole or a volume of gas, but his name is used to recognize the importance of the principle he proposed.

4.1.4 The molar mass of a substance

It would be quite impractical to count out the number of molecules needed, or to be found, in a solution but there is a way around this. We can define the **molar mass** of a substance as the mass (in grams) of one mole of the substance. So if we know the mass of a substance, we can work out the number of moles present and, hence, the number of molecules.

Note that **molar mass** does have units, of grams per mole (written g mol^{-1} or g/mol). Taking the example of two amino acids: 75.1 grams of glycine contains 1 mole (i.e. there are 6.02×10^{23} molecules of glycine in 75.1 grams of the substance), and thus the molar mass of glycine is 75.1 g mol^{-1}. Similarly, 1 mole of tryptophan has a mass of 204.2 grams, and so the molar mass of tryptophan is 204.2 g mol^{-1} (correct to 1 decimal place). But where did I get these numbers from?

For individual chemical elements, the molar mass comes from the atomic mass. You can look up the atomic mass of an element in a periodic table.[4] We defined one mole as the number of atoms in 12 grams of carbon, so the molar mass of carbon is 12 g mol^{-1}. Similarly, the molar mass of oxygen is 16 g mol^{-1}, and the molar mass of hydrogen is 1 g mol^{-1} (in round numbers).

But how do we find the molar mass of compound substances? As each molecule is made up of atoms, the mass required for one mole of them can easily be determined by addition of the masses of the individual atoms that make up the molecule. The chemical formula for common salt (sodium chloride) is NaCl: one molecule of salt consists of one atom of sodium and one atom of chlorine. The mass of one salt molecule is therefore found by adding atomic masses of sodium and chlorine together. The atomic mass of sodium is 22.990 and of chlorine is 35.453, so the molar mass of sodium chloride is 22.990 + 35.453 = 58.443 grams per mole.

[2] Initially, the mole was defined using the number of atoms in 16 grams of oxygen. There was, however, a complicating factor: chemists were using naturally occurring oxygen, which actually has isotopes oxygen-16, oxygen-17, and oxygen-18, but physicists were using pure oxygen-16 as their standard. This led to some small but potentially significant differences. So in 1961 they got together and set the new standard as carbon-12 and everyone now uses values relative to this. We'll perform calculations with isotope atomic masses in Chapter 10.
[3] Lorenzo Romano Amedeo Carlo Avogadro, conte di Quaregna e di Cerreto (1776–1856), Italian lawyer and scientist.
[4] There is a cool interactive periodic table at http://www.ptable.com/.

The same principle applies in more complicated examples:

EXAMPLE

The molecular formula for glycine is NH_2CH_2COOH. Each molecule of glycine thus consists of

- one atom of nitrogen (N);
- five atoms of hydrogen (H);
- two atoms of carbon (C);
- two atoms of oxygen (O).

Molar mass of nitrogen = 14.006 g mol^{-1}
Molar mass of hydrogen = 1.008 g mol^{-1}
Molar mass of carbon = 12.011 g mol^{-1}
Molar mass of oxygen = 15.999 g mol^{-1}
Therefore,

$$\text{the molar mass of } NH_2CH_2COOH = 1 \times (\text{molar mass of N}) + 5 \times (\text{molar mass of H}) + 2 \times (\text{molar mass of C}) + 2 \times (\text{molar mass of O})$$
$$= 1 \times (14.006 \text{ g mol}^{-1}) + 5 \times (1.008 \text{ g mol}^{-1}) + 2 \times (12.011 \text{ g mol}^{-1}) + 2 \times (15.999 \text{ g mol}^{-1})$$

(remembering the BEDMAS rules from Chapter 1)
$$= 14.006 \text{ g mol}^{-1} + 5.040 \text{ g mol}^{-1} + 24.022 \text{ g mol}^{-1} + 31.998 \text{ g mol}^{-1}$$
$$= 75.066 \text{ g mol}^{-1}.$$

The molar mass of glycine is thus 75.066 g mol^{-1}.

4.1.5 The molarity of a solution

Using the concept of moles allows us to define the concentration of solutions in terms of how many moles of a substance are contained in one litre of the solution, and this is referred to as the **molarity** (or molar concentration) of the solution.

A solution is **1 molar** when it contains one mole of the substance in 1 litre (or 1 dm^3) of that solution. The units of molarity are moles per litre, abbreviated to mol/l or mol l^{-1}. Alternatively, we may write the units as moles per cubic decimetre: mol/dm^3 or mol dm^{-3}. The symbol M is also commonly used for molarity, though its use in scientific literature is being discouraged, particularly in non-European countries. (Note that the capital distinguishes it from the symbol for metre, m.) Thus a 5 M solution of a substance will contain 5 moles of the substance per litre of solution and a 5 **millimolar solution** (written 5 mM) contains five thousandths of a mole per litre of solution.

This definition does not use the SI unit of volume, m^3, but a value that is 1000th of that. As we saw in Section 2.2.3, the litre (or dm^3) is a much more useful measure in the context of a bioscientist's work as it is easily measured, is more appropriate in size for the volumes of liquid likely to be needed in experiments, and can be easily stored and transported (1 m^3 of an aqueous solution would weight around 1000 kg!). However, some SI purists insist that molarity should be measured in moles per cubic metre (mol m^{-3}).

4.1.6 Application: measurements of cholesterol level

In the UK and Canada, a person's cholesterol level is measured in units of millimoles per litre (mmol l^{-1}). The normal cholesterol range is 3.6–6.5 mmol l^{-1}. In the USA, the units used are milligrams per decilitre (mg dl^{-1}). How are these two units related?

The chemical formula for cholesterol is $C_{27}H_{45}OH$. Adding up the atomic masses:

$$\text{molar mass} = 27 \times 12.011 + 46 \times 1.008 + 15.999$$
$$= 386.664 \text{ g mol}^{-1}.$$

Thus, 3.6 mmol = 3.6×10^{-3} mol, which weighs $3.6 \times 10^{-3} \times 386.664 = 1.392$ g.

So a concentration of 3.6 mmol l^{-1} is equivalent to 1.392 g l^{-1}

$$= 1.392 \times 10^3 \text{ mg } l^{-1}$$
$$= 1.392 \times 10^3 \times 10^{-1} \text{ mg } dl^{-1}$$
$$= 139.2 \text{ mg } dl^{-1}.$$

Similarly, 6.5 mmol l^{-1} equates to 251.3 mg dl^{-1}. In the USA, recommendations are that a cholesterol level under 200 is best, and 200–239 is 'borderline high'.

. . . .

4.2 Calculations involving moles and molarity

4.2.1 Calculating the number of moles in a sample

If we have a small sample of an element such as carbon, for which we know the atomic mass, and we weigh the sample, then we can work out how many moles of the substance are present. This is telling us the number of copies of the element that are present, useful if we are considering how much of another substance will interact with this.

The relationship is a simple linear one of direct proportionality:

> Mass of sample in grams = molar mass × number of moles present.   (4.1)

(Remember that the units of molar mass are grams per mole.)

This is similar to the general equation for direct proportionality, which we covered in Section 3.3.1:

$$y = mx \tag{3.1}$$

where in this case;

y is the mass of the sample in grams,

m is the constant value of the molar mass of the substance,

x is the number of moles present.

Manipulating this equation (by dividing each side by molar mass) in order to isolate the value we wish to determine, namely the number of moles present, produces $\dfrac{y}{m} = x$, or:

$$\frac{\text{Mass of sample in grams}}{\text{Molar mass}} = \text{number of moles present.} \tag{4.2}$$

Let's take an example of a sample of carbon. If we have 6 grams of carbon, how many moles of carbon do we have? From a periodic table, we can look up that the molar mass of carbon is 12 g mol^{-1} and we can substitute this, together with the mass of the sample, into the equation to give $\dfrac{6\,\text{g}}{12\,\text{g mol}^{-1}} = 0.5$ mol as the number of moles present.

If, however, we wish to do this kind of calculation with a compound substance, we first have to determine the molar mass of the molecule involved.

4.2.2 Calculating the molar mass of a compound

Suppose we now have 6 grams of common salt (sodium chloride). How many moles of salt are present?

As we saw in Section 4.1.4, the molecular formula for sodium chloride is NaCl, so each molecule of salt consists of one atom of sodium and one atom of chlorine. The atomic mass of sodium is 22.990, and the atomic mass of chlorine is 35.453, so the molar mass of sodium chloride is 22.990 + 35.453 = 58.443 grams per mole. Then a sample of 6 grams of salt will contain $\dfrac{6}{58.443} = 0.10266$ moles, or 102.66 millimoles.

(Note that the atomic mass values for sodium and chlorine are not whole numbers. This is because the larger atoms are not exact multiples of carbon-12, which is used to define the value of the atomic mass, and the naturally occurring elements contain atomic variants (isotopes) that are different in mass, though having the same chemical properties.)

With more complicated molecules involving more than one atom of an element, you must multiply the molar mass of the element by the number of copies present in a molecule. We illustrated this in Section 4.1.4.

Take our glycine amino acid example again. We saw that the molar mass of glycine is 75.064 g mol^{-1}. Thus 6 grams of glycine would contain $\dfrac{6}{75.064} = 0.080$ moles (to the nearest millimole). This can be written as 80 mmol. Just as 1000 ml = 1 litre, so 1000 millimoles = 1 mole (1 mol = 1000 mmol). We may even need to use micromoles (μmol): 1 mmol = 1000 μmol.

With more complicated biological molecules, this process can be quite a long one but is often easily bypassed by consulting tables, or reference material, for the compounds being used. Most manufacturers' catalogues will state the molar mass (though some may incorrectly refer to it as molecular weight) along with other key features of the compounds, and manufacturers' reagent bottles will also usually display this information.

4.2.3 Calculating the molarity of a solution

In Section 4.1.5 we defined the molarity (or molar concentration) of a solution as the number of moles of the chemical present per litre of solution. Writing this as an equation:

$$\text{molarity (mol l}^{-1}) = \frac{\text{number of moles (mol)}}{\text{volume (l)}}. \qquad (4.3)$$

The units of molarity may be written as M.

> **EXAMPLE**
>
> We mix 0.5 moles of uric acid ($C_5H_4N_4O_3$) into water and make up to 2.5 litres. What is the molarity of the solution?
>
> **Answer:** Molarity $= \dfrac{0.5\,\text{mol}}{2.5\,\text{litres}} = 0.2$ mol l^{-1}.

That is, it is a 0.2 M solution. Just as we have units of millilitres and millimoles, so we can talk about millimolar solutions. A millimolar solution is 10^{-3} mol l^{-1}, and is written as 1 mM. So the molarity in this example can be written as 0.2 M or 200 mM.

Notice also that giving you the chemical formula for uric acid in the question was a red herring: if we are dealing purely with volume, moles, and molarity, we don't need to know the molar mass of the chemical.

4.2.4 Calculating the moles present in a sample of solution

If we have a solution and we know its molarity, we can work out the number of moles present. Again this will help us with understanding how much of another substance will interact with this solution.

From (4.3) we can obtain a direct proportionality between the number of moles present and the volume of a solution of known molar concentration (the equation is again of the form $y = mx$):

$$\text{Number of moles (mol)} = \text{molarity (mol l}^{-1}) \times \text{volume (l)}. \qquad (4.4)$$

To be more consistent with SI units, you could replace litres (l) with cubic decimetres (dm^3).

Taking our glycine amino acid example again: If we have 100 ml of a 0.50 M solution of glycine, how many moles of glycine are present in the solution?

Entering the known values into the above equation gives us (remember that 100 ml is 0.100 litres):

$$\text{Number of moles (mol)} = 0.50 \text{ mol l}^{-1} \times 0.100 \text{ l}.$$

Here l^{-1} and l cancel, so this dimension disappears from the calculation

$$= 0.50 \times 0.100 \text{ mol}$$
$$= 0.050 \text{ mol}.$$

So there are 0.050 mol (or 50 millimoles) of glycine in 100 ml of a 0.50 M glycine solution.

4.2.5 Calculating the moles to add in making a solution

Making solutions of specific concentrations is a task that practising bioscientists do regularly and routinely. Thus it is important that you are able to carry out such calculations effectively. Fortunately they are quite straightforward and use the same equation (4.4) as we have seen in the previous section.

This time, however, we are not determining how much is present but how much we want to be present to have the desired final concentration and volume.

Thus, if we wish to make a solution of glycine with a molarity of 0.20 M and we are going to need about 300 ml (which is 0.300 dm^3) of this solution for our experiment, we put these values into (4.4) to give us:

$$\text{Number of moles (mol)} = 0.20 \text{ mol l}^{-1} \times 0.300 \text{ l} = 0.060 \text{ mol}.$$

So we would need to dissolve 0.060 mol (or 60 millimoles) of glycine in a total volume of 300 ml to have a 0.20 M solution of glycine.

4.2.6 Calculating the mass to add in making a solution

This is the next step in preparing a solution, as it is not possible to weigh out moles directly. We need to use equation (4.1).

We calculated the molar mass of glycine in Section 4.1.4 as 75.064 g mol^{-1}. For 500 ml of our 0.20 M solution we have determined that we need 0.060 mol of glycine.

Therefore:

Mass of sample in grams = 75.064 g mol^{-1} × 0.060 mol

(again, g mol^{-1} × mol^1 = g mol^0 = g)

Mass of sample in grams = 75.064 × 0.060 = 4.50 g.

In order to make our 300 ml of 0.2 M glycine solution, we need to obtain 4.50 g of our stock pure glycine powder and dissolve this so that the final volume of solution is 300 ml.

Problems: Now try the end-of-chapter problems 4.1–4.10.

4.3 Calculations for dilutions of solutions

4.3.1 Calculating the new concentration after diluting

When a solution is diluted, the total number of moles present remains the same; dilution means that the same amount of substance is dispersed in a greater volume. Using (4.4), this can be expressed as:

$$C_1V_1 = C_2V_2 \qquad (4.5)$$

where:

C_1 is the initial concentration;

V_1 is the initial volume;

C_2 is the final concentration;

V_2 is the final volume.

Let us consider the logic of this with a simple example. If we double the volume of a solution by adding water, it is obvious that the concentration will be half the initial value. This means that $V_2 = 2\,V_1$ but that $C_2 = \frac{1}{2}\,C_1$. Putting these into the equation (4.5) shows us that:

$$C_2V_2 = \frac{1}{2}\,C_1 \times 2\,V_1.$$

Multiplying $\frac{1}{2}$ by 2 gives 1 and thus:

$$C_2V_2 = C_1 \times V_1.$$

This shows that this works for this simple dilution, but it does also work for any dilution factor. It is an example of inverse proportion (Section 3.4): concentration is inversely proportional to volume, or $C \propto \dfrac{1}{V}$.

If we wish to know the final concentration after dilution, we need to manipulate the equation so that C_2 is isolated. This can be done by dividing each side of the equation by V_2:

$$\frac{C_1 V_1}{V_2} = \frac{C_2 V_2}{V_2}.$$

On the right-hand side, cancel the common factor of V_2 and thus this becomes:

$$\frac{C_1 V_1}{V_2} = C_2. \tag{4.6}$$

If we take the example of 300 ml of 0.20 M glycine solution from Sections 4.2.5 and 4.2.6 above, what would its concentration be if we added a further 100 ml of water?

C_1 is the initial concentration = 0.20 M.

V_1 is the initial volume = 0.300 l.

C_2 is the final concentration (the unknown).

V_2 is the final volume = 0.300 + 0.100 l = 0.400 l.

Hence:

$$\frac{0.20 \text{ M} \times 0.300 \text{ l}}{0.400 \text{ l}} = C_2.$$

As $\dfrac{0.300 \text{ l}}{0.400 \text{ l}} = 0.750$ (the dimensions cancel out),

$$0.20 \text{ M} \times 0.750 = C_2$$

$$0.15 \text{ M} = C_2.$$

After adding a further 100 ml to the 300 ml of 0.2 M glycine, the final concentration will be 0.15 M glycine.

4.3.2 Calculating how much to dilute to obtain a specific concentration

Frequently a bioscientist will wish to obtain a specific final concentration of a solution from a more concentrated stock solution. This time we need to rearrange the equation to isolate V_1, the volume of the initial concentration we need.

Starting with (4.5):

$$C_1 V_1 = C_2 V_2.$$

Divide both sides by C_1

$$\frac{C_1 V_1}{C_1} = \frac{C_2 V_2}{C_1}.$$

On the left-hand side $C_1/C_1 = 1$ and thus this becomes:

$$V_1 = \frac{C_2 V_2}{C_1}. \tag{4.7}$$

If we wish to have 100 ml (0.100 dm^3) of a solution of glycine that is 0.04 M, how much of the initial 0.2 M solution would be need? For a change, let's work in cubic decimetres (dm^{-3}).

C_1 is the initial concentration $= 0.20$ M.

V_1 is the initial volume $=$ the unknown.

C_2 is the final concentration $= 0.04$ M.

V_2 is the final volume $= 0.100$ dm^3.

$$V_1 = \frac{0.04 \ \cancel{mol \ dm^{-3}} \times 0.100 \text{ dm}^3}{0.2 \ \cancel{mol \ dm^{-3}}}.$$

Cancelling out the mol dm^{-3}:

$$V_1 = \frac{0.04 \times 0.100 \text{ dm}^3}{0.2}.$$

Then:

$$V_1 = \frac{0.004 \text{ dm}^3}{0.2}.$$

So:

$$V_1 = 0.02 \text{ dm}^3.$$

Converting between dm^3 and ml, 0.02 dm^3 is 20 ml (since 1 dm^3 = 1 litre = 1000 ml). Hence we would need to add 20 ml of our initial 0.2 M glycine and make the volume up to 100 ml in order to obtain a 0.04 M glycine solution. It may be obvious to you, and those who do this sort of calculation routinely, that the final concentration is one fifth of the starting concentration and so one fifth of the volume needs to be used (20 is one fifth of 100). It is, however, not always the case that such dilutions are this straightforward and in those cases we need to use this calculation.

Alternatively, we may be given the initial and final volumes, and need to work out the concentration of the solution after dilution. In this case, it is convenient to work out the **Dilution Factor.**

The Dilution Factor is defined as $DF = \dfrac{V_2}{V_1}$. Then starting from (4.5) again

$$C_1 V_1 = C_2 V_2$$

$$C_1 = C_2 \frac{V_2}{V_1} = C_2 \times DF$$

$$C_2 = \frac{C_1}{DF}. \tag{4.8}$$

The DF is a dimensionless quantity.

> **EXAMPLE**
>
> I take 20 ml of my 0.2 M glycine solution, and make up to 1 litre with water. What is the molarity of the final solution? $V_1 = 20$ ml, $V_2 = 1$ litre $= 1000$ ml
>
> So $DF = \dfrac{1000}{20} = 50$.
>
> Final concentration $C_2 = \dfrac{C_1}{DF} = \dfrac{0.2 \text{ M}}{50} = 0.004$ M, or 4 mM.

Problems: Now try the end-of-chapter problems 4.11–4.14.

4.3.3 Serial dilutions

It is not unusual to require a range of concentrations of a substance, or cells, in order to carry out a set of experiments to determine the relationship between that concentration and its effect. In these circumstances we would normally carry out serial dilutions rather than making a set of one-off dilutions. The main reason for this is that the accuracy of repeating the same type of dilution step is greater than when individual dilutions are made.

Serial dilutions[5] are in fact a special case of the standard dilution shown in Section 4.3.2 above. There is a repetition of the same dilution step and, therefore, the proportional change in concentration is the same at each step.

If we consider the case of the rate of an enzyme reaction that we wish to study at different concentrations of the material that the enzyme acts upon (the substrate), we would need to prepare a range of concentrations of the substrate for the enzyme to act upon. It is most useful if each concentration is half the concentration of its predecessor.

Starting with (4.5):

$$C_1V_1 = C_2V_2$$

and wishing to end up with $C_2 = \frac{1}{2}C_1$, then we would need $V_2 = 2V_1$, i.e. that the volume is doubled. We add 1 ml of solution from tube A to 1 ml of water in tube B.

If we repeat this, then each dilution will result in a solution half the concentration of that from which it started. Add 1 ml of solution from B to 1 ml of water in C, and so on.

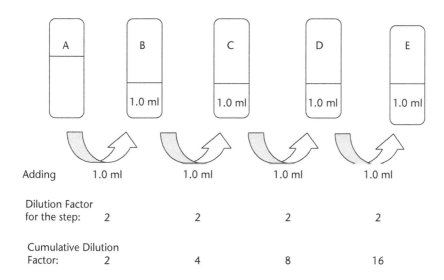

Hence the concentration in tube B is 1/2 of that in tube A, tube C is 1/4 of that in tube A, tube D is 1/8 of that in tube A and tube E is 1/16 of that in tube A.

Other situations may require greater dilutions and typically this may be by 10-fold dilution at each stage (or sometimes 100-fold). In those cases, such as when diluting suspensions of cells to obtain concentrations that contain only a few cells, the dilution is typically carried out by adding 1.0 ml of the previous solution to 9.0 ml of diluent (e.g. water).

---

[5] Not to be confused with cereal dilution, which is when you pour extra milk on your bowl of porridge . . .

	A	B	C	D	E
Dilution		1.0 ml A added to 9.0 ml	1.0 ml B added to 9.0 ml	1.0 ml C added to 9.0 ml	1.0 ml D added to 9.0 ml
Dilution Factor for the step	0	10	10	10	10
Cumulative Dilution Factor	0	10	100	1000	10 000
Concentration	A	A/10	A/100	A/1000	A/10 000

As an example, if the starting concentration of cells in tube A was 5×10^6 cells ml^{-1} then

- the concentration in tube B would be $5 \times 10^6/10 = 5 \times 10^5$ cells ml^{-1};
- in tube C $5 \times 10^6/100 = 5 \times 10^4$ cells ml^{-1};
- in tube D $5 \times 10^6/1000 = 5 \times 10^3$ cells ml^{-1}; and
- in tube E $5 \times 10^6/10\ 000 = 5 \times 10^2 = 500$ cells ml^{-1}.

Hence dividing the original concentration by the Dilution Factor gives the new concentration.

Problems: Now try the end-of-chapter problems 4.15–4.18.

4.3.4 Application: serial dilution in homeopathy

Homeopathy is a branch of alternative medicine based on the idea that a disease can be treated by administering a very dilute solution of an agent which would normally induce the symptoms of that disease.

Homeopathic notation uses X for a dilution step with a dilution factor of 10, and C for a dilution step with a dilution factor of 100. Thus, in the example of serial dilution at the end of the previous section, the solution in test tube E would be a 4X dilution, because it has been made by four consecutive 1:10 dilutions. The same result would be obtained by making two consecutive dilutions with dilution factor 100, denoted as 2C. So homeopaths would write 4X or 2C to mean a solution with a cumulative dilution factor of 10 000 (which is 10^4 or 100^2). In general, an nC solution has a cumulative dilution factor of 10^{2n}.

A major scientific criticism of homeopathy is that the dilutions used often result in all trace of the original agent being lost, so the homeopathic 'medicine' is in fact simply water. Let's quantify this. Suppose we start with 1 litre of a molar solution of the agent. We know that this contains 6.02×10^{23} molecules (Avogadro's number) of the agent. If we perform a sequence of n dilutions, each with a dilution factor of 100, the concentration is reduced by a factor of 10^{2n}, so the final litre of diluted solution will contain $(6.02 \times 10^{23})/10^{2n}$, or $6.02 \times 10^{23-2n}$ molecules. A 10C homeopathic solution would thus contain $6.02 \times 10^{23-20}$, or 6020 molecules. In a 12C solution, there will be 0.6 molecules, i.e. there is about a 50% chance of one molecule of the agent being present. A 15C dilution is equivalent to one molecule of agent in an Olympic-sized swimming pool (which contains 2.5×10^6 litres of water).

Samuel Hahnemann,[6] the founder of modern homeopathy, recommended using dilutions of 30C. And the active ingredient of *Oscillococcinum*, a common homeopathic remedy for

---

[6] Christian Friedrich Samuel Hahnemann (1755–1843), German physician.

influenza, is quoted[7] as 'Anas barbariae hepatis et cordis extractum 200CK HPUS'. This means a 200C dilution of extract of liver and heart of the Muscovy duck. The tablets will contain one molecule of extract for every 10^{400} molecules of sucrose and lactose, the other ingredients. Compare this with the number of atoms in the Universe, which you calculated in problem 2.19.

Homeopathic practitioners recognize that their high-dilution medicines will not contain any molecules of the active ingredients, but argue that the beneficial properties of the agent are transferred to the water by vigorous shaking after each dilution (or, for powders, by grinding with a pestle and mortar). Where the liver of the Muscovy duck comes in, is another story.

. . . .

4.4 Excel spreadsheets

The calculations we have made in the previous sections are simple: at most two multiplications or divisions. In fact, molarity and dilution are just an application of the principles of direct and inverse proportion we saw in Chapter 3. But simplicity is deceptive: are you sure you can recognize when to multiply and when to divide? One trick is to use dimensional analysis, to be sure that if the problem asks you to find a volume of a solution, the quantity you work out will have dimensions of litres or millilitres. You can memorize the three basic formulae:

Number of moles (mol) = concentration (mol l^{-1}) × volume (l)

Mass of sample in grams = molar mass × number of moles present

$C_1V_1 = C_2V_2$

but you're likely to make mistakes unless you understand the logic behind them. To show you the logical argument, I've created a set of Excel spreadsheets that each perform a standard type of calculation:

- SPREADSHEET 4.1: Calculate the number of moles in a solution.
- SPREADSHEET 4.2: Calculate the molarity of a solution.
- SPREADSHEET 4.3: Calculate a mass of solute.
- SPREADSHEET 4.4: Calculate the molarity of a made-up solution.
- SPREADSHEET 4.5: Calculate the mass of solute needed in making up a solution.
- SPREADSHEET 4.6: Calculate the molarity of a solution after dilution.
- SPREADSHEET 4.7: Calculate the volume of stock solution needed for a dilution.

SPREADSHEET 4.4 is shown in Figure 4.1, as an example.

Enter your own data in the yellow cells. The cell highlighted in light blue contains the answer. Each spreadsheet explains the logical reasoning being used, rather than simply

---

[7] See http://www.oscillo.com

	A	B	C	D	E	F	G	H	I	J	K	L	M	N
1	Calculating the molarity of a made-up solution												
2													
3	Question:												
4			Suppose we mix	7	grams of solute in water, and make up to			5	litres of solution. What is its molarity?				
5													
6	Answer:												
7			Molar mass of solute is	75.064	grams per mole (default solute is glycine)								
8													
9			So	75.064	grams of solute are equivalent to			1	mole				
10													
11			So	1	gram of solute is equivalent to	1 ÷	75.064	=	0.013322	moles			
12													
13			Then	7	grams of solute are equivalent to	7	×	0.013322	=	0.093254	moles		
14													
15			So	5	litres of solution contain	0.09325	moles of solute						
16													
17			Then	1	litre of solution would contain	0.09325	÷	5	=	0.018651	moles of solute		
18													
19					So the solution has a molarity of	0.01865	M						
20													

Figure 4.1 Calculating the molarity of a made-up solution (SPREADSHEET 4.4)

plugging numbers into an equation. Where a molar mass is needed, I have used the molar mass of glycine, but you can change this to the molar mass of your own chemical.

One complication which the spreadsheets don't handle is the use of SI prefixes on units. All calculations are done in moles, grams, and litres. You need to be able to obtain the right answer when millimoles, micrograms, or cubic centimetres are involved. For this, convert all quantities to standard units (grams, litres, moles), and convert the answer to a more appropriate unit at the end, if necessary:

EXAMPLE

25 mg of salt is dissolved in water, and the volume made up to 500 ml. What is the molarity of the solution?

Answer: The molar mass of salt (NaCl) is 58.443 grams/mol.

25 mg salt $= 25 \times 10^{-3}$ grams. So this amount contains

$$\frac{25 \times 10^{-3}}{58.443} = 4.28 \times 10^{-4} \text{ moles.}$$

500 ml water = 0.5 litres.
So molarity of the solution is $\dfrac{4.28 \times 10^{-4}}{0.5} = 8.56 \times 10^{-4}$ M , or 0.856 mM.

? PROBLEMS

In some of the problems you need to know atomic masses, as follows:

Element	Symbol	Atomic mass	Element	Symbol	Atomic mass
Hydrogen	H	1.008	Nitrogen	N	14.006
Calcium	Ca	40.080	Oxygen	O	15.999
Carbon	C	12.011	Potassium	K	39.098
Chlorine	Cl	35.453	Sodium	Na	22.990

Beware: Just because a chemical formula is given in the question, doesn't mean that you need to find the molecular mass in the calculation!

Don't cheat by using the Excel spreadsheets; but you can use them to check your answers afterwards. Give the answers in appropriate units.

4.1: How many moles of sucrose are in 2.5 litres of a 7 M solution?

4.2: We mix 7 moles of NaCl in water and make up to 4 litres. What is the molarity of the solution?

4.3: What is the mass of 7 moles of glycine NH_2CH_2COOH?

4.4: We mix 50 grams of Na_2CO_3 in water and make up to 250 millilitres. What is the molarity of the solution?

4.5: How many grams of Na_2CO_3 would be needed to make 0.5 litres of a 5 M solution?

4.6: How many moles of NaCl are in 20 ml of a 5 M solution?

4.7: We mix 25 mmol of glucose (molecular mass = 180.18) in water and make up to 2.4 litres. What is the molarity of the solution?

4.8: What is the mass of 2 mmol of $Ca(OH)_2$?

4.9: We mix 25 mg of $Ca(OH)_2$ in water and make up to 500 ml. What is the molarity of the solution?

4.10: How many grams of uric acid $C_5H_4N_4O_3$ are contained in 25 ml of a 50 mM solution?

4.11: We have 25 ml of a 2 M solution of glycine NH_2CH_2COOH, and we make it up to 2.5 l with water. What is the molarity of the resulting solution?

4.12: How much of a 10 M stock solution of Na_2CO_3 would be needed to produce 2 litres of a 6 M solution?

4.13: We have 250 ml of a 50 mM solution of $Ca(OH)_2$, and we add water to make it up to 1.5 litres. What is the final molarity?

4.14: How much of a 5 M stock solution of Na_2CO_3 would be needed to produce 25 ml of a 30 mM solution?

4.15: We take 10 ml of a 15 M solution of uric acid, and make up to 200 ml with water. We take 10 ml of this solution and make up to 200 ml with water. We repeat this process a further two times. What is the molarity of the final solution?

4.16: We take 250 ml of a 5 M solution of sucrose, and make up to 1 litre with water. We take 50 ml of this solution and make up to 250 ml. We take 10 ml of this solution and make up to 100 ml. What is the molarity of the final solution?

4.17: I mix 2.5 ml of a 10 M solution of NaCl with 50 ml of a 50 mM solution of NaCl. What is the molarity of the resulting solution?

4.18: We repeat the process of problem 4.16, but at the last stage, instead of using water to make up to 100 ml, we accidentally use the original 5 M solution. What is now the final molarity?

4.19: In problem 4.11, suppose that the volumes were measured correct to the nearest millilitre, and that the molarity of the glycine solution is correct to within 10 mM. What is the error in the final molarity?

4.20: The Nernst equation, used in modelling electrochemical cells, is

$$E = E_0 - \frac{RT}{nF} \ln\left(\frac{a_{red}}{a_{ox}}\right)$$

where

- E and E_0 are cell potentials (volts),
- R is the universal gas constant (J K^{-1} mol^{-1}),
- T is the cell temperature (K),
- n is the number of electrons transferred in the cell reaction, and
- a_{red} and a_{ox} are molar concentrations.

What are the dimensions of the Faraday constant F?

ln() denotes the natural logarithm, which we'll meet in Chapter 10.

5

Variables, functions, and equations

In Chapter 2 we described the units in which physical characteristics such as length, volume, and temperature can be measured. The mathematical concept we use to handle such physical characteristics is that of a **variable**. We use a single letter (upper-case or lower-case) to represent a variable: the height of a plant may be represented by h, the volume of a liquid by V, the temperature of an environment by T (since the lower-case letter v is normally used for a velocity and t for time). We may use subscripts if we have more than one variable of the same type, e.g. the concentrations of two substances A and B may be called c_A and c_B. As we saw in Chapter 1, you can treat variables just like numbers when applying the rules of arithmetic, and manipulating equations.

All the examples above are of **continuous variables**, where the variable can take any value on the real line (or an interval of it – all plant heights must be positive, for instance). There are also **discrete variables**, whose possible values are individual points on the real line. These usually arise from counting, e.g. the number of bacteria on a Petri dish can be represented by the discrete variable N, which is dimensionless and must be a natural number (a positive integer). You can't have 4.2 or $5\frac{1}{2}$ bacteria, so N, the number of bacteria, is a discrete variable. By contrast, a plant's height h can be any number on the positive real line, it isn't restricted to specific, discrete values with 'gaps' in between, so h is a continuous variable.

When we measure or calculate values of variables, we produce the data-points we discussed in Chapter 3. In that chapter, we talked about 'joining up the dots' with continuous lines or curves on the graph, expressing relationships between the variables x and y such as direct proportion, which can be expressed algebraically using an equation such as $y = m.x$. This brings us to the concept of a **function**: we say that y is a function of x. This is the basis for a huge body of mathematics, which has been developed since Euler[1] first used the word in its modern meaning, 100 years before Darwin and Mendel started laying the foundations of modern biology. Thus, when biologists started to analyse quantitatively the relations between physical variables, they found a fully formed mathematical universe all ready for them to apply to the biological universe. This combined universe, where the worlds of mathematics and biology meet, is what the rest of this book is all about. We start by defining what a function is, and explore the links between functions, equations, and graphs. Then Chapters 6–10 will form a systematic description of the different functions used in biology and mathematics.

This chapter revisits the ideas we introduced in Chapter 3, but from a 'function' point of view.

---

[1] Leonhard Euler, 1707–1783, German mathematician.

5.1 **What is a function?**

In an experiment, there may be several different variables involved. In growing plants under controlled conditions, some of the variables that can be directly controlled (and which we can give single-letter names) are:

- the temperature T of the environment;
- the amount of fertiliser in the soil, A;
- the daily duration of exposure to light, L;
- the amount of water applied daily, W.

These variables are completely independent of each other: any change in the amount of watering has no effect on the temperature of the environment, and *vice versa*. However, the rate of growth R of the plants will be affected by changes in any of these variables. We say that T, A, L, and W are **independent variables**, and that R is a **dependent variable**. The rate of plant growth R **depends on** the values of T, A, L, and W. We say that R is a **function** of T, A, L, and W; we can write this mathematically as

$$R = f(T, A, L, W). \tag{5.1}$$

T, A, L, and W are sometimes called **arguments** of the function R. (We used this term when introducing formulae in Excel, in Section 1.7.3.) Of course, there are other variables that can affect R which we haven't mentioned, such as the intensity of the light source. But in order to eliminate this from consideration, we use the same light source for all our experiments. That is, we keep the light intensity **constant**. So a variable can be turned into a constant by keeping its value fixed.

Of course, if we're sensible, we will remove some of the other independent variables above from consideration too, by keeping them constant. We may keep A, L, and W constant throughout our experiments, and just examine the effect on plant growth of changing the room temperature. In our mathematical model, we now have just one independent variable T that we can change; the plant growth R then depends on (is a function of) T:

$$R = f(T). \tag{5.2}$$

There is advanced mathematics associated with functions of several variables (as in equation 5.1). We will stick to the situation where we just have two variables x and y: x is the independent variable, and y is the dependent variable (i.e. y depends on x). The mathematical definition of saying that **y is a function of x** is that for *each* permitted value of x, there is a *single* corresponding value of y. Then we write this as

$$y = f(x).$$

The set of all permitted values of x is called the **domain** of the function. So if we choose any value of x in the domain, there will be a corresponding value of y; you can't have two different values of y for the same value of x, and you can't have a value of x in the domain, for which no value of y exists. The values of y lie in a set of numbers called the **range** of the function.

$f(x)$ is a function of x if, for each value of x in the domain, there is
precisely one value of y in the range, satisfying $y = f(x)$. $\qquad$ (5.3)

You can, however, have two different values of x producing the same value of y. In our example, the plant may grow at $R = 8$ mm/day when the temperature is 15 °C (too cold) and have the same growth rate when the temperature is 32 °C (too hot), but achieve a faster growth rate when the temperature is an optimal 25 °C.

You can think of the function f as being a '**black box**' with a keypad on the side for setting the value of the independent variable, and a screen on the front which displays the value of the dependent variable. You type a value of x into the keypad, and up pops the value of y on the screen. The 'black box' described by equation (5.1) would have four keypads on the side for T, A, L, and W to be fed in, but still a single screen on the front to display the value of R.

We can write '$f(3)$' to mean 'the value of $f(x)$ when $x = 3$'. We can also use letters which denote particular (constant) values of the variable x.

EXAMPLE

Let the function $s(t)$ indicate the distance s miles travelled by a car driving along a motorway after t hours, measured from the junction where it entered the motorway. Suppose the car passes Junction 8 of the motorway at time t_8; at that point it will have travelled $s(t_8)$ miles. When it passes Junction 9, at time t_9, it will have travelled $s(t_9)$ miles. The distance travelled between the two junctions is $s(t_9) - s(t_8)$ miles, which has been covered in a time interval of $(t_9 - t_8)$ hours, so the average speed of the car between the two junctions is given by $\dfrac{s(t_9) - s(t_8)}{t_9 - t_8}$ miles per hour.

Mathematicians also sometimes drop the use of f altogether, and write '$y = y(x)$' to mean that y depends on x, i.e. y is a function of x.

. . . .

5.2 Some simple functions

5.2.1 Functions on your calculator, and in Excel

We have already met some calculator functions in Chapter 1. Entering a value for x and pressing the x^2 key, produces on the screen the value of $f(x)$ for the function $f(x) = x^2$.

Similarly, the $\sqrt{}$ key produces the value of the function $f(x) = \sqrt{x}$. Notice that the function produces the positive square root. If we wrote $f(x) = \pm\sqrt{x}$ this would not be a function, as it would be associating two values of $f(x)$ with one value of x ($x = 4$ would produce $f(x) = 2$ and $f(x) = -2$).

The x^y key can be thought of as a function of two independent variables x and y: $f(x, y) = x^y$. See Section 1.5.1.

Another place where you may have used a function, perhaps without realizing it, is in an Excel spreadsheet. Look at SPREADSHEET 5.1.

Excel contains a library of standard functions which you can insert into your formulae without having to type them out. To insert a function in a cell in Excel, either click the f_x button, or choose Function ... from the Insert drop-down menu, or you can type it directly into the Formula Bar box. Functions are spelled out, not written with symbols; thus, $f(x) = \sqrt{x}$, where x is stored in cell A3, would be written as SQRT(A3).

SPREADSHEET 5.1

In part (a) you can see an example of the function $R = f(T, A, L, W)$ in equation (5.1) above. You can set values for T, A, L, and W in the yellow boxes in column D, and the resulting value of R appears in the 'black box' in cell H10. Notice that you get an error for the function if you try to put zero in the box for L. So points for which $L = 0$ are not in the domain of the function.

See Section 5.2.2 for descriptions of part (b) and part (c) of this spreadsheet.

As we said at the beginning of this chapter, functions can have more than one argument. An example of such a function in Excel was the IF() function used in SPREADSHEET 3.7. This has three arguments, and has the format

IF(logical_test, value_if_true, value_if_false).

The function IF(A3 > 0, A3, −A3) is equivalent to the mathematical function $|x|$, called the **absolute value** of x (where x is stored in cell A3). This was defined in Section 1.3.4. It means 'take the numerical value of x, discarding the − sign if x is negative'. We could write this mathematically as

$$|x| = \begin{cases} x & \text{if } x > 0 \\ -x & \text{otherwise} \end{cases} \tag{5.4}$$

since if $x = -3$, $|x| = -(-3) = 3$. This is the logic behind the Excel function IF(A3 > 0, A3, −A3).

There is also a standard Excel function for the absolute value: ABS(A3).

5.2.2 Direct and inverse proportionality functions

Mathematicians have developed many standard types of function over the centuries, which can be used to describe the relationships between continuous variables which are found in all fields of science including biology. The simplest relationships are of **proportionality**, which we looked at in Section 3.3.1. Let's recap what was said there, but in the context of a function.

If y is **directly proportional** to x, a doubling in the value of x will result in a doubling of the value of y. For example, with a fixed volume of solution, the molarity is proportional to the mass of chemical present in the solution. We have used the mathematical symbol $\propto$ to indicate a proportionality relation: we can write $y \propto x$ to mean that y is directly proportional to x.

There is an example of a proportionality function in SPREADSHEET 5.1(B). If you put zero in the yellow box for x in cell D27, the value of the function $f(x)$ in the 'black box' in cell H27 is also zero. If you try other values for x, you will find:

x	$f(x)$
0	0
0.5	2.1
1	4.2
2	8.4
3	12.6
10	42.0

We can write a direct proportionality relationship as

$$f(x) = mx. \tag{5.5}$$

Here, m is called the **constant of proportionality**. In the example in the spreadsheet, the constant of proportionality m is 4.2. The function is thus $f(x) = 4.2x$, and the formula for f, typed in cell H27, is = 4.2*D27, since the value of the independent variable x is in cell D27. To put a formula into a cell, highlight the cell by clicking on it, then type the formula, starting with an = sign. Just as with your calculator, you must type the multiplication symbol, which is an asterisk * in Excel. If you were to type = 4.2D27 and press Enter, you would see an error message. Look back at Section 1.7.3 if you're unsure.

If $y = f(x)$, the following are all equivalent ways of expressing the fact that when x has the value 2, y has the value 8.4:

- When $x = 2$, $y = 8.4$
- When $x = 2$, $f(x) = 8.4$
- y evaluated at $x = 2$, is equal to 8.4
- $f(x)$ evaluated at $x = 2$, is equal to 8.4
- $f(2) = 8.4$
- $y(2) = 8.4$.

An **inverse proportionality** relationship can be written as

$$f(x) = \frac{k}{x} \tag{5.6}$$

where k is the constant of inverse proportionality. Here, a doubling of x results in a halving of $f(x)$. Using the proportionality symbol, we could write $f(x) \propto 1/x$. In SPREADSHEET 5.1(C), y is inversely proportional to x, with the constant of proportionality being 4.2 again. Try entering different values of x into cell D37, and observe how $f(x)$ changes. The formula for this function, in cell H37, is = 4.2/D37. The forward-slash / is used in Excel for division. Notice that you are not allowed to put zero for x in cell D37; if you do, you see the error message #DIV/0!, which means 'You can't divide by zero!'.

EXAMPLE

Suppose we dissolve 10 g of salt in 20 ml of water, and that we then continue adding water. As the volume of solution V increases, the salt concentration c decreases. The concentration is inversely proportional to the volume. Here, the volume is the independent variable, as that is what we are directly controlling by adding water. The relationship is thus $c = k . \frac{1}{V}$. The constant of inverse proportionality k is the mass of salt (10 g), which stays unchanged whatever the dilution.

5.2.3 Quadratic functions, parameters

Here is a slightly more complicated example of a function.

If an object starts moving at time $t = 0$ with an initial velocity u metres per second (m s^{-1}), and travels with a constant acceleration a (measured in m s^{-2}), then after t seconds it will have travelled s metres according to the equation

$$s = ut + \tfrac{1}{2}at^2. \tag{5.7}$$

The involvement of a term in t^2 makes this an example of a **quadratic function**; we will examine these in Chapter 7. The quantities u and a are fixed values for a particular problem, so we treat them as constants whose values we know (even though in a different problem they will take different values). These sort-of-constants are called **parameters**. So the only variables in the equation are t (time, the independent variable) and s (distance travelled, the dependent variable), and we can describe the relationship as 's is a function of t', which we write mathematically as $s = s(t)$.

You can see this function implemented in Excel in SPREADSHEET 5.2 (see the left-hand side of Figure 5.1). In example 5.2(a) the values of the parameters u and a are set as 20 m s^{-1} and 10 m s^{-2} respectively, in cells D4 and D6 (shaded in light yellow), and you input the value of the independent variable t in cell D11 (shaded in bright yellow). The value of $s(t)$ appears in cell D14. With these values of the parameters, the equation could describe the motion of a stone thrown down from the top of a tall tower, with s being measured downwards from the top of the tower. If you click on cell D14 and look in the formula bar above the spreadsheet, you will see that equation (5.6) has been programmed into the cell as

$$f_x = D4*D11 + 0.5*D6*D11*D11 \tag{5.8}$$

Try entering different values for t. You should be able to obtain the values shown in the table below:

t	$s(t)$
0	0
1	25
2	60
3	105
4	160

You can then change the problem by altering the values of the parameters u and a. Remember from Section 1.7.3 that you can use the ^ symbol in Excel for taking exponents; for example, if we wanted to program the function $f(x) = x^4$ we would type = D11^4 instead of = D11* D11*D11*D11. Alternative formulae for (5.7), which will produce the same results as (5.8), are:

$$f_x = D11*(D4 + 0.5*D6*D11)$$
$$f_x = D4*D11 + 0.5*D6*D11^2$$
$$f_x = D11*(D4 + D6*D11/2)$$

In example 5.2(b) on the spreadsheet, the only difference is that the acceleration has been given a negative value: $a = -10$ m s^{-2}. This equation now describes the motion of an stone thrown – or an animal jumping – vertically upwards, against gravity. The behaviour of s as we change t now becomes more interesting:

t	$s(t)$
0	0
1	15
2	20
3	15
4	0

At $t = 2$ seconds, the distance $s(t)$ has reached its maximum value of 20, and at $t = 4$ seconds, $s(t)$ has returned to 0; in function notation, $s(2) = 20$ and $s(4) = 0$. These represent the times when the stone or animal has reached its maximum height, and when it has returned to the ground. Returning to our general function notation of $f(x)$, it is important to be able to find values of x for which $f(x)$ reaches a maximum or minimum value. These are called **turning points** of the function, and values of x for which $f(x) = 0$ are called **roots** of the function. We will see techniques for finding roots and turning points in Chapter 8.

A physical quantity may be treated as a constant, a parameter, or a variable, depending on the situation. For example, the value of acceleration due to gravity is treated as a constant ($g = 9.81$ m s^{-2}) in calculations for a stone dropped from a tower, or a shell fired from a gun. But if we are comparing such kinematics problems on different heavenly bodies, g will be a parameter with one value on Earth, another on the Moon, and a third value on Mars. And if we are calculating the speed of a rocket travelling from the Earth to the Moon, the acceleration due to gravity will be a continuously varying quantity, i.e. a variable.

5.3 Creating tables of values in Excel

We saw in Section 3.2 that if we have a set of data values, we can plot them as points and then draw a graph (a straight line or curve) through them. In Section 5.2 above we created tables of data values from a function, simply by choosing different values of the independent variable x, and calculating the corresponding values of $f(x)$. So if you are given a function, you should be able to plot its graph by drawing up a table of values.

Excel has features which make it easy to draw up a table of data values, especially if you choose values of the independent variable which increase by a fixed amount each time. For example, we have created in SPREADSHEET 5.3, shown in Figure 5.1, a table of t and $s(t)$ for

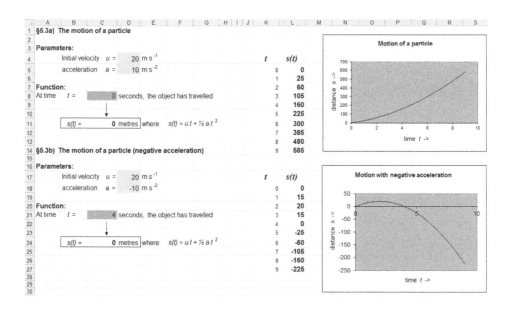

Figure 5.1 Motion of a particle (SPREADSHEET 5.3)

values of t starting at 0 and increasing by 1 each time, reproducing the table in Section 5.2.3 above. Here's how to do it:

SPREADSHEET 5.3

- Type the table headings `t` and `s(t)` in cells K4 and L4.
- Select the cells by dragging the mouse across them while holding down the left mouse key. Now use the formatting buttons in the toolbar, to make the headings bold, centred, and 16-point size. These buttons work just like in MS Word.
- Type 0 for the first value of t in cell K5, and 1 as the next value in cell K6.
- Select these two cells, then click on the 'handle' in the bottom right corner of the selection box. Holding the left mouse key down, drag the box downwards. You will see that Excel guesses you want the next numbers to be 2, 3, 4, etc., and puts these into the boxes K7, K8, K9, and so on.
- Now click on cell L5, the first value of $s(t)$. We are going to type in the formula (5.7) above, but in this case use cell K5 instead of D11 for the value of t. We are then going to click on the 'handle' of this cell and drag it down, to reproduce this formula in the other cells in the column. Excel automatically adjusts the references so that where we have K5 in the formula in cell L5, we will get K6 in the formula in cell L6, and so on. But we don't want the references to the parameter cells D4 and D6 to change. To fix them, we write them as $D$4 and $D$6, so the formula you type in L5 should be

    f_x = $D$4*K5 + 0.5*$D$6*K5*K5

Once you have grabbed the handle of cell L5 and dragged it down, you should see the table shown in **SPREADSHEET 5.3**. You should now be able to follow the procedure explained in Chapter 3, to select this table of values, `Insert a Chart...` and join the points with a smooth curve to produce the graph shown.

Finally, change the value of the acceleration parameter in cell D6 to −10, to see the graph of $s(t)$ when there is a negative acceleration.

CASE STUDY B5: Graphing the relationship in Excel

We can now draw up a table of values for the relationship between the angiogenic cancer cell line y and the non-angiogenic cell lines x_0, x_1. Writing $x = x_0 + x_1$, then y as a function of x is given by

$$y = \frac{-Q + \sqrt{Q^2 - 4rq[k(d-r) + pkx]}}{2rq}$$

where

$$Q = kq(d - r) + r + kp.$$

Adapting **SPREADSHEET 1.2** to produce a table of data-points for the function, we then insert a chart, as described earlier. The result is in **SPREADSHEET 5.4**.

For the values of the parameters d, r, k, p, q being used, the relationship is nearly linear.

5.4 Manipulating equations

5.4.1 What is an equation?

An **equation** is anything of the form A = B, where A (the left-hand side of the equation, abbreviated to l.h.s.) and B (the right-hand side or r.h.s.) are algebraic expressions. An **algebraic expression** is any valid combination of numbers (constants), parameters, and variables, together with arithmetic operators +, ×, etc. The '=' sign expresses the idea that the left-hand side is 'equal' to the right-hand side in some way, but confusion can arise because mathematicians use equations for several different purposes:

(1) **Axioms and identities:**   An equation like the distributive law: $a.(b + c) = a.b + a.c$ (which we saw in Section 1.4.3, equation 1.16) is stating a fact of algebra that is true for all values of a, b, and c. Mathematicians call these basic facts **axioms**; they are so simple and obvious that they can't be proved true, we just have to believe them. Other axioms for real numbers are $a + b = b + a$ and $a.1 = a$.

From the axioms we can deduce **identities** such as those in (1.20), e.g.

$$(x - a)(x + a) = x^2 - a^2 \quad \text{or} \quad (a + b)^2 = a^2 + 2ab + b^2,$$

which again are true for all values of the variables involved.

Mathematicians sometimes use the equality symbol ≡ ('is equivalent to') to indicate that an equation is an identity, e.g. $(x + 1)(x - 1) \equiv x^2 - 1$.

(2) **Relations between variables:**   We may write similar-looking equations which are only true for particular values of the variable(s) involved. For example, the equation $x^2 - 1 = 0$ is only true if $x = 1$ or $x = -1$. You spent many happy hours at school learning how to solve such equations. More generally, an equation like $x^2 + y^2 = 25$ is saying that the variables x and y are **related**, and there are only certain choices of x and y which **satisfy** this equation – meaning that if we plug in that choice of values, the equation will become an identity. For example, we could choose $x = 3$ and $y = 4$; then the equation becomes $9 + 16 = 25$, i.e. $25 = 25$ which is true, so that pair of values satisfy the equation. But if we choose $x = 1$ and $y = 4$, the equation becomes $1 + 16 = 25$, i.e. $17 = 25$, which is false, so that pair of values does not satisfy the equation.

(3) **Assignment of a value to a variable:**   In saying 'Let $x = 4$ and $y = 3$, and let $z = \sqrt{x^2 + y^2}$', we are simply considering the situation where a variable or a function has a particular value. We say that we are **assigning** the value 4 to x and assigning 3 to y. To find the value of z, you work out the expression on the right-hand side, and then assign this value to the variable z given on the left-hand side. This is the meaning of equations in computer programs: the assignment 'A = A + 1' means 'take the value in A, add 1 to it, and store the result back in A'. This is also the type of equation used in the formula bar in Excel.

(4) **Defining a function:**   When we write $s(t) = ut + \frac{1}{2}at^2$ we are **defining** the function $s(t)$.

If I write $s = ut + \frac{1}{2}at^2$, it is slightly ambiguous. I may be defining s as a function of t, with parameters u and a, as in (4), or I may be expressing a relationship between the variables s, u, a, and t, as in (2). The confusion arises because we don't have a recognized 'naming convention', where certain letters would only be used for functions and other letters only for variables. In most cases, a, b, c are used for constants or parameters, t, x, y, z are used for variables, and f, g are used for functions.

5.4.2 Linear equations and their graphs

We say that the variables x and y are **linearly related** if they satisfy an equation of the form

$$ax + by = c \qquad (5.9)$$

where a, b, c are constants (or parameters). For example, $2x - y = 7$ is a linear relation between x and y. If x increases by 1, y will have to increase by 2 so that the equation is still satisfied. If x increases by 2, y will have to increase by 4. In general, the increase in y will be double the increase in x, and the same proportion applies if x and y decrease. So in a linear relation, a change in one variable results in a proportional change in the other variable. Equation (5.9) is called a **linear equation** in x and y.

Linear relations are useful models for situations in the biosciences. For example:

- In a very simple linear predator–prey model, an increase/decrease in the number of predators over a short period of time will result in a proportional decrease/increase in the number of prey.

- When testing the absorbance of a sample containing a mixture of two chemicals in solution, the result provides a linear equation relating the two concentrations. This will be explained in the final example of this chapter.

In some situations it may be clear that x is the independent variable, and that y depends on x, but this is not always the case. In the predator–prey example, a change in the number of predators will affect the number of prey, but also a change in the number of prey (due to disease, for instance) will eventually have an effect on the number of predators the ecosystem can support. We'll model this using differential equations in Chapter 17.

The graph of a linear equation is a **straight line**. In SPREADSHEET 5.5 we have created a chart showing the graph of the linear equation

$$3x + 2y = 4. \qquad (5.10)$$

As this is a straight line, we only need to put two points in the table of data values that Excel uses to create the graph. If we want to produce a graph with x going from −10 to 10, we can use each of these x-values in the equation and solve for y:

When $x = -10$, the equation becomes $-30 + 2y = 4$, so $y = (4 + 30)/2 = 17$.

When $x = 10$, the equation becomes $30 + 2y = 4$, so $y = (4 - 30)/2 = -13$.

So we will plot the line between the points (−10, 17) and (10, −13). Excel will use the data table:

x	y
−10	17
10	−13

If you look at the formulae in the cells of SPREADSHEET 5.5, you will see that it is cleverer than it appears at first sight. The cells C8 and C9 for the y-values contain formulae that calculate y from the x-value in column B, and the coefficients 3, 2, and 4 in the cells in the equation in row 3. These cells have been highlighted in yellow, indicating that you can change the numbers in them. So we are treating x as the independent variable, and y as the dependent variable in this case.

SPREADSHEET 5.5

- The table of data values on the spreadsheet has the (x, y) values of the first point in cells B8 and C8, and the (x, y) values of the second point in cells B9 and C9.

- Select these four cells and click the `Chart` button (or choose `Insert > Chart...` from the drop-down menu bar).

- In the Chart Wizard which appears, choose the chart type '`XY (Scatter)`' and the sub-type 'Scatter by data-points connected by lines without markers'. Click the Next button.

- In step 2, on the Data Range tab choose the radio button to say that the series (i.e. the values of each variable) are in the columns of the table, not in rows. On the Series tab, you can enter a name for the line, for example '$3x + 2y = 4$' then click the Add button. Now highlight the name 'Series2' in the box and click Remove, so only your own name appears there.

- In step 3 you can remove the gridlines, change the option for where the legend will appear, etc. Then click Finish.

If you change your choice of x-values, the axes of the graph will change, e.g. if you put 0 instead of −10 in cell B8, the corresponding y-value in cell C8 becomes 2, and the graph will only be plotted from $x = 0$ to $x = 10$.

If you change any of the numbers in the yellow cells in the equation in row 3, the data values change and the graph shows the straight line for the new equation. In the next section we explain how these formulae were obtained.

5.4.3 Rearranging linear equations

A linear relation such as $3x + 2y = 4$, considered in the previous section, can easily be rearranged. If we want to think about x as the independent variable, and y as the dependent variable, we can rearrange the equation to write y as a function of x:

$$3x + 2y = 4.$$

Subtract $3x$ from both sides:

$$2y = -3x + 4.$$

Divide both sides by 2:

$$y = -1.5x + 2.$$

This type of function is called a **linear function**; we'll look at it in detail in Section 6.2.

If we rearrange the general linear relation $ax + by = c$ to solve for y as a function of x, we obtain:

$$y = (c - ax)/b.$$

This is the formula used in cells C8 and C9 in SPREADSHEET 5.5.

5.4.4 Equation of a circle

If you are given two points on a graph, with coordinates (x_1, y_1) and (x_2, y_2), how do you calculate the distance d between them? We could draw a right-angled triangle, as we did in Figure 3.4(a), and then use Pythagoras's theorem: the distance we want is the length of the hypotenuse. This is shown in Figure 5.2(a). The horizontal side of the triangle is of length $\Delta x = x_2 - x_1$, and the vertical side has length $\Delta y = y_2 - y_1$. If the distance between the two points is d, Pythagoras's theorem tells us that $d^2 = (\Delta x)^2 + (\Delta y)^2$. To find d we take square roots of both sides. Thus,

$$d = \sqrt{(\Delta x)^2 + (\Delta y)^2}$$
$$= \sqrt{(x_2 - x_1)^2 + (y_2 - y_1)^2}.$$

(5.11)

If one of the points is the origin $(0, 0)$ the formula becomes simpler:

The distance from the point (x_1, y_1) to the origin is $\sqrt{x_1^2 + y_1^2}$. See Figure 5.2(b).

EXAMPLE

In an experiment to observe ant movement, a large board is marked out with coordinate axes and gridlines (like graph paper), and an ants' nest is placed at the origin. After one minute, a photograph is taken from a camera positioned above the nest. The photograph is printed out, and the ants' coordinate positions can be read off.

Two of the ants are at coordinates $(5, 12)$ and $(-8, 14)$. How far is each ant from the nest, and how far are they from each other?

Distance of ant 1 from nest $= \sqrt{5^2 + 12^2} = \sqrt{25 + 144} = \sqrt{169} = 13$ cm.

Distance of ant 2 from nest $= \sqrt{(-8)^2 + 14^2} = \sqrt{64 + 196} = \sqrt{260} = 16.12$ cm.

Also, the distance of ant 1 from ant $2 = \sqrt{(-8 - 5)^2 + (14 - 12)^2} = \sqrt{(-13)^2 + (2)^2} = \sqrt{169 + 4} = \sqrt{173} = 13.15$ cm.

We can now write down the equation for a circle, with centre at the origin and radius r. The circle consists of all points (x, y) that are a distance r from the origin, that is, all points for which

$$x^2 + y^2 = r^2.$$

(5.12)

This is shown in Figure 5.3(a).

A circle centred at O and of radius 1, generated by the equation

$$x^2 + y^2 = 1$$

(5.13)

is called a **unit circle**.

It's worth noticing that we cannot write y as a function of x whose graph is a circle. If we rearranged (5.12) for a circle with radius $r = 5$, to find y:

$$y = \sqrt{25 - x^2}$$

this would only define the upper semicircle (above the x-axis) since $\sqrt{}$ means the positive square root. We can try to get round this by writing

$$y = \pm\sqrt{25 - x^2}$$

but then this is no longer a function. In Section 5.1 we said that a function is like a 'black box' which produces a single output value of y for each permitted input value of x. So if we put $x = 3$ into $y = \pm\sqrt{25 - x^2}$ and get out both $y = 4$ and $y = -4$, we are dealing with a relation between x and y, but y is not a function of x.

Now, think back to Section 1.8.1, where we talked about intervals and inequalities. If we replace the = sign in (5.13) by a < sign, then the inequality

$$x^2 + y^2 < 1$$

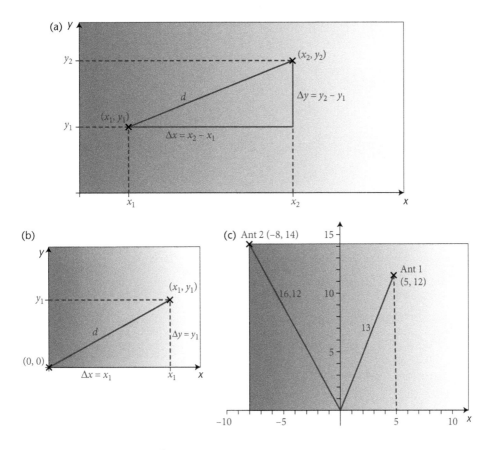

Figure 5.2 Distances on a graph

defines all points (x, y) that lie inside the unit circle; this is called the **unit disk**, shown in Figure 5.3(b).

Finally, take a fixed point (x_0, y_0) and a distance d. Then the inequality

$$\sqrt{(x - x_0)^2 + (y - y_0)^2} < d$$

defines all points which are within d of (x_0, y_0), i.e. a circular disk centred at (x_0, y_0) and of radius d. This is the two-dimensional counterpart of the one-dimensional interval $|x - a| < d$ of semi-width d and centre a (i.e. all points x on the real line that are within d of a) which we saw in equation (1.29) and Figure 1.3.

EXAMPLE

In our ants' nest example, we want to describe the distribution of ants around the nest, in our snapshot. We can define $N(d)$ to be the number of ants which are within a distance d from a nest located at (x_0, y_0). If we have all the ants' coordinates

$$(x_1, y_1), (x_2, y_2), (x_3, y_3), (x_4, y_4), \dots$$

entered into a database, then for a chosen value of d we can test each (x_i, y_i) to see if it satisfies

$$\sqrt{(x_i - x_0)^2 + (y_i - y_0)^2} < d,$$

(a)

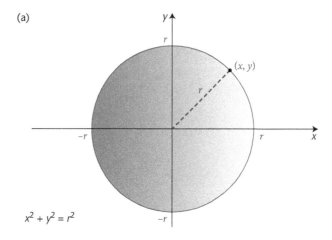

$$x^2 + y^2 = r^2$$

(b)

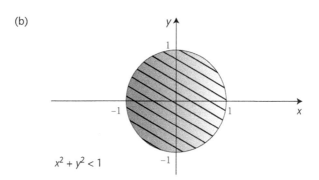

$$x^2 + y^2 < 1$$

Figure 5.3 Equations of (a) a circle and (b) unit disk

i.e. whether it lies within the disk of radius d centred around (x_0, y_0). The number that satisfies this inequality is $N(d)$; see Figure 5.4(a). Plotting a graph of $N(d)$ against d would produce a curve similar to that in Figure 5.4(b). We will see later how to extract parameters from such curves, so that we can describe them numerically.

5.4.5 Rearranging more complicated equations

Let's return to functions. In many situations it is clear which is the independent variable (or variables), and which is the dependent variable. This is especially true when time is a variable; it is always independent since the progress of time doesn't depend on anything else (If you are measuring the growth of a rat's tail over time, you can't make the rat younger by cutting a bit off its tail!) But there are other situations that are less clear-cut.

For a fixed body of gas, its pressure P, volume V and temperature T are related by the ideal gas law (see problem 2.33):

$$\frac{PV}{T} = k, \text{ a constant.} \tag{5.14}$$

In an experiment where we hold the temperature constant and compress the gas, causing an increase in pressure, T will be a parameter, V will be the independent variable, and the

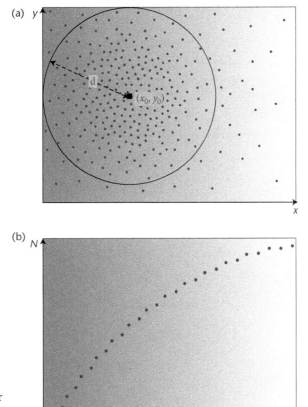

(a)

(b)

Figure 5.4 Graph of number of ants $N(d)$ that are within distance d from nest

dependent variable will be P. So we can write $P = P(V)$ and rewrite (5.14) (by multiplying both sides by T and dividing both sides by V), as

$$P = kT\frac{1}{V},$$

i.e. P is inversely proportional to V.

On the other hand, in an experiment where we heat the gas and allow it to expand at constant pressure, P is a parameter, T is the independent variable, and V the dependent variable: $V = V(T)$ and the relationship is

$$V = \frac{k}{P}T$$

which is a direct proportionality function.

You need to be able to rearrange fairly simple equations to make a particular variable the subject, i.e. to isolate it on the left-hand side. The key fact to remember is that algebraic variables represent numbers, so the rules of arithmetic summarized in Chapter 1 apply to variables and parameters as well. You can use the distributive law to expand brackets:

$$a.(b + c) \rightarrow a.b + a.c$$

or, conversely, to take out a common factor:

$$a.b + a.c \rightarrow a.(b + c).$$

You can perform the same sequence of operations on both sides of the equation, as discussed above. And finally you can swap over the l.h.s. and r.h.s. of the equation.

> **EXAMPLE**
>
> Given the equation $y = \dfrac{4}{x+3}$, make x the subject of the equation.
>
> **Solution:** Multiply both sides of the equation by $(x + 3)$, to get
>
> $$y.(x + 3) = 4.$$
>
> Expand the brackets:
>
> $$x.y + 3.y = 4.$$
>
> As we are trying to isolate x on the l.h.s., subtract $3y$ from both sides:
>
> $$x.y + 3y - 3y = 4 - 3y$$
> $$x.y = 4 - 3y$$
>
> (or 'move the $3y$ to the r.h.s., changing its sign', as we said in Section 1.3.6), and finally divide both sides by y:
>
> $$x = \frac{4-3y}{y}.$$

In the exercises you will be able to practise this sort of algebraic manipulation. Rearrangements can also be useful in planning experiments, as in the following example.

> **EXAMPLE**
>
> In an experiment where a small molecule binds to a protein across a membrane, the bound concentration b is related to the free concentration of the molecule a by
>
> $$b = \frac{P_0.a}{K+a} \qquad (5.15)$$
>
> where P_0 is the total protein concentration, and K is the dissociation constant. If we know P_0 and K and have a general idea of the size of b, we may want to work out the appropriate magnitude of the free concentration a to use. That is, we want to rearrange the equation to get a as a function of b, $a = a(b)$.
>
> This is very similar to the previous example. We can rearrange (5.15) by multiplying both sides by the denominator $(K + a)$:
>
> $$b.(K + a) = P_0.a.$$
>
> Multiply out the brackets using the distributive law:
>
> $$b.K + b.a = P_0.a.$$
>
> As we are trying to isolate a on the left-hand side, we collect the terms involving a on the l.h.s. and the other term on the right. Subtracting $P_0.a$ from both sides, and also subtracting $b.K$ from both sides, we get:
>
> $$b.a - P_0.a = -b.K.$$

Again using the distributive law, isolate a as a common factor on the l.h.s.:

$$a.(b - P_0) = -b.K$$

and finally divide both sides by $(b - P_0)$:

$$a = \frac{-b.K}{b - P_0}. \qquad (5.16)$$

5.5 Graphs of the direct and inverse proportion functions

SPREADSHEET 5.6, shown in Figure 5.5, has a chart showing the graphs of the direct proportion function $y = k.x$ and the inverse proportion function $y = \frac{k}{x}$. The constant of proportionality k is set in cell I3. The direct proportion function produces a straight-line graph, so we only need two data-points in the data table. We use the origin (0,0) as one point, and we have chosen the other point to be where $x = 10$.

The inverse proportion function is nonlinear, so we need a larger set of data-points, which we will join up with smooth curves. We cannot use 0 for x, as this is not in the domain (it would mean dividing by zero). We have started with 0.1 instead. We have also used unevenly spaced points, which are closer together in the part of the curve that has greatest curvature. You can try altering these points and see the effect on the graph.

The two functions were programmed into cell C8 (= $I$3*B8) and cell C15 (= $I$3/B15) and dragged down the column to fill up each data table. Notice that we must refer to the cell storing k as $I$3, so that the reference doesn't change as we drag down.

Now we can insert the chart. Here are the steps:

(1) Choose insert > chart... and in Step 1 of the Chart Wizard select the XY (scatter) chart type.

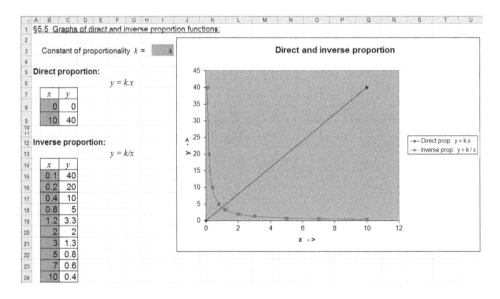

Figure 5.5 Graphs of direct and inverse proportion (SPREADSHEET 5.6)

(2) Choose the sub-type `Scatter with data points connected by smoothed lines`. Click the `Next` button.

(3) In Step 2, click the radio button for `Series in: Columns`, then click in the `Data range:` box and use the mouse to select the 2×2 block of cells for the first data table (B8 to C9). You should see an image of the straight-line graph.

(4) Click the `Series` tab at the top, and in the new panel, give this data table (Series 1) a name such as Direct Proportion.

(5) Then click the `Add` button. You can now define the new Series 2 for inverse proportion. Give it a name, define the `X Values` by selecting cells B15 to B24, and the `Y Values` by selecting cells C15 to C24. Then click `Next`.

(6) In Step 3, use the `Titles` tab to give the chart a title, and to put titles on each axis. On the `Gridlines` tab, remove the horizontal gridlines.

(7) Finally, click `Finish` and the chart appears, showing the line and the curve.

. . . .

5.6 Algebra of functions

The function notation is a very powerful tool in mathematics. It doesn't just have to sit on the left-hand side of an equation – it can be used like a variable, and has a couple of other tricks up its sleeve which can't be done by variables . . .

5.6.1 Changing the argument

As we said in Section 5.2.2, we can replace the argument x in $f(x)$ by a number to indicate the value of the function f evaluated at this value of x.

> **EXAMPLE**
> If $f(x) = x^3 - 5x - 2$ and $g(x) = \dfrac{x-1}{x+3}$, evaluate $f(0), g(0), f(4), g(4), f(-2), g(-2)$.
>
> **Solution:**
> $$f(0) = 0^3 - 5(0) - 2 = 0 - 2 = -2$$
> $$g(0) = \frac{0-1}{0+3} = -\frac{1}{3}$$
> $$f(4) = 4^3 - 5(4) - 2 = 64 - 20 - 2 = 42$$
> $$g(4) = \frac{4-1}{4+3} = \frac{3}{7}$$
> $$f(-2) = (-2)^3 - 5(-2) - 2 = -8 + 10 - 2 = 0$$
> $$g(-2) = \frac{-2-1}{-2+3} = \frac{-3}{1} = -3.$$

Remember to substitute the value for x everywhere that x occurs in the function definition.

We can extend this idea by replacing x by another variable. For example, if $f(x)$ is the 'squaring function' $f(x) = x^2$, I can write $f(t) = t^2$, using the argument t instead, or $f(z) = z^2$ or even $f(\xi) = \xi^2$. In all these cases, the essential function f is unchanged; it is the 'squaring' function on your calculator. We say that x is a **dummy argument**, because we can use any other name for it without changing the meaning.

What is less easy to get your head around, is that I can replace the argument x by an algebraic expression. For example, if $f(x) = x^2$, I can replace x by $3t + 1$, and then $f(3t + 1) = (3t + 1)^2$. I can even replace x by an algebraic expression involving x:

> **EXAMPLE**
>
> If $f(x) = x^3 - 5x - 2$ and $g(x) = \dfrac{x-1}{x+3}$, what are the functions $f(x + 1), g(x + 1), f(2x), g(2x),$ $f(x^2), g(x^2)$?
>
> **Solution:**
>
> $$f(x+1) = (x+1)^3 - 5(x+1) - 2 = (x^3 + 3x^2 + 3x + 1) - 5x - 7$$
> $$= x^3 + 3x^2 - 2x - 6$$
> $$g(x+1) = \frac{(x+1)-1}{(x+1)+3} = \frac{x}{x+4}$$
> $$f(2x) = (2x)^3 - 5(2x) - 2 = 8x^3 - 10x - 2$$
> $$g(2x) = \frac{2x-1}{2x+3}$$
> $$f(x^2) = (x^2)^3 - 5(x^2) - 2 = x^6 - 5x^2 - 2$$
> $$g(x^2) = \frac{x^2-1}{x^2+3}.$$

When we substitute $t - 1$ in place of the argument t in a function of time $f(t)$, what is the effect? Suppose that $f(3) = 7$, i.e. $f(t) = 7$ when $t = 3$ hours. Now consider $f(t - 1)$. We would now need to choose $t = 4$ in order to obtain the function value of 7: $f(4 - 1) = f(3) = 7$, i.e. $f(t - 1) = 7$ when $t = 4$. The effect is to shift the whole graph of $y = f(t)$ one unit to the right, and to make the function 'happen' one hour later than previously. We will see such 'shifted' graphs in Section 6.5.

> **EXAMPLE**
>
> When a patient takes an antibiotic capsule, the medicine's concentration in the bloodstream after t hours is $C(t)$. We assume that the medicine enters the bloodstream immediately, and produces an initial concentration of $C(0) = C_0$. We measure t from the time of taking the first capsule. A second capsule is taken 6 hours later. The concentration from the second capsule at time t will be $C(t - 6)$, so that $C(t - 6) = C_0$ when $t = 6$. The total concentration from the two tablets in the interval $6 < t < 12$ (i.e. between taking the 2nd and 3rd tablets) will be $C(t) + C(t - 6)$. Figure. 5.6 shows typical graphs of concentration against time.

5.6.2 Arithmetic operations with functions

Functions can also be treated just like variables, using the rules of algebra. They can be added and subtracted:

> **EXAMPLE**
>
> If $f(x) = x^3 - 2$ and $g(x) = 2x^3 - 3x + 1$, then
>
> - $f(x) + g(x) = (x^3 - 2) + (2x^3 - 3x + 1) = 3x^3 - 3x - 1,$
> - $f(x) - g(x) = (x^3 - 2) - (2x^3 - 3x + 1) = -x^3 + 3x - 3,$
> - $4f(x) = 4(x^3 - 2) = 4x^3 - 8.$

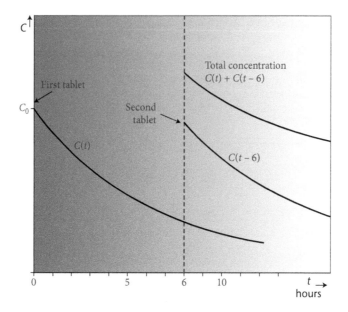

Figure 5.6 Antibiotic concentration over time

EXAMPLE

In modelling the progress of a non-fatal 'flu' epidemic through a fixed population of size N, we suppose that when a person is infected, they are ill for one week, then recover and are henceforth resistant to further infection. We can define three functions of time t:

- $I(t)$ = number of infected people at time t;
- $S(t)$ = number of susceptible people (who have not yet been infected);
- $R(t)$ = number of resistant people (who have been infected and recovered).

Initially, $I(0) = 0$, $S(0) = N$, $R(0) = 0$. Also, as the total population stays constant,

$$I(t) + S(t) + R(t) = N \quad \text{for all } t.$$

In Chapter 17 we will see a mathematical model that can predict the functions $I(t)$ and $S(t)$. Suppose the epidemic dies out after T weeks, i.e. $I(T) = 0$. Then we can find the number of people who went through the epidemic without being infected, by rearranging the above equation:

$$S(T) = N - R(T).$$

Functions can also be multiplied together. In Chapter 12 we will see rules for applying calculus to functions that can be written as products or quotients.

EXAMPLE

If $f(x) = (2x + 1)(x^4 - 5)$, we can write it as $f(x) = u(x).v(x)$, where $u(x) = 2x + 1$ and $v(x) = x^4 - 5$.

Similarly, if $g(x) = \dfrac{2x+1}{x^4-5}$ we could write it as $g(x) = \dfrac{u(x)}{v(x)}$.

5.6.3 Composition of functions (or 'function of a function')

In addition to adding, multiplying, etc. there is another operation that can be used to combine functions: **composition**. Suppose that $f(x) = 3x + 1$ and $g(x) = x^2 - 2$. Think again of a function $f(x)$ as a 'black box' that takes x as input and produces $f(x)$ as output. We could put x into f to produce $f(x)$, and then use this output as the input to g. The result would be written as $g(f(x))$:

$$g(f(x)) = g(3x + 1) = (3x + 1)^2 - 2.$$

Alternatively, we could apply g first, and then f:

$$f(g(x)) = f(x^2 - 2) = 3(x^2 - 2) + 1.$$

We have produced two new functions by composing f and g. The symbol for this operation of composition is $\circ$:

- The function $g \circ f$ is defined by $(g \circ f)(x) = g(f(x))$.
- The function $f \circ g$ is defined by $(f \circ g)(x) = f(g(x))$.

(5.17)

As we can see in the example, the composed functions $g \circ f$ and $f \circ g$ are not the same, in general.

It is possible to compose a function with itself:

$$(f \circ f)(x) = f(f(x)) = f(3x + 1) = 3(3x + 1) + 1 = 9x + 4.$$

This function is sometimes written as $f^2(x)$, but this can easily be confused with the square of $f(x)$.

> **EXAMPLE**
>
> A series of fields of different sizes contain milkweed plants, which are attractive to Monarch butterflies. The field areas are measured and the numbers of plants in the fields are counted. A trend line is plotted to the data in Excel to produce a linear function:
>
> - a field of size A m^2 contains $f(A)$ milkweed plants, where $f(A) = 25\,A - 3$.
>
> In a previous study, researchers reported a linear relationship between the number of milkweed plants in a field, and the number of Monarch butterflies:
>
> - a field with N milkweed plants attracts $g(N)$ butterflies, where $g(N) = 8N + 2$.
>
> Then the relationship between butterflies and field size is: $g(f(A)) = 8f(A) + 2 = 8(25A - 3) + 2 = 200A - 22$. This is the function $(g \circ f)(A)$: a field of area A will attract $(g \circ f)(A)$ butterflies.
>
> The alternative composition function $f \circ g$ is impossible. This would mean $f(g(N))$, but $g(N)$ is a count of butterflies, while f is expecting a field area as input.

Problems: Now try the end-of-chapter problems 5.25–5.32.

· ·

5.7 Extension: solving simultaneous linear equations

We have seen that if we have a single linear equation with a single unknown variable, for example $x + 3 = 5$, we can easily solve it to find x: $x = 5 - 3 = 2$. A linear equation involving two variables, such as $3x + 2y = 4$, does not have a unique solution for x and y. All the (x, y) points on the graph in SPREADSHEET 5.5 are solutions to this equation.

But if we know two linear equations involving two unknown variables, and both equations must be true simultaneously, then there is usually a unique solution for x and y. How do we find it?

EXAMPLE

Consider the two equations

$$3x + 2y = 4 \qquad\qquad (1)$$

and

$$7x + 12y = -5. \qquad\qquad (2)$$

We will refer to them as equation (1) and equation (2). I claim that there is just one pair of values of x and y that will satisfy both equations. Here's how to find it:

Multiply both sides of equation (1) by 7, and both sides of equation (2) by 3: we get

$$21x + 14y = 28 \qquad\qquad (3)$$

and

$$21x + 36y = -15. \qquad\qquad (4)$$

Call these new equations (3) and (4). Now we can subtract the left-hand side of (4) from the left-hand side of (3), and the right-hand side of (4) from the right-hand side of (3). The resulting equation will no longer have any term in x:

$$(21x + 14y) - (21x + 36y) = 28 - (-15).$$

Rearranging the left-hand side:

$$(21x - 21x) + (14y - 36y) = 28 - (-15).$$

The terms in x cancel, leaving

$$-22y = 43$$

so we find that

$$y = -43/22 = -1.95454545\ldots$$

We have found the value of y. To find the value of x, we simply substitute this value of y into one of the original equations (1) or (2), and solve for x. From equation (1):

$$x = \frac{1}{3}(4 - 2y)$$

so that

$$x = \frac{1}{3}(4 + 3.90909090\ldots)$$
$$= 2.636363\ldots$$

and so $x = 2.637$ (correct to 4 significant figures).

The solution is therefore: $x = 2.637$, $y = -1.955$ (correct to 4 significant figures).

Notice that, even if we know in advance that we want to quote the solution correct to 4 significant figures, we do not round off the value of *y* before using it to calculate *x*. Keep it in your calculator display, multiply it by 2, change the sign with the ± key, add 4 then divide by 3. Only then can you round *x* and *y* to 4 sig figs.

This technique is called **solution by elimination**, because we have chosen the multipliers of equations (1) and (2) in order to get the same coefficient of *x*, and hence to eliminate this variable when we subtract one equation from the other. We could equally well have chosen multipliers in order to eliminate *x* instead of *y*; can you see what these multipliers would be?

If you draw the graphs of equations (1) and (2) on the same set of axes, you get two straight lines. The point where these two lines intersect, is the solution for *x* and *y*. So an alternative way of solving this pair of simultaneous equations would be to carefully draw their graphs and read off the coordinates of the point of intersection.

SPREADSHEET 5.7, shown in Figure 5.7, demonstrates this elimination technique, and also draws the graphs. You can change the coefficients for the two equations (in the yellow cells in rows 3 and 4) and see how the elimination calculation changes, as well as how the graphs change.

SPREADSHEET 5.7

The table of data values has the left and right *x*-values in cells M22 and M23, the corresponding *y*-values for equation (1) in column N and for equation (2) in column O. To produce a chart with both lines on:

- Highlight the first two columns of the data table, and produce a chart with the graph for equation (1) as we did in Section 5.4.2. The legend for this line should be 'Equation (1)', and in Step 3 you can give the chart a title.

- Now with the chart still highlighted, look on the drop-down menu-bar at the top of the screen. As well as 'File Edit View' etc. there will be an option called `Chart`. Click on this and choose 'Add data...'

- Highlight cells O23 and O24 (containing the *y*-values for equation (2)). The new line will appear in the graph!

- The legend for the new line says Series2. To change it, have the chart highlighted and from the `Chart` item on the top menu-bar, choose 'Source data...' In the dialogue box with the series tab, you can highlight Series2 and enter the name Equation (2) and click Add. Then remove the name Series3, which also appears.

EXAMPLE

The **absorbance** A of a sample is defined from the ratio between the intensity of the light passing through a sample, and the light intensity without the sample.[2] As such, it is a dimensionless quantity. This is measured in a spectrometer, where the sample is placed in a cuvette of a certain length *d* and a light source with light of a single wavelength λ is used (the absorbance depends on the light wavelength). If the sample contains a solution of a given chemical at molar concentration *c*, the formula for absorbance is

$$A = A_\lambda.c.d$$

[2] Actually, physicists use the logarithm of this ratio.

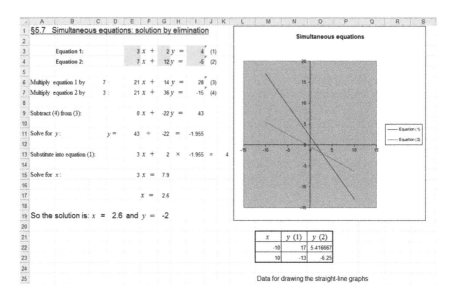

Figure 5.7 Simultaneous equations (SPREADSHEET 5.7)

where A_λ is the **molar absorbance** of that chemical for that particular light wavelength. By dimensional analysis (Section 2.4), you should be able to work out that this is in units of $M^{-1}\ mm^{-1}$, since the molar concentration is in units of M (see Section 4.1.5) and d is in millimetres.

If our sample is a mixture of two chemicals in solution, the absorbances add together. That is, if the molar concentrations of chemicals X and Y are c_X and c_Y, then the total absorbance of the sample will be given by

$$A = A_{\lambda X}.c_X.d + A_{\lambda Y}.c_Y.d.$$

As we have one linear equation in two unknowns (c_X and c_Y), we cannot find the individual concentrations from one measurement. But if we repeat the experiment with the same sample but using light of a different wavelength, we obtain a second linear equation with different coefficients. These two simultaneous equations can be solved.

For example, suppose the cuvette length is 10 mm. Using light of wavelength 450 nm, we obtain an absorbance of 0.72, and repeating the experiment with light of wavelength 500 nm we find an absorbance of 0.57. The molar absorbances of the two chemicals X and Y at the two wavelengths used are given by

Molar absorbance	chemical X	chemical Y
$\lambda = 450$ nm	112	476
$\lambda = 500$ nm	722	254

Find the concentrations of X and Y in the sample.

We therefore obtain the following two equations:

$$112 \times c_X \times 10 + 476 \times c_Y \times 10 = 0.72$$
$$722 \times c_X \times 10 + 254 \times c_Y \times 10 = 0.57$$

which can be written as:

$$112c_X + 476c_Y = 0.072$$
$$722c_X + 254c_Y = 0.057.$$

You can solve this yourself, following the previous example. Or you can cheat by looking at **SPREADSHEET 5.8**! This gives the solution as

$$c_X = 2.81\text{E-}05 \quad \text{and} \quad c_Y = 0.0001447.$$

As these are in units of M, they can be written more conveniently as

$$c_X = 0.028 \text{ mM} \quad \text{and} \quad c_Y = 0.145 \text{ mM}.$$

PROBLEMS

5.1–5.6: Rearrange the following equations to make x the subject, e.g. rearrange $dx + a = b$ to $x = \dfrac{b-a}{d}$.

5.1 $ax + b = cx - d$ 5.2 $y = \dfrac{3x}{x-2}$

5.3 $\dfrac{x+y}{x-y} = 3y+1$ 5.4 $\dfrac{5b}{x+3} = a(x-3)$

5.5 $y^2 + 3xy - 1 = 2x$ 5.6 $y^2 + 3xy - x^2 = 5.$

5.7–5.10: For the given pairs of functions $f(x)$ and $g(x)$, work out the functions: (a) $f(x) - g(x)$ (b) $f(x).g(x)$ (c) $f(g(x))$ (d) $g(f(x))$ (e) $2f(x) + 3g(x)$

5.7 $f(x) = 3x + 2, g(x) = 2x - 1$ 5.8 $f(x) = 3x + 2, g(x) = x^2 - 1$

5.9 $f(x) = 3x + 2, g(x) = \dfrac{1}{x}$ 5.10 $f(x) = 3x + 2, g(x) = \sqrt{x-1}.$

5.11–5.14: For the following functions $f(x)$, write down: (a) $f(0)$ (b) $f(3)$ (c) $f(-5)$ (d) $f(3x)$ (e) $(f(x))^2$ (f) $f(x^2)$ (g) $f(2t + 1)$ (h) $f(f(x))$.

Note: $f(f(x))$ may be written as $f^2(x)$, which is different from $(f(x))^2$.

Simplify expressions where possible.

5.11 $f(x) = 3x + 2$ 5.12 $f(x) = \dfrac{2}{x+5}$

5.13 $f(x) = x^2 + x + 1$ 5.14 $f(x) = x - \dfrac{1}{x}.$

5.15–5.19: Solve the following pairs of simultaneous linear equations. Give the solutions correct to 4 significant figures.

5.15 $x + y = 12$ 5.16 $2x + 3y = 0$
$3x - y = 8$ $x - 5y = 13$

5.17 $2x + 3y = 13$ 5.18 $2.5x - 1.3y = 10.0$
$5x - 2y = -1$ $-7.5x + 2.4y = 42.3$

5.19 $0.3x - 1.4y = 13.6$
$0.5x - 2.9y = 15.2$

5.20: Adam and Britney perform the absorbance experiment (Section 5.7) with a mixture of chemical Y with a new chemical A. They each look up the molar absorbances of the two chemicals, and measure the absorbance of the mixture at $\lambda = 450$ nm and 500 nm. The results are:

Molar absorbance	chemical A	chemical Y	Mixture absorbance
$\lambda = 450$ nm	324.5	476	0.069
$\lambda = 500$ nm	173	254	0.037

To deal with coefficients which are whole numbers, Adam decides to round up the absorbance of A at $\lambda = 450$ nm to 325. Britney also wants to use whole numbers, but rounds the absorbance down to 324. Do you think that this relative change of 0.15% in one coefficient value will make much difference to the final answer? Use SPREADSHEET 5.8 to solve the resulting pair of equations, once using 324 and again using 325 for the absorbance of A. Can you explain why their results for the mixture concentrations are so different? Which student would you give higher marks to?

Linear functions and curve sketching

In Chapter 5 we defined the concept of a function $f(x)$ and saw how it can be depicted by a line or curve (called its graph) with equation $y = f(x)$, using Cartesian x- and y-axes. This is important to experimental scientists, because they can produce such curves by 'joining up the dots' on a plot of data-points obtained from experiment. (We did this in Chapter 3.) If the resulting curve can be described by a mathematical function, we can use the theory of functions which has been developed by mathematicians, to analyse and make predictions about the physical process which is at work in the experiment.

We've seen that some basic functions such as $f(x) = \sqrt{x}$ are available as buttons on your calculator, and that functions can also be programmed into cells in an Excel spreadsheet. By these means we can work out the values of particular points (x, y) on the curve $y = f(x)$. But we need to see what the whole curve looks like, and that's what we'll be discussing from now on. For this, we need either a graphical calculator (too expensive!) or a graph-drawing software package. We will use a freely available (and free!) package called FNGRAPH to visualize the functions; we'll describe how to use FNGraph in the next section.

It would make life very easy if all physical relationships could be modelled by straight-line graphs. These are certainly a very important class of function, which we'll discuss in detail in this chapter. But real life is not that easy. Suppose we obtain data representing the monthly average temperature over the next decade, in a part of the world affected by a runaway 'greenhouse effect' on the climate. The overall trend is for temperature to rise at an increasing rate, but there will continue to be seasonal variations between winter and summer. A graph of the data-points will (with some exaggeration) look something like Figure 6.1. Can we find a mathematical function that will produce a curve like this?

Over the past 250 years, mathematicians have developed a library of functions with different characteristics, which can be shifted, stretched, and combined to produce almost any sort of curve we want (remember, though, that the curve of a function cannot turn back on itself, since a function is a 'black box' that produces only a single value of y for a particular chosen value of x). Here and in the remaining chapters forming Part I, we will explore this library of functions. For each function $f(x)$ we will want to know:

- the general formula for the function, involving x and various parameters;
- the shape of the graph of $y = f(x)$;
- useful points on the graph, such as where it crosses the x- or y-axes;

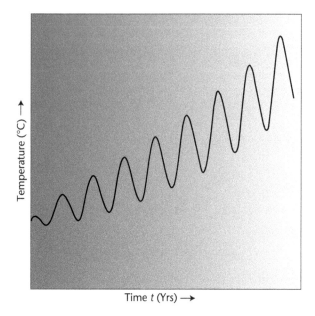

Figure 6.1 Temperature variation in 'greenhouse effect'

- how to use information we may know about our particular function, to work out the equation of the curve (that is, to find the values of the parameters);
- some of the physical processes which can be modelled by the function.

In the final chapter of Part I, we'll see some more advanced techniques for fitting functions to experimental data.

Our standard notation will be to take x as the independent variable, y as the dependent variable, and f as the name of the function, so that we will write $y = f(x)$ to mean that y is a function f of x. But we'll use more appropriate symbols for the variables when we show the functions used in different applications. It's important to remember that $y = x^2$, $f(x) = x^2$, $s = t^2, f(t) = t^2, g(u) = u^2, P = V^2$, etc. are all expressing the same 'squaring' function, although using different variables and function names.

In this chapter we'll study linear functions. In Chapter 7 we generalize to quadratic and polynomial functions; Chapter 8 looks at rational functions; and the final two chapters of Part I deal with trigonometric and exponential functions. To cope with the later material, it is essential that you completely understand simple linear functions. Sections 6.2–6.3 will cover this quite slowly, and you probably remember most of it from school anyway. Feel free to skim through these sections, but come back to them if you have problems in later chapters. But don't skip straight on to Chapter 7: in Sections 6.3–6.5 there is important material about asymptotes, limits, and linear transformations which will possibly be quite new to you.

6.1 Graph sketching with FNGraph

In Chapter 5 we used Excel to draw the graphs of some functions and relations, by constructing a table of data-points, and then plotting the line or curve through them (in the same way as we plotted data-points from an experiment in Chapter 3). But Excel can't sketch graphs directly from the function. If we want Excel to produce the graph of $y = x^2$, we can't simply give it this equation; we must work out a set of data-points (x, y), put them into a table of cells, and tell Excel to plot a chart with a smooth curve through the points, as we did for the inverse proportion function in Section 5.5. We have done this in some spreadsheets provided with this book, to create interactive demonstrations of graphical results, but it is a time-consuming process, and you need a lot of data-points to produce a smooth curve.

Instead, for graphs that we will encourage you to create or modify yourself, we will use a graph-drawing package called FNGraph. It is widely available as freeware; see the Technical Notes for information about downloading and installing it on your PC. We are very grateful to Alexander Minza, FNGraph's creator, for permission to use it in this textbook.

FNGraph files have the filename extension `.fng`. Each file contains a single graph. Once you have installed FNGraph, go to the textbook website and download the FNGraph files for Chapter 6; these are named `cm6-1.fng`, `cm6-2.fng`, `cm6-3.fng`, and so on. Open `cm6-1.fng`, and you will see a graph of a function we have built up from the library we are going to describe, and which mimics the 'greenhouse effect' temperature graph in Figure 6.1. Left-click anywhere on the window, hold and drag the cursor, and you can move the graph around. You use the + and − keys to zoom in or out. In the text, we will refer to the file `cm6-1.fng` as FNGRAPH 6.1, in the same way as we refer to the EXCEL SPREADSHEET 6.1 in file `cm-excel6.xls` as SPREADSHEET 6.1.

Creating the graph of your own function:

- Choose `File` > `New` from the menu, and you will see a blank graph with the x-axis and y-axis.
- In the top toolbar, click on `View` and choose the option `Graphs`.
- A window will appear, with a button labelled `Add`; click on this. An 'Add graph' window appears, with a blank box for the function.
- Type in your function as an algebraic expression using the variable x, in the same format as with Excel formulae: multiplication and division are indicated by * and /, and exponentiation (raising to powers) by ^. For example, to graph the function $y = 4x^3 − 2$ you type '4*x*x*x-2' or '4*x^3-2'. You don't type the = sign, as you do to enter a formula in Excel.
- Click OK and the graph appears as a blue curve, as shown in Figure 6.2.
- Save your graph using `File` > `Save`.

You can also use parameters in defining your functions:

- From the `View` option on the top toolbar, choose `Variables`.
- A new window appears. Click the `Add` button in it and set up a parameter with name a and value 4.
- Click `Add` again and set up a second parameter named b with value 3.
- Now go back to the `Graphs` window, click `Add` and create a new function '$a/x + b$'. By clicking the checkboxes against the functions, you can decide which curves are displayed.
- You can also edit the functions, and change the colours by selecting the function and clicking the `Properties` button.

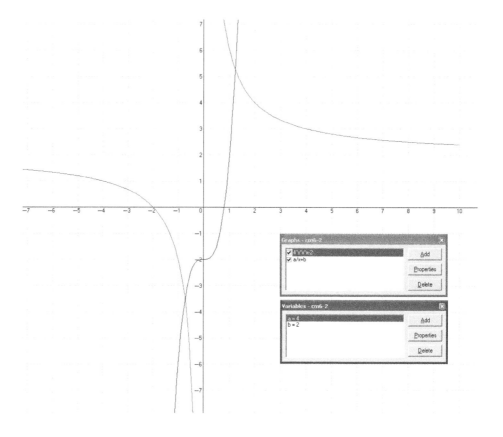

Figure 6.2 FNGraph sketching example

Check if the FNGraph file you have produced is the same as the file FNGRAPH 6.2 from our website. Now try creating your own graph showing the direct and inverse proportion functions, with the constant of proportionality k as a parameter.

Note: In FNGraph the independent variable is always x. Other names can be used for parameters, as with a and b above. FNGraph wrongly refers to these as 'variables'.

I claimed above that by scaling and combining functions from the library, we can generate functions with almost any desired properties. Open up FNGRAPH 6.1 again, right-click and choose Graphs. The function is written as

```
a*sqrt(x)*cos(c*x) + d*exp(k*x)
```

This represents the function

$$f(x) = a\sqrt{x}\cos(cx) + de^{kx}.$$

The notation, e.g. writing sqrt(x) for $\sqrt{x}$, is exactly how functions are programmed in Excel. So this function has been produced by combining a square root function, a cosine function, and an exponential function, together with appropriate scalings. We'll describe these in Chapters 8, 9, and 10, respectively.

As we work through the function library, each function will be illustrated by a FNGraph graph. You can download the FNGraph files from our website, and you should also try to construct each graph yourself. Then play around with the function by changing values of the parameters, and seeing the effect on the graph.

6.2 Constant functions $y = c$

In Chapter 5 we said that you could think of a function $f(x)$ as a 'black box': you type in a value of the (independent) input variable x on a keypad, and out comes the corresponding value of the (dependent) output variable y on a screen. With linear functions, there is a very clear link between the size of the input and the size of the output.

We first consider a function where, whatever value you type in as input, you always get the same output value. That is, y stays constant whatever the value of x. We could write this as $y = c$ or $f(x) = c$ where c is a constant. For example, $y = 5$ means that y has the value 5 whatever the value of x.

The graph of this function $y = c$, which we could also write as $f(x) = c$, is a horizontal line cutting the y-axis at c: whatever value of x you choose on the x-axis, if you move vertically up (or down) to the line, then horizontally across to the y-axis, you reach the value c. The point where a graph cuts the y-axis is called the **y-intercept**. The graph will never cross the x-axis (unless $c = 0$, in which case it lies along the x-axis and its equation is then $y = 0$).

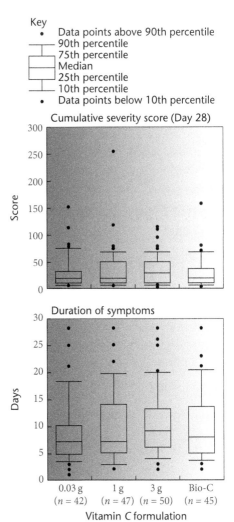

Figure 6.3 Results from trial of effectiveness of Vitamin C in treating colds (from Audera *et al.* 2001)

Create the graph of $y = c$ in FNGraph. Define a parameter c with the value 5 in the `Variables` window. Then, in the `Graphs` window, define a function which is simply c. (Notice that you will get an error message if you try to use the parameter c in a function definition before you have given it a value.) Try changing the value of c and see the effect on the graph.

You might think that graphs of constant functions are of no use to scientists. On the contrary, when we conduct an experiment and obtain such a graph, it tells us that the input variable had no effect on the output, which can be important, e.g. in showing that a drug is not effective at treating a particular illness.

> ### EXAMPLE
>
> In a 1999 double-blind trial of the effectiveness of daily doses of vitamin C in treating the common cold, staff and student volunteers at the Australian National University were given tablets and told to start taking them if they experienced symptoms of a cold. The tablets contained either 0.03 g (placebo), 1 g, 3 g, or 3-grams-plus-additives (Bio-C) of Vitamin C. The volunteers then recorded how long their symptoms lasted, and how severe they became. The results are summarized in Figure 6.3,[1] from Audera et al. (2001). If the median values were plotted on a graph against dosage amount, and a best-fit straight line fitted through the points, it would be almost horizontal in each case, indicating that the Vitamin C had no effect on the duration or severity of the symptoms.

. . . .

6.3 Linear functions

Linear functions are functions whose graphs are straight lines. We have already met linear functions. In Section 3.3 we drew a straight line through a set of data-points, and saw how to calculate its slope. Look back at this section, since we need to talk about the slope here. We will now discuss straight-line graphs from a function point of view.

6.3.1 The identity function $y = x$

The simplest relationship in which y actually changes as x changes, would be $y = x$, which we could also write as $f(x) = x$.

Figure 6.4(a) shows the graph of the identity function. The line now goes through the origin $(0, 0)$, and the points $(-1, -1)$, $(1, 1)$, $(2, 2)$, etc. It thus slants upwards at 45° (if you are using the same scale for both x- and y-axes). If you choose a value on the x-axis, go up to the line then across to the y-axis, you obtain the same value there; this is shown for the value 2 in Figure 6.4(a). At the origin, $y = x = 0$, so the line passes through this point.

The slope of this line is 1. If we constructed a right-angled triangle under the line, as in Figure 3.4(a) the side-lengths Δx and Δy would be equal, so the slope $m = \dfrac{\Delta y}{\Delta x} = 1$.

We have seen the usefulness of the line $y = x$ in drawing the 'cobweb diagram' for the population update equation in Case Study A6, and it will also be used to visualize the graph of an inverse function, in Section 8.5.

---

[1] From http://www.mja.com.au/public/issues/175_07_011001/audera/audera.html.

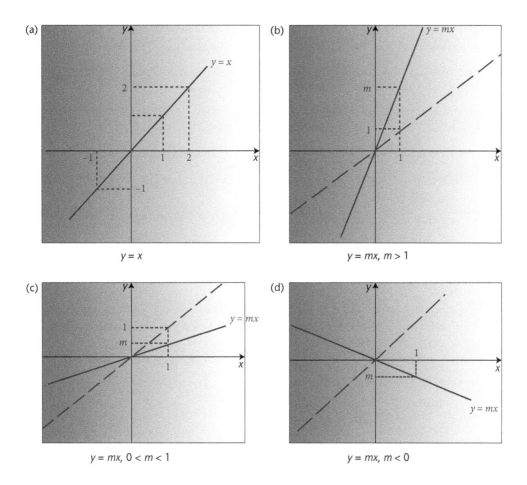

Figure 6.4 Proportionality functions

6.3.2 **Proportionality functions** $y = mx$

The identity relationship is important in 'cobwebbing' or drawing an inverse function, but of little practical use as a function. A much more common type of relationship is where y is a fixed multiple of x. For example, in the function $y = 3x$ the output value y will be 3 times as big as the input value x; the points $(1, 3)$, $(2, 6)$, $(-2, -6)$ will lie on this line. In general, if y is proportional to x, we can write the **proportionality function** as $y = mx$ or $f(x) = mx$; m is the **constant of proportionality** (and is also the slope of the line, which we saw how to calculate in Section 3.3.3). Modify the previous FNGraph by defining a parameter m with the value 3, and defining the function as m*x.

The line still passes through the origin, as with $y = x$, but the slope has increased so that when you choose 2 on the x-axis and go up to the line and then across, you reach 6 on the y-axis. So when m is greater than 1, the line slopes up more steeply (Figure 6.4(b)). Try changing the value of m in the Variables window, and see how the slope of the line changes. If you set $m = 0.5$, the line still slopes up, but less steeply than $y = x$ (Figure 6.4(c)). If you make m negative, the line slopes downwards (Figure 6.4(d)). The graphs of proportionality functions $y = mx$ always pass through the origin (in other words, the y-intercept is 0), because when x is zero then $y = m.0 = 0$ also.

Relationships of direct proportionality occur in all branches of science. There are direct proportion relationships between physical quantities, for example:

- For an object travelling with constant speed, distance travelled is proportional to elapsed time. The constant of proportionality is the speed.

- For different quantities of water in test tubes, the mass of the water is proportional to its volume. The constant of proportionality is the density.

Proportionality also occurs in conversion factors between different units:

- The length of an object in millimetres is proportional to its length in metres. The constant of proportionality is 1000.

- The length of an object in centimetres is proportional to its length in inches. The constant of proportionality is approximately 2.54, since 1 inch = 2.54 cm.

EXAMPLE

When they are chasing food or escaping enemies, some sharks can swim very fast, up to forty miles per hour. One mile is equivalent to 1.61 km, so we could write (length in km) = 1.61 × (length in miles).

Thus if (length in miles) is 40, the equivalent length in km is 1.61 × 40 = 64.4 km.

So a speed of 40 miles per hour is equivalent to 64.4 km/hour. And 64.4 km/hour = 64400 m/hour = $\frac{64400}{3600}$ or 17.9 m/second.

At this speed, in 5 seconds a shark could cover a distance of 17.9 × 5 = 89.4 metres. In general, (distance in metres) = 17.9 × (time in seconds) at this speed.

To reach a fish which is 100 metres away, the shark would take $\frac{100}{17.9}$ = 5.6 seconds.

Direct proportion crops up in many physical situations, in all sorts of fields. In Section 3.3 we cited the example of load being proportional to extension for a stretched elastic skin sample. Some other examples, taken from the online 'Encyclopedia of Life Sciences'[2] include:

- In the lungs of mammals, the amount of oxygen that is actually utilized is directly proportional to the difference in partial pressures between the air breathed in and the carbon dioxide breathed out.

- The species diversity in a region is directly proportional to the size of available habitat. So when a tropical forest is fragmented by cutting down swathes for timber extraction, species of primate in the separated fragments will die out.

- When a moustached bat hunts, it emits ultrasonic calls towards its prey, and its ears detect the echo. The bat then determines the distance to its prey by using the fact that this distance will be proportional to the call–echo delay time.

- The calcium ion in cells is a messenger which controls processes such as gene transcription and muscle contraction. When hepatocytes (cells in the liver) are stimulated with hormones, increased calcium levels occur in pulses rather than a sustained increase (which would be toxic). The frequency of the pulses is directly proportional to the concentration of hormone applied.

[2] http://www.els.net.

CASE STUDY C4: Equations of motion 1

From the schoolboy formula $\text{speed} = \dfrac{\text{distance}}{\text{time}}$, we can deduce the equation expressing distance travelled s as a function of elapsed time t, for an animal travelling at constant velocity v:

$$s = v.t$$

expressing the distance $s(t)$ as a function of t. Here, v is the constant of proportionality, i.e. the rate of change of distance with time.

If the velocity is not constant, it means the animal is accelerating or decelerating (speeding up or slowing down). Acceleration is the rate of change of velocity with time. If the acceleration a is constant, we can write

$$v = a.t$$

to express the velocity $v(t)$ as a function of t.

Dimensionally, v is in units of m s^{-1}, so the dimension of acceleration is m s^{-2} (metres per second squared).

EXAMPLE

A lion can accelerate from rest at 6.25 m s^{-2}. How long will it take him to reach his maximum velocity of 14 m s^{-1}?

Solution: Setting $a = 6.25$ and $v = 14$, we find $t = \dfrac{v}{a} = \dfrac{14}{6.25} = 2.24$ seconds.

Note: The equation $s = v.t$ applies when v is constant, i.e. acceleration a is zero. The equation $v = a.t$ applies when a is constant and v is changing.

Do not try to use both equations in the same situation!

6.3.3 General linear functions $y = mx + c$

We already met this idea of linearity in Section 5.4, where we considered linear relations between x and y, which we wrote as $ax + by = c$. We also saw how this equation could be manipulated to express y as a linear function of x, namely:

$$y = -\frac{a}{b}x + \frac{c}{b}. \tag{6.1}$$

That is, y is equal to a multiple of x plus a constant term. This is equivalent to the equation for straight-line graphs, which we saw in Section 3.3. Thus, the general form of this function is

$$y = mx + c, \tag{6.2}$$

where m and c are constants. Both of these constants still have the geometric interpretations we saw above:

- m is the slope of the line;
- c is the y-intercept (the point at which the line crosses the y-axis).

Use FNGraph to see that the graph of $y = mx + c$ is again a straight line. In the `Variables` window you should have the parameters m and c defined, with values 5 and 3, respectively.

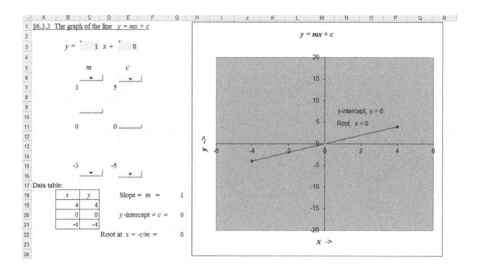

Figure 6.5 The graph of $y = mx + c$ (SPREADSHEET 6.1)

Now in the Graphs window, click Add and define a function m*x + c. The file FNGRAPH 6.3 is programmed with the parameters m and c, and the graphs of $y = c$, $y = x$, $y = mx$, and $y = mx + c$.

The graph of $y = mx + c$ is also drawn in an Excel chart, in SPREADSHEET 6.1, which is shown in Figure 6.5. You can adjust the values of the slope m and the y-intercept c using the vertical scrollbars,[3] which we'll call sliders in this book, on the sheet. Click on the slider button and drag it up or down. You can also click on the buttons at top and bottom of the slider, to adjust the values by small increments.

The graph of $y = 3x + 5$ has the same slope as that of $y = 3x$, but has been moved up by 5 units so that it crosses the y-axis at $y = 5$. See Figure 6.6(a).

In proportionality functions, the value of y is proportional to the value of x. An important generalization of this is that in linear functions **a change in x produces a proportionate change in y**. This is an important property of many real-life relationships.

To see this, consider $y = mx + c$. Remember from Section 3.3.3 that we use Δx and Δy to represent the distances in the x and y directions in moving from (x_1, y_1) to (x_2, y_2). Then y increases from $y_1 = mx_1 + c$ to $y_2 = mx_2 + c$. The change in y is therefore

$$\begin{aligned} \Delta y = y_2 - y_1 &= (mx_2 + c) - (mx_1 + c) \\ &= mx_2 + c - mx_1 - c \\ &= mx_2 - mx_1 \\ &= m(x_2 - x_1) = m.\Delta x. \end{aligned}$$

(6.3)

Try changing the values of m and c in SPREADSHEET 6.1, and see the effect. Changing the value of m changes the slope. Changing the value of c moves the line up or down. The line $y = mx + c$ crosses the y-axis at $y = c$ (in other words, the y-intercept is c), since, when $x = 0$, then $y = m.0 + c = 0 + c = c$.

---

[3] Note: When you try to open a spreadsheet that contains controls such as sliders, you may get a security warning message about macros. Try clicking the 'Enable macros' button. If the spreadsheet still doesn't open, see the Technical Notes at the front of this book.

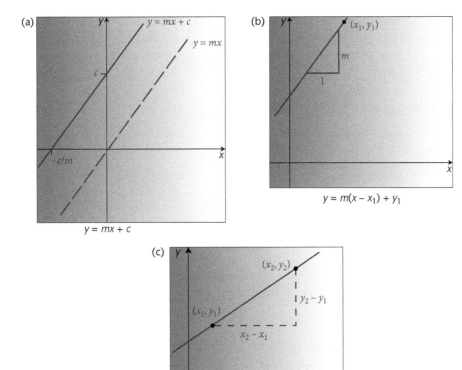

Figure 6.6 General linear functions

Where does the line cross the x-axis? We need to substitute $y = 0$ and then solve for x:

$$mx + c = 0$$

$$mx = -c$$

$$x = \frac{-c}{m}.$$

This crossing-point is also shown on Figure 6.6(a). It is called the **root** of the function $f(x) = mx + c$.

A **root** of the function $f(x)$ is a value of x at which $f(x) = 0$.   (6.4)

EXAMPLE

Describe the graph of $y = 4x - 9$.

Solution: The function is linear, with $m = 4$ and $c = -9$, so its graph will be a straight line with slope 4 (positive slope, so it is sloping upwards), and crossing the y-axis at $y = -9$. It will cross the x-axis when $4x - 9 = 0$, i.e. when $x = 9/4 = 2.25$.

You only need to locate two points in order to draw the graph of a linear function. If you are given the slope m and the y-intercept c, you immediately know one point: the point $(0, c)$ on the y-axis. To find a second point, simply choose a convenient value of x, x_0 say, substitute this into the equation $y = mx + c$ and calculate y_0; then mark the point (x_0, y_0) on the graph, and draw a line passing through the two points. A neat choice of x_0 would be $x_0 = -c/m$, the point at which $y = 0$.

But what if you are given different information, and asked to find the equation of the line? There are two common situations:

(1) If you know the slope m and one point (x_1, y_1), you can find c by solving the equation $y_1 = mx_1 + c$ for c: $c = y_1 - mx_1$. Now put this into $y = mx + c$: $y = mx + (y_1 - mx_1)$, which we can simplify to

$$y = m(x - x_1) + y_1. \tag{6.5}$$

See Figure 6.6(b). You can either try to memorize this formula, or you can remember the algorithm for finding c: Substitute (x_1, y_1) into $y = mx + c$ and solve for c: $c = y_1 - mx_1$. Then draw the graph of $y = mx + c$.

We can draw the graph of the function as a straight line through the two points (x_1, y_1) and $(0, c)$. The second point is the y-intercept, on the y-axis.

EXAMPLE

A tree grows at a constant rate of 0.7 metres per year. Three years after it was planted, its height h is 4.5 m. What was its height when it was planted, and how tall will it be after a further two years of growth?

Solution: We can write the height h as a linear function of time t: $h = mt + c$.

Here, the slope $m = 0.7$, and we know the data-point $(3, 4.5)$. Using (6.5), the height h (in metres) as a function of time t (in years) is therefore

$$h = m(t - t_1) + h_1 = 0.7(t - 3) + 4.5 = 0.7t - 2.1 + 4.5 = 0.7t + 2.4.$$

The data, and equation (6.5), are used to draw the graph, in FNGRAPH 6.4.

The constant term is $c = 2.4$, which is the initial height at $t = 0$. After a further two years, i.e. when $t = 5$, the height will be $0.7 \times 5 + 2.4 = 3.5 + 2.4 = 5.9$ metres.

Alternatively, we could have calculated $c = h_1 - m t_1 = 4.5 - 0.7 \times 3 = 4.5 - 2.1 = 2.4$. Then we could draw the line through $(3, 4.5)$ and $(0, 2.4)$. This is illustrated in SPREADSHEET 6.2.

(2) If you know two points (x_1, y_1) and (x_2, y_2) that lie on the line, we can find the slope using equation (6.3) above:

$$m = \frac{\Delta y}{\Delta x} = \frac{y_2 - y_1}{x_2 - x_1}. \tag{6.6}$$

Since the line passes through (x_1, y_1), we can substitute this into (6.5) and simplify:

$$y = \frac{y_2 - y_1}{x_2 - x_1}(x - x_1) + y_1 \tag{6.7}$$

or

$$y - y_1 = \frac{y_2 - y_1}{x_2 - x_1}(x - x_1). \tag{6.8}$$

See Figure 6.6(c). Again, you can memorize this formula, or you can remember an algorithm for finding m and c: Substituting each point's coordinates into (6.2), we obtain

$$y_1 = mx_1 + c$$
$$y_2 = mx_2 + c$$

which we can write as two simultaneous equations in m and c:

$$x_1 m + c = y_1$$
$$x_2 m + c = y_2.$$

Use the technique in Section 5.7 to solve them. To eliminate c, subtract the first equation from the second:

$$(x_2 m + c) - (x_1 m + c) = y_2 - y_1$$
$$x_2 m + c - x_1 m - c = y_2 - y_1$$
$$x_2 m - x_1 m = y_2 - y_1$$
$$(x_2 - x_1)m = y_2 - y_1$$
$$m = \frac{y_2 - y_1}{x_2 - x_1}.$$

Finally, find c by $c = y_1 - mx_1$, as in the previous situation.

EXAMPLE

Three years after it was planted, a tree's height h is 4.5 m. After a further two years of growth, its height is 6.8 m. What was its height when it was planted, and how tall will it be after ten years of growth? Assume a constant rate of growth.

Solution: Here, we know the two data-points (3, 4.5) and (5, 6.8). Inserting in equation (6.8):

$$h - h_1 = \frac{h_2 - h_1}{t_2 - t_1}(t - t_1)$$
$$h - 4.5 = \frac{6.8 - 4.5}{5 - 3}(t - 3)$$
$$h - 4.5 = \frac{2.3}{2}(t - 3)$$
$$h - 4.5 = 1.15(t - 3)$$
$$h - 4.5 = 1.15t - 3.45$$
$$h = 1.15t - 3.45 + 4.5$$
$$h = 1.15t + 1.05.$$

So the slope $m = 1.15$, and $c = 1.05$. When it was planted, the tree was 1.05 metres tall. After 10 years of growth, the tree's height will be $1.15 \times 10 + 1.05 = 12.55$ m.

The data, and equation (6.8), are used to draw the graph, in FNGRAPH 6.5.

The constant term is $c = 2.4$, which is the height at $t = 0$. After a further two years, i.e. when $t = 5$, the height will be $0.7 \times 5 + 2.4 = 3.5 + 2.4 = 5.9$ metres.

Alternatively, we have two simultaneous equations:

$$3m + c = 4.5$$
$$5m + c = 6.8.$$

Subtracting the first from the second, we get $5m - 3m = 6.8 - 4.5 = 2.3$, so $m = 2.3/2 = 1.15$.

Then $c = h_1 - mt_1 = 4.5 - 1.15 \times 3 = 4.5 - 3.45 = 1.05$. Then we could draw the line through $(3, 4.5)$ and $(5, 6.8)$. This is illustrated in SPREADSHEET 6.3.

CASE STUDY C5: A constant-velocity chase

A lion is running at 14 m s^{-1}, chasing a young zebra. The zebra is 30 m ahead, and only running at a speed of 10 m s^{-1}. However, the lion can only maintain its speed for a further 6 seconds; if it has not caught the zebra by that time, it will have to give up. Will the zebra escape?

As the animals are travelling at constant velocity, we use the equation $s = v.t$, which we saw in the previous instalment, and write two equations of motion: one for the lion and one for the zebra. We first need to choose the location of the origin, from which we will measure distances s, and the point in time from which to measure time t. We take the origin to be the spot where the lion is, at the time in the question, which will be $t = 0$.

So, for the lion, $s(0) = 0$ and its speed $v = 14$.

For the zebra, which is 30 m ahead, $s(0) = 30$ and $v = 10$.

The equation of motion for the lion is $s_1 = 14t$.

For the zebra, we must generalize the equation by adding an initial distance from the origin, which in this case is 30 metres. The zebra's equation of motion is thus $s_2 = 10t + 30$. The lion would catch the zebra when $s_1 = s_2$. That is,

$$14t = 10t + 30.$$

Solving for t, we find that $4t = 30$, or $t = 7.5$ seconds. As the lion has to give up after 6 seconds, it will not catch the zebra.

In fact, the maximum distance d which the lion would need to be from the zebra at time $t = 0$ in order to catch it, is given by

$$14t = 10t + d \text{ when } t = 6,$$

from which $d = 24$ metres.

Summary: Linear function: $y = mx + c$

The graph is a straight line, with slope m, crossing the y-axis at $y = c$ (the y-intercept).

It crosses the x-axis at $x = -\dfrac{c}{m}$ (the root).

Other forms of the equation:

$$y = m(x - x_0) + y_0$$

$$y - y_0 = \frac{y_1 - y_0}{x_1 - x_0}(x - x_0).$$

6.3.4 Trend lines revisited: goodness of fit

If you have only two data-points, you can draw a unique straight line through them. But in most experiments or observations you obtain more than two data-points, and they won't all lie on a single straight line. Resist the temptation to slightly 'adjust' some of the values until they do so! Small random errors are inevitable. You may be justified in fitting a 'least squares' or 'best-fit' straight line (called a Trend Line in Excel) to your data-points, but there are some things you should check when doing this.

In Section 3.3.3 we explained how to fit a 'best-fit' trend line through the points, in an Excel chart, and to find its equation:

- Type the data-point values into a table in an Excel worksheet.
- With the data values highlighted, click on `Insert` on the top toolbar, and choose `Chart...` from the dropdown menu.
- Select the chart type 'XY (Scatter)' and sub-type 'Scatter'.
- To draw a 'best-fit' straight line through the points, highlight the `Scatter` chart. On the top toolbar there should appear a new option called `Chart`. Click on this and select `Add Trendline...` Choose the `Linear` option.
- You can obtain the line's equation, with the values of the parameters m and c, by right-clicking on the line. Choose the menu item 'Format Trendline...' In the window that appears, click the Options tab, and check the box 'Display equation on chart'. The equation now appears on the chart.

There is a practical question you should ask yourself before fitting the trend line: 'Am I convinced that the relationship revealed by the data-points is a linear one (with some random experimental errors), or does it really represent a nonlinear relationship, which would be better represented by fitting a curve?' We will deal with fitting curves to data in Section 8.3. You were prompted to think about this in problem 3.21.

Suppose you are satisfied that a linear relationship is appropriate, and you have fitted the trend line and noted its equation (see the example in SPREADSHEET 3.4). You should now ask yourself 'How good a fit is this trend line to the data?' Partly this is a question of whether the data-points all lie close to the line. It often happens that one or two points are **outliers**: they lie a relatively long way from the line, and from the other points. If there are outliers, it doesn't necessarily mean that you made a mistake when reading or recording the values. If we were plotting a class of students' marks in maths against their marks in chemistry, there should be in general a linear relationship with positive slope, but there may be one exceptional student who has a natural talent for maths but who cannot remember the first thing about chemistry. His data-point will be a long way from the others, and this means it will have a big influence on the trend line. It may be appropriate to remove him from the data-set. The question of what to do about outliers is discussed in statistics textbooks.

Ideally, all the data-points should lie reasonably close to the line. They should also be scattered evenly along the line, as in Figure 3.13, so that we are not proposing a linear relationship in an area of the graph where there is little or no evidence of one. Now look at the graphs in Figure 3.14. In both cases there is a mass of points showing no coherent relationship at all, plus one or two points in a faraway part of the graph, which have really determined the location of the trend line.

There is a statistical measure of how closely the trend line fits the data – sometimes called the 'goodness of fit'. It is denoted by R^2 and is usually called 'the R^2 value', though its proper name is the **coefficient of determination**. You can obtain this value on your graph, in the same way as you obtained the equation in the final step above. In the `Format Trendline`, `Options` tab, check the box `Display R-squared value on chart`. R^2 is a number between 0 and 1. The closer it is to 1, the better the fit. If you display R^2 in the chart on SPREADSHEET 3.4, you find it is 0.9685, indicating a good fit to the data. Try displaying R^2 for the trend lines you drew in problems 3.1–3.6.

We will show how the coefficients m and c in the equation of the trend line are calculated, in Section 8.7. The process of calculating this equation is called **linear regression**.

Problems: Now try the end-of-chapter problems 6.1–6.24.

. . . .

6.4 Limits and asymptotes

To sketch the curve of a given function $f(x)$, we find out where it crosses the x- or y-axes, where it has turning-points – and if necessary we can evaluate it at particular values of x to find points on the curve. This information will tell us what the curve looks like within the central graph area itself. But we may also need to know how the function behaves as it approaches the edges of the graph. An **asymptote** is a line we can draw on the graph, which the curve approaches but never actually meets or crosses as it heads out beyond the edges of the graph. One way of finding an asymptote is to work out the **limit** of the function $f(x)$ as x approaches a particular value, or as x heads off to infinity.

Polynomial functions (including linear functions) do not have asymptotes. But consider the function $f(x) = \dfrac{1}{x}$. If we make x larger and larger, then $\dfrac{1}{x}$ becomes smaller and smaller, and closer and closer to zero. But however large a value of x you choose, $\dfrac{1}{x}$ will never actually equal zero.

6.4.1 Limits of functions

Consider again the function $f(x) = \dfrac{1}{x}$, where $f(x)$ becomes closer and closer to zero as we make x larger and larger. In fact, we can make $f(x)$ as close to zero as we wish, by taking a sufficiently large value of x. For example, if we want $f(x)$ to be smaller than 10^{-6}, we would take a value of x which is greater than 1,000,000. The concept of **infinity**, whose mathematical symbol is ∞, refers to the far right-hand end of the real line. Note that ∞ is a concept not a number; we can't do arithmetic with ∞. Instead of talking about 'x getting larger and larger', we can now say 'x approaches infinity', and we write this mathematically as '$x \to \infty$'. We can then write '$f(x) \to 0$ as $x \to \infty$'. We say that 0 is the **limit** of $f(x)$ as x approaches infinity. The mathematical notation for a limit is 'lim', and we write the previous statement mathematically as

$$\lim_{x \to \infty} f(x) = 0$$

or, using the definition of $f(x)$:

$$\lim_{x \to \infty} \frac{1}{x} = 0. \tag{6.9}$$

Mathematicians tend to say 'tends to' rather than 'approaches': '$f(x)$ tends to 0 as x tends to ∞'. We may also say that $f(x)$ **converges to** the limit 0.

We can also let x approach the far left-hand end of the real line, which is 'minus infinity'; we write this as '$x \to -\infty$'. If we do this with the function $f(x) = \dfrac{1}{x}$, the values of $f(x)$ will be negative, but will still be getting smaller and smaller, so that $\lim\limits_{x \to -\infty} f(x) = 0$.

We could write $\dfrac{1}{x}$ using exponent notation as x^{-1}. The same behaviour for large x will also occur for the functions $\dfrac{1}{x^2}, \dfrac{1}{x^3}, \ldots$. In fact, we can write down the following limits, generalizing (6.9):

$$\text{For } k = 1, 2, 3, \ldots, \quad \lim_{x \to \infty} \left(x^{-k} \right) = 0 \text{ and } \lim_{x \to -\infty} \left(x^{-k} \right) = 0. \tag{6.10}$$

You can see these functions as the blue curve in FNGRAPH 6.6, shown in Figure 6.7. You can change the value of k in the Variables window.

6.4.2 Horizontal asymptotes

A horizontal asymptote of $f(x)$ is a horizontal line on the graph indicating a value of y which $f(x)$ approaches as $x \to \infty$ or $x \to -\infty$. Geometrically, the curve of $y = f(x)$ will approach this line as it passes across the left-hand or right-hand edge of the graph. Thus, the line $y = 0$, i.e. the x-axis, is a horizontal asymptote of the function $f(x) = \dfrac{1}{x}$, since $\lim\limits_{x \to \infty} \dfrac{1}{x} = 0$. From (6.10) we can conclude that the functions $\dfrac{1}{x^2}, \dfrac{1}{x^3}, \ldots$ will also have the horizontal asymptote $y = 0$. The graphs of these functions, which we could write as x^{-k}, for $k = 1, 2, 3, \ldots$, are in FNGRAPH 6.6.

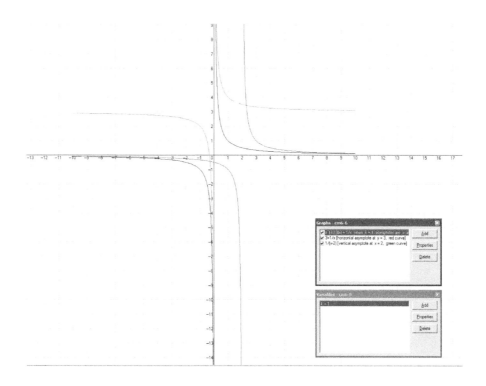

Figure 6.7 Horizontal and vertical asymptotes (FNGRAPH 6.6)

The function $g(x) = 3 + \dfrac{1}{x}$ will have its horizontal asymptote at $y = 3$, since

$$\lim_{x \to \infty}\left(3 + \frac{1}{x}\right) = 3 + \lim_{x \to \infty}\frac{1}{x} = 3 + 0 = 3.$$

The graph of $g(x)$ is the same as that of $f(x)$, but shifted vertically upwards by 3 units; see FNGRAPH 6.6.

6.4.3 Vertical asymptotes

A vertical asymptote of $f(x)$ is a vertical line on the graph indicating an isolated value of x for which no value of $f(x)$ exists. Geometrically, the curve of $y = f(x)$ will approach this line as it passes across the top or bottom edges of the graph. Vertical asymptotes occur at values of x that would involve dividing by zero in calculating $f(x)$. This would happen if we tried to substitute $x = 0$ into the function $f(x) = \dfrac{1}{x}$. Thus, the line $x = 0$, i.e. the y-axis, is a vertical asymptote of the function $f(x) = \dfrac{1}{x}$, and also of the functions $\dfrac{1}{x^2}, \dfrac{1}{x^3}, \ldots$ and of $g(x) = 3 + \dfrac{1}{x}$. See FNGRAPH 6.6 and 6.7.

The function $h(x) = \dfrac{1}{x - 2}$ will have a vertical asymptote at $x = 2$, since this value of x makes the denominator zero. If you evaluate $h(x)$ at a value of x very close to 2, you will be dividing by a very small denominator, which will make $h(x)$ very large and positive (if $x > 2$) or very large and negative (if $x < 2$). This is why the curve heads off the top or bottom of the graph close to a vertical asymptote. The graph of $h(x)$ is the same as that of $f(x)$, but shifted to the right by 2 units; see FNGRAPH 6.6.

Vertical asymptotes can occur when the function involves division by a polynomial in x. These are called rational functions, and we will look at them in Chapter 8. They can also occur with trigonometric functions, which we'll look at in Chapter 9. The tangent function, you may remember, is defined as the sine divided by the cosine, so it will have vertical asymptotes at values of the angle which make the cosine zero.

6.4.4 Limits of sequences

The concept of a limit can also be applied to sequences of numbers. In Case Study A we have been generating sequences of numbers, of the form

$$N_0, N_1, N_2, \ldots, N_p, \ldots$$

where N_p is the size of a population at the pth generation, starting from an initial population N_0. The simplest model was of geometric growth (see Study A1–A3), where the new population is defined from the previous one by the update equation

$$N_{p+1} = \lambda . N_p.$$

A sequence produced by such an update equation is called a **geometric sequence**. In mathematics, a general geometric sequence is defined using two parameters: the first term a and the common ratio r. Each successive term in the sequence is obtained by multiplying the previous term by r. The sequence is therefore

$$a, ar, ar^2, ar^3, \ldots, ar^n, \ldots \tag{6.11}$$

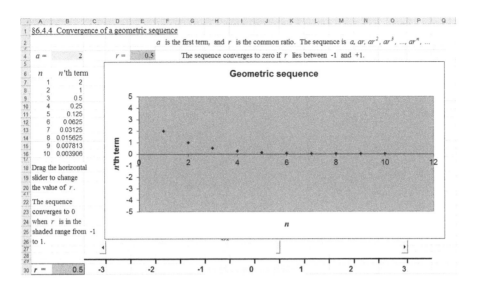

Figure 6.8 Convergence of a geometric sequence (SPREADSHEET 6.4)

The $(n + 1)$th term of the sequence is ar^n. What is the limit of this sequence, as $n \to \infty$? You can experiment with the sequence in SPREADSHEET 6.4 (see Figure 6.8). Type a value for a in cell B4, and vary the value of r by moving the horizontal slider; you can change r between −3 and 3. From the chart you can see which values of r produce a sequence whose values are **converging to zero**. You will find that this happens when r lies between −1 and +1 (this interval of the real line is shaded − cells H30 to K30). And this is true whatever positive or negative value of a you set. We could write this condition for convergence as $-1 < r < 1$, or as $|r| < 1$, meaning that the absolute value of r is less than 1. We can state this result as

$$\lim_{x \to \infty} \left(ar^n \right) = 0 \text{ provided } |r| < 1. \tag{6.12}$$

If the absolute value of r is greater than 1 (i.e. r is greater than 1 or less than −1), the terms of the sequence grow unboundedly. In Section 7.3.1 we will see that when $|r| < 1$ it is possible to calculate the sum of all the terms in the sequence.

Problems: Now try the end-of-chapter problems 6.25−6.33.

. . . .

6.5 Extension: linear transformations

We mentioned in Section 5.6.1 that the graph of the function $y = f(x − 1)$ will be the same as that of $y = f(x)$, but shifted horizontally by one unit to the right, and we saw further examples of shifted graphs in the previous section. There are other ways of transforming the basic function $f(x)$, which have the effect of stretching or squashing, flipping, or shifting the graph. The transformations involve multiplying x or $f(x)$ by a coefficient, or adding a constant. They are called **linear transformations**, because these are exactly the operations involved in forming the general linear function $mx + c$ from x. They are important because each type of transformation can be applied to a general function $f(x)$, and will always have the same effect

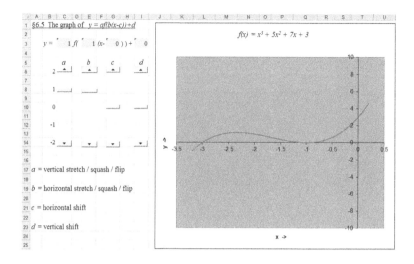

Figure 6.9 The graph of $y = x^3 + 5x^2 + 7x + 3$ (SPREADSHEET 6.5)

on its curve. We will be using them to transform the polynomial, trigonometric, and exponential functions we will study in the next few chapters, so it is worth summarizing them here. We will illustrate them by applying them to the cubic function:

$$f(x) = x^3 + 5x^2 + 7x + 3 \tag{6.13}$$

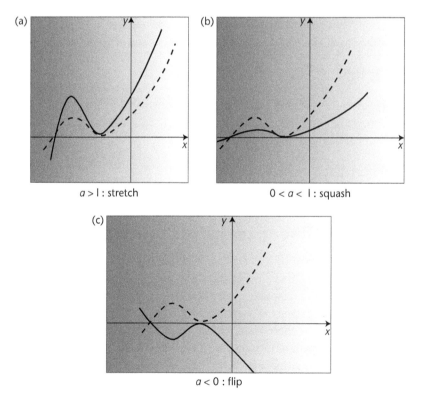

Figure 6.10 Vertical stretch/squash/flip: $y = a.f(x)$

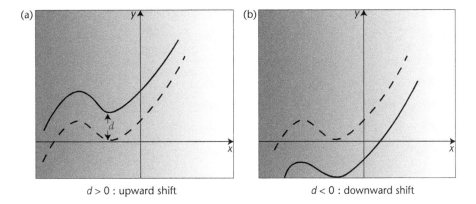

Figure 6.11 Vertical shift: $y = f(x) + d$

whose graph is drawn in FNGRAPH 6.7 and in SPREADSHEET 6.5 (see Figure 6.9). On the spreadsheet there are sliders which you can use to adjust the values of the parameters a, b, c, d, and see the effects which are described in the rest of this section. There will be further examples, starting in Section 7.1, so we suggest you use the current section as reference, and look back at it as you see examples of linear transformations in Chapters 7–10.

6.5.1 Vertical stretch/squash/flip: $y = a.f(x)$

If we multiply $f(x)$ by a coefficient a, we are scaling the y-values of the curve. This stretches or squashes it in the y-direction. When $a > 1$ we stretch the curve away from the x-axis, and when $a < 1$ we squash it towards the x-axis. If a is negative, then positive y-values become negative and vice versa, meaning that the curve is flipped vertically about the x-axis. The flip will also involve a stretch if $a < -1$, or a squash if $-1 < a < 0$. These effects are illustrated in Figure 6.10.

6.5.2 Vertical shift: $y = f(x) + d$

Adding a positive constant d to the y-values, shifts the whole curve vertically upwards by a distance of d units. If d is negative the curve is shifted downwards. See Figure 6.11.

6.5.3 Horizontal stretch/squash/flip: $y = f(bx)$

If we replace x by $b.x$ wherever it occurs in the formula for $f(x)$, we are scaling in the x-direction. When $b > 1$ we squash the curve towards the y-axis, and when $0 < b < 1$ we stretch it away from the y-axis. If b is negative, then positive x-values become negative and vice versa, meaning that the curve is flipped horizontally about the y-axis. The flip will also involve a squash if $b < -1$, or a stretch if $-1 < b < 0$. These effects are illustrated in Figure 6.12.

6.5.4 Horizontal shift: $y = f(x - c)$

Subtracting a positive constant c from the x-values, shifts the whole curve horizontally to the right by a distance of c units. If c is negative the curve is shifted to the left. See Figure 6.13.

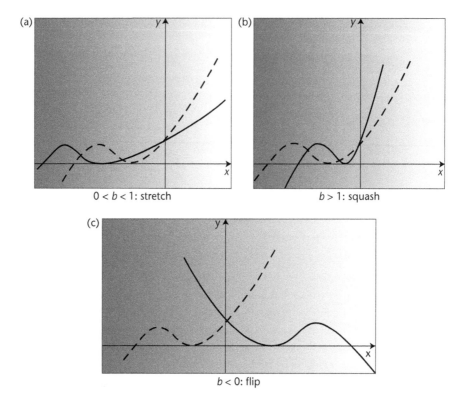

Figure 6.12 Horizontal stretch/squash/flip: $y = f(b.x)$

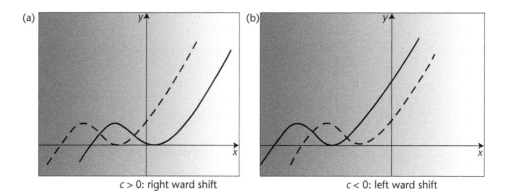

Figure 6.13 Horizontal shift: $y = f(x - c)$

These individual transformations can be combined; experiment with this using the sliders in SPREADSHEET 6.5. By this means a standard curve can be distorted and shifted to produce the best fit to the set of data-points we have obtained from experiment. You can see examples in subsequent chapters; the next chapter starts by applying these transformations to a simple 'squaring' function.

6.1–6.6: Draw the graphs of the following functions on the same set of axes. In each case, evaluate two points on the graph and join them with a straight line. Use the same scales for the x- and y-axes. Check your answers by drawing the graphs in FNGraph.

6.1 $y = -3$ 6.2 $y = 5x$

6.3 $y = -7x$ 6.4 $y = 5x + 1$

6.5 $y = -5x - 2$ 6.6 $y = -0.2x + 3$.

6.7–6.10: Describe the graphs of the following linear functions:

- Does the line slope upwards or downwards?
- Where does it cross the y-axis?
- Where does it cross the x-axis?

6.7 $y = 5x + 2$ 6.8 $y = 5x - 4$

6.9 $y = 3 - x$ 6.10 $y = -25x$.

6.11–6.14: Find the y-intercepts of the following graphs, and hence find the equation of the graph.

6.11 slope 7, passing through (1, 10)

6.12 slope –2, passing through (2, 1)

6.13 slope 0.3, passing through (–3, 1.6)

6.14 slope –1/3, passing through (–9, –2).

6.15–6.18: Find the equations of the lines passing through the points:

6.15 (1, 2) and (3, 4)   6.16 (0, 0) and (5, –7)

6.17 (1, –2) and (–2, –4)   6.18 (–1, 0) and (2.5, 4.5).

6.19: Americans (and the older generation in the UK) use the Fahrenheit system for measuring temperature. In this, water freezes at 32 °F and boils at 212 °F. Produce a linear function $y = f(T)$ that converts a temperature T °C to y °F. Draw the function in FNGraph.

What are 0 °F and 100 °F in Celsius?

Is there any temperature which has the same numerical value in both Celsius and Fahrenheit? If so, calculate it, and also find it graphically by drawing the line $y = T$ on your FNGraph plot.

6.20: In a maths examination, the teacher forgot to give out a formula sheet to the students. This disadvantaged the weaker students more than the top students (who knew all the formulas by heart anyway). The teacher decides to scale up the marks, and wants to increase the marks of the weaker students by more than the marks of the better students. He applies a rotation scaling; this is a linear transformation, in which a mark of 100% stays at 100%, but he decides that a raw mark of 30% should be increased to 40%. Produce a linear function $y = f(x)$, where a raw mark of x% is increased to y%, which satisfies these requirements.

What is a raw mark of 50% increased to? What raw mark results in a scaled mark of 50%? Give your answers to the nearest %.

Write an equation which would enable a student to work out his or her raw mark, knowing their scaled mark.

Produce a general function for a rotation scaling where 100% stays at 100% but a raw mark of x_0% is increased to y_0%.

6.21–6.24: On a graph with the same scale for x- and y-axes, two lines with slopes m_1 and m_2 will be perpendicular (i.e. they meet at an angle of 90°), if $m_1 m_2 = -1$. Using this fact, find the equation of the line that is perpendicular to the given line, and passes through the given point. Draw both lines in FNGraph.

   6.21 perpendicular to $y = 5x + 2$, and passing through $(1, 7)$

   6.22 perpendicular to $y = -2x + 5$, and passing through $(3, -1)$

   6.23 perpendicular to $y = 0.1x + 2.5$, and passing through $(0, 0)$

   6.24 perpendicular to $y = -x - 2.8$, and passing through $(4, 0)$.

6.25–6.28: Find the vertical and horizontal asymptotes, and hence sketch the curves. Check your graphs by sketching the functions in FNGraph:

   6.25 $y = \dfrac{3}{x} - 5$ 6.26 $y = 3 + \dfrac{1}{x-2}$

   6.27 $y = \dfrac{1}{x} + \dfrac{1}{x-2}$ 6.28 $y = 3 + \dfrac{1}{x^2 - 4}$.

6.29–6.33: Rearrange the equations to find y as a function of x, then find the asymptotes (if any) and sketch the curves. Check your graphs using FNGraph:

   6.29 $x - 4y = 6$ 6.30 $1 - 4xy = 6$ 6.31 $x - 4xy = 6$

   6.32 $y - 4xy = 6$ 6.33 $y - 4x^2 y = 6$.

6.34–6.37: Sketch the graph of $y = f(x)$ where $f(x) = \sqrt{x}$. (Use FNGraph to see what it looks like, if you're not sure.) Note that the domain of this function is only the non-negative numbers $x \geq 0$. Use the theory in Section 6.5 to deduce what the following curves look like, and sketch them on the same graph. Also say what the domain of each function is.

   6.34 $y = \sqrt{x-2} + 1$ 6.35 $y = 3\sqrt{x+4} - 1$

   6.36 $y = \sqrt{2x-1}$ 6.37 $y = 5\sqrt{\dfrac{x}{2}} + 1$.

6.38–6.40: The graph of $y = x^2$ is a parabola, with a minimum at $(0, 0)$. See Figure 7.1. By applying linear transformations, write the equations of the form $y = a(x - c)^2 + d$ that will produce the following curves:

   6.38 Minimum at $(4,1)$, passing through $(6,9)$

   6.39 Minimum at $(-3,-1)$, passing through $(0,8)$

   6.40 Maximum at $(1,5)$, passing through $(2,3)$.

6.41–6.46: The graph of $y = f(x)$ where $f(x) = x^2 - 4$ is a parabola with a minimum at $(0, -4)$, and roots (points where it crosses the x-axis, i.e. where $y = 0$) at $x = -2$ and $x = 2$. Express each of the functions below as linear transformations of $f(x)$. Using the theory of Section 6.5, say where the turning-point (maximum or minimum) and the roots (if any) lie, for the following curves. Check your answers using FNGraph.

    6.41 $\;y = (x - 6)^2 - 4$ 6.42 $\;y = x^2 + 4$

    6.43 $\;y = 9x^2 - 4$ 6.44 $\;y = 4x^2 - 19$

    6.45 $\;y = -x^2 - 4$ 6.46 $\;y = -x^2 + 4.$

6.47–6.52: Find the points of intersection with $y = f(x) = x^2 - 4$ (if any), of each of the curves in Problems 6.41–6.46. Sketch the graphs.

6.53: If you want to install a slider control (a scrollbar) on your own Excel spreadsheets, here's how. We will create a copy of SPREADSHEET 6.2, with a scrollbar controlling the value of m in cell B4. The slider will adjust this value from 0 to 5 in steps of 0.025.

- Create a copy of the sheet. To do this, select the sheet by clicking its tab, then right-click on the tab and choose `Move or Copy . . .` from the menu. Choose `(move to end)` and check the box for `Make a copy` before clicking OK.

- Open the new sheet. If the sheet is protected, unprotect it. Choose `Tools > Protection > Unprotect sheet. . .`

- Bring up the Control Toolbox. Choose `View > Toolbars > Control Toolbox`

- In the Toolbox, click the button for a scrollbar (in my Excel, it's the 7th button down on the left). The cursor should change to a + sign.

- Left-click down, and using the cursor, draw the toolbar, horizontal or vertical, e.g. from cell A12 down to cell A20. Release the mouse button and the scrollbar appears! While it is highlighted, you can drag it into position, or resize it. In the Control Toolbox, the button in top left is highlighted, showing you are in Design mode. While in Design mode, you can modify the scrollbar, but it won't operate.

- Right-click on the scrollbar and choose `Properties` from the menu. A Properties window appears.

- There is a property called `LinkedCell`. In the right-hand column for this property, enter the address of a blank cell, e.g. C16. For the property Max, change its value from 32767 to 200.

- Close the Properties window. Click the Design mode button in the Toolbox. You have now exited the Design mode; now you can't modify the scrollbar but it will function. Drag the scrollbar button down, and as you do so, numbers from 0 to 200 will appear in cell C16.

- We will now program cell B4 so that values from 0 to 5 appear there as the values in C16 change from 0 to 200. The formula is: = 0.025*C16.

You should now have a spreadsheet like that in SPREADSHEET 6.1.

    To practise this, create a copy of SPREADSHEET 5.6, with a scrollbar changing the value of k in cell I3, from −4 to 4 in steps of 0.1. To program cell I3, you need to work out a linear function $f(x)$ with the property that $f(0) = -4$ and $f(\text{max}) = 4$, where max is the Max value you enter in the scrollbar properties.

7

Quadratic and polynomial functions

When we have enough information about the independent variable x and a dependent variable y to allow us to define two data-points, we can plot them as two points on a graph, fit a straight line through them, and find the equation of the line. This expresses y as a linear function of x, which may or may not be a realistic model of the relationship; for example, we may start with a population of $N = 100$ bacteria on a Petri dish at time $t = 0$, and after $t = 1$ hour there are $N = 200$ bacteria, but is it realistic to model the population growth over time as a linear relationship $N = 100t + 100$, passing through the two data-points (0, 100) and (1, 200)? To test this we should perform further experiments or observations, and obtain more data-points to plot on our graph. Only if these new points all lie on the straight line, could we confidently report that we have a linear relationship.

Not all physical relationships are linear. Suppose we have data defining three points on the graph, and they are not in a straight line – can we fit a smooth curve through them, and find its equation? The answer is 'Yes', and the function can be written by adding an 'x squared' term to the linear function:

$$y = ax^2 + bx + c. \tag{7.1}$$

The function in the equation above is called a **quadratic function**. In this chapter we will see different forms of the quadratic function, and their graphs.

· ·

7.1 Simple quadratic functions

In the last chapter we built up to the linear function $y = mx + c$ by starting with the identity function $y = x$, and then moving to the proportion function $y = mx$. We take the same approach now, starting with the 'squaring' function $y = x^2$.

7.1.1 The squaring function: $y = x^2$

We saw the squaring function

$$y = x^2 \tag{7.2}$$

in Chapter 5. An important feature of the function is that $f(x) = x^2$ is always positive (or zero, when $x = 0$), since for example $f(-2) = (-2)^2 = 4$. We could make a table of values:

x	-4	-3	-2	-1	0	1	2	3	4
$f(x) = x^2$	16	9	4	1	0	1	4	9	16

The graph is sketched in Figure 7.1(a). The shape of the curve is a **parabola**, concave upwards. If you imagine looking down on the curve from above, you would see a concave shape; we will refer to this as **concave up**. If we follow the path of the curve from left to right, it heads down until it reaches (0, 0), where it turns round and moves up again. We say that the curve has a **turning-point** when $x = 0$. In this case the turning-point is a **minimum** (i.e. $f(x)$ reaches its minimum value).

Parabolas occur in real life. Parabolic reflectors and antennas focus incoming rays of light or waves of sound onto a single point – the feathers around the ears of a barn owl form a paraboloid, to focus all incoming sound waves onto a small flap of skin (the pre-aural flap) which reflects them into the ear canal. Projectiles follow parabolic curves: when an artillery shell is fired from a gun, or a ski-jumper flies off the ramp, they follow a parabolic curve through the air, rising up to a maximum altitude before curving down to ground level again. This is a parabola concave downwards, which involves a maximum; we'll see the equation of such a curve in the next section.

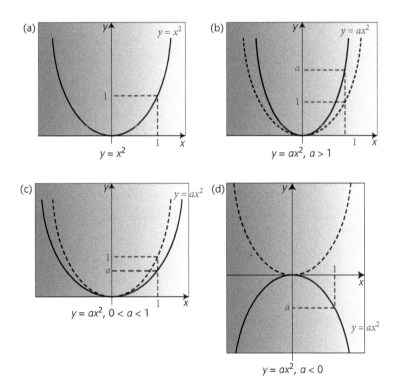

Figure 7.1 The 'proportional to the square' function

7.1.2 The 'proportional to the square' function: $y = ax^2$

Let's now consider what happens when we include a constant of proportionality a with the squaring function:

$$y = ax^2. \tag{7.3}$$

There are important physical relationships of this type:

- When a bungee jumper jumps, the distance she falls is proportional to the square of the elapsed time (until the elastic becomes taut).
- The kinetic energy of a moving body is proportional to the square of its velocity.
- The surface area of a Petri dish (and hence the number of bacteria that can be grown on it) is proportional to the square of the dish diameter.

EXAMPLE

A bungee jumper of mass m jumps from a high bridge over a canyon, at time $t = 0$. As she falls, she is pulled down by the force of gravity, meaning she has a downward acceleration of $g = 9.8$ m s^{-2}. The equations of motion describing her velocity $v(t)$ and vertical distance travelled $s(t)$ after t seconds, are

$$v(t) = u + gt$$
$$s(t) = ut + \frac{1}{2}gt^2.$$

We will derive these equations of motion in Case Study C6, later in this chapter. Here, u is the initial velocity, so if she simply topples from the bridge rather than jumping, $u = 0$ and the equations simplify to

$$v(t) = gt$$
$$s(t) = \frac{1}{2}gt^2.$$

The kinetic energy E of a body of mass m kg travelling at a constant velocity v m s^{-1} is

$$E = \frac{1}{2}mv^2.$$

The unit of kinetic energy is the joule, dimensionally equivalent to kg m^2 s^{-2} (see Table 2.2). So if our bungee jumper weighs 60 kg, then after 3 seconds we have:

- distance fallen: $s(3) = 0.5 \times 9.8 \times 3^2 = 44.1$ metres;
- instantaneous speed: $v(3) = 9.8 \times 3 = 29.4$ m s^{-1};
- kinetic energy: $E(3) = 0.5 \times 60 \times 29.4^2 = 25.93$ kJ.

(In the last calculation we divided by 1000 to express the answer in kilojoules.)
    If we double the elapsed time to 6 seconds, we have:

- distance fallen: $s(6) = 0.5 \times 9.8 \times 6^2 = 176.4$ metres;
- instantaneous speed: $v(6) = 9.8 \times 6 = 58.8$ m s^{-1};
- kinetic energy: $E(6) = 0.5 \times 60 \times 58.8^2 = 103.72$ kJ.

So as t doubles from 3 to 6: $s(t)$ quadruples (i.e. proportional to t^2); $v(t)$ doubles (i.e. proportional to t); and $E(t)$ quadruples (i.e. proportional to t^2).

Compared to the graph of $y = x^2$ – the dotted curve in Figure 7.1(b)–(d) – the parabolic curve of $y = ax^2$ has been stretched vertically, away from the x-axis, making it narrower, if $a > 1$: see Figure 7.1(b). If $0 < a < 1$ the curve is squashed down towards the x-axis, making it broader: see Figure 7.1(c). So this constant of proportionality squashes or stretches the parabolic curve vertically.

If a is negative, then $f(x) = ax^2$ will always be negative, reaching a maximum at $(0, 0)$. The curve lies below the x-axis: see Figure 7.1(d). So a negative constant of proportionality 'flips' the curve around the x-axis (as well as stretching or squashing it). The curve is now **concave down** (it would appear convex looking from above, but concave when looking from below). Figure 13.5 illustrates the ideas of 'concave up' and 'concave down'. You can use FNGRAPH 7.1 to see the curve of $y = ax^2$ for different values of a, positive and negative. Try values such as $a = 0.5, 3, -3$ and compare with the curve of $y = x^2$ (the dotted curve). Notice that the parabola of $y = -3x^2$ is equivalent to 'flipping' or reflecting the curve of $y = 3x^2$ around the x-axis.

7.1.3 Adding a constant term: $y = ax^2 + d$

The effect of adding a constant term, for example, $y = 3x^2 + 5$, is that 5 is added to each value of y on the curve of $y = 3x^2$. This means that the shape of the parabola is unchanged, but it is shifted vertically upwards by a distance of 5 units. We saw the same behaviour in Section 6.3.3, where the graph of $y = 3x + 5$ was found by moving the line $y = 3x$ vertically upwards. This is illustrated in Figure 7.2(a), and in FNGRAPH 7.2. So the graph of

$$y = ax^2 + d \tag{7.4}$$

is a parabolic curve, concave upwards if $a > 0$ or concave downwards if $a < 0$, and with its turning-point at the point $(0, d)$.

When a is positive and d is negative, Figure 7.2(b), or vice versa Figure 7.2(c), an important new feature of the graph appears. Take, for example, the function $f(x) = 2x^2 - 18$, whose curve will be similar to the parabola in Figure 7.2(b). This curve cuts the x-axis in two places. These are called the **roots** of $f(x)$, as we defined in (6.4): they are the values of x at which $f(x) = 0$. To calculate the values of x at these points, set $f(x) = 0$ and solve for x. Here are the calculations for the particular function $f(x) = 2x^2 - 18$, and for the more general function $f(x) = ax^2 + d$:

Particular example	General method
$f(x) = 2x^2 - 18 = 0$	$f(x) = ax^2 + d = 0$
$2x^2 = 18$	$ax^2 = -d$
$x^2 = 9$	$x^2 = -\dfrac{d}{a}$
$x = \pm\sqrt{9} = +3 \text{ or } -3$	$x = \pm\sqrt{-\dfrac{d}{a}}$ (7.5)

Here, the $\pm$ symbol means 'plus or minus', indicating that there are two values, a positive root, $x_1 = +\sqrt{\dfrac{d}{a}}$, and a negative root, $x_2 = -\sqrt{\dfrac{d}{a}}$. In our example $f(x) = 2x^2 - 18$, we interpret $x = \pm\sqrt{9}$ to mean x is either the positive *or* the negative of the square root of 9. Since $\sqrt{9} = 3$, this means that $x = +3$ or -3.

Remember that if d and a have opposite signs, d/a will be negative, so $-d/a$ will be positive and we can take the square root. If d and a have the same sign, the curve will not cut the x-axis, and the function will have no roots; this is the case for the curve in Figure 7.2(a).

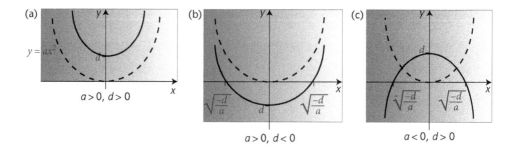

Figure 7.2 Graphs of $y = ax^2 + c$

We can apply the idea of roots to linear graphs. We saw in Section 6.3.3 that the linear function $y = mx + c$ crosses the x-axis at $x = -c/m$. So this is the root of the linear function; a linear function always has just one root, whereas a quadratic function may have no roots (as in Figure 7.2(a)), one root (which is also the turning-point, as in Figure 7.1), or two roots (as in Figure 7.2(b) and (c)).

We may sometimes talk about the roots of the equation $y = f(x)$, meaning the roots of the function $f(x)$, i.e. the values of x for which $y = 0$.

In order to sketch the graph of a function $f(x)$, we should try to find the location of any turning points, and also the location of any roots. The roots can have important physical meaning. If $P(t)$ is a model for the population at time t, then finding the root of $P(t)$ will predict the time at which the population becomes extinct. If $h(t)$ describes the height of a rocket after t seconds, solving $h(t) = 0$ gives us the time when the rocket will land.

7.1.4 The 'shifted squaring' function: $y = a(x - \alpha)^2$

Let's modify the 'proportional to the square' function $y = ax^2$ in another way. Suppose we replace x by $(x - \alpha)$. Here, α (the Greek letter a, pronounced 'alpha') is a parameter, just like a and c. The effect of this change is to shift the parabola of $y = ax^2$ horizontally, a distance of α units to the right. You can see this if we construct a table of values for $f(x) = x^2$ and $f(x) = (x - 3)^2$:

x	-2	-1	0	1	2	3	4	5	6
$f(x) = x^2$	4	1	0	1	4	9	16	25	36
$f(x) = (x - 3)^2$	25	16	9	4	1	0	1	4	9

The turning-point has been moved from $(0, 0)$ to $(3, 0)$. The minimum value of $(x - \alpha)^2$ is zero, which occurs when $x = \alpha$. In that case, $y = a(x - \alpha)^2 = a(\alpha - \alpha)^2 = a.0^2 = 0$. So this is also the root of the function.

This is illustrated in Figure 7.3(a), and in FNGRAPH 7.3, where the graphs of $f(x) = 2x^2$ and $f(x) = 2(x - 3)^2$ are sketched.

So the graph of

$$y = a(x - \alpha)^2 \tag{7.6}$$

is a parabolic curve, concave up if $a > 0$ or concave down if $a < 0$, and with its turning-point at the point $(\alpha, 0)$. The turning-point is a minimum if $a > 0$, and a maximum if $a < 0$. Its

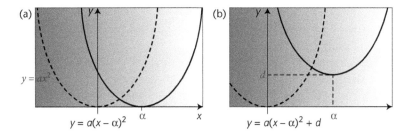

Figure 7.3 Graphs of 'shifted squaring' functions

y-intercept (the place where it cuts the y-axis) is found by putting $x = 0$, which produces $y = a(0 - \alpha)^2 = a.\alpha^2$.

EXAMPLE

The function $y = 3(x - 5)^2$ is a quadratic function with a root and turning-point at $(5, 0)$. Its graph is a parabola concave upwards, and its y-intercept is at $(0, 75)$.

. . . .

7.2 General quadratic functions

In the previous section, we started with the basic squaring function $y = x^2$, and applied a number of changes, called **linear transformations**:

- Multiplying by a constant of proportionality, $y = ax^2$, makes the parabola squatter or taller by squashing or stretching it in the vertical direction. It also flips the curve to be 'concave downwards' if a is negative.

- Adding a constant term, $y = ax^2 + d$, shifts the curve vertically upwards (d positive) or downwards (d negative).

- Replacing x by $x - \alpha$, to give $y = a(x - \alpha)^2$, shifts the curve horizontally, to right (α positive) or left (α negative).

These illustrate the linear transformations that we defined in Section 6.5.

By applying all of these transformations simultaneously, we can produce any parabolic curve we wish, located anywhere in the graph. That is, we can produce the equation for the general quadratic function. In this section we'll look at two forms of the equation for a general quadratic, and find their turning-points, and where they cross the x- and y-axes.

7.2.1 The 'completed square' form: $y = a(x - \alpha)^2 + d$

Applying all three transformations, we obtain the most useful form of the general quadratic:

$$y = a(x - \alpha)^2 + d. \tag{7.7}$$

The turning-point of the curve (which is a minimum if a is positive, and a maximum if a is negative) will occur when $x = \alpha$, as in Section 7.1.4. At this point, we will have $y = a(x - \alpha)^2 + d = a(\alpha - \alpha)^2 + d = a.0^2 + d = 0 + d = d$. So the turning-point is located at (α, d). This is illustrated in Figure 7.3(b), and in FNGRAPH 7.3 you can experiment with different values of the three parameters a, α, and d.

For example, the graph of the function $y = 2(x - 3)^2 - 7$ will be the parabola of $y = 2x^2$ shifted 3 units to the right and 7 units downwards. Thus, the curve will be a parabola that is concave up, with its minimum located at $(3, -7)$.

We can calculate the roots of the quadratic (if there are any) by setting $f(x) = 0$ and solving for x. Here are the calculations for the particular function $f(x) = 2(x - 3)^2 - 7$, and for the general function $y = a(x - \alpha)^2 + d$:

Particular example	General method
$f(x) = 2(x-3)^2 - 7 = 0$	$f(x) = a(x - \alpha)^2 + d = 0$
$2(x-3)^2 = 7$	$a(x - \alpha)^2 = -d$
$(x-3)^2 = \dfrac{7}{2} = 3.5$	$(x - \alpha)^2 = \dfrac{-d}{a}$
$x - 3 = \pm\sqrt{3.5}$	$x - \alpha = \pm\sqrt{\dfrac{-d}{a}}$
$x = 3 \pm \sqrt{3.5}$	$x = \alpha \pm \sqrt{\dfrac{-d}{a}}$
$x = 1.129$ or $x = 4.871$	

$$(7.8)$$

Check that in FNGRAPH 7.3 – shown in Figure 7.4 – the roots lie at 1.13 and 4.87 on the x-axis. There will be two roots if the quantity $-d/a$ under the square root sign is positive. This will happen when d and a are of opposite sign.

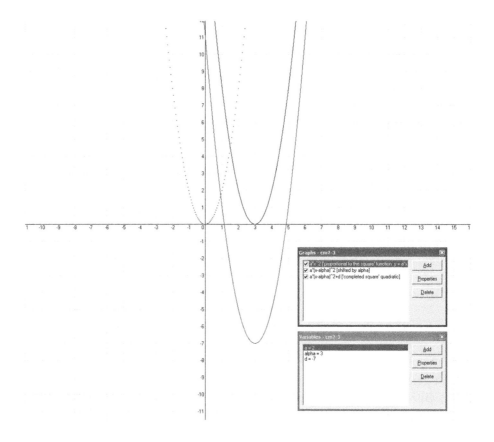

Figure 7.4 The graph of $y = 2(x - 3)^2 - 7$ (FNGRAPH 7.3)

The y-intercept of the curve is found by setting $x = 0$:

$$f(0) = a(0 - \alpha)^2 + d = a(-\alpha)^2 + d = a.\alpha^2 + d.$$

So for our function $f(x) = 2(x - 3)^2 - 7$, the y-intercept is at

$$y = 2 \times 3^2 - 7 = 18 - 7 = 11.$$

Figure 7.5 shows the curve, with the locations of the turning-point, roots and y-intercept labelled, for the cases where

(a) $a > 0$ and $d < 0$

(b) $a < 0$ and $d > 0$.

Even if you don't remember the formulae for the roots and y-intercept, you should be able to work them out for a particular example.

EXAMPLE

An Olympic diver jumps from a springboard projecting 0.5 m over the edge of the swimming pool, at a height of 3 m above the water surface. He reaches a maximum height of 5 m above the pool, and at this point he is above a point on the water surface

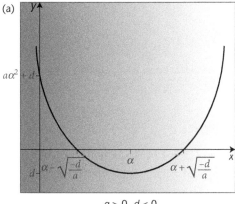

(a)

$a > 0, d < 0$

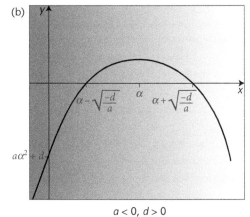

(b)

$a < 0, d > 0$

Figure 7.5 The 'completed square' function $y = a(x - \alpha)^2 + d$, with two roots

that is 1.5 m in from the edge of the pool. How far in from the pool edge is he when he hits the water?

Solution: The diver follows a parabolic curve in (x, y) space, where x is the horizontal distance in from the edge of the pool, and y is his vertical height above the water. So he leaves the springboard at the point $(0.5, 3)$. He reaches his maximum height at the point $(1.5, 5)$. So in (7.7) we have $\alpha = 1.5$ and $d = 5$. The equation is

$$y = a(x - 1.5)^2 + 5.$$

To find a we use the fact that he starts at $(0.5, 3)$. Substituting into the equation:

$$3 = a(0.5 - 1.5)^2 + 5$$
$$3 = a(-1)^2 + 5$$
$$3 = a + 5$$
$$a = 3 - 5 = -2$$

so the equation describing his flight is

$$y = -2(x - 1.5)^2 + 5.$$

Plot the curve using FNGRAPH 7.4.

He hits the water when $y = 0$. Inserting this value for y, we solve for the roots x:

$$0 = -2(x - 1.5)^2 + 5$$
$$2(x - 1.5)^2 = 5$$
$$(x - 1.5)^2 = 2.5$$
$$x - 1.5 = \pm\sqrt{2.5}$$
$$x = 1.5 \pm \sqrt{2.5}$$
$$x = 1.5 + 1.581 = 3.081 \quad \text{or} \quad x = 1.5 - 1.581 = -0.081.$$

We are expecting a positive value of x, so we discard the negative root. The diver hits the water 3.08 metres in from the edge.

7.2.2 The polynomial form: $y = ax^2 + bx + c$

We said that the 'completed square' form of the general quadratic was the most useful form, because it's easy to obtain the turning-point, roots, and y-intercept of the curve from it. But the most usual way of obtaining a quadratic function is by adding an x^2 term to a linear function, as we did in the Introduction to this chapter. That is, we write the quadratic as

$$y = ax^2 + bx + c. \tag{7.9}$$

The numbers represented by a, b, c are called the **polynomial coefficients**.

Now it's not obvious where the turning-point is, or whether there exist any roots. It is however easy to see the y-intercept: putting $x = 0$ means that $y = a.0^2 + b.0 + c = c$. So the y-intercept is at $(0, c)$ on the y-axis.

As an example, consider the quadratic $y = x^2 + 6x + 7$. Comparing with (7.9), the polynomial coefficients are $a = 1$, $b = 6$ and $c = 7$. The curve will cut the y-axis at $y = 7$. To find its turning-point and roots, we need to manipulate the polynomial function to get it into the 'completed square' form; the process of doing this is called **completing the square**.

Consider the first two terms: $x^2 + 6x$. We need to add a constant term to this, so that it becomes a square, i.e. something that we can write in the form $(x + k)^2$. Recall from (1.20) in Section 1.5 that $(x + k)^2 = x^2 + 2kx + k^2$. Comparing the first two terms of this expansion with

$x^2 + 6x$, we see that we need to take $k = 3$ because we need $2k$ in the term $2kx$ to have a value of 6. Then $(x + 3)^2 = x^2 + 6x + 9$. But our original equation is $y = x^2 + 6x + 7$. So our constant term is too large; to adjust it, we need to subtract 9 and then add 7, i.e. to subtract 2.

We can perform the process of completing the square as:

$$y = x^2 + 6x + 7$$
$$= (x^2 + 6x + 9) - 9 + 7$$
$$= (x + 3)^2 - 2.$$

Compare this with the 'completed square' form (7.7). In this case $a = 1$, $\alpha = -3$ (which makes $x - \alpha = x - (-3) = x + 3$) and $d = -2$. So the turning point is at $(-3, -2)$ and the roots are at

$$-3 \pm \sqrt{-\left(\frac{-2}{1}\right)} = -3 \pm \sqrt{2} = -3 + 1.4142 \text{ or } -3 - 1.4142 = -1.586 \text{ or } -4.414.$$

Notice that the value of x at the turning-point, namely at $x = -3$, is midway between the two roots.

Completing the square is more tricky when there's a coefficient multiplying the x^2 term. Here's an example:

$$y = 3x^2 + 6x + 7$$
$$= 3(x^2 + 2x) + 7$$
$$= 3((x^2 + 2x + 1) - 1) + 7$$
$$= 3((x + 1)^2 - 1) + 7$$
$$= 3(x + 1)^2 - 3 + 7$$
$$= 3(x + 1)^2 + 4$$

which is in the 'completed square' form $y = a(x + k)^2 + d$.

In this case, the turning-point is at $(-1, 4)$ and there are no roots since $a = 3$ and $d = 4$ have the same sign.

You can see the completing-the-square process demonstrated in Excel SPREADSHEET 7.1, reproduced in Figure 7.6. Don't worry if you can't follow all that was going on – though you should be able to check that the algebra is correct. What's important is to understand that the completing-the-square process can be applied to any quadratic in polynomial form, and thus the roots of the quadratic, if any, can be found.

To see how the process works graphically, look at SPREADSHEET 7.2,[1] which is reproduced in Figure 7.7. The polynomial form of the equation (7.9) is set in Row 8, and drawn as the red curve in the chart. The 'completed square' form $y = a(x + k)^2 + d$ is in Row 15, and the values of the parameters a, k, d can be adjusted using the slider controls. Initially they are set to give the simple squaring function $y = x^2$, by taking $a = 1$, $k = 0$, $d = 0$. The function is the blue curve on the chart. The 'completing the square' process involves choosing the parameters so that the two equations define identical functions; graphically, they should produce the same curve. First adjust d (which shifts the blue curve vertically) and k (which shifts the blue curve horizontally), until the minimum point matches that of the red curve. Then adjust a (which changes the curvature) until the two curves match. The resulting values of a, k, d should be the same as those found algebraically in SPREADSHEET 7.1. Try repeating the

---

[1] Note: When you try to open a spreadsheet that contains controls such as sliders, you may get a security warning message about macros. Try clicking the 'Enable macros' button. If the spreadsheet still doesn't open, see the Technical Notes at the front of this book.

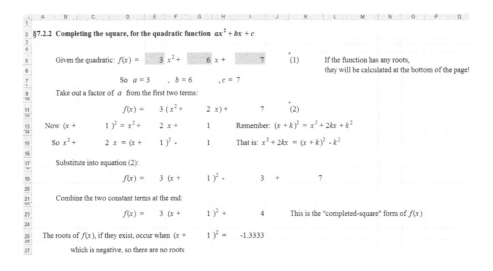

Figure 7.6 Completing the square (SPREADSHEET 7.1)

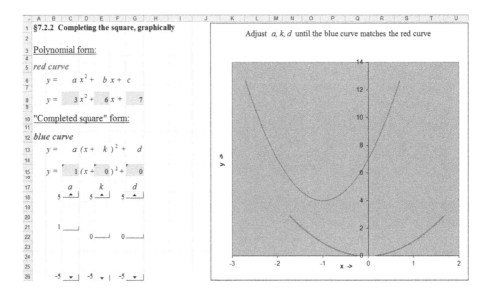

Figure 7.7 Completing the square, graphically (SPREADSHEET 7.2)

process for different values of a, b, c (but don't make the turning-point too far from the origin, as the sliders can only choose values between −5 and +5).

If the completing-the-square process is applied to the general quadratic (7.9), and then the roots are found, the result is a very important formula for the roots of the quadratic, which you do need to memorize:

The roots of the quadratic $y = ax^2 + bx + c$, i.e. the values of x for which $ax^2 + bx + c = 0$, are given by

$$x = \frac{-b \pm \sqrt{b^2 - 4ac}}{2a}.$$

(7.10)

(i) There will be two roots if the quantity under the square root sign is positive, that is, if $D > 0$, where

$$D = b^2 - 4ac. \qquad (7.11)$$

This quantity D is called the **discriminant**. See Figure 7.8(a). We say that the roots are **distinct**, meaning that they occur at two different values of x.

(ii) If $D = 0$, there will be just one root, at $x = \dfrac{-b}{2a}$ (which is also the turning-point). This is called a **double root**, because you can imagine the two different roots in (7.10) getting closer and closer together as D approaches zero, until they both lie in the same place. Mathematicians say that the two roots **coincide** (are no longer distinct). As the

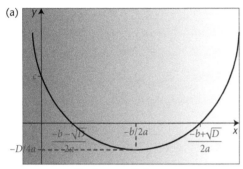

(a)

$D > 0$: two distinct roots

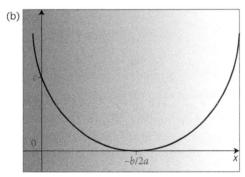

(b)

$D = 0$: a double root

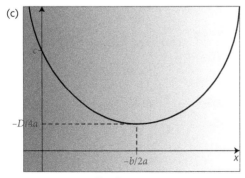

(c)

$D < 0$: no roots

Figure 7.8 Graphs of $y = ax^2 + bx + c$
$(a > 0, b > 0, c > 0)$

turning-point is sandwiched between the roots, it will also lie at the point where the two roots coincide. See Figure 7.8(b).

(iii) If $D < 0$, there are no roots (because you can't take the square root of a negative number). That is, either the whole curve lies above the x-axis, or the whole curve lies below the x-axis. See Figure 7.8(c).

In each of the three cases above, the turning-point of the parabola will occur when $x = \dfrac{-b}{2a}$. The y-value of the turning-point is $y = \dfrac{-D}{4a}$.

To summarize:

The graph of the quadratic function $y = ax^2 + bx + c$, is a parabola, concave up if $a > 0$ or concave down if $a < 0$. It cuts the y-axis at $y = c$. The turning point of the curve is at $x = \dfrac{-b}{2a}$, and is a minimum if $a > 0$, or a maximum if $a < 0$. Calculate the discriminant $D = b^2 - 4ac$.

If $D > 0$ the curve cuts the x-axis at the two roots

$$x = \frac{-b \pm \sqrt{b^2 - 4ac}}{2a}.$$

If $D = 0$ the curve touches the x-axis at the double root

$$x = \frac{-b}{2a} \text{ (which is also the turning-point).}$$

If $D < 0$ there are no roots, i.e. the curve does not touch the x-axis.   (7.12)

If the function $y = ax^2 + bx + c$ does have two roots, we could write them as α and β, where

$$\alpha = \frac{-b - \sqrt{b^2 - 4ac}}{2a} \quad \text{and} \quad \beta = \frac{-b + \sqrt{b^2 - 4ac}}{2a}.$$

SPREADSHEET 7.3 demonstrates the calculation of the roots (if any) using this formula, and FNGRAPH 7.4 draws the curve. To see the link between the calculated roots (if any) and the points where the curve cuts the x-axis (if any), you should experiment with several choices of the coefficients a, b, c. Type them into the yellow cells in SPREADSHEET 7.3, and see the calculation. Then type them into the Variables box in FNGRAPH 7.4, and see the graph. The quadratic that is provided for you is $y = 3x^2 - 7x + 2$, for which $D = 25$, and there are two roots, at $\alpha = 0.33$ and $\beta = 2$.

The 'shifted square' and 'completed square' functions can be written in polynomial form by multiplying out the term in brackets:

For the example in Section 7.1.4,

$$y = 3(x - 5)^2 = 3(x^2 - 10x + 25) = 3x^2 - 30x + 75.$$

For the example in Section 7.2.1,

$$y = 2(x - 3)^2 - 7 = 2(x^2 - 6x + 9) - 7 = 2x^2 - 12x + 18 - 7 = 2x^2 - 12x + 11.$$

Examples of finding the roots:

(i) $y = 2x^2 - 5x + 1$. Here, $a = 2$, $b = -5$, and $c = 1$.

The discriminant $D = (-5)^2 - 4 \times 2 \times 1 = 25 - 8 = 17$.

$D > 0$ so there are two roots:

$$\alpha = \frac{-(-5) - \sqrt{17}}{2 \times 2} = \tfrac{1}{4}(5 - \sqrt{17}) = 0.2192 \text{ and}$$

$$\beta = \frac{-(-5) + \sqrt{17}}{2 \times 2} = \tfrac{1}{4}(5 + \sqrt{17}) = 2.2808.$$

The y-intercept is at $(0, 1)$.

The turning-point occurs when $x = -\frac{-5}{2 \times 2} = 1.25$. At this value of x,
$y = 2(1.25)^2 - 5(1.25) + 1 = 3.125 - 6.25 + 1 = -2.125$. The coefficient of x^2 is positive, so the graph of the function is a parabola concave upwards, with a minimum at $(1.25, -2.125)$.

(ii) $y = -x^2 + 6x - 9$. Here, $a = -1$, $b = 6$, and $c = -9$.

The discriminant $D = (6)^2 - 4 \times (-1) \times (-9) = 36 - 36 = 0$.

$D = 0$ so there is just one root, which is also a turning-point, at

$$x = \frac{-6}{2 \times (-1)} = \frac{-6}{-2} = 3.$$

The y-intercept is at $(0, -9)$.

The coefficient of x^2 is negative, so the graph of the function is a parabola concave downwards, with a maximum at $(3, 0)$.

(iii) $y = x^2 - 2x + 3$. Here, $a = 1$, $b = -2$, and $c = 3$.

The discriminant $D = (-2)^2 - 4 \times 1 \times 3 = 4 - 12 = -8$.

$D < 0$ so there are no roots.

The y-intercept is at $(0, 3)$.

The turning-point occurs when $x = -\frac{-2}{2 \times 1} = 1$. At this value of x,
$y = (1)^2 - 2(1) + 3 = 2$. The coefficient of x^2 is positive, so the graph of the function is a parabola concave upwards, with a minimum at $(1, 2)$.

Problems: Now try the end-of-chapter problems 7.1–7.23.

CASE STUDY C6: Equations of motion 2

In the last instalment we saw a general equation of motion $s = v.t + d$, where the moving animal will be at the point $s(t)$ on the real line at time t. Its initial position is $s(0) = d$. This assumes that it is moving at a constant velocity v. What if v is changing, e.g. because the animal is accelerating?

We consider an animal that is travelling with constant acceleration a, and at time $t = 0$ its position is $s(0) = d$ and its initial velocity is u. Then its velocity $v(t)$ at time t is given by generalizing the equation $v = a.t$, which assumes that $v(0) = 0$. We have to add u, the initial velocity:

$$v = u + a.t. \tag{7.13}$$

But we also need an equation for $s(t)$, the animal's position at time t. We derive this as follows:

At time $t = 0$ the animal's velocity is u.

After t seconds, its velocity is $u + a.t$.

So its average velocity over that time period is $v_{ave} = \frac{1}{2}\left(u + (u + at)\right)$.

The distance travelled will be

$$v_{ave}.t = \frac{1}{2}\left(u + (u + at)\right).t = \frac{1}{2}\left(2ut + at^2\right) = ut + \frac{1}{2}at^2.$$

If it starts at position $s(0) = d$, its position at time t is given by

$$s = d + ut + \frac{1}{2}at^2. \tag{7.14}$$

By solving (7.13) for t and inserting in (7.14), you should be able to derive the third equation of motion, expressing v as a function of s:

$$v^2 = u^2 + 2a(s - d). \tag{7.15}$$

Example: Returning to the lion chasing the young zebra in Part C5, suppose now that the lion is lying in wait when the zebra runs past. The lion then accelerates after it. How close would the zebra have to be when the lion starts to move, so that the lion will catch it? Use the information about the lion and zebra from Parts C4 and C5.

Solution: We saw in Part C4 that the lion has an acceleration of 6.25 m s^{-2}, and using (7.13) we found that it will take 2.24 seconds for it to reach its maximum velocity of 14 m s^{-1}. We measure distances from the lion's hiding-place, so to find the distance it travels in that time, we use (7.14):

$$s = 0 + 0 \times 2.24 + 0.5 \times 6.25 \times 2.24^2 = 15.68 \text{ metres}.$$

It is then able to run at this velocity for 6 seconds, covering a further $14 \times 6 = 84$ metres. When it would have to give up the chase, it would be at the position $s_1 = 15.68 + 84 = 99.68$ metres, after $2.24 + 6 = 8.24$ seconds.

In that time, the zebra, running at a constant speed of 10 m s^{-1}, will cover $8.24 \times 10 = 82.4$ metres. Suppose the lion started moving when the zebra was d metres away from it. Then the position of the zebra will be $s_2 = d + 82.4$ metres.

For the lion to catch the zebra before it gives up, we require $s_1 > s_2$, i.e. $99.68 > d + 82.4$. The lion must start the chase before the zebra has gone more than 17.28 metres away.

Problems: Now try the end-of-chapter problems 7.22–7.24.

CASE STUDY A7: Fibonacci's Rabbits: geometric growth in a population structured by age

In Part A2 of this Case Study, back in Chapter 1, we stated that the geometric growth model, exemplified by the growth of a population of bacteria cells on a Petri dish, can be written as

$$N_p = \lambda^p.N_0, \tag{7.16}$$

where N_0 is the initial population size, N_p is the population size after p generations, each cell divides into λ cells at each generation, and no cells die (so that the update equation is $N_{p+1} = \lambda N_p$). But more complex organisms have a more complicated reproductive behaviour. In 1202, the mathematician Leonardo Pisano Fibonacci (1170–1250) published the following problem:

> 'A man puts a pair of baby rabbits into an enclosed garden. Assuming each pair of rabbits bears a new pair every month, which from the second month on itself becomes productive, how many pairs of rabbits will there be in the garden after one year?'

In this model, each pair of newborn rabbits will first mate at the end of one month, and produce two offspring after a further month – and every month thereafter. We assume that no rabbits die.

Let N_p denote the number of pairs of rabbits in the population after p generations. So $N_0 = 1$ pair, called Alice and Arnold, say. After 1 month they are old enough to mate (so N_1 is still 1), and after a further month they produce a baby pair: Betty and Basil, so now $N_2 = 2$ pairs. In month 3, pair A produce more offspring (Charles and Camilla) while pair B mate, so $N_3 = 3$. In month 4 we have pair A producing Dan and Dina, pair B producing Eugene and Erika, while pair C mate, so $N_4 = 5$. Continuing this process, you will find that the sequence of numbers

$$N_0, N_1, N_2, N_3, N_4, \ldots \quad \text{is} \quad 1, 1, 2, 3, 5, 8, 13, 21, \ldots$$

This is the famous **Fibonacci sequence**. To find the update equation, we argue as follows:

The population in the $(p + 1)$th generation will be made up of:

- all the N_p pairs from the previous generation, plus
- new pairs from each of the pairs that are at least two months old. The population two months previously was N_{p-1}, so there will be N_{p-1} of these.

The update equation is therefore:

$$N_{p+1} = N_p + N_{p-1}. \tag{7.17}$$

That is, each new member of the sequence is the sum of the two previous ones. The initial conditions are that $N_0 = N_1 = 1$, and we can use (7.17) repeatedly to calculate $N_2, N_3, N_4, \ldots$

But we would like a way of calculating N_{12} (the population after one year) without having to work out all the intermediate populations; such an explicit formula is called a **closed-form solution**. We can do this if we assume that this is still some sort of geometric growth. That is, we look for a solution resembling equation (7.16): $N_p = r^p.N_0$ for some value of r. Substituting into (7.17) we get $r^{p+1}.N_0 = r^p.N_0 + r^{p-1}.N_0$.

Take all the terms onto the left-hand side, and note that there is a common factor of N_0, and also of r^{p-1} (since $r^p = r.r^{p-1}$ and $r^{p+1} = r^2.r^{p-1}$):

$$r^{p-1}.N_0(r^2 - r - 1) = 0.$$

Since N_0 and r are not zero by (1.21), we must have:

$$r^2 - r - 1 = 0. \tag{7.18}$$

This is a quadratic equation, and the formula (7.10) tells us that the roots are:

$$r_1 = \frac{1 - \sqrt{5}}{2} \text{ and } r_2 = \frac{1 + \sqrt{5}}{2}.$$

But an equation such as $N_p = N_0 \left(\dfrac{1 + \sqrt{5}}{2} \right)^p$ will not produce our Fibonacci sequence.

The trick is to take a linear combination of the two solutions:

$$N_p = A r_1^p + B r_2^p \tag{7.19}$$

and to find the coefficients A and B so as to create the initial conditions $N_0 = N_1 = 1$. You will see how to do this in the problems, and also how to program an Excel spreadsheet to produce the Fibonacci sequence using (7.17), and also to evaluate individual terms using (7.19).

The Fibonacci sequence appears in many seemingly unrelated situations in Nature, such as the arrangements of leaves, petals, and seeds, e.g. on sunflowers and fir cones. There is a comprehensive and readable exploration of Fibonacci numbers at Dr Ron Knott's multimedia website.[2]

Problems: Now try the end-of-chapter problems 7.25–7.30.

7.2.3 Calculating the polynomial coefficients

In the introduction to this chapter, we claimed that if we are given three data-points, we can fit a quadratic curve (7.9) through the points. That is, given the data-points $(x_1, y_1), (x_2, y_2)$, (x_3, y_3), there is a quadratic curve $y = ax^2 + bx + c$ that passes through them. But how do we find this curve, i.e. find the values of a, b, and c?

If we put the x and y values from each data-point in turn into the quadratic equation, we obtain three simultaneous equations involving the three unknowns a, b, c:

$$ax_1^2 + bx_1 + c = y_1$$
$$ax_2^2 + bx_2 + c = y_2$$
$$ax_3^2 + bx_3 + c = y_3. \tag{7.20}$$

We saw in Section 5.7 how to solve two simultaneous linear equations involving two unknowns. It is possible to reduce the system (7.20) by subtracting one equation from another. If you subtract the first equation from the second, the two '+ c' terms will cancel and disappear:

$$(ax_2^2 + bx_2 + c) - (ax_1^2 + bx_1 + c) = y_2 - y_1$$
$$ax_2^2 + bx_2 + c - ax_1^2 - bx_1 - c = y_2 - y_1$$
$$ax_2^2 - ax_1^2 + bx_2 - bx_1 + c - c = y_2 - y_1$$
$$(x_2^2 - x_1^2)a + (x_2 - x_1)b = y_2 - y_1.$$

---

[2] http://www.maths.surrey.ac.uk/hosted-sites/R.Knott/Fibonacci/fib.html.

We could do the same thing by subtracting the second equation in (7.20) from the third. So we will have two equations involving a and b, and we know how to solve this. Once we know the values of a and b, we can substitute them into one of the original equations to find c. Here is a practical example, from a topic close to my own heart – quite literally!

7.2.4 Application: Reduction of cholesterol level

At the start of May 2008, I had a blood test that showed that my cholesterol level was 8.1 millimoles/litre. This is well into the high-risk zone for a heart attack; the 'normal' range in the UK (see Section 4.1.6) should be 3.6–6.5 mmol l^{-1}. I was put onto a daily course of statin tablets. After 8 weeks a second blood test was performed, and my cholesterol level had dropped to 7.3 mmol/l. A third blood test, at the start of October, showed a level of 6.4 mmol/l. Although this was now in the 'normal' range, my GP told me I need to get it down to 5.5 mmol/l if possible. By fitting a quadratic function to this data, predict when my cholesterol level will fall to (i) 6.0 mmol l^{-1}, (ii) 5.5 mmol l^{-1}.

We will measure elapsed time t in months from 1 January 2008. The dependent variable $y(t)$ is the cholesterol level at time t. So the three data-points are (4, 8.1), (6, 7.3), and (9, 6.4). The system of equations (7.20) becomes

$$16a + 4b + c = 8.1$$
$$36a + 6b + c = 7.3$$
$$81a + 9b + c = 6.4.$$

Subtract the first equation from the second, and the second equation from the third:

$$(36 - 16)a + (6 - 4)b + c - c = 7.3 - 8.1$$
$$(81 - 36)a + (9 - 6)b + c - c = 6.4 - 7.3.$$

That is,

$$20a + 2b = -0.8$$
$$45a + 3b = -0.9.$$

We can divide through the first equation by 2, and the second equation by 3:

$$10a + b = -0.4$$
$$15a + b = -0.3.$$

Subtracting the first of these equations from the second, we see that $5a = -0.3 - (-0.4) = 0.4 - 0.3 = 0.1$, so $a = 0.02$. Substitute this into one of the equations and solve for b: $b = -0.4 - 10a = -0.4 - 0.2 = -0.6$.

Finally, substitute the values of a and b into one of the original equations to find c:

$$c = 8.1 - 16a - 4b = 8.1 - 16(0.02) - 4(-0.6) = 8.1 - 0.32 + 2.4 = 10.18.$$

So the equation of our quadratic curve is

$$y = 0.02t^2 - 0.6t + 10.18. \tag{7.21}$$

(i) To predict when my cholesterol level would fall to 6.0 mmol l^{-1}, we need to find the value of t when $y = 6$. That is, we have to solve

$$0.02t^2 - 0.6t + 10.18 = 6.$$

Subtracting 6 from both sides gives us the quadratic equation

$$0.02t^2 - 0.6t + 4.18 = 0.$$

We can apply the important formula (7.10) to solve this equation:

$$t = \frac{-(-0.6) \pm \sqrt{(-0.6)^2 - 4(0.02)(4.18)}}{2(0.02)}$$
$$= \frac{0.6 \pm \sqrt{0.36 - 0.3344}}{0.04} = \frac{0.6 \pm \sqrt{0.0256}}{0.04} = \frac{0.6 \pm 0.16}{0.04} = \frac{0.76}{0.04} \text{ or } \frac{0.44}{0.04}.$$

The roots are at $t = 19$ or $t = 11$ months. It is the earlier time which we are looking for, i.e. the level should fall to 6.0 after 11 months from 1 January 2008, i.e. around 1 December.

(ii) By the same argument, the time when the level will fall to 5.5 mmol l^{-1} will be the solution of:

$$0.02t^2 - 0.6t + 10.18 = 5.5.$$

Subtracting 5.5 from both sides gives us the quadratic equation

$$0.02t^2 - 0.6t + 4.68 = 0.$$

The discriminant for this quadratic is

$$D = (-0.6)^2 - 4(0.02)(4.68) = 0.36 - 0.3744 = -0.0144.$$

The discriminant is negative, meaning that the equation has no roots. By this model, my cholesterol level will never fall to 5.5.

In fact, the graph of the function (7.21) is a parabola concave upwards, which lies above the x-axis, as shown in Figure 7.9. The curve has a minimum at $t = 15$, $y = 5.68$.

So this model predicts that my cholesterol level will drop to a minimum of 5.68 mmol/l after 15 months and then start rising again. This is not realistic (I hope), if I keep taking the statins. So the quadratic model is not suitable for making longer-term predictions (a process called **extrapolation**). I hope that a more realistic curve is the dashed one in Figure 7.9, and we will see in Chapter 10 the type of function that can produce this model. However, the quadratic model is valid for estimating cholesterol levels at times within the range of our data-points. For example, it says that after 5 months my level was $0.02(5)^2 - 0.6(5) + 10.18 = 0.5 - 3.0 + 10.18 = 7.68$ mmol/l. This process is called **interpolation**. We saw linear interpolation in Section 3.6, and we present quadratic interpolation in the final section of this chapter, which shows how to obtain the interpolated value directly from the data-points, without needing to find the equation of the curve.

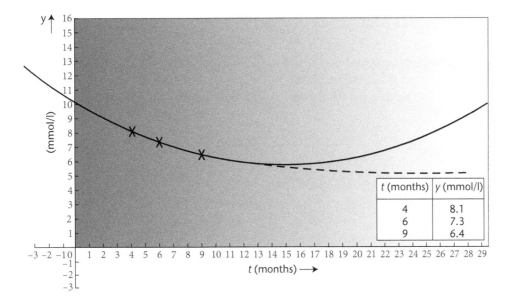

Figure 7.9 Reduction of cholesterol level: $y = 0.02t^2 - 0.6t + 10.18$

CASE STUDY B6: Derivation of the equilibrium density

Look back to part B2 of this case study, in Section 1.1. We said that the rate of increase of the angiogenic cancer cells is given by

$$r.y.\left(1 - \frac{y}{k}\right) - d.y - \frac{p.y.(x_0 + x_1 + y)}{q.y + 1}.$$

Here, the variables x_0, x_1, y represent respectively the densities of healthy cells, non-angiogenic cancer cells, and angiogenic cancer cells in the tumour. We went on to claim that the equilibrium density of angiogenic cells will be given by

$$y = \frac{-Q + \sqrt{Q^2 - 4rq[k(d - r) + pk(x_0 + x_1)]}}{2rq}$$

where

$$Q = kq(d - r) + r + kp.$$

We are now able to see how this expression is derived. You have probably noticed that the expression above does look similar to the formula for the roots of a quadratic (7.10).

The equilibrium state will occur when the rate of increase of angiogenic cells is zero, i.e. when

$$r.y.\left(1 - \frac{y}{k}\right) - d.y - \frac{p.y.(x_0 + x_1 + y)}{q.y + 1} = 0.$$

This doesn't look like a quadratic function of y, but we need to remove the expression involving y, which is the denominator in the last term. To do this we will have to multiply through the whole equation by $q.y + 1$. Notice also that there is a common factor of y on the left-hand side, and the right-hand side is zero. So by (1.21) either $y = 0$ (meaning that the angiogenic cell line does not develop), or we can divide through by y to get:

$$r(q.y + 1)\left(1 - \frac{y}{k}\right) - d(q.y + 1) - p(x_0 + x_1 + y) = 0.$$

Multiplying out the brackets in each term:

$$r.q.y - \frac{rq}{k}y^2 + r - \frac{r}{k}y - dqy - d - p(x_0 + x_1) - py = 0.$$

We can now write this as a quadratic in y, by collecting up terms:

$$-\frac{rq}{k}y^2 + \left(rq - \frac{r}{k} - dq - p\right)y - d + r - p(x_0 + x_1) = 0.$$

To remove the fractions, multiply both sides by k, and also multiply by -1 to change negative terms to positive (the right-hand side remains zero, as $-k.0 = 0$), and then a bit of rearranging:

$$rqy^2 + (-rkq + r + dkq + kp)y + kd - kr + pk(x_0 + x_1) = 0$$

$$rqy^2 + (kq(d - r) + r + kp)y + k(d - r) + pk(x_0 + x_1) = 0.$$

Compare the coefficients of y^2, y and constant term in this last equation with the polynomial form of the quadratic in y:

$$ay^2 + by + c = 0$$

and we have

$$a = rq$$

$$b = kq(d - r) + r + kp$$

$$c = k(d - r) + pk(x_0 + x_1).$$

Notice how b is the same expression as Q – that is, $b = Q$. Therefore, if we write $Q = b = kq(d - r) + r + kp$ and apply the formula (7.10) for the roots of the quadratic, we obtain:

$$y = \frac{-Q \pm \sqrt{Q^2 - 4rq[k(d - r) + pk(x_0 + x_1)]}}{2rq}.$$

The $\pm$ symbol indicates that there are two roots. We are looking for the solution that is smallest in magnitude, as y starts out as 0 (no angiogenic cells). We therefore want the two terms on the numerator to partially cancel. We saw in the case study part B3 in Chapter 1 that Q is positive, so we take the $+$ sign in front of the square root. This gives us the formula for y at the top of this box!

7.2.5 The factorized form: $y = a(x - \alpha)(x - \beta)$

Consider for a moment the simple quadratic function $f(x) = x^2 + 2x - 3$. Using (7.12) we can write its roots as

$$\alpha = \frac{-2 - \sqrt{2^2 - 4(-3)}}{2} = \frac{-2 - \sqrt{4 + 12}}{2} = \frac{-2 - \sqrt{16}}{2} = \frac{-2 - 4}{2} = -3$$

$$\beta = \frac{-2 + \sqrt{2^2 - 4(-3)}}{2} = \frac{-2 + \sqrt{4 + 12}}{2} = \frac{-2 + \sqrt{16}}{2} = \frac{-2 + 4}{2} = 1.$$

These are values of x for which $f(x) = 0$, i.e. $f(-3) = 0$ and $f(1) = 0$.

But this quadratic can also be written as a product of two factors: $f(x) = (x + 3)(x - 1)$. Check this by multiplying out (see Section 1.5.3). By (1.21) this product will be zero if either factor is zero:

either $(x + 3) = 0$ or $(x - 1) = 0$,

meaning that either $x = -3$ or $x = 1$.

This demonstrates an important correspondence between factors and roots, for polynomial functions $f(x)$:

> If α is a root of $f(x)$, i.e. $f(\alpha) = 0$, then $(x - \alpha)$ is a factor of $f(x)$.
> Conversely, if $(x - \alpha)$ is a factor of $f(x)$, then $f(\alpha) = 0$. $\qquad$ (7.22)

This can sometimes help us to find factors of a polynomial. For example, I can't easily factorize $g(x) = 7x^2 - 20x + 13$, but I can see that if I put $x = 1$, then $g(1) = 7 - 20 + 13 = 0$. So I know that $(x - 1)$ is a factor of $g(x)$.

Sometimes we may need to sketch a quadratic that is given in factorized form, e.g. $f(x) = 3(x - 5)(x + 2)$. We immediately know that the roots are $x = 5$ and $x = -2$. These are where the curve $y = f(x)$ crosses the x-axis.

To find the y-intercept, put $x = 0$: $f(0) = 3(-5)(2) = -30$.

The turning-point will occur when x is midway between the roots, i.e. when $x = \dfrac{5 + (-2)}{2} = 1.5$.

Finally, the term in x^2 is $3x^2$, with positive coefficient, so the curve is a parabola concave upwards, as sketched in Figure 7.10.

In general, the curve of the quadratic in factorized form:

$$f(x) = a(x - \alpha)(x - \beta) \qquad (7.23)$$

has roots α and β, so crosses the x-axis at $x = \alpha$ and $x = \beta$.

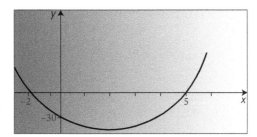

Figure 7.10 Graph of $y = 3(x - 5)(x + 2)$

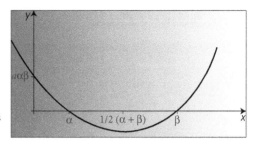

Figure 7.11 The quadratic function
$y = a(x - \alpha)(x - \beta)$, $a > 0$

Multiplying out, we get $f(x) = ax^2 - a(\alpha + \beta)x + a\alpha\beta$. So the y-intercept is at $y = a\alpha\beta$, and the turning-point occurs when $x = \frac{1}{2}(\alpha + \beta)$.

The curve is a parabola concave upwards if $a > 0$, or concave downwards if $a < 0$. The curve is sketched in Figure 7.11.

7.3 Cubic and higher-degree polynomials

We have now looked in considerable detail at linear functions and quadratic functions (which we obtained by adding a term in x^2 to the linear function). We could continue this process by adding further terms involving higher powers (or exponents) of x (x^3, x^4, and so on). These functions are called **polynomials** in x. The **degree** of a polynomial is the largest exponent of x that occurs in it. The term that has this largest exponent is called the **leading term**. So, for example, the function $p(x) = 3x^5 - 7x^4 + 2x^2 - 1$ is a polynomial of degree 5. Its leading term is $3x^5$. Here is a summary of the first few classes of polynomials, which have special names:

degree	name	general form of $p(x)$
0	constant	c
1	linear	$mx + c$
2	quadratic	$ax^2 + bx + c$
3	cubic	$ax^3 + bx^2 + cx + d$
4	quartic	$ax^4 + bx^3 + cx^2 + dx + e$
5	quintic	$ax^5 + bx^4 + cx^3 + dx^2 + ex + f$

In this notation, the leading term of the general polynomial of degree n would be ax^n.

7.3.1 Power functions and geometric series

The power functions are polynomials with just one term. The power function of degree n is the polynomial

$$P_n(x) = x^n, \quad \text{where } n = 1, 2, 3, \ldots \tag{7.24}$$

Thus, $P_1(x) = x^1 = x$, $P_2(x) = x^2$, $P_3(x) = x^3$,   and so on.

We have already seen the graphs of P_1 and P_2. The graphs of P_2, P_3, P_4, and P_5 are shown on the same axes in Figure 7.12, taken from **FNGRAPH 7.5**. When n is an even number (2, 4, 6, …)

then x^n is always positive (or zero at the origin), and the curve will be a U-shape (a concave upwards parabola). When n is an odd number (3, 5, …) then x^n is negative when x is negative, and positive when x is positive, so the curve snakes from bottom left to top right of the graph.

Now consider the functions $T_n(x)$ defined by

$$T_n(x) = P_n(x) - 1,$$

so that $T_1(x) = x - 1$, $T_2(x) = x^2 - 1$, $T_3(x) = x^3 - 1$, $T_n(x) = x^n - 1$.

If you put $x = 1$ you see that $T_n(1) = 0$, so $(x - 1)$ will be a factor of each of these polynomials. We have already seen this factorization in (1.20) in Section 1.5.3, for the first two polynomials:

$$T_2(x) = x^2 - 1 = (x - 1)(x + 1)$$
$$T_3(x) = x^3 - 1 = (x - 1)(x^2 + x + 1).$$

In general, the factorization is

$$T_n(x) = x^n - 1 = (x - 1)(x^{n-1} + x^{n-2} + \ldots + x + 1). \tag{7.25}$$

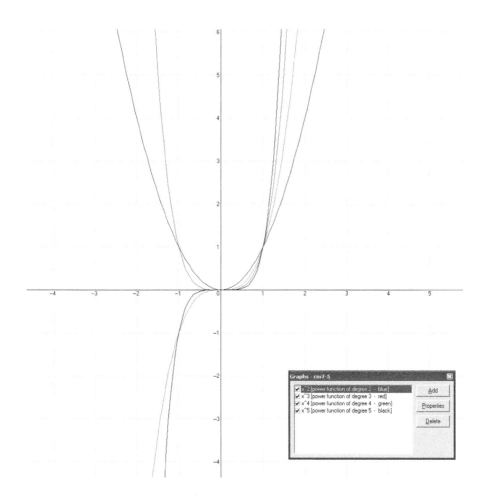

Figure 7.12 Graphs of the power functions (FNGRAPH 7.5)

You can check this by multiplying out. Dividing both sides by $(x - 1)$, we get:

$$x^{n-1} + x^{n-2} + \cdots + x + 1 = \frac{x^n - 1}{x - 1}.$$

The left-hand side of this equation is called a **geometric series**, and the right-hand side is its sum. You may remember it from school, in the following form:

$$a + ar + ar^2 + \cdots + ar^{n-1} = \frac{a(1 - r^n)}{1 - r} \tag{7.26}$$

where you used it to work out compound interest. The geometric series is the result of adding together the terms of the geometric sequence in (6.11). We are interested in the situation where the absolute value of r is less than 1, i.e. $|r| < 1$, meaning that r lies between -1 and $+1$. Then all the powers of r in (7.26) will be smaller than 1. As we take n larger and larger, the value of r^n becomes smaller and smaller (provided $|r| < 1$), and we saw in Section 6.4.4 that

$$\lim_{n \to \infty} (ar^n) = 0 \text{ provided } |r| < 1. \tag{6.12}$$

So, taking an infinite number of terms in our geometric series (i.e. letting $n \to \infty$) and using (6.12) in the right-hand side sum, we obtain a useful identity:

$$a + ar + ar^2 + \cdots + ar^n + \cdots = \frac{a}{1 - r} \quad \text{provided } |r| < 1. \tag{7.27}$$

On the left-hand side we have an **infinite geometric series**, meaning a geometric series with an infinite number of terms. Despite having an infinite number of terms added together, its sum is finite, provided that $|r| < 1$. For example, taking $r = \frac{1}{2}$, we get

$$1 + \frac{1}{2} + \frac{1}{4} + \cdots + \frac{1}{2^n} + \cdots = \frac{1}{1 - \frac{1}{2}} = 2.$$

If we take $r = -\frac{1}{2}$, the series becomes $1 + \left(-\frac{1}{2}\right) + \left(-\frac{1}{2}\right)^2 + \left(-\frac{1}{2}\right)^3 + \cdots$, and multiplying out and using (7.27) we get

$$1 - \frac{1}{2} + \frac{1}{4} - \frac{1}{8} + \cdots + \frac{(-1)^n}{2^n} + \cdots = \frac{1}{1 - \left(-\frac{1}{2}\right)} = \frac{1}{\left(\frac{3}{2}\right)} = \frac{2}{3} = 0.66667$$

We use a factor of $(-1)^n$ in writing the nth term as a neat way of saying that the terms are alternately positive and negative. Check that $(-1)^n$ is $+1$ when n is even (or zero), and is -1 when n is odd. If you don't believe that the geometric series converges to $\frac{2}{3}$, write an Excel spreadsheet to work out the sum of the first n terms, for increasing values of n – see the Problems. You should find that as you add more and more terms, the sum converges to the right-hand side of (7.27).

Problems: Now try the end-of-chapter problems 7.31–7.34.

7.3.2 General properties of polynomial functions

A polynomial of degree n may have up to n roots (i.e. it may cross the x-axis up to n times), with up to $n-1$ turning-points. The correspondence between roots α and factors $(x-\alpha)$ explained in Section 7.2.4, applies to polynomials in general. For example, the cubic polynomial

$$p(x) = x^3 + 7x^2 + 11x + 5 \tag{7.28}$$

can be written in factorized form as $p(x) = (x+5)(x+1)^2$, so it will have a root at $\alpha = -5$ and a double root (which is also a turning-point) at $\beta = -1$. You can draw the function in FNGraph by typing in the function as either `x^3 + 7*x^2 + 11*x + 5` or `(x + 5)*(x + 1)^2` in the Graphs window.

Notice that when x is large, the size of the leading term ax^n will be greater than that of the other terms. For example, in the polynomial (7.28), the leading term is x^3, and when $x = 1,000$ the leading term will be 10^9, while the other terms add up to less than 10^7. When x is large and negative, x^3 is also large and negative: $(-10^3)^3 = -10^9$. So the behaviour of the curve at the left-hand and right-hand ends of the graph can be identified by looking at the leading term. When x is large, the sign of ax^n depends on:

- the sign of x (positive or negative),
- the sign of a, and
- whether the exponent n is odd or even (since a negative number raised to an odd power is negative, but raised to an even power it will be positive).

This enables us to tell whether the graph of the polynomial will be heading up or down at each end of the curve. The results are summarized in Figure 7.13.

The last term in the expression for $p(x)$ tells us the y-intercept (the point where the curve crosses the y-axis). The polynomial in (7.28) will cross the y-axis at $y = 5$. But there is no easy way to identify the locations of the roots (where the curve crosses the x-axis) or the turning-points of a higher-degree polynomial, unless we can factorize it. The quest for such formulae has engaged mathematicians over centuries, and forms a remarkable story which we'll relate in a 'historical interlude' at the end of Part I. We'll also look at this problem again in Part II, using the more powerful techniques of calculus developed there.

7.3.3 Algebraic long division

To look for the roots of a higher-degree polynomial, you can try substituting values of x into it, looking for one that makes the polynomial zero. For instance, trying $x = -5$ in the polynomial in (7.28) produces

$$p(x) = x^3 + 7x^2 + 11x + 5$$
$$p(-5) = (-5)^3 + 7(-5)^2 + 11(-5) + 5$$
$$= -125 + 175 - 55 + 5$$
$$= 180 - 180 = 0$$

so we know that $(x+5)$ is a factor. Sketching the graph of the polynomial in FNGraph will help to identify likely values to try. Of course, this only works if there is a root that is a nice whole number.

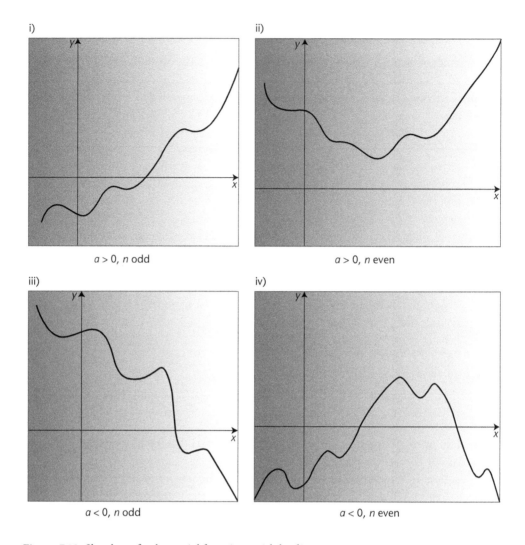

Figure 7.13 Sketches of polynomial functions with leading term ax^n

Once you have identified one factor, you can find remaining factors by a process called algebraic long division, which you hopefully remember from school. It works just like long division in arithmetic; I'll give a couple of examples to jog your memory. I've also produced SPREADSHEET 7.4 which performs the process; look at the formulae used in the cells for the coefficients. Here is algebraic long division for the example we are considering:

$$
\begin{array}{r}
x^2 + 2x + 1 \\
x+5 \enclose{longdiv}{x^3 + 7x^2 + 11x + 5} \\
\underline{x^3 + 5x^2 } \\
2x^2 + 11x \\
\underline{2x^2 + 10x } \\
x + 5 \\
\underline{x + 5} \\
0
\end{array}
$$

So $p(x) = (x + 5)(x^2 + 2x + 1)$. Here's a description of the first few steps:

- The x in $(x + 5)$ needs to be multiplied by x^2 to create x^3, the first term in $p(x)$, so write x^2 on the top line.

- $x^2(x + 5) = x^3 + 5x^2$; write this below the first two terms of $p(x)$.

- Subtract: $(x^3 + 7x^2) - (x^3 + 5x^2) = 2x^2$. Write this on the fourth line and bring down the third term ($11x$) from $p(x)$.

- The x in $(x + 5)$ needs to be multiplied by $2x$ to create $2x^2$, so write $2x$ on the top line.

And so on.

EXAMPLE

Show that $(x - 2)$ is a factor of $p(x) = 3x^3 + x^2 - 13x - 2$, and find the other factor.

Solution: Evaluate $p(2)$: $p(2) = 3 \times 2^3 + 2^2 - 13 \times 2 - 2 = 24 + 4 - 26 - 2 = 28 - 28 = 0$. As $p(2) = 0$, $(x - 2)$ is a factor of $p(x)$. Now divide out:

$$
\begin{array}{r}
3x^2 + 7x + 1 \\
\hline
x - 2 \,\big|\, 3x^3 + x^2 - 13x - 2 \\
3x^3 - 6x^2 \\
\hline
7x^2 - 13x \\
7x^2 - 14x \\
\hline
x - 2 \\
x - 2 \\
\hline
0
\end{array}
$$

So $p(x) = (x - 2)(3x^2 + 7x + 1)$. We could find the roots of the other factor using the formula for the roots of a quadratic.

If we divide a polynomial by a polynomial of smaller degree that is not a factor, we get left with a **remainder**.

EXAMPLE

Divide $p(x) = x^4 + 2x^3 - 3x^2 + x - 50$ by $(x - 3)$.

Solution:

$$
\begin{array}{r}
x^3 + 5x^2 + 12x + 37 \\
\hline
x - 3 \,\big|\, x^4 + 2x^3 - 3x^2 + x - 50 \\
x^4 - 3x^3 \\
\hline
5x^3 - 3x^2 \\
5x^3 - 15x^2 \\
\hline
12x^2 + x \\
12x^2 - 36x \\
\hline
37x - 50 \\
37x - 111 \\
\hline
61
\end{array}
$$

The remainder is 61, and we can write $p(x) = (x - 3)(x^3 + 5x^2 + 12x + 37) + 61$. It is this example which is initially programmed into SPREADSHEET 7.4. Try different problems by changing the coefficients in the yellow squares, and check the arithmetic yourself.

We could have found this remainder without performing the long division. If the remainder is R, so that $p(x) = (x - 3).q(x) + R$, where $q(x)$ is the unknown other factor, then $p(3) = (3 - 3).q(3) + R = 0.q(3) + R = R$, so simply evaluate $p(3)$:

$$p(3) = 3^4 + 2 \times 3^3 - 3 \times 3^2 + 3 - 50$$
$$= 81 + 54 - 27 + 3 - 50$$
$$= 138 - 77 = 61.$$

7.4 Logistic growth

7.4.1 The logistic function

When life scientists need to model a behaviour in which the dependent variable initially increases, but soon levels off and then drops back down towards its starting value, they reach for a quadratic function. You have probably experienced this sort of behaviour in real life: the harder you study for an exam, the better you will do in it – but every extra hour of study results in a smaller pay-off, and beyond a certain point, further study can be harmful: if you overdo it and study all night, you will be so exhausted in the morning that you fail the exam! A simple quadratic model would be a parabola concave downwards, with one root at the origin (zero study results in zero marks) and another root at some positive value k, as in Figure 7.14 (studying for k hours results in total exhaustion, and zero marks). We can write the equation of such a parabola as:

$$f(x) = ax\left(1 - \frac{x}{k}\right) \tag{7.29}$$

where a and k are positive coefficients. Here, x represents hours of study, and $f(x)$ is the performance achieved. The factors of $f(x)$ are x and $(1 - x/k)$, which are zero respectively when $x = 0$ and when $1 - x/k = 0$, i.e. when $x = k$. Multiplying out,

$$f(x) = ax\left(1 - \frac{x}{k}\right) = ax - ax.\frac{x}{k} = -\frac{a}{k}x^2 + ax.$$

The coefficient of x^2 is $-a/k$, which will be negative, so the parabola opens downwards, and it passes through the origin $(0, 0)$. The curve is sketched in Figure 7.14, and in FNGRAPH 7.6.

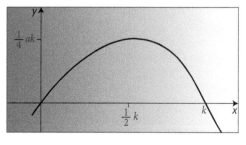

Figure 7.14 The logistic function $y = ax\left(1 - \dfrac{x}{k}\right)$

In this section we will describe two examples of this behaviour. The main application is in models of population growth, and from this we get the name of this function: the logistic function. The logistic function is also used in economic models; in Case Study C7 (predator–prey relationships) we describe a model that predicts the effects on fish stocks of over-fishing.

CASE STUDY C7: Fisheries management

From lions hunting prey on the African savannah, we now turn to a very different predator–prey situation. We will describe the Gordon–Schaefer bioeconomic model of fisheries management (see Anderson 1977, Chapter 3). Consider a fish stock in the ocean, which humans are going to exploit by fishing. The yield Y (measured as value of annual catch) will depend on the amount of effort E that the people put into the fishing enterprise. Effort is a single variable that combines factors such as the man-hours spent, the equipment used (boats, nets), the maintenance of the equipment, the associated infrastructure, etc. What will the yield-effort curve look like? When little effort is put in, the yield will be small. Increased effort will result in larger catches. But beyond a certain point, intensive effort will mean over-fishing which will reduce the fish population faster than it can breed, leading to smaller catches. Ultimately, over-fishing can drive the fish stock to extinction so the yield drops to zero.

The Gordon–Schaefer model depicts the yield-effort curve $Y = Y(E)$ as a parabola, concave downwards. It passes through the origin (zero effort results in zero yield), as shown in Figure 7.15(a).

The equation of the yield-effort curve will be

$$Y = aE\left(1 - \frac{E}{k}\right).$$
(7.30)

In this model, the fishing cost is taken to be directly proportional to the level of effort. Also shown in Figure 7.15(a) is the cost-effort line $C = mE$, where C is the cost for a level of fishing effort E. We assume that cost is directly proportional to effort, and the constant of proportionality m is the dollar cost per unit of effort.

The turning-point of the curve, at effort E_2, is where the greatest yield is obtained; this is called the Maximum Sustainable Yield (MSY). But this yield is not going to lead to the maximum profit. To find the net annual profit we have to subtract the cost from the gross yield; this is the vertical distance from the curve down to the cost-effort line, for a particular level of effort. This distance is a maximum at some lower level of effort E_1, and the yield at this point is called the Maximum Economic Yield (MEY). Unfortunately, in an open-access fishing regime, the fishing effort will increase as more people come in from outside in search of a profit, until the effort ceases to be profitable. This is where the line cuts the curve, at effort E_3, called the Equilibrium Point (EP). As the fishing industry becomes unprofitable, the government may seek to ensure its survival by providing subsidies. This will reduce the unit cost of fishing effort m, which rotates the cost-effort line clockwise as indicated by the arrow in Figure 7.15(b), increasing the over-fishing and driving the fish stock closer to extinction at effort E_4. In contrast, the challenge of fisheries management is to introduce regulations that will reduce the fishing effort from the free-market E_3 back to the more profitable E_2, or ideally to E_1.

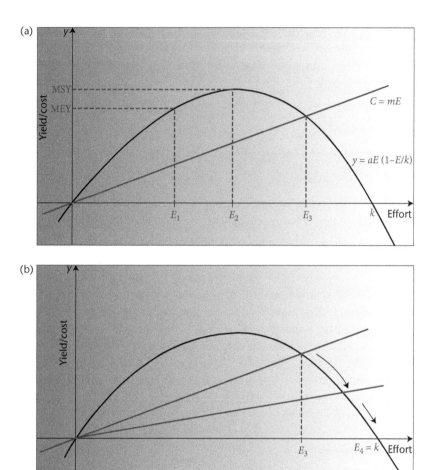

Figure 7.15 The Gordon–Schaefer model of fisheries management

7.4.2 Logistic growth of populations

In our Case Study A: Models of population growth, we have so far considered the geometric growth model. Starting from an initial population of size N_0, we write the population size after p generations as N_p. In geometric growth, the population after $p + 1$ generations is a fixed multiple of the previous size:

$$N_{p+1} = N_p + R.N_p = (R + 1)N_p. \tag{7.31}$$

Here, R is the growth rate per head of population, so that $R.N_p$ is the net growth of a population of size N_p, over one generation. Adding this to the original population produces the total new population. Look back at the Case Study A boxes in Chapter 1.

This geometric growth model is quite realistic for the early stages of growth of a colony of cells placed on a Petri dish, but – as we noted in Part 5 of the case study, at the end of Chapter 2 – growth will not continue unboundedly. The available supply of nutrient is limited, and this limits the number of cells that the Petri dish can support. The same situation occurs in the wild: the size of a population of animals is limited by the available food and other resources.

In order to model this, we use a growth rate that is not constant, but that depends on the population size. This is called **density-dependent growth**, and the particular model we will look at is called **logistic growth** ('logistics' refers to the management of food, equipment and other resources by an organization such as the Army). In this model, we assume that the environment has sufficient resources to support a maximum size of population of K individuals. The parameter K is called the **carrying capacity** of the environment. When the population is small (i.e. N is much less than K) this does not restrict growth, which occurs at a maximum growth rate r_m (the equivalent of the growth rate R in the geometric growth model). But as the population grows, the growth rate r steadily reduces until when $N = K$ there is zero growth. In terms of functions, we can say that growth rate r is a function of population N, written $r = r(N)$, with the conditions that $r(0) = r_m$ and $r(K) = 0$. A linear function satisfying these conditions, is

$$r(N) = r_m\left(1 - \frac{N}{K}\right). \tag{7.32}$$

The graph of this function is shown in Figure 7.16, and Figure 7.17(a) shows the sort of population growth over time that we hope this model will produce. We can now look at how this growth model is used in Case Studies A and B.

Problems: Now try the end-of-chapter problems 7.35–7.38.

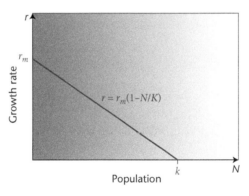

Figure 7.16 The logistic growth rate

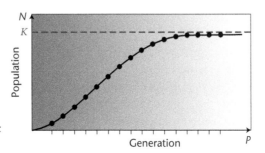

Figure 7.17 Logistic population growth

CASE STUDY B7: Logistic growth of cancer cells

The rate of increase of the angiogenic cancer cells in a tumour is given by

$$r.y.\left(1 - \frac{y}{k}\right) - d.y - \frac{y.p.(x_0 + x_1 + y)}{q.y + 1}.$$

Here, y is the density of angiogenic cells, r is the replication rate, and d is the death rate of the angiogenic cells. So, as we noted in Part B3 of this case study (Section 1.7.2), the first two terms describe the net increase ('births minus deaths') in the angiogenic cell population.

But whereas the death rate d is constant, the birth rate b is density-dependent, following equation (7.32): $b = r\left(1 - \frac{y}{k}\right)$. Recall that k is the maximum density of angiogenic cells, i.e. the carrying capacity.

CASE STUDY A8: 'Cobwebbing' of logistic growth model

We can find out whether the logistic growth model really produces the population growth curve in Figure 7.13, by using equation (7.32) in place of the growth rate R in the geometric growth model update equation

$$N_{p+1} = N_p + R.N_p. \tag{7.31}$$

The update equation for logistic growth is thus

$$N_{p+1} = N_p + r_m.N_p\left(1 - \frac{N_p}{K}\right). \tag{7.33}$$

If we have values for the initial population N_0, and the fixed parameters r_m and K, we could produce a table of data-points $(0, N_0)$, $(1, N_1)$, $(2, N_2)$, $(3, N_3)$, etc., and we could thus produce a graph of N_p against p, as we did for geometric growth in Part A6 of this case study (in Chapter 3).

   Recall from Part A6 that we can also produce a 'cobweb graph' illustrating the growth of the population from one generation to the next. In this case we plot a graph with the old population N_p along the x-axis, and the new population N_{p+1} along the y-axis, then the points (N_0, N_1), (N_1, N_2), (N_2, N_3), etc. will lie on the curve of the update equation. For the logistic growth model, this update curve will be:

$$y = x + r_m.x\left(1 - \frac{x}{K}\right). \tag{7.34}$$

This is a quadratic function, whose curve is a parabola concave downwards (because the coefficient of the x^2 term is negative), crossing the x-axis at the origin $(0, 0)$ and at a positive value of x, as in the yield-effort curve of Figure 7.15. We also draw the diagonal line $y = x$. We can now draw the 'cobweb' by starting at N_0 on the x-axis, moving up to the curve, across to the line, up to the curve, and so on. The graph will look like Figure 7.18, with the cobweb converging to the point where the line and curve intersect. This represents the equilibrium point when the population has reached the carrying capacity K.

This logistic growth model is programmed into SPREADSHEET 7.5, where the graph of N_p against p, and a cobweb graph are also plotted. We have used a carrying capacity of $K = 50$, an initial population of $N_0 = 1$, and a maximum growth rate $r_m = 0.8$. With these parameter values, we do indeed obtain graphs matching Figures 7.17 and 7.18. The spreadsheet is reproduced in Figure 7.19.

But some very strange phenomena occur if we increase the growth rate r_m in cell B6. We will explore these phenomena in Chapter 18. Figure 18.9 shows how the population behaves for values of r_m between 1.8 and 2.85. Looking at the lower graph in SPREADSHEET 7.5 as you increase r_m, you realise why it's called a cobweb graph. There's an example in Figure 18.10.

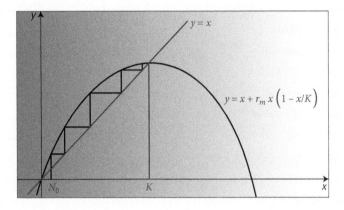

Figure 7.18 'Cobwebbing' of logistic growth

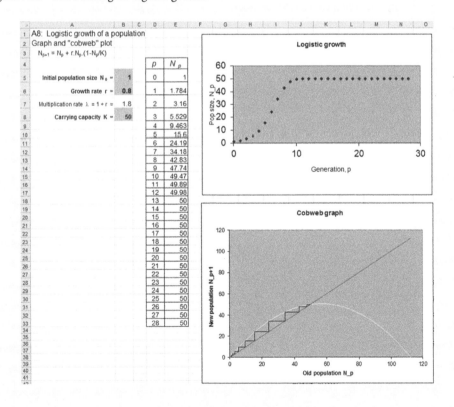

Figure 7.19 Logistic growth, for $r = 0.8$ (SPREADSHEET 7.5)

7.5 **Extension: quadratic interpolation**

In Section 3.6.2 we introduced the technique of interpolation. If we have a set of data-points (x_i, y_i) and are given an intermediate x-value, the task of interpolation is to estimate the corresponding y-value. We saw how to perform linear interpolation, which works by joining up the data-points on the graph by a series of straight lines. So if the interpolating x-value x lies between two data-point x-values x_0 and x_1, the corresponding value of y is calculated from the straight line joining (x_0, y_0) and (x_1, y_1). If the interval width is h, we first calculate the interpolating factor s by

$$s = \frac{x - x_0}{h} \tag{7.35}$$

and the first difference

$$\Delta y_0 = y_1 - y_0, \tag{7.36}$$

then the interpolated y-value is given by

$$y = y_0 + s.\Delta y_0. \tag{7.37}$$

The algorithm was demonstrated in SPREADSHEETS 3.6 AND 3.7.

These are simple calculations, but the resulting estimate of y will not be very accurate (unless the relationship between x and y is nearly linear in this region of the graph). To make a better estimate, we could use a third data-point and fit a quadratic curve through the three points, instead of a line through two data-points. This new algorithm is called **quadratic interpolation**.

Suppose that we have the same problem as in Section 3.6: to estimate y corresponding to an x-value x lying between x_0 and x_1. In addition to the data-points (x_0, y_0) and (x_1, y_1), we also use the next data-point (x_2, y_2). As before, we work out the first difference Δy_0 using equation (7.36). But now we can also work out a first difference Δy_1 by:

$$\Delta y_1 = y_2 - y_1. \tag{7.38}$$

And the difference of the two first differences, is called the **second difference** $\Delta^2 y_0$:

$$\Delta^2 y_0 = \Delta y_1 - \Delta y_0. \tag{7.39}$$

Now the interpolated y-value, using quadratic interpolation, is given by

$$y = y_0 + s.\Delta y_0 + \frac{1}{2}s(s-1)\Delta^2 y_0. \tag{7.40}$$

Here, the interpolating factor s is exactly the same as was used in linear interpolation, i.e. found from equation (7.35). The first two terms in the formula for y are the same as in the linear interpolation formula (7.37); the third term is a correction using the third data-point.

The algorithm is illustrated in SPREADSHEET 7.6, shown in Figure 7.20. You can change the values in the yellow cells: the y-values, the first x-value x_0, and the interval width h. Note that the width of the interval between x_1 and x_2 must be the same as that between x_1 and x_0, so x_2 is calculated as $x_2 = x_0 + 2h$. The quadratic curve through the three data-points is shown in the chart.

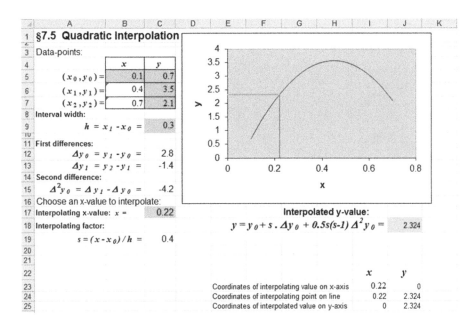

Figure 7.20 Quadratic interpolation (SPREADSHEET 7.6)

Here is the calculation made in Section 3.6.2, but this time using quadratic interpolation. The data table is

x	y
1.00	3.92
1.50	8.26
2.00	11.45
2.50	13.51
3.00	14.42
3.50	14.20
4.00	12.83
4.50	10.33
5.00	6.68

We want to estimate the value of y when $x = 1.8$.

General method	Particular example
x is the interpolating x-value. The data x-values on the left and right of x are x_0 and x_1.	We want to interpolate at $x = 1.8$. This lies between the data x-values $x_0 = 1.5$ and $x_1 = 2.0$.
The interval width is $h = x_1 - x_0$.	The interval width is $h = 2.0 - 1.5 = 0.5$.
The interpolating factor is $$s = \frac{(x - x_0)}{h}.$$	The interpolating factor is $$s = \frac{(1.8 - 1.5)}{0.5} = \frac{0.3}{0.5} = 0.6.$$

The data y-values corresponding to x_0, x_1 and x_2 are y_0, y_1, and y_2.	The data y-values corresponding to $x_0 = 1.5$, $x_1 = 2.0$, $x_2 = 2.5$ are $y_0 = 8.26$, $y_1 = 11.45$, $y_2 = 13.51$.
The first differences are: $$\Delta y_0 = y_1 - y_0$$ $$\Delta y_1 = y_2 - y_1.$$	The first differences are: $$\Delta y_0 = 11.45 - 8.26 = 3.19$$ $$\Delta y_1 = 13.51 - 11.45 = 2.06.$$
The second difference is: $$\Delta^2 y_0 = \Delta y_1 - \Delta y_0.$$	The second difference is: $$\Delta^2 y_0 = 2.06 - 3.19 = -1.13.$$
The interpolated value is then $$y = y_0 + s.\Delta y_0 + \frac{1}{2}s(s-1)\Delta^2 y_0.$$	The interpolated value is then $$y = 8.26 + 0.6 \times 3.19 + \frac{1}{2}.0.6(-0.4)(-1.13)$$ $$= 8.26 + 1.914 + 0.1356$$ $$= 10.3096.$$

So when $x = 1.8$, we now estimate y as 10.31 (to 2 decimal places).

In SPREADSHEET 7.7 (see Figure 7.21) we see the quadratic interpolation algorithm applied to this table of data-points. See Section 3.6.3 for more information on how to use the spreadsheet. The only extra feature of the data table is that column I contains the second differences. The interpolated y-values in column J are calculated using equation (7.40). For a small extra calculation, we obtain much more accurate interpolated values than from linear interpolation.

Problems: Now try the end-of-chapter problems 7.39–7.45.

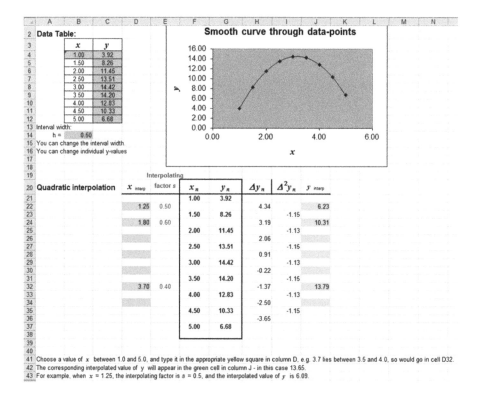

Figure 7.21 Piecewise quadratic interpolation (SPREADSHEET 7.7)

7.1–7.4: Try to find the 'completed square' form of the following quadratic functions, by entering the coefficients in SPREADSHEET 7.2 and adjusting the sliders to match the curve:

7.1 $f(x) = x^2 - 12x + 21$ 7.2 $f(x) = 3x^2 + 12x + 8$

7.3 $f(x) = x^2 - 7x + 15$ 7.4 $f(x) = 2x^2 - 9x + 6$.

7.5–7.8: Now convert the functions in 7.1–7.4 to 'completed square' form algebraically. Compare your result with the values you estimated graphically. Find the roots of the function, if any. Check your answers using SPREADSHEET 7.1.

If you have not already done so, try problems 6.38–6.52.

7.9–7.16: Find the roots (if any) of the following quadratic functions, using the formula for the roots of a quadratic. Express them as surds, and also evaluate them correct to 4 significant figures. Check your answers by drawing the parabolas in FNGraph:

7.9 $x^2 - 5x + 4$ 7.10 $x^2 - 2x + 4$

7.11 $x^2 + 6x + 9$ 7.12 $2x^2 + 3x - 9$

7.13 $3x^2 + 2x - 9$ 7.14 $3x^2 - 2x + 9$

7.15 $7 - 2x - x^2$ 7.16 $-x^2 - 7x - 2$.

7.17–7.20: Find the equations, in the form $ax^2 + bx + c$, for the quadratic functions passing through the three points:

7.17 $(-2, 12), (0, -2), (2, 24)$ 7.18 $(1, 0), (4, 30), (6, 60)$

7.19 $(-1, -6), (1, 0), (5, -12)$ 7.20 $(1, -12), (4, 3), (5, 12)$.

7.21: The end of a ski-jump track is 6 metres above the ground. The ground is sloping downhill at a gradient of 1 in 2; See Figure 7.22. A ski-jumper speeds down the inrun and once he shoots off the lip at the end of the track, he describes a parabolic curve until he lands, further down the slope. At his maximum height, he is at a point 2 metres vertically above the lip, and 20 metres horizontally forward of it. How far down the slope does he land (the distance d in Figure 7.22)? Take the origin of (x, y) coordinates at the point O on the slope directly beneath the track lip.

7.22–7.23: A lion can accelerate at 6.25 m s⁻², until he reaches a top speed of 14 m s⁻¹. A gazelle can accelerate at 3.75 m s⁻², until it reaches a top speed of 25 m s⁻¹. In the following situations, will the lion catch the gazelle or not?

7.22: The lion is 6 m away from the gazelle, and they both start running at the same instant.

7.23: The lion starts running when the gazelle is 10 m away, and after 0.6 seconds the gazelle sees him and starts running.

7.24: Using equations (7.13) and (7.14), derive (7.15).

7.25–7.30: Case Study A7: Fibonacci's Rabbits
In Case Study A7 we derived the Fibonacci update equation for the rabbit colony:

$$N_{p+1} = N_p + N_{p-1} \tag{7.17}$$

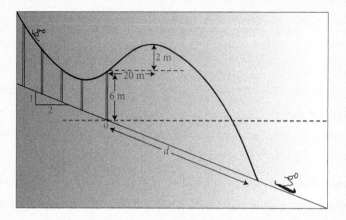

Figure 7.22 Ski-jump

and by substituting $N_p = N_0 r^p$ we obtained the quadratic equation

$$r^2 - r - 1 = 0. \tag{7.18}$$

7.25: Use the formula (7.10) to find the roots; you should find that

$$r_1 = \frac{1 - \sqrt{5}}{2} \text{ and } r_2 = \frac{1 + \sqrt{5}}{2}.$$

We theorize that the closed-form solution for N_p will be of the form

$$N_p = A r_1^p + B r_2^p \tag{7.19}$$

for some values of A and B. By requiring (7.19) to produce the first two Fibonacci numbers ($N_0 = 1$, $N_1 = 1$) obtain two simultaneous equations involving A and B. Solve them, and you should find that

$$A = \tfrac{1}{2}\left(1 - \frac{1}{\sqrt{5}}\right), \quad B = \tfrac{1}{2}\left(1 + \frac{1}{\sqrt{5}}\right).$$

Using these and the values of r_1 and r_2, all written as surds, evaluate N_2 using (7.19) and check that it produces $N_2 = 2$.

If you're really keen, you can try to check that (7.19) also produces $N_3 = 3$.

7.26: Create an Excel spreadsheet that works out the Fibonacci sequence using the update equation (7.17), to answer Fibonacci's original problem, i.e. to find N_{12}. Make a table with the generation number p in column A: $p = 1, 2, 3, \ldots, 11, 12$. The Fibonacci numbers N_p will be in the corresponding cells of column B: type 1 into the cells for N_0 and N_1. Then program the calculation (7.17) in the cell for N_3. Select the cell, grab its handle, and drag down the column to create the rest of the Fibonacci sequence. You should find that $N_{12} = 233$ pairs. See if you also obtain 233 by evaluating (7.19) for N_{12}. You will have to convert r_1, r_2, A, and B to decimals, and evaluate r_1^{12} and r_2^{12} using the [x^y] key on your calculator.

7.27: The Fibonacci numbers $N_0, N_1, N_2, N_3, \ldots$ grow larger and larger, unboundedly. But what happens if we look at the ratio of each Fibonacci number to its predecessor?

$$\frac{N_2}{N_1} = \frac{2}{1} = 2, \ \frac{N_3}{N_2} = \frac{3}{2} = 1.5, \ \frac{N_4}{N_3} = \frac{5}{3} = 1.667, \ \frac{N_5}{N_4} = \frac{8}{5} = 1.6, \ \frac{N_9}{N_8} = \frac{13}{8} = 1.625, \ldots$$

The ratios appear to be converging, to a number between 1.635 and 1.667. This number is called the **Golden Section**, and commonly written as Φ (a capital Greek letter f, pronounced phi). Using the notation of limits, we could define Φ as

$$\Phi = \lim_{p \to \infty}\left(\frac{N_{p+1}}{N_p}\right) \text{ where } N_p \text{ is the } p\text{th Fibonacci number.}$$

Modify your Excel spreadsheet from problem 7.26 to calculate this ratio sequence. In column C, in the cell for $p = 2$, program the formula to calculate $\frac{N_2}{N_1}$ using the numbers in column B. Then drag this cell down the column to produce the ratio sequence. The sequence should converge to $\Phi = 1.6180339887\ldots$ This number crops up in nature, and was used by the Ancient Greeks to define the proportions of their temples. The ratio Φ:1 is called the **Golden Ratio**. You can learn more about it at Dr Knott's website, quoted in the Case Study. We will meet it again in Section 8.6.

This mysterious number is actually a root of a simple quadratic equation. To find it, start with (7.17) and divide both sides by N_p:

$$\frac{N_{p+1}}{N_p} = 1 + \frac{N_{p-1}}{N_p}.$$

Now take the limit of each term as $p \to \infty$. Then $\frac{N_{p+1}}{N_p} \to \Phi$, but also $\frac{N_p}{N_{p-1}} \to \Phi$, so the equation becomes $\Phi = 1 + \frac{1}{\Phi}$. Rearrange this to obtain a quadratic equation in Φ, find its roots, and decide which one we should use. You should find that $\Phi = \frac{1}{2}\left(1 + \sqrt{5}\right)$.

Extension: On the spreadsheet, find $\lim_{p \to \infty}\left(\frac{N_{p+1}}{N_{p-1}}\right)$ and $\lim_{p \to \infty}\left(\frac{N_{p+1}}{N_{p-2}}\right)$. Can you work out what these limits represent?

7.28: There are many other Fibonacci-like sequences. Here are two examples:

(i) The Lucas numbers use the Fibonacci update equation (7.17) but start with $N_0 = 2$, $N_1 - 1$. Try changing the first two terms in your spreadsheet, to these values and to other pairs of starting-values; does the ratio sequence still converge to Φ?

(ii) The Pell numbers are defined by the update equation $N_{p+1} = 2N_p + N_{p-1}$, starting with $N_0 = 0$, $N_1 = 1$. Find the equivalent of (7.18) for this case, work out r_1 and r_2, and find the new A and B in (7.19). Hence show that a closed-form solution for the pth Pell number N_p is: $N_p = \frac{1}{2\sqrt{2}}\left[\left(1 + \sqrt{2}\right)^p - \left(1 - \sqrt{2}\right)^p\right]$. Adapt your Fibonacci spreadsheet to produce the Pell numbers, and the ratio sequence.

7.29: In your spreadsheet from problem 7.27, program a blank cell with the formula for $\sqrt{5}$. Then program four more cells to use this value to calculate A, B, r_1, r_2. Use another blank cell for the value of p, and program two more cells to calculate Ar_1^p and Br_2^p. Program a final cell to use these to calculate N_p in (7.19). Does it produce 233 when you enter $p = 12$?

Now replace the formula for $\sqrt{5}$ in the first cell, with 2.24, which is $\sqrt{5}$ correct to 3 significant figures. Perform an error analysis (Section 2.6) to find the uncertainty interval for the final calculation of N_{12}. Does it include the true value 233?

7.30: We return to Fibonacci's rabbit colony, but we now suppose that each rabbit pair only starts producing offspring in their third month, rather than their second month. Write down the update equation, and modify your Excel spreadsheet to produce the new sequence. Look at the new ratio sequence: is it converging? You may need to extend the table beyond $p = 12$. Using the same technique as we have followed in problem 7.27, find a cubic equation of which the limit Φ of the ratio sequence $\left(\dfrac{N_{p+1}}{N_p}\right)$ should be a root.

You should find that it is

$$\Phi^3 - \Phi^2 - 1 = 0. \tag{7.41}$$

This can be rearranged to $\Phi = 1 + \dfrac{1}{\Phi^2}$. Find out if the limit of your ratio sequence does indeed satisfy this equation.

Use a bracketing method (Section 8.6) to find a root of the cubic equation (7.41), in the interval $1 < \Phi < 2$. Is this the same as the limit of the ratio sequence?

We can also find a root of the cubic equation (7.41) using the closed-form solution we derived in problem 1.66. We first have to make a linear transformation from Φ to a new variable x. Write $x = \Phi - d$, so that $\Phi = x + d$. Substitute this into (7.41), multiply out the terms in Φ^3 and Φ^2, and collect together the terms in x^3, x^2, x, and the constant term. We want the polynomial to be of the form $x^3 + ax - b$, with no term in x^2. Check that this happens if we take $d = \frac{1}{3}$. Use this value of d to evaluate the coefficient of x and the constant term. You should find that the equation in x is:

$$x^3 - \tfrac{1}{3}x - \tfrac{29}{27} = 0. \tag{7.42}$$

Now use the values $a = -\frac{1}{3}$ and $b = \frac{29}{27}$ in the formula for the root given in problem 1.66. You need to be very careful in the arithmetic. Work systematically:

$\dfrac{a}{3} =$ 　　　　　 $, \dfrac{b}{2} =$

$\left(\dfrac{a}{3}\right)^3 =$ 　　　　 $,\left(\dfrac{b}{2}\right)^2 =$

$\left(\dfrac{a}{3}\right)^3 + \left(\dfrac{b}{2}\right)^2 =$

$z = \sqrt{} =$

$\dfrac{b}{2} + z =$ 　　　 $,\dfrac{b}{2} - z =$

$\sqrt[3]{\dfrac{b}{2} + z} =$ 　　 $,\sqrt[3]{\dfrac{b}{2} - z} =$

Hence $x = \sqrt[3]{} + \sqrt[3]{}$

Then $\Phi = x + \frac{1}{3} =$

7.31–7.32: For the following geometric series, find a and r in equation (7.27), and hence find the sum of the series. Check your answer by programming an Excel spreadsheet; in column A it should work out the individual terms in the series, and in column B the partial sums.

Thus, the formula in cell B6 (holding the sum of the first 6 terms) would be ' = B5 + A6', meaning 'sum of the first 5 terms, plus the 6th term'.

7.31 $2 + \dfrac{1}{2} + \dfrac{1}{8} + \dfrac{1}{32} + \cdots$

7.32 $1 - \dfrac{1}{3} + \dfrac{1}{9} - \dfrac{1}{27} + \cdots$

7.33: Use equation (7.26), for the sum of a geometric series, to find the total number of grains of wheat on the chessboard, in problem 1.45. If a grain of wheat weighs 50 mg, calculate the total mass of wheat, to the nearest metric tonne (1000 kg).

7.34: *'A cheetah is chasing a rabbit. The cheetah can run much faster than the rabbit, but it never catches it. How is this possible?'*

'Well, suppose that initially the rabbit is 20 metres ahead. The cheetah runs up to the point where the rabbit was – but in that time the rabbit has moved forward. So the cheetah runs to the point where the rabbit now is – but in the time this takes, the rabbit has now moved forward again. And so on: every time the cheetah reaches the rabbit's previous position, the rabbit has moved forward – so the cheetah can never catch the rabbit.'

Can you explain what is wrong with this logic?[3]

7.35: In a Pacific coastal fishing area, the value Y of the annual catch (in thousands of dollars) is related to the level of fishing effort E (in 'units of effort') by the Gordon–Schaefer model:

$$Y = aE\left(1 - \frac{E}{k}\right) \tag{7.30}$$

where $a = 0.7$ and $k = 80$. The cost-effort relationship is $C = mE$, where $m = 0.2$. What are the units of a, k, m and C?

Referring to Figure 7.15(a), calculate:

 (i) the effort E_2 required to generate the Maximum Sustainable Yield, and the value of that yield $Y(E_2)$;

 (ii) the effort E_3 at the Equilibrium Point, and the value of the Equilibrium Point yield;

 (iii) the effort E_1 that will generate the Maximum Economic Yield, the value of that yield, and the net profit of that yield, i.e. $Y(E_1) - C(E_1)$;

 (iv) the net profit for the level of effort E_2, i.e. $Y(E_2) - C(E_2)$.

7.36: A population of N Takahe birds on an island off New Zealand is in geometric decline, described by the update equation $N_{p+1} = aN_p$, where N_p is the population at the pth generation, and $0 < a < 1$. In a conservation programme, a steady stream of new Takahe are introduced from the mainland, equivalent to b new individuals at each generation. The population model then becomes

$N_{p+1} = aN_p + b$.

---

[3] This is a paradox which was proposed by the Greek philosopher Zeno of Elea (490 BC–430 BC). He described a race between Achilles and a tortoise. There is an animation of the race on YouTube.

We could write this update equation in function notation as $N_{p+1} = f(N_p)$, where $f(x) = ax + b$. Let N_0 be the initial population. Then:

after 1 generation, $N_1 = f(N_0) = aN_0 + b$;

after 2 generations, $N_2 = f(N_1) = f(f(N_0)) = a(aN_0 + b) + b = a^2N_0 + b(a + 1)$; and so on. The nth generation is given by $N_p = f^n(N_0)$, where f^n means the function f composed with itself n times (see Section 5.6.3).

Find N_3, and deduce that N_p is given by

$$N_p = a^p N_0 + b(a^{p-1} + a^{p-2} + \cdots + a + 1).$$

Using the theory of limits of functions (Section 6.4.1) and of geometric series (Section 7.3.1), show that as $n \to \infty$ the population will approach the equilibrium level $N_\infty = \dfrac{b}{1-a}$.

7.37: Could we have obtained the result of problem 7.36 with a lot less work? Given the update equation $N_{p+1} = aN_p + b$, when the sequence $N_0, N_1, N_2, \ldots$ defined by $N_{p+1} = f(N_p)$ converges to its equilibrium value N_∞, successive values will be equal. So if $N_{p+1} = N_p = N_\infty$ we have $N_\infty = f(N_\infty)$, i.e. $N_\infty = aN_\infty + b$.

Solving this equation for N_∞ we get $N_\infty = \dfrac{b}{1-a}$.

Or is there something extra which we have proved in problem 7.36, which is not proved here?

7.38: Use the argument of the previous problem to show that if the population generated by the logistic growth equation (7.33) reaches an equilibrium, it will be either extinction ($N_\infty = 0$) or the carrying capacity $N_\infty = K$.

7.39–7.43: Repeat the calculations of problems 3.13–3.17, but using quadratic interpolation. Compare your answers with the linear-interpolation estimates. Which do you think are more realistic?

Year	Atmospheric CO_2 (ppm)
1959	315.98
1969	324.62
1979	336.78
1989	352.90
1999	368.14
2009	387.35

Source: National Oceanic and Atmospheric Administration, quoted on CO2now.org website.

7.39 1972

7.40 1987

7.41 2007

7.42 2012

7.43 2020.

7.44–7.46: The following table gives the population of Sweden from 1780 to 1960, collected from national censuses at 20-year intervals. Use quadratic interpolation to estimate the population in the following years:

Year	Population (thousands)	Year	Population (thousands)
1780	2104	1880	4572
1800	2352	1900	5117
1820	2573	1920	5876
1840	3123	1940	6356
1860	3824	1960	7480

Source: Hoppensteadt and Peskin 1992.

7.44 1890

7.45 1970

7.46 2000.

We will see this data again in Section 10.4.1.

8

Fitting curves; rational and inverse functions

In this chapter we will look at two concepts which can provide us with some useful new functions: the reciprocal of a function and the inverse of a function. These two ideas often get confused in people's minds – mainly because of the inconsistent names and notation which have developed. We'll deal with reciprocals in Section 8.1 and inverses in Section 8.5. We'll also study a class of functions called rational functions, which have important applications in the biosciences, e.g. in biochemistry.

This is also a convenient point at which to discuss how the functions we are producing can be used with real data: how do we use the data-points obtained from experiment or observation to determine the values of the parameters in the function? We have already done this for linear functions $y = mx + c$: in Section 6.3.3 we saw how to find the values of the parameters m and c from two data-points, and if we have more than two data-points Section 6.3.4 showed how to put a best-fit straight line through them. Section 8.3 will show how to extend this technique to fit more general curves to data, and in the final, extension Section 8.7 we reveal how the equation of the best-fit straight line is actually calculated in Excel.

. . . .

8.1 Reciprocal functions

Recall from Section 1.3.3 that the reciprocal of a number a is $\frac{1}{a}$. So the reciprocal of 2 is $\frac{1}{2}$ or 0.5, the reciprocal of 10 is $\frac{1}{10}$ or 0.1, and the reciprocal of -2 is $\frac{1}{-2} = -\frac{1}{2} = -0.5$. If you take the reciprocal of the reciprocal, you get back to the original number:

$$\frac{1}{1/a} = \frac{1 \times a}{\frac{1}{a} \times a} = \frac{a}{1} = a.$$

You will also remember from the laws of exponents in Section 1.5.2 that raising a number to the power of -1 produces the reciprocal:

$$a^{-1} = \frac{1}{a}.$$

If you're unsure of the rules for exponents you should look back to that section now, as we'll be using those rules a lot in the remainder of Part I.

8.1.1 Definition of the reciprocal of $f(x)$

The reciprocal of a function $f(x)$ is defined in just the same way as for a number:

> The reciprocal of the function $f(x)$ is the function $\dfrac{1}{f(x)}$. $\qquad(8.1)$

So the reciprocal of the function $y = x$ is the function $y = \dfrac{1}{x}$.

Notation warning: This is where the inconsistent notation starts. We can write $\dfrac{1}{a}$ as a^{-1}, when a is a number, but we **cannot** write $\dfrac{1}{f(x)}$ as $f^{-1}(x)$. This notation is used for the inverse of f, which is something different (Section 8.5).

Problems: Now try the end-of-chapter problems 8.1–8.4.

8.1.2 Rational functions $y = \dfrac{1}{x}$, $y = \dfrac{1}{ax+b}$, $y = \dfrac{x}{ax+b}$

Recall from Chapter 1 that rational numbers are numbers that can be expressed as a fraction (or ratio) of two integers (or whole numbers). Similarly, rational functions are fractions where the numerator (the expression on top) and the denominator (the expression on the bottom) are polynomial functions.

The simplest rational function is $y = 1/x$, whose graph is shown in Figure 8.1(a). The shape of the curve is called a **rectangular hyperbola**. Notice that it doesn't cross the x-axis or y-axis, because if $x.y = 1$ it is impossible for either x or y to be zero. The axes are **asymptotes** to the curve (lines which the curve approaches but never reaches), which we introduced in Section 6.4. Notice also that there are two separate parts of the curve, coming from x, y both positive or both negative. We saw experimental data producing such a rectangular hyperbola in SPREADSHEET 3.5: in Section 5.2.2 we identified it as the inverse proportion function

$$y = \frac{k}{x}. \qquad (8.2)$$

A more general rational function comes from taking the reciprocal of the linear function $y = ax + b$ which we discussed in Section 6.3.3. We now call the parameters a, b rather than m, c since in the reciprocal function they no longer represent the slope and y-intercept. The graph of

$$y = \frac{1}{ax+b} \qquad (8.3)$$

is again a hyperbola. To sketch it, we investigate whether it crosses the axes, and whether it has any asymptotes.

To find the y-intercept, set $x = 0$. Then $y = \dfrac{1}{a.0+b} = \dfrac{1}{b}$. So the curve crosses the y-axis at the point $(0, 1/b)$.

It is impossible for the denominator $ax + b$ to be zero, or we would be dividing by zero. As $ax + b = 0$ is equivalent to $x = -b/a$, there is no point on the curve for which x equals $-b/a$. This means that the vertical line at $x = -b/a$ is an asymptote.

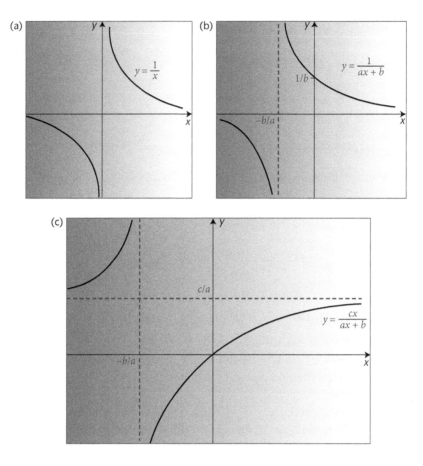

Figure 8.1 Rectangular hyperbolae

Since $\lim\limits_{x\to\infty}\dfrac{1}{ax+b}=0$ (see Section 6.4.1), the x-axis will be a horizontal asymptote.

The graph of the function (8.3) is sketched in Figure 8.1(b). FNGRAPH 8.1 shows the graph for the case when $a=2$ and $b=5$, in which case the curve crosses the y-axis at $y=1/5=0.2$, and the vertical asymptote is at $x=-5/2=-2.5$. Try entering other values of a, b to see the effect; use negative as well as positive values. Notice that the function must be entered in FNGraph as 1/(a*x + b) using brackets. Because of the BEDMAS order of operations, typing 1/a*x + b would produce the graph of $y=\dfrac{1}{a}x+b$.

A rational function of practical importance in the biosciences is obtained by putting an x on the numerator, that is

$$y=\frac{cx}{x+b}.\tag{8.4}$$

This curve will now pass through the origin, as when $x=0$ on the numerator then the whole fraction will become zero. It now has the vertical asymptote at $x=-b$, but in practical applications we are usually only interested in the graph for positive values of x. To find the horizontal asymptote, we need to know what happens to the right-hand side of (8.4) when x becomes large. It's not immediately obvious from looking at the equation, because both the numerator and denominator will be large, so what is the ratio? The trick is to look at the leading terms of each expression (see the introduction to Section 7.3). The leading term is the term with the highest exponent of x, and this is the term that will become largest when x is large. On the numerator there's only one term, but on the denominator there are two. The leading term is x. When x is very large while b stays constant, we can conclude that x will be much larger than b. If we form the ratio with only the leading terms, we see that when x is large

$$y = \frac{cx}{x+b} \approx \frac{cx}{x} = c.$$

The curly equals sign $\approx$ means 'is approximately equal to'.

So the horizontal asymptote is at $y = c$. The graph of the more general function $y = \dfrac{cx}{ax+b}$ is sketched in Figure 8.1(c). Take $a = 1$ to obtain the function in (8.4).

EXAMPLE

The graph of the function $y = \dfrac{3x}{x+5}$ passes through the origin, has a vertical asymptote at $x = -5$, and a horizontal asymptote at $y = 3$. The graph is sketched in FNGRAPH 8.2.

Problems: Now try the end-of-chapter problems 8.5–8.12.

CASE STUDY C8: A hyperbolic model of animal speed

In previous instalments, we have modelled the movement of lions and gazelles by assuming that they accelerate from stationary with a constant acceleration a until they reach their maximum speed, at which time their acceleration suddenly drops to zero. On a graph of velocity v against time t, this linear-speed model consists of a line $v = at$, then a horizontal line $v = v_{\max}$, as shown in Figure 8.2(a). It would be more realistic if the animal's acceleration $a(t)$ were time-dependent, gradually reducing from its initial value a_0 as the animal's speed neared its maximum value $v_{\max}$. This would produce a smooth curve on a velocity–time graph, starting at the origin and with horizontal asymptote $v = v_{\max}$. The curve is sketched in Figure 8.2(b). We want the slope of the curve at the origin to be the initial acceleration a_0. In Chapter 12 we will prove that for the rational function $y = \dfrac{cx}{x+b}$, the slope of the curve at the origin is c/b. The horizontal asymptote is c, so for our model of animal acceleration we want $c = v_{\max}$ and $c/b = a_0$. Solving for b:

$$b = \frac{v_{\max}}{a_0}.$$

The equation of the curve is thus

$$v = \frac{v_{\max} \cdot t}{t+b}. \tag{8.5}$$

where $b = \dfrac{v_{\max}}{a_0}$. Check that b has units of $(\text{m s}^{-1}) \div (\text{m s}^{-2}) = \text{s}$.

The lion that we encountered in Part C6 could accelerate at $a_0 = 6.25$ m s^{-2} and had a maximum speed of $v_{\max} = 14$ m s^{-1}. So $b = 14/6.25 = 2.24$ s. Using this value of b, plot the graph of (8.5) in FNGraph; also plot the line $v = a_0 . t$, and you should see that this line through the origin has the same slope as the curve at that point. We say that the line is

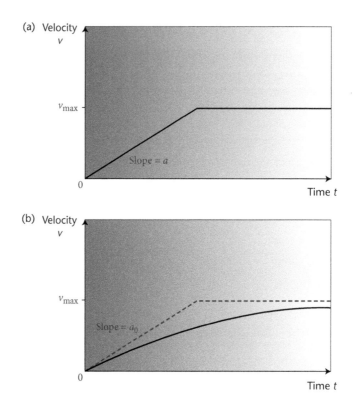

Figure 8.2 Models of animal acceleration

a tangent to the curve. This concept of 'the tangent to the curve' is the basis of the theory of Calculus, which we'll study in Part II. We'll introduce it more fully in Chapter 9.

In this model, where acceleration is not constant but time-dependent, equations (7.13)–(7.15) in Part C6 no longer apply. To find the distance travelled, we will need to use calculus.

Problems: Now try the end-of-chapter problems 8.13–8.14.

8.2 General rational functions $\dfrac{p(x)}{q(x)}$

All the functions in the preceding section are examples of the general rational function $y = f(x) = \dfrac{p(x)}{q(x)}$, where $p(x)$ and $q(x)$ are polynomials. To analyse a rational function, and thereby sketch its curve, there are four pieces of information we need:

(1) points where the curve cuts the x-axis (if any), i.e. x-intercepts;

(2) a point where the curve cuts the y-axis (if there is one), i.e. the y-intercept;

(3) any horizontal asymptotes (see Section 6.4.2);

(4) any vertical asymptotes (see Section 6.4.3).

Here we've defined the information geometrically. We can also describe it algebraically:

(1) values of x that make $f(x) = 0$;

(2) the value of $f(x)$ when $x = 0$, i.e. $f(0)$;

(3) the behaviour of $f(x)$ when x is large (either positive or negative);

(4) values of x for which $f(x)$ cannot be evaluated.

We already saw in Chapter 6 how to determine these for simple functions. We now extend this to general rational functions.

8.2.1 Finding the *x*-intercepts

If $y = f(x) = \dfrac{p(x)}{q(x)}$, the values of x that make $f(x) = 0$ will be those that make the numerator $p(x)$ zero. That is, they are roots of $p(x)$.

For example, the curve of $y = \dfrac{x^2 - 2}{x^2 + 1}$ will cut the x-axis when $x^2 - 2 = 0$, i.e. when $x = -\sqrt{2}$ and $x = \sqrt{2}$.

If $u(x)$ is a constant, for example when $f(x) = \dfrac{6}{3 - x}$, then there will be no x-intercepts since the numerator can never be zero, so the curve will not cut the x-axis.

8.2.2 Finding the *y*-intercept

To find the point where the curve of $y = f(x) = \dfrac{p(x)}{q(x)}$ cuts the y-axis, simply set $x = 0$. For example, the curve of $y = \dfrac{x^2 - 2}{x^2 + 1}$ cuts the y-axis at $y = \dfrac{0^2 - 2}{0^2 + 1} = -2$.

In general, the y-intercept occurs at the point $(0, f(0))$ where $f(0) = \dfrac{p(0)}{q(0)}$.

If setting $x = 0$ in $f(x) = \dfrac{p(x)}{q(x)}$ makes the denominator $q(0) = 0$, then there is no such point (as we can't divide by zero), and the value $x = 0$ is not in the domain of the function $f(x)$; we have seen this with $y = \dfrac{1}{x}$. Geometrically, the curve does not cross the y-axis in this case.

8.2.3 Finding the horizontal (and sloping) asymptotes

We said in Section 6.4.2 that a horizontal asymptote of $f(x)$ is a horizontal line on the graph indicating a value of y that $f(x)$ will approach as $x \to \infty$ or $x \to -\infty$. We could apply the technique of taking limits, which we used there, for general rational functions if we first divide top and bottom by the highest power of x occurring.

EXAMPLE

Find the horizontal asymptote of $y = \dfrac{3x^3 + 2x - 5}{6x^3 - x^2 + 1}$.

Solution: The highest power of x occurring is x^3. Dividing top and bottom by x^3:

$$y = \frac{3 + \dfrac{2}{x^2} - \dfrac{5}{x^3}}{6 - \dfrac{1}{x} + \dfrac{1}{x^3}}.$$

Now take the limits as $x \to \infty$, using the result in (6.10): $\lim\limits_{x \to \infty}\left(\dfrac{1}{x^k}\right) = 0$.

As $x \to \infty$, $y \to \dfrac{3 + 0 - 0}{6 - 0 + 0} = \dfrac{3}{6} = 0.5$.

An easier approach is to use the concept of the **leading term** of a polynomial (see Section 7.3). When x is large, only the leading terms of each polynomial are important, so in our example

$$y \approx \frac{3x^3}{6x^3} = \frac{3}{6} = 0.5.$$

In either method, we conclude that the line $y = 0.5$ is a horizontal asymptote of the curve. If the degree of the polynomial on the numerator is less than that of the denominator polynomial, then $y = 0$ will be a horizontal asymptote:

EXAMPLE

Find the horizontal asymptote of $y = \dfrac{3x^2 + 2x - 5}{6x^3 - x^2 + 1}$.

Solution: The leading term of the numerator is $3x^2$, and the leading term of the denominator is $6x^3$. So when x is large,

$$y \approx \frac{3x^2}{6x^3} = \frac{3}{6x} = \frac{1}{2x} \to 0 \text{ as } x \to \infty.$$

So the horizontal asymptote is $y = 0$.

On the other hand, what happens if the degree of the numerator polynomial is greater than that of the denominator?

EXAMPLE

Find the behaviour of $y = \dfrac{3x^3 + 2x - 5}{6x^2 - x + 1}$ when x is large.

Solution: The leading term of the numerator is $3x^3$, and the leading term of the denominator is $6x^2$. So when x is large,

$$y \approx \frac{3x^3}{6x^2} = \frac{3x}{6} = 0.5x.$$

This tells us that at the left-hand and right-hand edges of the graph, the curve approaches the **sloping asymptote** $y = 0.5x$. Draw the graph of this function in FNGraph to see what happens geometrically.

8.2.4 Finding the vertical asymptotes

From Section 6.2.4, a vertical asymptote is an isolated value of x for which no value of $f(x)$ exists. This will happen when the denominator of the rational function becomes zero. So we will need to find the roots of $q(x)$.

8.2.5 Example of graph sketching

Here is an illustration of the arguments to use in order to sketch the graph of a rational function. Suppose we want to sketch the graph of the function $y = \dfrac{3x^2 - 5}{x^2 - 4}$.

Here, $f(x) = \dfrac{3x^2 - 5}{x^2 - 4}$, $p(x) = 3x^2 - 5$, $q(x) = x^2 - 4$. For the x-intercepts, find the roots of $p(x)$:

$$3x^2 - 5 = 0$$
$$x^2 = \frac{5}{3}$$
$$x = \pm\sqrt{\frac{5}{3}} = \pm 1.291.$$

So the x-intercepts are at $(1.291, 0)$ and $(-1.291, 0)$.

The y-intercept is found by setting $x = 0$:

$$f(0) = \frac{3 \times 0^2 - 5}{0^2 - 4} = \frac{-5}{-4} = \frac{5}{4} = 1.25.$$

Thus, the y-intercept is at $(0, 1.25)$.

For the horizontal asymptote, take the leading terms of $p(x)$ and $q(x)$: When x is large, $y \approx \dfrac{3x^2}{x^2} = 3$, so the horizontal asymptote is $y = 3$.

For the vertical asymptotes, find the roots of $q(x)$:

$$x^2 - 4 = 0$$
$$x^2 = 4$$
$$x = \pm 2.$$

so the vertical asymptotes are at $x = -2$ and $x = 2$.

Next, mark the intercepts and draw the asymptotes on a blank graph, as shown in Figure 8.3(a). We know that the curve doesn't cross the axes anywhere else. Finally, we have to draw the curve, to fit with this information. It is easy to sketch the curve between the two vertical asymptotes, passing through the three intercepts. To find what happens when $x > 2$, evaluate another point: when $x = 3$, then $f(3) = \dfrac{3 \times 3^2 - 5}{3^2 - 4} = \dfrac{22}{5} = 4.4$, so the curve passes through the point $(3, 4.4)$. This tells us that it lies in the upper right-hand quadrant.

As $f(x)$ only involves x^2, the graph will be symmetric about the y-axis. So the curve passes through $(-3, 4.4)$, and for $x < -2$ the curve will lie in the upper right-hand quadrant. The final curve is shown in Figure 8.3(b).

8.3 Fitting curves to data

Suppose that we want to investigate a physical behaviour that can be modelled by equation (8.4), i.e. $y = \dfrac{cx}{x + b}$. We can perform an experiment and obtain a set of n data-points (x_1, y_1), (x_2, y_2), (x_3, y_3), ..., (x_n, y_n); how do we use these data to estimate the coefficients b, c in the equation? The technique is to find a transformation that converts the equation to a linear equation between new variables X, Y. We can then apply the technique of Section 6.3.4 to fit a trend line through the transformed points. We'll start with a simple example.

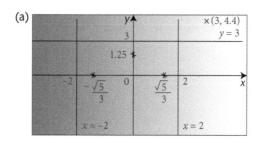

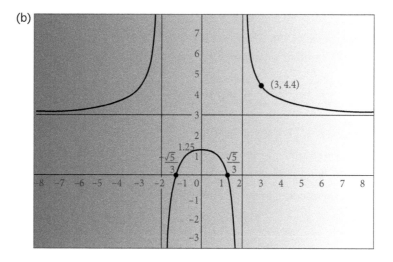

Figure 8.3 Graph of $y = \dfrac{3x^2 - 5}{x^2 - 4}$

8.3.1 Inverse proportion

SPREADSHEET 3.5 showed a data table of volume V (independent variable) and pressure P (dependent variable), from an experiment where a fixed body of gas was compressed at constant temperature. Here is the data table again (see Section 3.2):

Volume V	Pressure P
1.00	4.3
2.00	2.2
5.00	0.9
10.00	0.4
20.00	0.2

We expect that V and P are related by Boyle's law: $P.V = $ constant. We can write this in the form of a function $y = f(x)$ as an inverse proportion relation:

$$P = \frac{k}{V} \tag{8.6}$$

and our task is to find the value of k. To transform this equation to linear form, write it as

$$P = k.\frac{1}{V}.$$

This can be seen as a direct proportion relation between $1/V$ (independent variable) and P (dependent variable), i.e. it is a special form of linear relationship $Y = k.X$ where the slope is k and the y-intercept is zero (see Section 6.3.2). Our transformed variables are $X = 1/V$ and $Y = P$, and by transforming the data values (taking the reciprocals of the V values) we get the table:

$X = 1/V$	$Y = P$
1.00	4.3
0.50	2.2
0.20	0.9
0.10	0.4
0.05	0.2

Plot these transformed data-points and we find they lie (roughly) along a straight line. Fit a best-fit straight line, find its slope (by hand or in Excel) and this gives us the value of k as 4.3225. The tables and graphs are shown in SPREADSHEET 8.1, shown in Figure 8.4. Remember that when you have produced the chart for the transformed variables, you add the best-fit line by highlighting the chart and choosing Add trendline.. from the Chart menu. In the Trendline dialog box, the line Type should be chosen as linear. Then click the Options tab, and check the box for Display equation on chart. Notice that there is another checkbox for Set intercept = 0. We want to check this too, as we know that in this case the line $y = k.x$ has y-intercept zero. The result for k is 4.3225.

The procedure of fitting a best-fit straight line through a set of data-points is called **linear regression**.

As well as the Display equation on chart checkbox, there is one labelled Display R-squared value on chart. You may remember this from Section 6.3.4. The R^2 value,

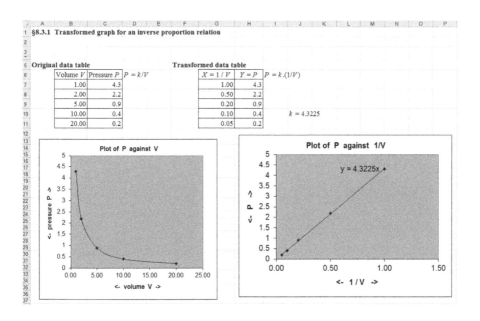

Figure 8.4 Transformed graph for inverse proportion relation (SPREADSHEET 8.1)

called the Coefficient of Determination, is a number between 0 and 1, which indicates how closely the transformed data-points lie to the trend line. The closer the R^2 value is to 1.0, the better is the fit. If you bring up the Trendline dialog box in SPREADSHEET 8.1 and check both these boxes, you will see that the R^2 value is 0.9996, indicating an extremely good fit to the data.

To bring up the dialog box, highlight the chart and then move the mouse cursor until the arrow tip is touching the line, and the text 'Series 1 Trendline 1' appears. Then right-click, and choose the option Format Trendline. . . . This brings up the Trendline dialog window.

8.3.2 Rational function $y = \dfrac{1}{ax+b}$

The previous section showed how to deal with functions of the form (8.2). The function in (8.3), namely $y = \dfrac{1}{ax+b}$, is transformed by taking the reciprocal of both sides:

$$\frac{1}{y} = ax + b. \tag{8.7}$$

So plot $\dfrac{1}{y}$ against x, i.e. our transformed variables are $X = x$ and $Y = \dfrac{1}{y}$, so that (8.7) can be written as

$$Y = aX + b.$$

The best-fit line through the transformed data-points will have slope a and y-intercept b. In this case we do not check the Set intercept = 0 box in the Add Trendline.. dialog window (if we did, it would be forcing b to be zero).

If you try to fit equation (8.7) to the data-points in the previous section, i.e. not assuming that $b = 0$, you should find that the equation of the trend line is $Y = 0.253X - 0.0634$, with an R^2 value of 0.9993, again a very good fit.

8.3.3 Quadratic functions

At first sight, it seems impossible to apply this process to the quadratic function $y = ax^2 + bx + c$, since there are three parameters a, b, c. But the last parameter c is the y-intercept of the curve, and in practice we often know this from the experiment (if x represents time, then c is the initial value of y, which will usually be the first of our data-points). Even if we don't know c from the data, we can plot the points, sketch the curve, and see where it crosses the x-axis, to estimate c.

Assuming we know c, take it to the left-hand side:

$$y - c = ax^2 + bx.$$

Now divide both sides by x:

$$\frac{y - c}{x} = ax + b. \tag{8.8}$$

So the transformed variables are x (independent variable) and $\dfrac{y - c}{x}$ (dependent variable); that is, set $X = x$ and $Y = \dfrac{y - c}{x}$. Plot $\dfrac{y - c}{x}$ against x, fit the best-fit line, and again it will have slope a and y-intercept b. Note that if you have used the first data-point (x_1, y_1) to fix the value of c, you cannot use it again in fitting the best straight line. Fit the trend line through $(x_2, y_2), \ldots, (x_n, y_n)$.

8.3.4 Rational function $y = \dfrac{a}{x} + b$

We shall work through this example by hand, using just two data-points. Suppose I had obtained the (x, y) data-points $(2, 4.5)$ and $(5, 3)$ from an experiment, and suspected that these lie on a curve of the form

$$y = \frac{a}{x} + b. \tag{8.9}$$

Here, we again transform the variables into new variables X, Y that will be related linearly: $Y = mX + c$.

In this case, notice that (8.9) can be written as

$$y = a.\frac{1}{x} + b,$$

so if we take $Y = y$ and $X = \frac{1}{x}$ this becomes $Y = aX + b$. We now need to convert the (x, y) data-points to (X, Y) data-points:

- when $x = 2$, $X = \dfrac{1}{2} = 0.5$, so the data-point $(2, 4.5)$ becomes $X = 0.5$, $Y = 4.5$;
- when $x = 5$, $X = \dfrac{1}{5} = 0.2$, so the data-point $(5, 3)$ becomes $X = 0.2$, $Y = 3$.

Plugging these values into $Y = aX + b$ we obtain the simultaneous equations

$$0.5a + b = 4.5$$
$$0.2a + b = 3.0.$$

Subtract the second equation from the first, and we get $0.3a = 1.5$, so $a = \dfrac{1.5}{0.3} = \dfrac{15}{3} = 5$. Then $b = 3 - 0.2 \times 5 = 3 - 1 = 2$. The equation of the curve is thus

$$y = \frac{5}{x} + 2.$$

Note again that often we know data-points at special values that immediately tell us one of the parameters. For example, if we knew that the speed v of an animal varies with time t according to the relation $v = \dfrac{a}{1 + bt}$, and that the animal's initial speed is 7 m s^{-1}, then putting $t = 0$, $v = 7$ in the equation we get $7 = \dfrac{a}{1 + 0}$, so we see immediately that $a = 7$.

Problems: Now try the end-of-chapter problems 8.15–8.22.

· ·

8.4 Application: enzyme kinetics

We have demonstrated the technique of transforming variables for several simple rational functions – but not for the function $y = \dfrac{cx}{x + b}$ which we introduced in Section 8.1.2. We claimed there that this function has important applications in the biosciences. Here, then, is the most important one.

8.4.1 The Michaelis–Menten equation

In a biochemical enzyme-catalysed reaction, the initial rate v of the reaction depends on the concentration s of the substrate; we can write $v = v(s)$. If there is no substrate to be converted, there will be no reaction, so the graph of v against s will start at the origin. For low substrate concentrations, the rate v is approximately proportional to s; when double the amount of substrate is present to be converted, the reaction rate is twice as fast. However, when a larger concentration of substrate is present, the enzyme does not work so efficiently, and there is a theoretical maximum rate of reaction v_{max} that is achievable with the amount of enzyme present. The resulting type of graph of v against s (called a v–s plot) is shown in Figure 8.5(a). The curve is a rectangular hyperbola, passing through the origin and approaching the horizontal asymptote at $v = v_{max}$. This is exactly the type of curve we saw in Figure 8.1(c), whose equation is (8.4). The equation relating v and s is of this form, and was proposed by Michaelis and Menten in 1913. It is now known as the Michaelis–Menten equation:

$$v = \frac{v_{max} \cdot s}{K_m + s}. \tag{8.10}$$

This function is a rational function of the form $y = \dfrac{cx}{x+b}$ which we saw in Section 8.1.2. By the argument we used at the end of Section 8.1, we can see that as s becomes large,

$$v = \frac{v_{max} \cdot s}{K_m + s} \approx \frac{v_{max} \cdot s}{s} = v_{max}$$

so the horizontal asymptote is at $v = v_{max}$. The other parameter involved is K_m, and this also has a physical significance: it is the concentration that produces an initial rate of reaction

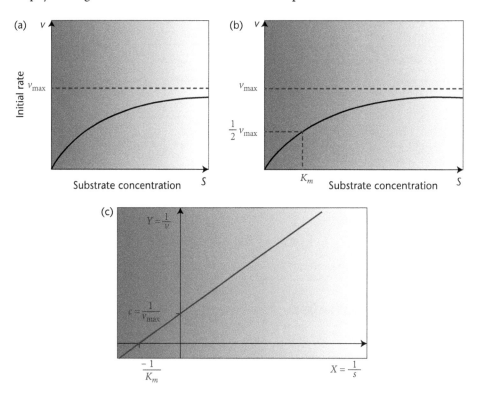

Figure 8.5 The Michaelis–Menten equation and the Lineweaver–Burk plot

that is exactly 50% of the theoretical maximum. To see this, put $v = \frac{1}{2}v_{max}$ into (8.10) and solve for K_m:

$$\frac{1}{2}v_{max} = \frac{v_{max}.s}{K_m + s}$$

> Cancel the common factor of v_{max}

$$\frac{1}{2} = \frac{s}{K_m + s}$$

> To remove the fraction, multiply both sides by $(K_m + s)$.

$$\frac{1}{2}(K_m + s) = s$$

$$\frac{1}{2}K_m + \frac{1}{2}s = s$$

$$\frac{1}{2}K_m = s - \frac{1}{2}s$$

$$\frac{1}{2}K_m = \frac{1}{2}s$$

$$K_m = s.$$

K_m is called the **Michaelis constant**. The physical significance of K_m is illustrated in Figure 8.5(b).

The reaction rate is typically measured in μM s^{-1}, and the concentration (and thus also the Michaelis constant) in mM or μM.

8.4.2 The Lineweaver–Burk transformation

We can apply the transformation-of-variables technique of Section 8.3 to the Michaelis–Menten equation (8.10). If we invert both sides of the equation, we get

$$\frac{1}{v} = \frac{K_m + s}{v_{max}.s}.$$

The right-hand side can be written as separate fractions (the factor of s on the top and bottom of the second fraction cancels):

$$\frac{1}{v} = \frac{K_m + s}{v_{max}.s}$$

$$= \frac{K_m}{v_{max}.s} + \frac{s}{v_{max}.s}$$

$$= \frac{K_m}{v_{max}.s} + \frac{1}{v_{max}}.$$

So we have rewritten (8.10) as

$$\frac{1}{v} = \frac{K_m}{v_{max}}.\frac{1}{s} + \frac{1}{v_{max}}, \tag{8.11}$$

which expresses a linear relationship between $\frac{1}{v}$ and $\frac{1}{s}$. Equation (8.11) is equivalent to $Y = mX + c$, where the transformed variables are $X = \frac{1}{s}$, $Y = \frac{1}{v}$, and the transformed parameters are $m = \frac{K_m}{v_{max}}$, $c = \frac{1}{v_{max}}$.

Figure 8.5(c) shows the straight-line graph of the transformed equation $Y = mX + c$; this is called a Lineweaver–Burk plot. If you compare this with the graph of $y = mx + c$ in Figure 6.6(a), you see that the intercept on the vertical axis is at $Y = c = \dfrac{1}{v_{max}}$, and the intercept on the horizontal axis is at $X = -\dfrac{c}{m} = -\dfrac{1}{v_{max}} \cdot \dfrac{v_{max}}{K_m} = -\dfrac{1}{K_m}$.

Thus, if you are drawing the best-fit straight line on graph paper, the values of v_{max} and K_m can be obtained directly from the X-intercept and Y-intercept. If you have calculated the coefficients m and c, you can deduce v_{max} and K_m by:

$$v_{max} = \frac{1}{c}, \qquad K_m = m.v_{max}.$$

(8.12)

The whole process is programmed in SPREADSHEET 8.2, a printout of which is in Figure 8.6.

8.4.3 Error analysis

There is a practical problem with using the Lineweaver–Burk transformation. In experiments, it is difficult to obtain precise values of s and v at low substrate concentrations, and the Lineweaver–Burk calculation of v_{max} and K_m is particularly sensitive to errors in these points. We can illustrate this using the example in SPREADSHEET 8.2. The table of data is

s	v
1.00	2.00
2.00	3.00
4.00	3.50
8.00	4.30
12.00	4.50

and suppose that each measurement is correct to 2 decimal places. Both s and v are measured quantities, so this is a more complicated situation than previous error analyses where we have considered the effect of error in one variable only. Thus, for the first data-point, s could lie between 0.995 and 1.005, and v could lie between 1.995 and 2.005. Instead of an interval on the real line, this defines a rectangle in two-dimensional (s, v) space. For each possible value of s and each possible value of v, we could work out a value of K_m by SPREADSHEET 8.2, i.e. assuming all the other data-points are exact. Provided this rectangle is small enough, the extreme (greatest and smallest) values of K_m will occur at one of the corners. I have tried changing the values of s and v in cells B6 and C6, and noting the value of K_m that is produced. Here is what I found:

$s = 0.995$ $v = 2.005$ produces $K_m = 1.4855$		$s = 1.005$ $v = 2.005$ produces $K_m = 1.5059$
	$s = 1.00$ $v = 2.00$ produces $K_m = 1.5045$	
$s = 0.995$ $v = 1.995$ produces $K_m = 1.5031$		$s = 1.005$ $v = 1.995$ produces $K_m = 1.5237$

So the least value of K_m is $K_m = 1.4855$, and the greatest value of K_m is $K_m = 1.5237$. The uncertainty interval for K_m due to the precision of the first data-point is $1.4855 < K_m < 1.5237$, which we can write as $K_m = 1.5046 \pm 0.0191$. Now repeat this analysis considering the same precision in the final data-point. Here is the table:

$s = 11.995$ $v = 4.505$ produces $K_m = 1.5060$		$s = 12.005$ $v = 4.505$ produces $K_m = 1.5059$
	$s = 12.00$ $v = 4.50$ produces $K_m = 1.5045$	
$s = 11.995$ $v = 4.495$ produces $K_m = 1.5031$		$s = 12.005$ $v = 4.495$ produces $K_m = 1.5030$

The uncertainty interval for K_m due to the precision of the last data-point is thus $1.5030 < K_m < 1.5060$, which we can write as $K_m = 1.5045 \pm 0.0015$. You can see that the uncertainty interval for K_m when we consider the possible error in the data-point at the smallest value of s is over ten times greater than the uncertainty when we consider the same level of precision in the data-point at the greatest value of s. The calculated value of K_m is excessively sensitive to the data-points for small values of substrate concentration. Unfortunately it is just these points that are more susceptible to experimental error.

There are two alternative transformations which are used:

- In the **Eadie–Hofstee** plot, we take $X = \frac{v}{s}$, $Y = v$.
- In the **Hanes–Woolf** plot, we take $X = s$, $Y = \frac{s}{v}$.

Both of these plots produce a linear relationship $Y = mX + c$. You can explore this in the Problems.

Another approach to the curve-fitting task is to use the regression technique to fit a curve, rather than a line, directly on the nonlinear $v = v(s)$ relationship (8.10). To find out about this, look in the index of a statistics textbook for 'nonlinear regression' or 'curvilinear regression'.

Problems: Now try the end-of-chapter problems 8.23–8.25.

8.4.4 Allosteric regulation

There are certain enzymes in the cell that are able to regulate their level of activity according to signals, which may include the substrate concentration. These are called allosteric enzymes. This regulation occurs through the enzyme existing in two different states: a less active conformation (the T state) and a more active one (the R state). At low substrate concentrations almost all the enzyme is in the T state, but as the concentration increases more of the enzyme converts to the R state; this is called *positive cooperativity*. This causes the v–s curve to become sigmoidal (S-shaped). In Figure 8.7, taken from the article by Traut (2007) in the *Encyclopaedia of Life Sciences*, a normal v–s curve is shown on the left, and a v–s curve for a so-called K-type allosteric enzyme in the middle graph. Under certain experimental conditions it is also possible to produce V-type allosteric enzymes, where a contrary regulation occurs (*negative cooperativity*), producing a v–s curve as in the graph on the right.

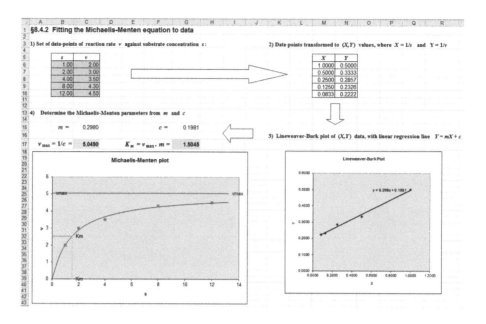

Figure 8.6 Fitting the Michaelis–Menten equation to data (SPREADSHEET 8.2)

These types of curve can be produced by modifying equation (8.10) to

$$v = \frac{v_{max} \cdot s^{\eta}}{K_m + s^{\eta}}. \tag{8.13}$$

That is, we raise the independent variable s to a power η (a Greek letter e, pronounced 'eta'), which is a third parameter, called the **Hill coefficient**, to be determined along with v_{max} and K_m. For normal enzymes, $\eta = 1$. For enzymes exhibiting positive cooperativity, $\eta > 1$, and for those with negative cooperativity $\eta < 1$. The Hill coefficient is referred to as n_H in the graphs of Figure 8.6, where it takes the values of 1, 2 and 0.7 respectively. Negative cooperativity is associated with a large increase in the value of v_{max}; this is not visible in the graphs of Figure 8.7, which plot $\frac{v}{v_{max}}$ on the y-axis (so the horizontal asymptote is always at $\frac{v}{v_{max}} = 1$).

The curve of (8.13) is plotted in FNGRAPH 8.3, where the Hill coefficient is called eta; try changing its value to see the effect on the curve. The Michaelis–Menten curve ($\eta = 1$) is also available as a graph. The question of how to determine the parameters in (8.12) from experimental data, will have to wait until Chapter 10 (Section 10.4.3).

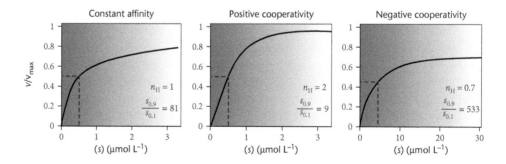

Figure 8.7 Allosteric regulation (from Traut 2007)

8.5 Inverse functions

8.5.1 Definition of the inverse of $f(x)$

If f is a function that takes x to y, as in $y = f(x)$, then the inverse of f is a function that takes y back to x. We saw some inverse functions in Chapter 1 when we explored the calculator buttons. For example, the squaring function $\boxed{x^2}$ is the inverse of the square-root function $\boxed{\sqrt{}}$, and vice versa. Applying one and then the other takes you back to where you started:

    9 $\boxed{x^2}$ 81 $\boxed{\sqrt{}}$ 9 (start with 9, square it to get 81, take the square root and you get back to 9)

or

    9 $\boxed{\sqrt{}}$ 3 $\boxed{x^2}$ 9 (start with 9, take the square root and get 3, then square it and you get back to 9).

Notation: we write the inverse of $f(x)$ as $f^{-1}(x)$. This does **not** mean $\dfrac{1}{f(x)}$. This is a special, one-off notation for the inverse of a function. If you wanted to square the inverse of f, you'd have to write it as $(f^{-1}(x))^2$, not $f^{-2}(x)$. So if $f(x) = \sqrt{x}$, then its inverse is $f^{-1}(x) = x^2$.

If you are given a function $y = f(x)$, you can find the inverse of f by solving the equation to express x in terms of y. The resulting equation gives you $x = f^{-1}(y)$. If you now want to write the inverse function as $y = f^{-1}(x)$, simply replace the x on the left-hand side by y, and replace y by x wherever it occurs on the right-hand side.

For example, to find the inverse of the function $f(x) = \sqrt{x}$, write it as $y = \sqrt{x}$, then solve for x by squaring both sides and writing as $x = y^2$. This is $x = f^{-1}(y)$. Now replace x by y on the left-hand side and y by x on the right-hand side to get $y = x^2$, so $f^{-1}(x) = x^2$.

To summarize: if we know $f(x)$, we can define $f^{-1}(x)$ by:

$$y = f^{-1}(x) \text{ means that } x = f(y). \tag{8.14}$$

That is, $f^{-1}(x)$ is the value that, when you then apply f to it, produces x again.

The inverse of a linear function is another linear function. To see this, take the general linear function $f(x) = mx + c$, with slope m and y-intercept c. Set $y = f(x)$ and solve for x:

$$y = mx + c$$
$$mx = y - c$$
$$x = \frac{1}{m} y - \frac{c}{m}$$

now interchange x and y, to obtain $y = f^{-1}(x)$:

$$f^{-1}(x) = \frac{1}{m} x - \frac{c}{m}$$

which is a linear function with slope $\dfrac{1}{m}$ and y-intercept $\dfrac{-c}{m}$.

Thus, the inverse of $y = 2x + 5$ is $y = \dfrac{1}{2} x - \dfrac{5}{2}$. What is the relationship between the graphs of these two functions? They are sketched in FNGRAPH 8.4 as the blue and red lines respectively. Also drawn on the graph is the green line $y = x$. You should see that the red line would be obtained by reflecting the blue line through the green line acting as a 'double mirror'. This is the geometric equivalent of switching x and y around, so that the point (a, b) on the graph

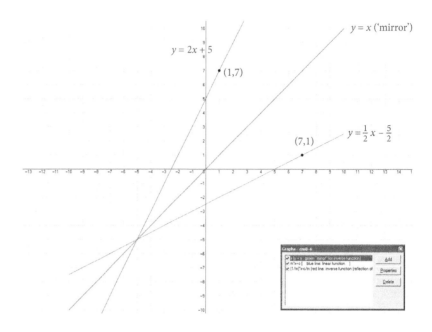

Figure 8.8 Graphs of a linear function and its inverse (FNGRAPH 8.4)

of $y = f(x)$ will produce the point (b, a) on the graph of $y = f^{-1}(x)$. This is illustrated in Figure 8.8 with the point $(1, 7)$ on the graph of $f(x) = 2x + 5$ corresponding to the point $(7, 1)$ on the graph of the inverse.

> **EXAMPLE**
> In FNGraph, draw the graphs of $y = x^2$ and $y = \sqrt{x}$ (which you code as `sqrt(x)`), together with the 'mirror' line $y = x$, and see how one curve is the reflection of the other. What will be the inverses of (i) $f(x) = 4x^2 + 3$ (ii) $g(x) = \sqrt{x - 5}$?
>   Answers: (i) $f^{-1}(x) = \frac{1}{2}\sqrt{x - 3}$ (ii) $g^{-1}(x) = x^2 + 5$.

8.5.2 The inverse of rational functions

The inverse of a rational function is another rational function. We shall show this by finding the inverse of a function similar to equation (8.4):

$$y = \frac{cx}{ax + b}.$$

First rearrange the equation to produce x as a function of y:

$$y = \frac{cx}{ax + b}$$

> Multiply both sides by $ax + b$ to remove the fraction

$$axy + by = cx$$

$$axy - cx = -by$$

> Collect the terms involving x on the left-hand side

$$x(ay - c) = -by$$

> Take out x as a factor

$$x = \frac{-by}{ay - c}.$$

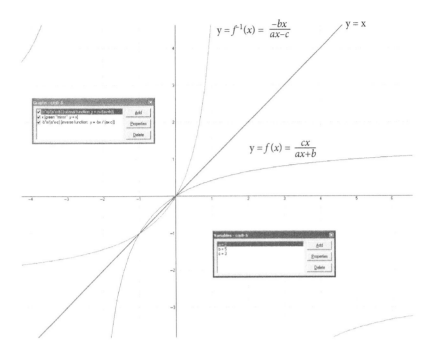

$$y = f^{-1}(x) = \frac{-bx}{ax-c}$$

$$y = x$$

$$y = f(x) = \frac{cx}{ax+b}$$

Figure 8.9 Graphs of a rational function and its inverse (FNGRAPH 8.5)

Thus, interchanging x and y, we have found that the inverse of $f(x) = \dfrac{cx}{ax+b}$ is $f^{-1}(x) = \dfrac{-bx}{ax-c}$. The two functions, together with the 'mirror' $y = x$, are drawn in FNGRAPH 8.5, and printed in Figure 8.9.

Inverse functions become more important in the next two chapters, where we apply the idea to trigonometric and exponential functions.

Problems: Now try the end-of-chapter problems 8.26–8.31.

8.6 Bracketing methods

We saw in Section 8.2.1 that, to find the points where a rational function $\dfrac{p(x)}{q(x)}$ cuts the x-axis, we look for the roots of the numerator polynomial $p(x)$. In Chapter 7 we explored this task for quadratic and higher-degree polynomial functions. In (7.12) we found the locations of the roots and turning-point (maximum or minimum) for quadratic functions, but analysing cubics and other polynomials is much more problematic. In fact, as we shall see in the Historical Interlude at the end of Part I, it is impossible to find a method that produces solutions to the general quintic equation.

But there is another approach to this whole problem of finding roots and turning-points. The methods we have been looking for produce complicated mathematical expressions for the roots, involving square roots and cube roots; in problem 1.66 you saw such a formula for a root of a cubic polynomial. They define the values of the roots exactly; we say they are

analytic or **closed-form solutions**. But if we want to use them to get a value for the root, we need to evaluate the square roots on a calculator or computer, and this can only be done to a certain level of precision – so the resulting value will be correct to a certain number of decimal places. Instead of developing complicated closed-form solutions that can only be evaluated to a certain precision, there are methods that work as follows:

- Start with an approximate value of the root.
- Plug this value into a relatively simple algorithm that produces a more accurate approximation.
- Now start with this improved approximation, plug it into the same algorithm, and produce an even better approximation.
- Repeat this process as many times as necessary, until you achieve an approximation that is sufficiently accurate.

Methods of this kind are called **iterative methods**.[1] Each repetition of the algorithm to produce an improved estimate is called an **iteration**. These methods are also classed as **numerical methods**, or numerical algorithms, because what they produce is not a complicated algebraic expression, but just a numerical value for the solution. The algorithms for interpolation which we saw in Sections 3.6 and 7.5 are also numerical methods. One drawback of these methods is that they don't produce an equation such as (7.10) defining all the roots of $f(x)$; they will find a root that is close to the approximate value we start with. Another root could be found by the same process, starting with a different initial guess – but first you need to know that it's out there.

Numerical methods are easy-to-use and extremely powerful. In particular, they are ideal for writing as computer programs, and we will see them programmed into Excel spreadsheets. In this section we will describe some simple iterative methods for solving the two problems mentioned at the start:

- Find a root of $f(x)$, i.e. find a value of x for which $f(x) = 0$.
- Find a value of x at which $f(x)$ has a maximum or minimum.

One great advantage of these methods is that they can be applied to any function $f(x)$; not just polynomials but also trigonometric and exponential functions which we'll describe in Chapters 9 and 10.

The algorithms we'll describe in this section are all **bracketing methods**. Rather than starting with a single approximation to the solution, they start with an interval on the real line in which the solution lies. Then by eliminating part of the interval at each iteration, they home in on the solution; once the interval is small enough, taking its midpoint gives us the final approximation to the answer. We will describe two root-finding algorithms (the methods of bisection and *regula falsi*), and a minimization algorithm (golden section search) which can find turning-points.

We assume that the function $f(x)$ is continuous, i.e. that it changes gradually as x changes, without any sudden jumps in value.

---

[1] The name comes from the Latin 'iter' meaning 'again'.

8.6.1 Root-finding algorithms

The first algorithm we'll describe is called the **method of bisection**. It's a very simple idea – in fact, you may have already thought of it yourself, if you ever played the following children's game:

> I think of a number between 1 and 100 and you have to guess what it is. Each time you make a guess, I tell you if it is too low or too high, then you guess again. Once you've guessed my number, we swap roles and I try to guess your number; if I get your number in fewer guesses than you took, I win.

Children soon learn that the most reliable strategy is to make their first guess 50. If that's too high then their next guess is 25, if it's too low they guess 75. At each stage they guess a number in the middle of the interval of possible numbers, and the answer enables them to eliminate half of that interval. The method of bisection works just like this.

We are trying to find a root of the function $f(x)$, which we'll denote as α (Greek letter 'alpha'). So we want to find a number α with the property that

$$f(\alpha) = 0. \tag{8.15}$$

To start the algorithm, we need to find two numbers a and b that lie one either side of the root. Assuming that a is the smaller number, we need to find numbers a, b such that $a < \alpha < b$. We say that a and b **bracket the root α**, just like 0 and 100 bracket my number in the game. It's actually quite easy to do this, if you think about the graph of $y = f(x)$ around the point where we suspect α to be. As shown in Figure 8.10, the curve cuts the x-axis at $x = \alpha$, so either it is moving from negative values of y to positive ones, as in Figure 8.10(a), or from positive values of y to negative ones, as in Figure 8.10(b). If we can find values of a and b with the property that either

- $f(a) < 0$ and $f(b) > 0$, as in Figure 8.10(a); or
- $f(a) > 0$ and $f(b) < 0$, as in Figure 8.10(b);

then we know that there must be a point somewhere between a and b at which $f(x) = 0$. There's a neat way to check both these conditions at the same time. Recall from Chapter 1 that if the product of two numbers is negative, then one must be positive and one negative. So we calculate the product $f(a).f(b)$, and if

$$f(a).f(b) < 0 \tag{8.16}$$

there must be a root α between a and b.

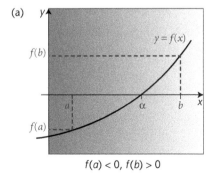
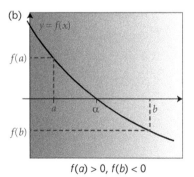

(a) $f(a) < 0$, $f(b) > 0$

(b) $f(a) > 0$, $f(b) < 0$

Figure 8.10 Bracketing the root

In real-life problems we usually have an idea whereabouts on the real line the root we are interested in should lie, and we can then try evaluating $f(x)$ at convenient points in this region, until we find two points a, b that satisfy (8.16).

As an example, we'll try to find a root near 0 of the quintic function:

$$f(x) = x^5 - 12x + 5. \tag{8.17}$$

Evaluating this function at convenient points either side of 0 gives:

x	$f(x)$
−1	$(-1)^5 - 12(-1) + 5 = -1 + 12 + 5 = 16$
0	$0 - 0 + 5 = 5$
1	$(1)^5 - 12(1) + 5 = 1 - 12 + 5 = -6$

Moreover, we know from the graph in Figure 7.13(i) that when x is large and negative the leading term of (8.17), which is x^5, is negative and so $f(x) < 0$, and that when x is large and positive the leading term is positive and so $f(x) > 0$. This indicates that there must be roots lying further out in each direction, and in fact:

$$f(-2) = (-2)^5 - 12(-2) + 5 = -2^5 + 24 + 5 = -32 + 29 = -3 \text{ and}$$
$$f(2) = (2)^5 - 12(2) + 5 = 2^5 - 24 + 5 = 32 - 19 = 13.$$

So $f(-2) < 0$, $f(-1) > 0$, $f(0) > 0$, $f(1) < 0$, and $f(2) > 0$.

We can deduce that there must be a root lying between −2 and −1, another root lying between 0 and 1, and a third root lying between 1 and 2. Let's try to find the root closest to zero, by taking $a = 0$ and $b = 1$. So we have

$$f(a) = f(0) = 5, \text{ and}$$
$$f(b) = f(1) = -6.$$

(We could have cheated by using FNGraph to sketch the curve x^5 − 12*x + 5. You should try this, in case there are other roots; a quintic can have up to five roots.)

Now, as in the game, we evaluate the function at the midpoint of the interval $a < x < b$. That is, we work out

$$c = \tfrac{1}{2}(a + b) \tag{8.18}$$

and evaluate $f(c)$. In our example, $c = \dfrac{1}{2}(0 + 1) = 0.5$, and $f(c) = f(0.5) = (0.5)^5 - 12(0.5) + 5 = -0.9688$ (to 4 decimal places).

Since $f(0.5)$ is negative, we know that the root must lie between $f(0)$, which is positive, and $f(0.5)$. So 0.5 becomes the new right-hand end b of our interval, and at the next step we will evaluate $f(x)$ at the new midpoint $c = \dfrac{1}{2}(0 + 0.5) = 0.25$. And so on. The algorithm can be summarized as:

Method of bisection

To find a root of $f(x)$:

1. Pick a, b satisfying $a < b$ and $f(a).f(b) < 0$.

2. Set $c = \tfrac{1}{2}(a + b)$ and evaluate $f(c)$. If $f(c) = 0$, stop; c is the root.

3. If $f(a).f(c) < 0$, the root must lie between a and c; replace b with c.
   Otherwise, the root must lie between c and b; replace a with c.   (8.19)

4. Re-evaluate $f(a)$ and $f(b)$.

5. Return to line 1.

This algorithm has been programmed into SPREADSHEET 8.3, reproduced in Figure 8.11. You should enter your initial choices of a and b into cells B6 and E6. Reading along row 6, once we have calculated $f(a)$, $f(b)$, c, and $f(c)$, we test the condition in line 3 of the algorithm, in cell S6. The crucial step is in cells V6 and W6, where we select the new values of a and b according to this criterion. The formula programmed into cell V6 is

= IF($H6*$P6<0, B6, N6)

The format of the IF function is:

IF(*test, value _ if _ true, value _ if _ false*)

so this formula in V6 is saying: if $f(a).f(c) < 0$ take the value in B6 (which is the old a), otherwise take the value in N6 (which is c). This sets the value of the left-hand end for the new interval (the new a) according to line 3 of the algorithm. This value is used in cell B7, for the next iteration of the iteration, on row 7. Check how the formula in cell W6 sets the value for the right-hand end of the new interval (the new b), and that this value is used in cell E7. I have colour-coded the cells to indicate which ones are linked.

Each following row is a new iteration of the algorithm. Ten iterations have been programmed, down to row 16. You can perform further iterations by selecting all of row 16, from A16 to W16, then holding the cross on the bottom right corner of the selection and dragging the whole row down.

For the approximation to the solution at each iteration, we use the midpoint c of the interval. So if you look down column N you can see a sequence of numbers, the **iterates**, which should be converging towards the true solution. The usual practice is to stop when successive values of c agree correct to the number of decimal places of precision we require, but if the sequence is converging slowly this may not produce the correct answer. After ten iterations, the last two values of c, 0.41699 and 0.41748, both round to 0.417 correct to 3 decimal places, so that would be our estimate of the root that lies between 0 and 1. If you continue the

	Left value		Right value		Function values at a and b				Evaluate f at mid-point $c = (a+b)$			Test if a,c bracket the root:	New a	New b
§8.6.1 Finding a root of $f(x)=x^5-12x+5$ by bisection							Error checks:							
							i) Check $a < b$:	OK						
Pick a,b: values of x which bracket the root and $a < b$;							ii) Check $f(a).f(b) < 0$:	OK						
$a =$	0	$b =$	1	$f(a) =$	5	$f(b) =$	-6	$c =$	0.5	$f(c) =$	-0.97	$f(a).f(c) < 0$ so root between a and c	0	0.5
$a =$	0	$b =$	0.5	$f(a) =$	5	$f(b) =$	-0.969	$c =$	0.25	$f(c) =$	2.001	$f(a).f(c) > 0$ so root between c and b	0.25	0.5
$a =$	0.25	$b =$	0.5	$f(a) =$	2.001	$f(b) =$	-0.969	$c =$	0.375	$f(c) =$	0.507	$f(a).f(c) > 0$ so root between c and b	0.375	0.5
$a =$	0.375	$b =$	0.5	$f(a) =$	0.5074	$f(b) =$	-0.969	$c =$	0.4375	$f(c) =$	-0.23	$f(a).f(c) < 0$ so root between a and c	0.375	0.438
$a =$	0.375	$b =$	0.4375	$f(a) =$	0.5074	$f(b) =$	-0.234	$c =$	0.4063	$f(c) =$	0.136	$f(a).f(c) > 0$ so root between c and b	0.406	0.438
$a =$	0.406	$b =$	0.4375	$f(a) =$	0.1361	$f(b) =$	-0.234	$c =$	0.4219	$f(c) =$	-0.05	$f(a).f(c) < 0$ so root between a and c	0.406	0.422
$a =$	0.406	$b =$	0.4219	$f(a) =$	0.1361	$f(b) =$	-0.049	$c =$	0.4141	$f(c) =$	0.043	$f(a).f(c) > 0$ so root between c and b	0.414	0.422
$a =$	0.414	$b =$	0.4219	$f(a) =$	0.0434	$f(b) =$	-0.049	$c =$	0.418	$f(c) =$	-0	$f(a).f(c) < 0$ so root between a and c	0.414	0.418
$a =$	0.414	$b =$	0.418	$f(a) =$	0.0434	$f(b) =$	-0.003	$c =$	0.416	$f(c) =$	0.02	$f(a).f(c) > 0$ so root between c and b	0.416	0.418
$a =$	0.416	$b =$	0.418	$f(a) =$	0.0203	$f(b) =$	-0.003	$c =$	0.417	$f(c) =$	0.009	$f(a).f(c) > 0$ so root between c and b	0.417	0.418
$a =$	0.417	$b =$	0.418	$f(a) =$	0.0087	$f(b) =$	-0.003	$c =$	0.4175	$f(c) =$	0.003	$f(a).f(c) > 0$ so root between c and b	0.417	0.418

Figure 8.11 Finding a root of $f(x)$ by the method of bisection (SPREADSHEET 8.3)

iterations, however, you find the root more accurately, as 0.41773 …, which would round to 0.418 correct to 3 decimal places. It is important to continue the iterations until the sequence of iterates has really converged.

To find the root of the quintic that lies between −2 and −1, and the root that lies between 1 and 2, you just need to change the values of a and b defining the initial interval, in cells B6 and E6. Try this!

If you want to use this spreadsheet to find the roots of other functions $f(x)$, you need to unprotect the sheet (using `Tools > Protection > Unprotect Sheet...`), then program the new formula for $f(x)$ into cells H6 (using B6 for x), K6 (using E6 for x), and P6 (using N6 for x). Then highlight each cell and drag it down the column. You will only see the correct calculations after you have dragged down all three columns.

The method of bisection really is used by research scientists to solve equations, partly because of its simplicity, but also because of two great advantages

- provided the initial interval contains a root, the method is guaranteed to find it;
- if you know how big your initial interval is, and it get gets halved at each iteration, you can work out in advance how big an interval you will have, and hence how accurate an answer you are going to produce, after 100 iterations, say; see problem 8.35.

There is a modification of bisection that can greatly improve the speed of convergence of the iterates in most cases. We have laboriously calculated $f(a)$, $f(b)$, and $f(c)$ at each iteration, but the only use made of them is to see if they are positive or negative. Can't we also make use of the function values themselves? If you consider the root between −2 and −1 in our quintic example, we know that $f(−2) = −3$ and $f(−1) = 16$. So it seems very likely that the root α, $−2 < \alpha < −1$, for which $f(\alpha) = 0$, will lie much closer to −2 than to −1, because −3 is a lot closer to 0 than 16 is. Looking at Figure 8.12, it seems like a neat idea to choose our point c, not at the midpoint between a and b, but at the point where the line through the points $(a, f(a))$ and $(b, f(b))$ cuts the x-axis. A straight line joining two points on a curve is called a **secant line**. We would be using the secant line as a **linear approximation** to the curve between a and b.

To find the equation of the line, look back to Section 6.3.3, equation (6.7). It will be

$$y - f(a) = \frac{f(b) - f(a)}{b - a}(x - a). \tag{8.20}$$

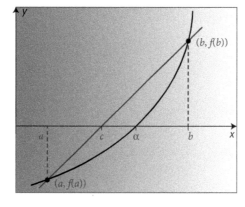

Figure 8.12 The method of *regula falsi*

But if you go back to equation (3.2) you see that the ratio $\dfrac{f(b)-f(a)}{b-a}$ is the slope of the line, which we write as m. So (8.20) can be written as

$$y - f(a) = m(x - a). \tag{8.21}$$

$x = c$ is the value at which $y = 0$:

$$0 - f(a) = m(c - a).$$

Solving this equation for c, you should obtain

$$c = a - \frac{f(a)}{m}. \tag{8.22}$$

Our modified root-finding algorithm becomes:

Method of regula falsi

To find a root of $f(x)$:

1. Pick a, b satisfying $a < b$ and $f(a).f(b) < 0$.

2. Evaluate $m = \dfrac{f(b) - f(a)}{b - a}$.

3. Set $c = a - \dfrac{f(a)}{m}$ and evaluate $f(c)$.

4. If $f(a).f(c) < 0$, the root must lie between a and c; replace b with c.

   Otherwise, the root must lie between c and b; replace a with c.

5. Re-evaluate $f(a)$ and $f(b)$.

6. Return to line 1.

(8.23)

The algorithm is known by the Latin name of ***regula falsi***, or the rather less impressive English translation: the method of false position. It is programmed in SPREADSHEET 8.4. You can see that it finds a good approximation to the root very quickly, though it then seems to get 'stuck' for several iterations before the interval is significantly reduced in size. There exist further modifications that are designed to overcome this problem; nevertheless, *regula falsi* is the main practical method for scientists to use in solving equations numerically.

If it is possible to use calculus with the function, then another powerful iterative method can be used: the Newton–Raphson algorithm, which we'll see in Chapter 13. Finally, it's worth noting that these algorithms can also be used to find when a function $f(x)$ attains a certain prescribed value d, say. If we want to find the value of x that makes $f(x) = 25$, for example, we look for a root of the function $g(x)$ where $g(x) = f(x) - 25$. Then $g(x) = 0$ will mean that $f(x) = 25$.

Similarly, if you have two functions $f_1(x)$ and $f_2(x)$, and you want to find a value of x at which they are equal, look for a root of the function $g(x)$ defined by $g(x) = f_1(x) - f_2(x)$. Then $g(x) = 0$ will mean that $f_1(x) = f_2(x)$.

8.6.2 Minimization algorithms

Here, we consider the problem of finding the turning-points of a function $f(x)$. A turning-point can be a minimum or a maximum. We will only describe minimization algorithms. If you wanted to find a local maximum of $f(x)$, simply flip it around the x-axis (Section 6.5.1) and use a minimization method on the function $-f(x)$. As these methods really cover 'maximization' as well as minimization, the field of study has the general title of **optimization**. We'll discuss optimization in more detail in Chapter 13.

In a bracketing method for finding minima, we will define an interval, and look for a local minimum within that interval. We will be talking about several different intervals, so we'll use a notation which we introduced in Section 1.8.1: we write $[a, b]$ to mean the interval between a and b, that is, the values of x that satisfy $a \leq x \leq b$. Such intervals are called **closed intervals**, because they include their end-points a and b. Open intervals are written with normal brackets, i.e. (a, b) means those x satisfying $a < x < b$, but this can get confused with Cartesian coordinates, so we'll stick to closed intervals and the notation $[a, b]$. Thus, $[0, 1]$ means the closed interval of all values of x between 0 and 1 (including 0 and 1 themselves).

> The closed interval $[a, b]$ means all real numbers lying between a and b. That is, it includes all values of x satisfying $a \leq x \leq b$.

You might think that we could adapt the method of bisection to find local minima, but life isn't that easy. If we take an interval $[a, b]$ and pick a point c inside the interval, and if $f(c)$ is smaller than both $f(a)$ and $f(b)$ (as in Figure 8.13) – and if the graph of $y = f(x)$ is a smooth continuous curve – then we can be sure that a minimum lies somewhere in the interval from a to b. We'll assume that we have chosen our interval well enough, so that there is a single turning-point between a and b. But this minimum could lie between a and c, or between c and b. In the figure, we show this with two possible curves through the three points. In order to narrow down the interval, we need to evaluate $f(x)$ at **two** points c and d within the interval: $a < c < d < b$. Then:

(i)   if $f(c) < f(d)$ we know that a minimum lies between a and d, or

(ii) if $f(c) > f(d)$ we know that a minimum lies between c and b.

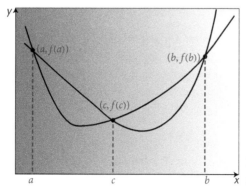

Figure 8.13 Three points are not enough to locate the minimum

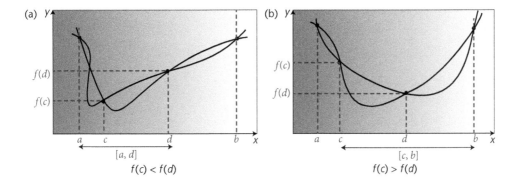

Figure 8.14 Locating the minimum

These two cases are illustrated in Figure 8.14. In either case, we can narrow down the interval (to $[a, d]$ or to $[c, b]$), then pick two more sampling points inside the new interval, and repeat the process. But how do we choose these interior points c, d?

We don't want our method to be biased towards the left or the right end of the interval, so we should pick c and d to lie symmetrically either side of the midpoint. To do this, we choose a fraction τ (Greek letter t, pronounced 'tau') that lies between $\frac{1}{2}$ and 1. Then we say that d lies at a point that is τ of the distance from a to b, and c will lie at the point that is $(1 - \tau)$ of the distance from a. For example, if we pick τ to be $\tau = 0.6$, then d will lie 0.6 of the way from a to b, and c will lie 0.4 of the distance from a. As an example, if $a = 3$ and $b = 8$, then the length of the interval is $8 - 3 = 5$ units, and so d will lie at $a + \tau \times 5 = 3 + 0.6 \times 5 = 3 + 3 = 6$, and c will lie at $a + (1 - \tau) \times 5 = 3 + 0.4 \times 5 = 3 + 2 = 5$. So c and d lie symmetrically a distance of 0.5 units either side of the interval midpoint at 5.5. The general formulae for the locations of the interior points are:

$$c = a + (1 - \tau)l \quad \text{and} \quad d = a + \tau.l \tag{8.24}$$

where $l = b - a$ is the length of the interval $[a, b]$. This is illustrated in Figure 8.15.

But how to choose τ? This is the clever part: we choose τ so that when we narrow down the interval as in Figure 8.14, when one interior point becomes a new end-point, the other interior point will become one of the interior points of the new interval. Thus, we will only need to make one new evaluation of $f(x)$, at the second, new interior point!

Figure 8.16 illustrates what we mean. In each case we start with the interval $[a_1, b_1]$, of length $l_1 = b_1 - a_1$, and with interior points c_1, d_1. Figure 8.16(a) shows how we form the new interval $[a_2, b_2]$ and its interior points c_2, d_2, if $f(c_1) < f(d_1)$. The left-hand end stays the same ($a_2 = a_1$) and the point d_1 becomes the new right-hand end ($b_2 = d_1$). The length of the new interval is then

$$l_2 = b_2 - a_2 = d_1 - a_1 = \tau.l_1.$$

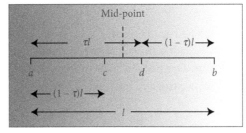

Figure 8.15 Definition of sampling points

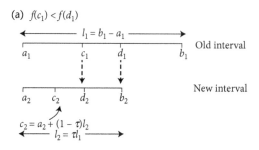

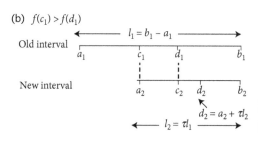

Figure 8.16 The method of golden section search

If we choose τ cleverly, the point c_1 on the old interval becomes point d_2 on the new interval, so that we only need to evaluate f at one new point. In this case,

$$a_1 + (1 - \tau)l_1 = a_2 + \tau . l_2.$$

As $a_2 = a_1$ and $l_2 = \tau . l_1$, this equation reduces to

$$(1 - \tau)l_1 = \tau . \tau l_1,$$

and cancelling the common factor of l_1 we find that τ must satisfy the quadratic equation

$$\tau^2 + \tau - 1 = 0. \tag{8.25}$$

Using the formula in (7.10) the roots of this equation are

$$\tau = \frac{-1 \pm \sqrt{5}}{2} = -1.61803.... \quad \text{or} \quad 0.61803....$$

We are looking for a value of τ between 0.5 and 1, so we take the positive root:

$$\tau = \frac{\sqrt{5} - 1}{2} = 0.6180339887.... \tag{8.26}$$

The case where $f(c_1) > f(d_1)$ is shown in Figure 8.16(b). You should work through the algebra; the same value of τ will be obtained.

The ratio $\tau : 1$ is called the **golden section**[2] or **Golden Ratio**. It is closely related to the Fibonacci sequence of numbers, which you saw in the Case Study A7 (Section 7.2.2), where we solved a very similar quadratic equation (7.18).

---

[2] There is more about the golden section on Dr Ron Knott's website, at http://www.maths.surrey.ac.uk/hosted-sites/R.Knott/Fibonacci/fib.html.

We can summarize our minimization algorithm as:

> **Method of golden section search**
>
> To find a local minimum of $f(x)$:
>
> 1. Set $\tau = \dfrac{\sqrt{5}-1}{2} = 0.6180339887....$
> 2. Pick an interval $[a, b]$ that contains a local minimum of $f(x)$.
> 3. Evaluate $f(a)$, $f(b)$ and the interval length $l = b - a$.
> 4. Evaluate two interior points: $c = a + (1 - \tau)l$ and $d = a + \tau.l$
> 5. Evaluate $f(c)$, $f(d)$.
> 6. Set new interval length l to be $\tau.l$
> 7. Set new values of a, b, c, d and function values as follows:
>
> If $f(c) < f(d)$:   replace the values of $b, f(b)$ with $d, f(d)$
>
>   replace the values of $d, f(d)$ with $c, f(c)$
>
>   set new $c = a + (1 - \tau)l$ and evaluate $f(c)$
>
> If $f(c) < f(d)$:   replace the values of $a, f(a)$ with $c, f(c)$
>
>   replace the values of $c, f(c)$ with $d, f(d)$
>
>   set new $d = a + \tau.l$ and evaluate $f(d)$
>
> 8. If l is still too large, return to step 6.

$$(8.27)$$

At each iteration the length of the interval reduces by a factor of τ, so as with the method of bisection it is possible to predict the interval length after, say, 100 iterations (see problem 8.36). Using logarithms we will also be able to find how many iterations will be needed to achieve a desired level of precision (see Chapter 10). We take our final approximation to α to be the midpoint of the final interval.

The method of golden section search has been programmed into SPREADSHEET 8.5, reproduced in Figure 8.17. In order to keep the structure of the spreadsheet simple, we have not taken advantage of the feature that only one new function evaluation needs to be made at each iteration; in fact, the function values at all four new points a, b, c, d (in columns A, B, C, D) are evaluated at the start of each iteration, in columns H, I, J, K. In order to re-program the spreadsheet to minimize a different function, you just need to redefine the function in cell H12. Highlight it, click on the handle, and drag it across into cells I12, J12, K12. Then, with all four cells highlighted, click on the handle and drag down the sheet. In the example in the spreadsheet the function programmed is the quadratic $f(x) = -(24x(1 - 0.125x) - 7x)$, and the iteration is converging to the minimum at $f(x) = -24.0833$ when $x = 2.8333$. This means that the function $g(x) = -f(x) = 24x(1 - 0.125x) - 7x$ has a maximum at $g(x) = 24.0833$ when $x = 2.8333$. This example is described in Case Study C9.

In Problem 7.27, you showed that the ratios of successive numbers in the Fibonacci sequence:

$$\frac{1}{2}, \frac{2}{3}, \frac{3}{5}, \frac{5}{8}, \cdots$$

get closer and closer to the golden section 0.618 ... There is a related search algorithm called **Fibonacci Search**, in which the value of τ changes at each iteration, and is taken as one of these ratios.

1	**§8.6.2 Finding the Max Economic Yield by Golden Section search**								Error checks:						
2	**Case Study C9: we maximise *P(E)* in equation (8.28)**								i) Check *a* < *b*:		OK				
3	See cell H12 for the function being evaluated								i) Check *f(c)* < *f(a)*:		OK				
4	Pick *a*,*b*: values of *x* which bracket the minimum, and *a* ≤ *b*:								ii) Check *f(d)* < *f(b)*:		OK				
5															
6	*a* =	0			b =	8			Golden Section ratio τ =		0.61803				
7															
8	Interval length: *l* = *b* - *a* =				8		Interior points: *c* = *a* + (1-*t*)*l* =		3.05573		*d* = *a* + *t.l* =		4.9443		
9															
10		Points on interval:			Interval		Function values:			Test:	New	Points on new interval:			
11	*a*	*c*	*d*	*b*	length	*f(a)*	*f(c)*	*f(d)*	*f(b)*		length:	*a*	*c*	*d*	*b*
12	0.0000	3.0557	4.9443	8.0000	8.0000	0.0000	-23.9350	-10.7151	56.0000	f(c) < f(d)	4.9443	0.0000	1.8885	3.0557	4.9443
13	0.0000	1.8885	3.0557	4.9443	4.9443	0.0000	-21.4055	-23.9350	-10.7151	f(c) > f(d)	3.0557	1.8885	3.0557	3.7771	4.9443
14	1.8885	3.0557	3.7771	4.9443	3.0557	-21.4055	-23.9350	-21.4113	-10.7151	f(c) < f(d)	1.8885	1.8885	2.6099	3.0557	3.7771
15	1.8885	2.6099	3.0557	3.7771	1.8885	-21.4055	-23.9336	-23.9350	-21.4113	f(c) > f(d)	1.1672	2.6099	3.0557	3.3313	3.7771
16	2.6099	3.0557	3.3313	3.7771	1.1672	-23.9336	-23.9350	-23.3395	-21.4113	f(c) < f(d)	0.7214	2.6099	2.8854	3.0557	3.3313
17	2.6099	2.8854	3.0557	3.3313	0.7214	-23.9336	-24.0752	-23.9350	-23.3395	f(c) < f(d)	0.4458	2.6099	2.7802	2.8854	3.0557
18	2.6099	2.7802	2.8854	3.0557	0.4458	-23.9336	-24.0749	-24.0752	-23.9350	f(c) > f(d)	0.2755	2.7802	2.8854	2.9505	3.0557
19	2.7802	2.8854	2.9505	3.0557	0.2755	-24.0749	-24.0752	-24.0422	-23.9350	f(c) < f(d)	0.1703	2.7802	2.8452	2.8854	2.9505
20	2.7802	2.8452	2.8854	2.9505	0.1703	-24.0749	-24.0829	-24.0752	-24.0422	f(c) < f(d)	0.1052	2.7802	2.8204	2.8452	2.8854
21	2.7802	2.8204	2.8452	2.8854	0.1052	-24.0749	-24.0828	-24.0829	-24.0752	f(c) > f(d)	0.0650	2.8204	2.8452	2.8606	2.8854
22	2.8204	2.8452	2.8606	2.8854	0.0650	-24.0828	-24.0829	-24.0811	-24.0752	f(c) < f(d)	0.0402	2.8204	2.8357	2.8452	2.8606

Figure 8.17 Finding a minimum of $f(x)$ by the method of golden section search (SPREADSHEET 8.5)

CASE STUDY C9: Fisheries management: finding the Maximum Economic Yield

The problem solved in SPREADSHEET 8.5 is from the fisheries management model of instalment C7 (Section 7.4). The yield-effort curve (7.30) is

$$Y = 24E\left(1 - \frac{E}{8}\right)$$

and the cost-effort line is $C = 7E$. The profit P for a given level of effort is defined as 'profit = yield − cost', or

$$P(E) = Y(E) - C(E). \tag{8.28}$$

This is the function we want to maximize, to find the Maximum Economic Yield. Our method is a minimization algorithm, so we actually use it to minimize the function $-P(E)$. The formula typed in cell H12 is thus

```
= –(24*A12*(1-A12/8)-7*A12)
```

I have taken the initial interval as from 0 to 8, which are the two roots of $Y(E)$. One nice feature of iterative methods is that there is no need to be very precise in choosing the initial approximation or initial interval.

The sheet shows ten iterations of the search; to perform more iterations, highlight row 22 from A22 to T22, click on the handle and drag down. The cells have been set to display numbers rounded to 4 decimal places. You should find that to this precision the Maximum Economic Yield is 24.0833, occurring when $E = 2.8333$.

You should be able to check this result by writing $P(E)$ as a quadratic function of E, and finding its roots using (7.10), and hence where it has its turning-point.

Problems: Now try the end-of-chapter problems 8.32–8.36.

8.7 Extension: finding the equation of a trend line

If we know two data-points, we can fit a straight line through them, and we saw in Chapter 6 how to find its equation. We may, however, know more than two points. If we're lucky, they will all lie on the straight line, as in Experiment (a) of Chapter 3. In this case, we can choose any two points to use in the calculation. (The points should not be close together, otherwise small errors in the data will cause a large error in the calculation of the slope.)

In Experiment (b) of Chapter 3, there were more than two data-points, and although there appeared to be a linear relationship, the points did not all lie exactly on a straight line. In Excel we used the Add trend line... feature to fit a best-approximation line through the points (SPREADSHEET 3.2), and in Sections 3.3.3 and 6.3.4 we saw how to make Excel reveal its equation (SPREADSHEET 3.4), as well as the R^2 measure of goodness-of-fit. We can now explain how the equation of the trend line is worked out. The technique is called **linear regression**. A warning before we begin: this is quite a complex and challenging piece of algebra. Most students of statistics simply learn to apply the final formula (8.33) without worrying about where it comes from. We are presenting the derivation here as an Extension, partly to satisfy your curiosity as to where Excel obtained its trend-line equation from – but mainly as an example of the practical use of some of the mathematical theory we have studied: completing the square, minimization, and simultaneous equations.

Suppose that we have n data-points: $(x_1, y_1), (x_2, y_2), (x_3, y_3), \ldots, (x_n, y_n)$.

We wish to fit a straight line $y = mx + c$ that passes as close as possible to the data-points. In Figure 8.18 we show the data-points and the line, and also the vertical distance d_i from the line to each data-point (x_i, y_i). Our task is to find values of the parameters m and c, so that the sum total of all the vertical distances from the data-points to the straight line is as small as possible.

To refer to the sum of all the d_i, we will use the mathematical notation Σ for summation. The symbol Σ (a Greek capital S, pronounced 'sigma') is not a variable but a shorthand for 'the sum of'. If we had four values of d:

$d_1 = 4 \quad d_2 = -3 \quad d_3 = 0 \quad d_4 = 4.$

Then $\Sigma d_i = d_1 + d_2 + d_3 + d_4 = 4 + (-3) + 0 + 4 = 5$, and $\Sigma d_i^2 = d_1^2 + d_2^2 + d_3^2 + d_4^2 = 4^2 + (-3)^2 + 0^2 + 4^2 = 16 + 9 + 16 = 41$.

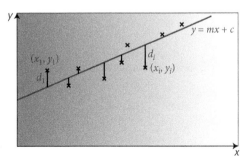

Figure 8.18 Linear regression

You may be familiar with this notation in statistics: the **mean** (or average) of a set of data values $x_1, x_2, x_3, \ldots, x_n$ is given by

$$\bar{x} = (x_1 + x_2 + \cdots + x_n)/n = \frac{1}{n}\sum x_i.$$

Each d_i is a distance in the y-direction, from the y-coordinate of the line $y = mx + c$ to the y-coordinate of the data-point (x_i, y_i). That is,

$$d_i = (mx_i + c) - y_i$$
$$= mx_i - y_i + c.$$

We could add up all the d_i, but there's as snag: positive and negative distances (from points that lie below and above the line) will partially cancel out. This is the situation we saw in Section 2.6.3 when discussing how errors add up. Our solution there was to use the absolute values of the errors. Here, we do something a bit different: we add up the squares of the d_i. As you know, the square of a number is always positive.

So our task becomes: find values of m and c, which will make

$$\sum d_i^2 = d_1^2 + d_2^2 + d_3^2 + \cdots + d_n^2 = \sum (mx_i - y_i + c)^2 \tag{8.29}$$

as small as possible. This is called a **least-squares** problem.

If we multiply out d_i^2 using the distributive law (Section 1.4) we get

$$d_i^2 = (mx_i - y_i + c)^2$$
$$= (mx_i - y_i + c).(mx_i - y_i + c)$$
$$= m^2x_i^2 - 2mx_iy_i + y_i^2 + 2cmx_i - 2cy_i + c^2.$$

So

$$\sum d_i^2 = m^2 \sum x_i^2 - 2m\sum x_iy_i + \sum y_i^2 + 2cm\sum x_i - 2c\sum y_i + nc^2. \tag{8.30}$$

Here, we have used the fact that 2, m, and c are constants. So, for example,

$$\sum 2cmx_i = 2cmx_1 + 2cmx_2 + 2cmx_3 + \cdots + 2cmx_n$$
$$= 2cm(x_1 + x_2 + x_3 + \cdots + x_n)$$
$$= 2cm\sum x_i.$$

In the last term, $\sum c^2 = c^2 + c^2 + c^2 + \cdots + c^2 = n.c^2$

There are five sums involved in calculating the right-hand side of (8.30):

$$A = \sum x_i^2 \quad B = \sum x_i.y_i \quad C = \sum y_i^2$$
$$D = \sum x_i \quad E = \sum y_i$$

so we can write (8.30) using A, B, C, D, E as:

$$\sum d_i^2 = Am^2 - 2Bm + C + 2Dcm - 2Ec + nc^2. \tag{8.31}$$

For example, if we had three data-points $(-2, -3)$, $(0, -1)$ and $(2, 5)$, we will calculate

$$A = \sum x_i^2 = (-2)^2 + 0^2 + 2^2 = 4 + 4 = 8$$

$$B = \sum x_i . y_i = (-2)(-3) + (0)(-1) + (2)(5) = 6 + 0 + 10 = 16$$

$$C = \sum y_i^2 = (-3)^2 + (-1)^2 + 5^2 = 9 + 1 + 25 = 35$$

$$D = \sum x_i = -2 + 0 + 2 = 0$$

$$E = \sum y_i = -3 - 1 + 5 = 1.$$

Substituting into (8.30) we find that

$$\sum d_i^2 = 8m^2 - 32m + 35 - 2c + 3c^2.$$

Now we are going to use the trick of **completing the square**, from Section 7.2.2. We can rewrite the right-hand side as

$$\sum d_i^2 = (8m^2 - 32m + 32) + 3c^2 - 2c + 3 = 8(m - 2)^2 + 3\left(c - \frac{1}{3}\right)^2 + 2\frac{2}{3}.$$

You should check that this is correct, by multiplying out the squared terms.

Remember, we are trying to find values of m and c that will make $\sum d_i^2$ as small as possible. The smallest that each of the squared terms can be is zero, so the smallest value of $\sum d_i^2$ will be $2\frac{2}{3}$, which occurs when $m = 2$ and $c = \frac{1}{3}$. The equation of our trend line is therefore $y = 2x + \frac{1}{3}$.

We were lucky that the coefficient in this example, $D = \sum x_i$, worked out to be zero. (Using our data, $D = \sum x_i = -2 + 0 + 2 = 0$). This meant that there was no term in mc in (8.31), so we were able to separate out the terms in m from the terms in c, and do the completing-the-square trick with each variable separately. In the general case, this will not work, and we have to use a more powerful piece of mathematics. This is the theory of calculus, which is the subject of Part II of this book. Using calculus to minimize $\sum d_i^2$ (you will see how this is done in the Extension to Chapter 16), produces two simultaneous equations:

$$Am + Dc = B$$
$$Dm + nc = E \tag{8.32}$$

which we must solve for m and c. Using the method of Section 5.7 to eliminate c, you should find that

$$m = \frac{DE - nB}{D^2 - nA},$$

and substituting this into the second equation and solving for c:

$$c = \frac{1}{n}(E - mD).$$

Writing this solution in terms of the summations in (8.31), the equation of the linear regression (trend) line is $y = mx + c$, where

$$m = \frac{\left(\sum x\right)\left(\sum y\right) - n\sum xy}{\left(\sum x\right)^2 - n\left(\sum x^2\right)} \quad \text{and}$$

$$c = \frac{1}{n}\left(\sum y - m\sum x\right) = \frac{\left(\sum x\right)\left(\sum xy\right) - \left(\sum x^2\right)\left(\sum y\right)}{\left(\sum x\right)^2 - n\sum x^2}. \tag{8.33}$$

Check that if you use the data from our example, you again obtain $m = 2$ and $c = \frac{1}{3}$.

Notice that the coefficient $C = \sum y_i^2$ does not appear in the formulae; can you see why not, by looking at our three-point example?

The calculation is embodied in SPREADSHEET 8.6, shown in Figure 8.19. Check the formulas for the cells in row 14; they use the Excel function SUM(). Also check how (8.33) is programmed in the cells in row 16. You can use your own data, for up to 6 data-points. You can use SPREADSHEET 3.4 (reproduced as SPREADSHEET 8.7) with the same data, to see Excel fit the trend line; it should come up with the same equation! The Excel functions SLOPE() and INTERCEPT() produce the regression coefficients m, c for a set of data-points. The R^2-vaue is produced by RSQ().

The use of the formulae in (8.33) to fit a least-squares straight line through a set of data, is a standard technique in statistics textbooks; look up 'linear regression' in the index. But statistics books do not generally derive the formulae, and may even omit the formulae altogether, treating the linear regression feature of Excel or Minitab as a 'black box'.

Problems: Now try the end-of-chapter problems 8.37–8.38.

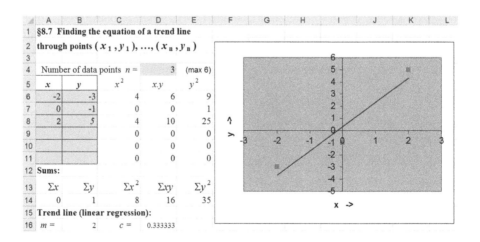

Figure 8.19 Calculating the equation of the trend line (SPREADSHEET 8.6)

8.1–8.4: Write down the reciprocals of the following functions, simplifying where possible:

8.1 $f(x) = \dfrac{x}{ax + b}$

8.2 $f(x) = a + \dfrac{b}{x}$

8.3 $f(x) = \dfrac{kx}{ax^2 + bx + c}$

8.4 $f(x) = \dfrac{x}{1 + \dfrac{1}{x}}.$

8.5–8.12: Find where the following functions

(i) cross the x-axis,

(ii) cross the y-axis, and also find

(iii) any horizontal or sloping asymptotes

(iv) any vertical asymptotes.

Using this information and evaluating individual points if necessary, sketch the function. Check your graph by sketching it in FNGraph:

8.5 $y = \dfrac{2}{3x - 2}$ 8.6 $y = \dfrac{x - 3}{3x - 2}$

8.7 $y = \dfrac{3x - 2}{2x - 3}$ 8.8 $y = \dfrac{x^2 - 9}{x^2 + 6}$

8.9 $y = \dfrac{x^2 - 9}{x - 9}$ 8.10 $y = \dfrac{3x^3 + 24}{x^2 - 8}$

8.11 $y = \dfrac{x - 9}{x^2 - 9}$ 8.12 $y = \dfrac{2x + 9}{x^2 + 9}.$

8.13: In Case Study C8 we wrote a hyperbolic-curve model for animal speed $v(t)$, equation (8.5), given its initial acceleration a_0 and maximum velocity v_{max}. Consider the lion for which $a_0 = 6.25$ m s^{-2} and $v_{max} = 14$ m s^{-1}, and a gazelle for which $a_0 = 3.75$ m s^{-2} and $v_{max} = 25$ m s^{-1}. If the two animals accelerate from rest at the same time $t = 0$, at what time t^* will their velocities be equal? Give t^* correct to 3 significant figures.

8.14: Show that the general solution for Problem 8.13 is

$$t^* = \frac{V_1 V_2}{a_1 a_2} \cdot \frac{(a_2 - a_1)}{(V_1 - V_2)}$$

where V_1, a_1 are the maximum velocity and initial acceleration of the lion, and V_2, a_2 are the maximum velocity and initial acceleration of the gazelle.

8.15: Fit a curve of the form $y = a.\dfrac{1}{x} + b$ to the following data-set:

x	y
3	2.23
5	3.75
8	5.17
12	6.38
20	7.92

8.16: Fit a curve of the form $y = \dfrac{cx}{x + b}$ to the following data-set:

x	y
3	1.30
5	1.97
8	2.75
12	3.49
20	4.32

8.17: Fit a curve of the form $y = \dfrac{1}{ax + b}$ to the data-set of problem 8.15. Which of the two curves is the better fit to the data?

8.18: Fit a quadratic curve to the Swedish population data in problems 7.44–7.46, for the census years 1880–1960. Make the curve go through the first data-point (the 1880 census). Hence estimate the population in the year 2000.

8.19: Now fit a quadratic curve to the data-set in problem 8.18, which goes through the last data-point (i.e. the 1960 census). Again estimate the size of the population in 2000.

8.20: We expect that the following data-set fits a curve of the form $Y(s) = as^n + bs + c$, for some positive integer value of n. By using Excel to fit curves using different values of n, and seeing how good a fit they are, find out what the value of n most probably is:

s	Y
0.0	5.0
1.0	3.8
2.0	5.7
3.0	17.7
4.0	54.0

8.21: The volume V (in litres) and pressure P (in pascals) for a fixed body of gas, are related by Boyle's law (equation 8.6) where the proportionality constant $k = 37.1$ (correct to 3 significant figures). Suppose that the volume is measured correct to the nearest millilitre. Using error analysis, what is the uncertainty interval for P when (i) $V = 0.01$ litres, (ii) $V = 1$ litre, (iii) $V = 10$ litres?

8.22: In the UK, a car's fuel consumption is measured in units of miles per gallon (mpg). Elsewhere in Europe it is measured in litres per 100 km. As we saw in Problem 1.53, the rule for conversion is:

$$g = \frac{282.36}{f}$$

where f is the consumption in mpg and g is the consumption in l/100 km. The constant 282.36 is correct to two decimal places.

My car can achieve 42.2 mpg on the motorway, but only 12.7 mpg when driving around town (to the nearest 0.1 mpg). Convert each of these to litres/100 km, and find the absolute error in each case. Which has the greater error?

8.23: Apply the Lineweaver–Burk transformation of Section 8.4.2 (i.e. plot Y against X, where $X = \frac{1}{s}$, $Y = \frac{1}{v}$, and fit the best straight line) to estimate the values of v_{max} and K_m in the Michaelis–Menten equation (8.10), from the following set of experimental data:

s	v
20	25.3
40	33.0
60	36.8
80	39.0
100	40.8
120	41.8

8.24: As an alternative to the Lineweaver–Burk transformation, the Eadie–Hofstee plot for the Michaelis–Menten equation takes $X = \frac{v}{s}$, $Y = v$ and also obtains a linear relationship $Y = mX + c$. Find out what m and c represent in this case. Hence show how v_{max} and K_m can be derived from m and c. Apply this plot to estimate the values of v_{max} and K_m in the data of Section 8.4.3. Perform the error analysis of that section to see if the drawback of the Lineweaver–Burk plot also applies here.

8.25: A third way of transforming the Michaelis–Menten equation is to take $X = s$, $Y = \frac{s}{v}$. This again gives a linear relationship between X and Y; it is called the Hanes–Woolf plot. Use this plot to determine the values of v_{max} and K_m in the data of Section 8.4.3, and again perform the error analysis of that section.

8.26–8.28: Find the inverse functions of the functions in Problems 8.5–8.7. Find where these inverse functions cross the axes, and their asymptotes. Draw them alongside the original function.

8.29–8.30: Find and sketch the inverse functions of the functions in problems 8.1 and 8.2.

8.31: In Problem 6.19 you found a function $y = f(T)$ to convert a temperature T in °C to y °F (degrees Fahrenheit). What is the inverse function, to convert Fahrenheit to Celsius?

8.32: Perform four iterations of the method of bisection (8.19) on your calculator to find the root between 1 and 2 of the function $f(x) = x^5 - 3$. Note that in this algorithm, you don't need to work out the values of $f(x_n)$, only whether it is positive or negative, i.e. whether x^5 is greater or smaller than 3. Write down the final interval enclosing the root x. Check by evaluating $\sqrt[5]{3}$ on your calculator, using the x^y key.

8.33: In Excel, use the methods of bisection and of *regula falsi* (8.23) to solve problem 8.32. Does *regula falsi* take fewer iterations, and less work?

8.34: In Excel, use the method of golden section search (8.27) to find the minimum of $f(x) = x^5 - x - 3$ between 0 and 1, correct to 4 decimal places.

8.35: In the method of bisection (8.19) for finding the root of a function, the length of the interval is halved at each iteration. Suppose we started with an interval of length 1.0 containing the root. Use your calculator to find the length of the interval after 5 iterations. Continue halving, to find out how many iterations would be needed before the interval length is less than 0.001. We could then quote the midpoint of this final interval as being the solution, with a precision of ±0.0005.

8.36: In the method of golden section search (8.27), by what factor does the interval decrease at each iteration? Suppose we started with an interval of length 1.0 containing the minimum of a function. Use your calculator to find the length of the interval after 5 iterations. How many iterations would be needed before the interval length is less than 0.001?

8.37: Follow the algebra of Section 8.7, to show that if we require the trend line to pass through the origin, its equation will be given by $y = mx$, where

$$m = \frac{\sum xy}{\sum x^2}.$$

8.38: In 'Prey capture by the African lion' (Elliott et al. 1977), the authors consider the effect of hunger on resumption of hunting. They propose a direct-proportion relationship between E, the caloric value of the last meal eaten, and h, the length of time before the lion starts hunting for its next prey. Use the result of problem 8.37 to find using your calculator a trend line $h = mE$ for the following data-set. Give m correct to 4 significant figures. Check your answer using Excel.

E, caloric value of last meal eaten (kJ)	h, time before next hunt (hours)
120	36
180	42
50	15
15	4
90	21
150	35

Periodic functions

Very many natural processes are **cyclical** in nature:

- air temperature rises and falls on a daily cycle;
- the rate of tree growth varies on an annual cycle;
- the heart beats about 70 times per minute (heart rate of a normal male at rest).

Polynomial functions are not suitable for modelling cyclical processes. We need functions that have a certain basic pattern of behaviour (a **cycle**) that repeats. They are called **periodic** functions, and the extent of a single cycle is called a **period**. The independent variable doesn't have to be time t, but in life sciences applications this is virtually always the case.

So if a heart beats around 70 times per minute, a single cycle lasts 1/70 minutes, or about 850 milliseconds. An electrocardiogram recording of the electrical activity (in millivolts) of the heart in a human patient, plotted against time (in seconds) is shown in Figure 9.1. If we decided to define a typical cycle as running from the minimum at $t_1 = 0.9$ seconds to the minimum at $t_2 = 1.8$ seconds, then the period would be

$$T = t_2 - t_1 = 1.8 - 0.9 = 0.9 \text{ seconds.}$$

The most commonly used and relevant periodic functions are the basic **trigonometric** functions: the sine and cosine. These are functions of an angle, so we'll have to look at how angles are measured. However, we will start with a simple Excel function that can be used to produce periodic graphs with sudden changes in value.

9.1 Sawtooth functions

Sawtooth functions are used in sound waves created in music synthesizers, and also describe the horizontal scanning function in cathode-ray-tube televisions and monitors. We will use them to illustrate how to design functions with a given period, amplitude and phase, before we get into the complications of trigonometry.

9.1.1 Basic sawtooth function

Let's define a function called int(x). This function int rounds the number x to the first integer to the left of x on the real line.

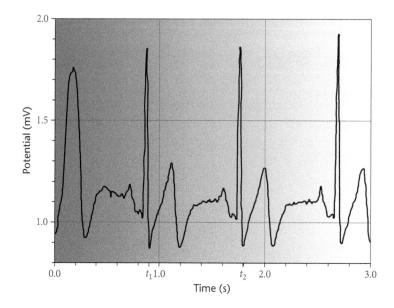

Figure 9.1 Heartbeat graph

So int(3.142) = 3, and int(−5.23) = −6. This function exists in Excel as INT(). Then if cell B3 contains the value 3.142, typing =INT(B3) into cell A5 will display the value 3 in the cell.

The function int(x) also exists in FNGraph. If you use FNGraph to plot the function int(x) the result is not very useful: an ascending 'staircase' of horizontal steps (Figure 9.2(a)). But now try producing the graph of the function

$$y = x - \text{int}(x). \tag{9.1}$$

This is sketched in Figure 9.2(b). This is a periodic function, where each cycle consists of:

- a 'ramp', which is a line sloping upwards with slope 1, i.e. at 45°, from $y = 0$ to $y = 1$, followed by
- a 'cliff' where the graph suddenly drops back down to $y = 0$.

This is called a **sawtooth function**, because its profile resembles the teeth on a saw. The staircase and sawtooth functions are plotted in **FNGRAPH 9.1**.

A sawtooth function might be a first rough approximation to our heartbeat graph in Figure 9.1, if we could transform it to have the same maximum, minimum, and period.

The quantity $x - \text{int}(x)$ is actually the fractional part which is thrown away from x to get int(x), e.g. if $x = 3.142$ then int(x) = 3 and $y = x - \text{int}(x) = 3.142 - 3 = 0.142$. So as x increases from 3 to 4, $y = x - \text{int}(x)$ will increase from 0 to 0.999 . . . before suddenly dropping back to 0 again when $x = 4$.

The length in the x-direction of each cycle is 1 unit. This is the **period** of the function. The **amplitude** of the function is half the vertical distance from the minimum function value over a cycle to the maximum value. This distance is from $y = 0$ to $y = 1$, so the amplitude is $\frac{1}{2}(1 - 0) = 0.5$. The **average** is the average of minimum and maximum values; here, the average is $\frac{1}{2}(1 + 0) = 0.5$.

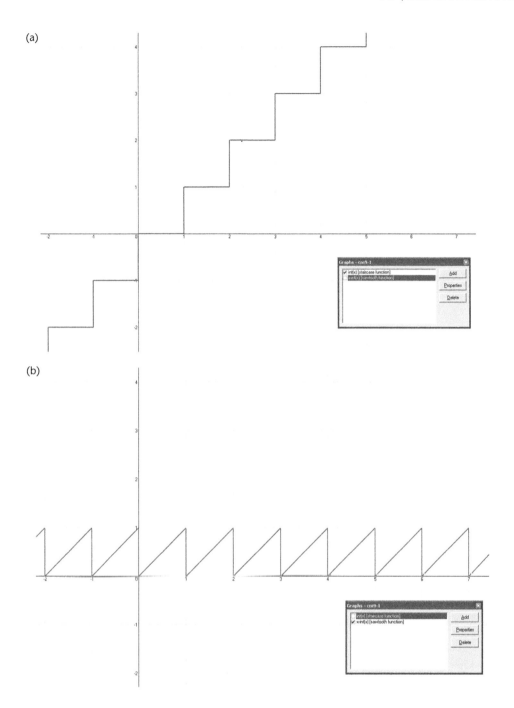

Figure 9.2 Staircase function $y = \text{int}(x)$, and sawtooth function $y = x - \text{int}(x)$

9.1.2 Specifying the period and amplitude

If we want to create a function with period T and amplitude A, we need to perform a horizontal stretch and a vertical stretch; see Section 6.5. The resulting function will be

$$y = 2A\left(\frac{1}{T}x - \text{int}\left(\frac{1}{T}x\right)\right). \tag{9.2}$$

The first cycle now starts when $x = 0$ and ends when $x = T$ (making $\frac{1}{T}x = 1$), so the period is T. The vertical stretch by a factor of $2A$ means that the amplitude becomes $0.5 \times 2A = A$. This function is plotted in FNGRAPH 9.2, with T and A initially set to 4 and 3.

9.1.3 Specifying the vertical shift and phase

The function (9.2) goes from a minimum value of $y = 0$ to a maximum value of $y = 2A$. The average is then $\frac{1}{2}(2A + 0) = A$. If we want different minimum and maximum values (but keeping the same amplitude), we apply a **vertical shift** (see Section 6.5.2), by adding d to the right-hand-side expression; this will shift the graph vertically upwards by d units (thus increasing the minimum, maximum, and average values by d).

Finally, if we wanted the cycle to start when $x = t_1$ rather than $x = 0$, we would replace x by $(x - t_1)$ in the definition; see Section 6.5.4. This horizontal shift is called the **phase** of the periodic function. So the final general sawtooth function, with period T, amplitude A, vertical shift d, and phase t_1, is

$$y = 2A\left(\frac{1}{T}(x - t_1) - \text{int}\left(\frac{1}{T}(x - t_1)\right)\right) + d. \tag{9.3}$$

Let's now model the heartbeat cycle in Figure. 9.1. We need to create a sawtooth function with period T = 0.9. Also, the electrical potential rises from a minimum of 0.85 mV at the start of the cycle to a maximum of 1.85 mV at the end, before dropping back down to 0.85 mV. The difference is 1 mV, so the amplitude is $0.5 \times 1 = 0.5$ millivolts. The vertical shift will be $d = 0.85$, making the average value $\frac{1}{2}(1.85 + 0.85) = 1.35$ mV, and the phase will be $t_1 = 0.9$ seconds. Our function, as a function of time, would be

$$y = \left(\frac{1}{0.9}(t - 0.9) - \text{int}\left(\frac{1}{0.9}(t - 0.9)\right)\right) + 0.85. \tag{9.4}$$

The graph of (9.3), with the parameter values in (9.4), is plotted in FNGRAPH 9.3, shown in Figure 9.3.

Note that (9.4) can be simplified to $y = \frac{1}{0.9}t - \text{int}\left(\frac{1}{0.9}t - 1\right) - 0.15$.

· ·

9.2 Revision of school trigonometry

For the rest of this chapter we will consider functions whose graphs are smooth oscillations; these curves are called **sinusoidal**. We start by reminding you of the basic trigonometry you learnt at school.

At school you measured angles using a protractor, marked in **degrees**. A right-angle or quarter-turn is 90°, and four right-angles make up a complete turn of 360° (Figure 9.4(a)). The circumference c of a circle is proportional to its diameter d. The factor of proportionality is the number written as π (Greek letter p, pronounced 'pi'), which is approximately 3.142 or

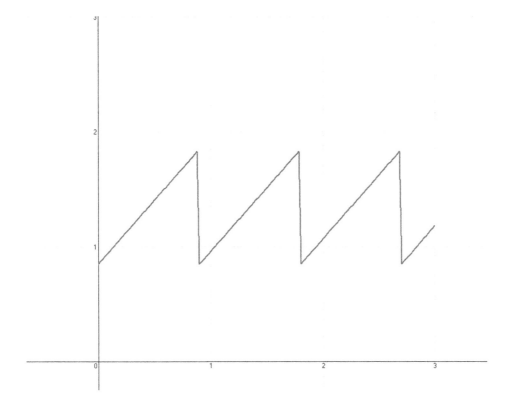

Figure 9.3 Model of heartbeat using sawtooth function (9.4) (FNGRAPH 9.3)

$\frac{22}{7}$. It should be programmed into a button on your calculator. If the radius of the circle is r (so $d = 2r$) this means that

$$c = \pi d = 2\pi r. \tag{9.5}$$

The length of one quarter of a circle will be $\frac{c}{4} = \frac{\pi}{2}.r$ units, see Figure 9.4(b). The area of the circle is πr^2.

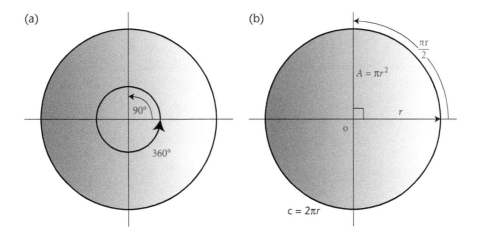

Figure 9.4 (a) Angles of rotation, and (b) length of circumference

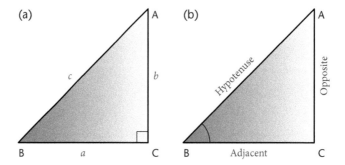

Figure 9.5 Pythagoras's theorem

Figure 9.5(a) shows a right-angled triangle ABC with the right-angle at C. Let a be the length of the side facing A, etc. The side facing C, of length c, is called the **hypotenuse**. The lengths of the three sides are related by **Pythagoras's theorem**:

$$a^2 + b^2 = c^2. \tag{9.6}$$

We will use A, B, C to denote the angles at the corners A, B, C. Then the **sine** of the angle B is defined as

$$\sin B = \frac{AC}{AB} = \frac{b}{c} \tag{9.7}$$

and the **cosine** of B is

$$\cos B = \frac{BC}{AB} = \frac{a}{c}. \tag{9.8}$$

For the angle at B, the side AC is called the **opposite** side, and the other side BC is the **adjacent** side (Figure 9.5(b)). The sine is given by the ratio of lengths $\dfrac{\text{opposite}}{\text{hypotenuse}}$, and the cosine by the ratio $\dfrac{\text{adjacent}}{\text{hypotenuse}}$. As a and b are smaller than c, the sine and cosine of B are numbers between 0 and 1.

A third function, the **tangent** of B is

$$\tan B = \frac{\sin B}{\cos B} = \frac{AC}{BC} = \frac{b/c}{a/c} = \frac{b}{a}. \tag{9.9}$$

That is, tangent $= \dfrac{\text{opposite}}{\text{adjacent}}$.

The angles of a triangle add up to 180°, and the angle at C is 90°, so $A = 90° - B$. Also, for the angle at A, the side BC is the opposite side, and AC the adjacent, so

$$\sin A = \frac{a}{c} = \cos B = \cos(90° - A)$$
$$\cos A = \frac{b}{c} = \sin B = \sin(90° - A). \tag{9.10}$$

There are two standard right-angled triangles, shown in Figure 9.6: (a) an isosceles triangle with two sides of length 1 unit and hypotenuse of length $\sqrt{2}$, and (b) half of an equilateral

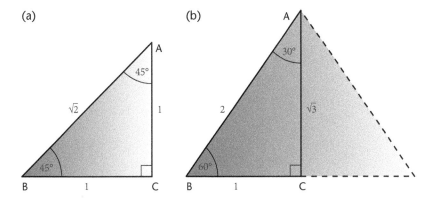

Figure 9.6 Standard right-angled triangles

triangle, with sides of length 1 and $\sqrt{3}$, and hypotenuse of length 2. You should check that in both cases equation (9.6) is satisfied. From these we can deduce the values of the sine and cosine of particular angles:

$$\cos 30° = \sin 60° = \frac{\sqrt{3}}{2} = 0.8660$$

$$\cos 60° = \sin 30° = \frac{1}{2} = 0.5000$$

$$\cos 45° = \sin 45° = \frac{1}{\sqrt{2}} = 0.7071$$

$$\tan 45° = 1.0000$$

(9.11)

where the values are given correct to 4 decimal places.

The big deficiency in this way of defining the sine and cosine is that it only applies to angles between 0° and 90°; these are called **acute** angles. We need a definition that will work for all angles, including negative ones; but first we'll need to define what negative angles are! And while we're defining angles, we'd like a more scientific and rational unit of measurement than the degree. (If you have skipped ahead to read the Historical Interlude, you'll realize that having 360 degrees making up a complete turn is a relic from the Babylonian predilection for counting in multiples of 60.)

. . . .

9.3 Measurement of angles in radians

We will be treating angle as a continuous dimension, like distance or time, so we should use proper variable names for angles. The most common name which indicates to mathematicians that the variable is an angle is the Greek letter θ. This is pronounced 'theta' and represents a 'th' sound in Greek.[1] Other Greek letters commonly used for angles are φ ('phi') and ψ ('psi').

---

[1] As in AΘINAI, the Greek name for Athens.

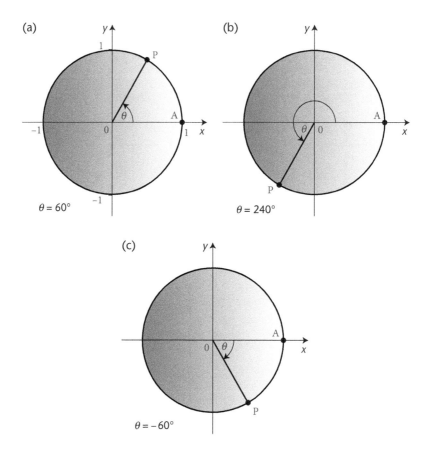

Figure 9.7 Examples of angles of rotation

As listed in Chapter 3, the SI unit of angle is the **radian**. This is defined as follows:

We start with a **unit circle**, as sketched in Figure 9.7. Recall from Section 5.4.4 that this is a circle centred at the origin O of a Cartesian graph, and with radius 1 unit. It is the graph of all points (x, y) that satisfy the equation

$$x^2 + y^2 = 1. \tag{9.12}$$

We draw an angle θ by taking a radius OP of the circle from the origin O at $(0, 0)$, along the positive x-axis to the point A at $(1, 0)$, and rotating it anticlockwise through the angle. If the angle is greater than 360° we will make one full revolution and start a second one. If the angle is negative, we rotate the radius OP clockwise.

Figure 9.7 shows this done for angles of (a) 60°, (b) 240°, (c) −60°. As we rotate OP, the point P travels from A a certain distance around the circle. The distance that P travels is directly proportional to the size of the angle. It is this distance AP going anticlockwise around the unit circle, which defines the measurement of the angle in radians. An angle of 1 radian is that angle that involves travelling an arc of length 1 unit around the unit circle. This is shown in Figure 9.8(a).

One complete turn of 360° will mean the point P travelling all around the circumference c of the circle, and we know from the previous section that $c = 2\pi r = 2\pi$ units, since the radius of the unit circle is 1 unit. So

an angle of 360° is equivalent to 2π radians.

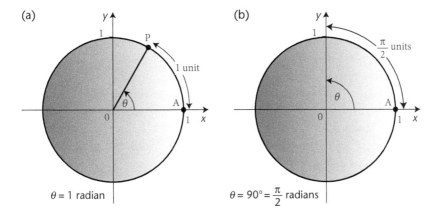

Figure 9.8 Angles in degrees and radians

Similarly, for a right-angle of 90° P travels around one quarter of the circumference, i.e.
$\frac{c}{4} = \frac{2\pi}{4} = \frac{\pi}{2}$ units. So

an angle of 90° is equivalent to π/2 radians.

See Figure 9.8(b).

To convert an angle in degrees into radians, multiply the degrees by $\frac{2\pi}{360}$. We could cancel a factor of 2 in this fraction, and get the rules:

To convert from degrees to radians, multiply by $\frac{\pi}{180}$.

To convert from radians to degrees, multiply by $\frac{180}{\pi}$.

(9.13)

Here is a table of conversions:

Angle in degrees	Angle in radians
360°	2π
180°	π
90°	$\pi/2$
60°	$\pi/3$
45°	$\pi/4$
30°	$\pi/6$
1°	$\pi/180$

In EXCEL SPREADSHEET 9.1 you can experiment with an angle-conversion machine. Type your angle in degrees into cell C7, and you see it converted to radians in cell C15. The unit circle with the radius OP is drawn in the graph. You can also vary the angle continuously from 0° up to 450° using the slider control.[2]

[2] Note: When you try to open a spreadsheet that contains controls such as sliders, you may get a security warning message; you can ignore this. See the Technical Notes.

We said that the value of π is approximately 3.142, or $\frac{22}{7}$. A more accurate decimal expansion is $\pi = 3.14159265\ldots$ This expansion of π never terminates, and never starts endlessly repeating a digit or set of digits. You can check this by going to the website http://www.piday.org/million.php, where you can see the first million digits of π.

We say that π is an irrational number (see Section 1.6.3) because it cannot be written as a fraction. The approximation $\frac{22}{7}$ evaluates to $3.1428571\ldots$, which has a relative error of 0.04%. This may look insignificant, but if an approximate value of π is used repeatedly in calculations, errors can build up (as we discussed in Section 2.6). For this reason, most software that uses π has an accurate value programmed in. In Excel you can use the function PI() which gives π accurate to 15 digits. There is a constant named pi in FNGraph too.

The radian is a dimensionless unit. It is properly defined as the length around the circle divided by the radius, so it is a ratio of two lengths. It is the official SI unit of angle, so although it may be easier to visualize an angle of 45° we should henceforth refer to it primarily as $\pi/4$ radians. Also, from now on if we write $\sin\theta$ (or $\sin x$ or $\sin t$) we mean that θ (or x or t) is measured in radians. You can switch your calculator from working in degrees to working in radians; the key may be labelled DRG, meaning Degrees/Radians/Grads.[3] So if your calculator is set on radians, typing 1 and pressing the ⌈sin⌉ key produces 0.8414709. This is the sine of 1 radian, which is equivalent to $180/\pi = 57.29578$ degrees (to 5 decimal places).

9.4 **The sine and cosine functions**

Figure 9.9 again shows the unit circle, the angle θ, and the radius OP (of length 1 unit). We will write the coordinates of P as (x, y). But now we draw a vertical line down from P to C on the x-axis. The triangle OPC looks very much like the triangle BAC in Figure 9.5, which we

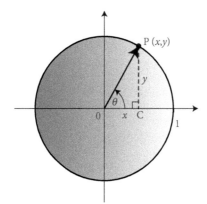

Figure 9.9 Definition of sine and cosine functions

[3] The grad was an early attempt at metrication for angles: 100 grads make a right-angle. It is hardly ever used nowadays.

used to define the sine and cosine of the angle at B. This is now the angle at O, called θ. Following the definitions in (9.7)–(9.9) we have

$$\sin\theta = \frac{PC}{OP} = \frac{y}{1} = y$$

$$\cos\theta = \frac{OC}{OP} = \frac{x}{1} = x \tag{9.14}$$

$$\tan\theta = \frac{\sin\theta}{\cos\theta} = \frac{PC/OP}{OC/OP} = \frac{PC}{OC} = \frac{y}{x}.$$

So the Cartesian coordinates of the point P can be written $(\cos\theta, \sin\theta)$. This is how we define the sine and cosine of any angle θ:

> If the point P on the unit circle, obtained by rotating the radius through
> an angle θ, has coordinates (x, y), then $\cos\theta = x$ and $\sin\theta = y$. $\qquad$ (9.15)

Look at how the y-coordinate of P (i.e. its height above the x-axis) changes as the radius OP makes one complete revolution (i.e. θ increases from 0 to 2π radians). The height starts at 0.0, increases to 1.0 when $\theta = \frac{\pi}{2}$ (i.e. 90°), and drops back to 0.0 again when $\theta = \pi$ (i.e. 180°). It now becomes negative as P falls below the x-axis, reaching a minimum of −1.0 when $\theta = \frac{3\pi}{2}$ (i.e. 270°), before climbing back up to 0.0 at $\theta = 2\pi$ (i.e. 360°). If θ continues to increase, P repeats its path around the circle, giving us a periodic function. In Figure 9.10(a) this sine function is plotted against θ; we see that it produces a sinusoidal graph.

The x-coordinate of P shows the same sinusoidal behaviour, but out-of-phase by 90°. It starts at 1.0 when $\theta = 0$, drops down to a minimum of −1.0 when $\theta = \pi$ (i.e. 180°), then climbs back up to 1.0 again. The graph of $y = \cos\theta$ is shown in fig 9.10(b). This behaviour is summarized in Figure 9.11 and the table below:

θ in degrees	θ in radians	$x = \cos\theta$	$y = \sin\theta$	Figure 9.11
0°	0	1	0	
0° < θ < 90°	0 < θ < π/2	0 < x < 1	0 < y < 1	(a)
90°	π/2	0	1	
90° < θ < 180°	π/2 < θ < π	−1 < x < 0	0 < y < 1	(b)
180°	π	−1	0	
180° < θ < 270°	π < θ < 3π/2	−1 < x < 0	−1 < y < 0	(c)
270°	3π/2	0	−1	
270° < θ < 360°	3π/2 < θ < 2π	0 < x < 1	−1 < y < 0	(d)
360°	2π	1	0	

You can see an animation of this process in SPREADSHEET 9.2, reproduced in Figure 9.12. Use the slider control to set the angle θ (in degrees) into the yellow cell A7, and in column A you will see the value of θ in radians, and of $\cos\theta$ and $\sin\theta$. The chart with the unit circle will

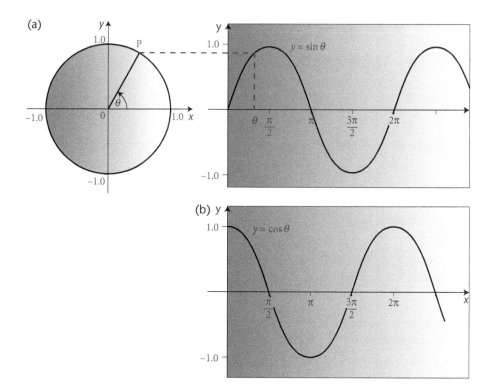

Figure 9.10 Plots of sine and cosine functions

show the radius in the appropriate position. The vertical yellow line is of length $y = \sin \theta$, and the horizontal purple line is of length $x = \cos \theta$. The yellow line is reproduced on the graph of $y = \sin \theta$, at the appropriate value of θ along the horizontal axis.

As you saw in **SPREADSHEET 9.1**, once we have rotated through 360° or 2π radians, we are back where we started, and continuing to rotate anticlockwise will repeat the same cycle. So the cosine and sine functions are periodic, with a period of 2π. This can be represented mathematically as

$$\cos (\theta + 2\pi) = \cos \theta \quad \text{and} \quad \sin (\theta + 2\pi) = \sin \theta. \tag{9.16}$$

If you have to find the cosine or sine of an angle greater than 360°, simply deduct multiples of 360° (or 2π) until you get an angle between 0 and 360°:

$$\sin 765° = \sin(765 - 360)° = \sin 405° = \sin(405 - 360)° = \sin 45° = \frac{1}{\sqrt{2}}$$
$$\cos (19\pi/3) = \cos (19\pi/3 - 6\pi) = \cos (\pi/3) = \frac{1}{2}$$

We can also define the sine and cosine of negative angles, by starting at the positive x-axis but rotating the radius clockwise. Figure 9.11(e) shows the cosine and sine angles for a negative angle.

As well as a period of 2π (360°), both functions $\sin \theta$ and $\cos \theta$ have an amplitude of 1.0 and an average value of 0.0. The cosine function is equivalent to the sine function after a horizontal shift of $\pi/2$ to the left:

$$\cos \theta = \sin \left(\theta + \frac{\pi}{2} \right). \tag{9.17}$$

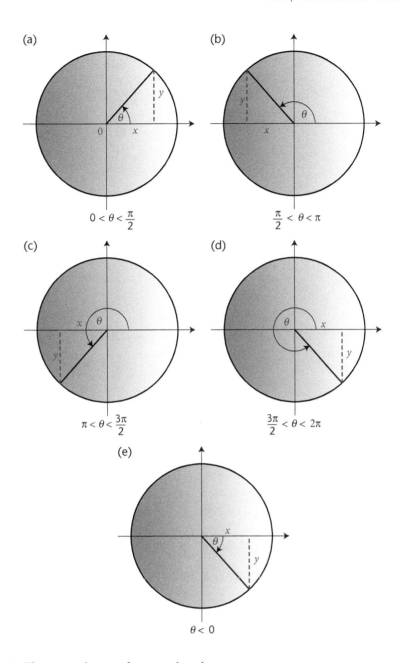

Figure 9.11 The sine and cosine for general angles

The graph of $y = \sin x$ crosses the x-axis at $x = \ldots, -3\pi, -2\pi, -\pi, 0, \pi, 2\pi, 3\pi, \ldots$ We could write this as $x = n\pi$ where n is any integer.

Finally, there is a very important **trigonometric identity** linking sine and cosine of any angle. First note that we can treat $\cos\theta$ and $\sin\theta$ just like variables; in particular we can square them. We have seen that $\cos\dfrac{\pi}{6} = \dfrac{\sqrt{3}}{2}$, so $\left(\cos\dfrac{\pi}{6}\right)^2 = \left(\dfrac{\sqrt{3}}{2}\right)^2 = \dfrac{3}{4}$, for example. We write $(\cos\theta)^2$ as $\cos^2\theta$.

Remember that an **identity** is an equation that is true whatever values you put in for the variables. An example would be $(a+b)(a-b) = a^2 - b^2$.

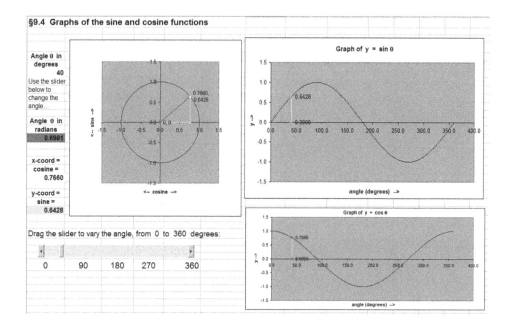

Figure 9.12 Graphs of the sine and cosine functions in Excel (SPREADSHEET 9.2)

The point P with coordinates (x, y) lies on the unit circle whose equation is (9.12). And as (9.14) tells us $\cos \theta = x$ and $\sin \theta = y$, we conclude that for any angle θ:

$$\cos^2 \theta + \sin^2 \theta = 1. \tag{9.18}$$

To sketch the graphs of these functions using Cartesian coordinates, use the FNGraph package. For the cosine function, click Add in the Graphs window and type the function as $\cos(x)$; for the sine function type $\sin(x)$. This is done in FNGRAPH 9.4, reproduced in Figure 9.13. The cosine curve is in red, and the sine curve in blue.

Both functions oscillate sinusoidally between a maximum value of +1 and a minimum value of −1. Thus the **amplitude** of the functions is half the difference between the maximum and minimum values, namely $(1 − (−1))/2 = 2/2 = 1$ unit.

Notice that the blue curve is the same as the red curve, but shifted to the left by a distance of $\pi/2$, or 1.571 units. This illustrates (9.17).

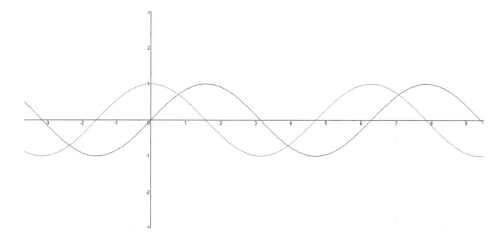

Figure 9.13 Graphs of $\sin x$ and $\cos x$ (FNGRAPH 9.4)

The **period** of the functions is the distance along the x-axis for one complete cycle, which is 2π (which is approximately 6.283). To see just one cycle of each function, highlight the function in the `Graphs` window, click `Properties`, and set the `Min` and `Max` x-values of the graph to be 0 and 6.283.

9.5 Periodic functions of time

We want to use the sine and cosine to produce sinusoidal functions that can model real data from natural processes (such as temperature variation over the seasons) where time t is the independent variable. For this we need to use the linear transformations of Section 6.5 to obtain the correct period (e.g. 24 hours, or 365 days), amplitude, phase, and average.

9.5.1 General sine and cosine functions

One cycle of the basic function $\cos t$, from its maximum value of 1.0 down to -1.0 and back to the maximum, has a period of 2π. To change the period, we apply a horizontal stretch. Consider the function $\cos \omega t$ (ω is a Greek letter named omega). The angle θ is now ωt. When $t = 0$, $\theta = \omega.0 = 0$ so the cycle starts at $t = 0$. The next cycle starts when the angle $\theta = 2\pi$. This occurs when $t = \frac{2\pi}{\omega}$ (making $\omega t = 2\pi$). The period is then $\frac{2\pi}{\omega} - 0 = \frac{2\pi}{\omega}$.

To obtain a cosine function with a period of T (measured in seconds, hours or days) we want $\frac{2\pi}{\omega} = T$, and solving for ω, $\omega = \frac{2\pi}{T}$, so we use $\cos\left(\frac{2\pi}{T}t\right)$. For this function, one cycle starts at $t = 0$ and ends at $t = T$, for then $\cos 0 = 1$ and $\cos\left(\frac{2\pi}{T}t\right) = \cos\left(\frac{2\pi}{T}T\right) = \cos(2\pi) = 1.0$.

We can apply the other linear transformations from Section 6.5, as we did for the sawtooth function earlier in this chapter, to produce the general cosine function with period T, amplitude A, phase t_0, and average value d:

$$f(t) = A\cos\left(\frac{2\pi}{T}(t - t_0)\right) + d. \tag{9.19}$$

Here, the angle $\theta = \frac{2\pi}{T}(t - t_0)$.

One cycle for $f(t)$ runs from $t = t_0$ to $t = T + t_0$, the values of t that make the angle 0 and 2π respectively. You should check that

$$f(t_0) = A\cos\left(\frac{2\pi}{T}(t_0 - t_0)\right) + d = A\cos 0 + d = A + d$$

$$f(\tfrac{1}{2}T + t_0) = A\cos\left(\frac{2\pi}{T}(\tfrac{1}{2}T + t_0 - t_0)\right) + d = A\cos\pi + d = -A + d$$

$$f(T + t_0) = A\cos\left(\frac{2\pi}{T}(T + t_0 - t_0)\right) + d = A\cos 2\pi + d = A + d$$

which are the function values (maximum, minimum, maximum) at the start, midpoint, and end of the cycle. The function is sketched in Figure 9.14 and plotted in FNGRAPH 9.5. Unfortunately in FNGraph we must use x for the independent variable, not t. Also, FNGraph

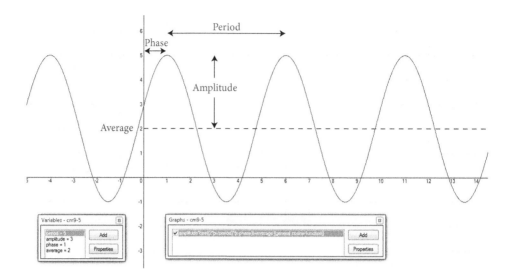

Figure 9.14 General cosine function (FNGRAPH 9.5)

doesn't use capital letters for parameters, so we've called them 'period', 'amplitude', 'phase' and 'average' instead of T, A, t_0 and d.

Similarly, the general sine function is

$$g(t) = A\sin\left(\frac{2\pi}{T}(t - t_0)\right) + d. \tag{9.20}$$

In modelling a set of real data, it generally doesn't matter whether we use the sine or cosine function; only the phase t_0 will be different depending on our choice. But often we can see that one is more convenient, e.g. if the data starts at a maximum value we would use the cosine function. In the next section there is an example of fitting some real data.

CASE STUDY C10: A simple model of predator–prey population dynamics

Imagine an isolated ecosystem, such as a small island or the Ngorongoro Crater, which includes a population of predators and a population of the animals that are its main prey, e.g. lions and gazelles, or wolves and moose. Let $P_1(t)$ and $P_2(t)$ be the sizes of the populations of predators and prey respectively, at time t. These two populations are not independent of each other. When the number of predators is high, we would expect the size of the prey population to decrease. Conversely, when there are small numbers of prey available, the predator population would decrease due to lack of food. Because of this interdependence, each population tends to vary in cycles. We can describe one cycle, using wolves as predators and moose as prey, as follows:

We start when the moose population is at a minimum. The wolves cannot find enough moose to catch, and so start to die out. As the number of wolves declines, the moose population has a chance to recover. Increasing numbers of moose will provide the food resources that allow the wolf population to expand. But at a certain point the wolf population will grow so big that the rise in moose numbers is halted. As moose numbers drop, wolves find it increasingly difficult to catch food, and so the cycle repeats.

A simple model of this process would be to represent each population by a sinusoidal function (9.19), with the predator oscillation lagging behind the prey oscillation by a quarter-period. Here is a practical example.[4]

Isle Royale is an island in Lake Superior, which has resident populations of wolves and moose. A biologist named Rolf Peterson has been observing the populations since 1960, and writing annual reports. He noted that 'for the period from 1959 to 1980 the wolf and moose population appeared to cycle in tandem, with wolves peaking about a decade after moose' (Peterson 2007). In 1980 the wolf population reached a peak of 50, with about 1000 moose. A decade previously, there were about 1250 moose and 30 wolves. Let us use this data in our simple model.

If the wolf population peak lags behind the moose population peak by 10 years, which represents a quarter-period, then the period $T = 40$ years. We will measure time t in years from the point when recording started in 1960. Looking at (9.19), the peak population is $P_1(t_0) = A + d$, occurring when $t = t_0$. So for the wolf population, the peak of 50 in 1980 means $t_0 = 20$ and $A + d = 50$. Ten years (a quarter-period) earlier, we have $P_1(t) = A = 30$ wolves. So $d = 20$ wolves. The function for the predator population is thus

$$P_1(t) = 30\cos\left(\frac{2\pi}{40}(t - 20)\right) + 30. \tag{9.21}$$

The moose population reached its peak of 1250 ten years earlier than the wolves, in 1970. By a similar set of calculations, check that the function for the moose population will be

$$P_2(t) = 250\cos\left(\frac{2\pi}{40}(t - 10)\right) + 1000. \tag{9.22}$$

The functions are graphed in FNGRAPH 9.6; the red curve represents the wolves, and the blue curve the moose. To fit the two curves on the same graph, different vertical scalings have been applied. A value of 12 on the vertical scale corresponds to populations of 60 wolves or 2500 moose.

The graph is printed in Figure 9.15 together with a plot of the populations from 1960 to 1995. You can see that both populations ceased to correspond to this simple sinusoidal model after 1980. Peterson suggests that this was due to a disease introduced by visitors affecting the wolves; the low numbers of wolves then allowed the moose to more than double in number.

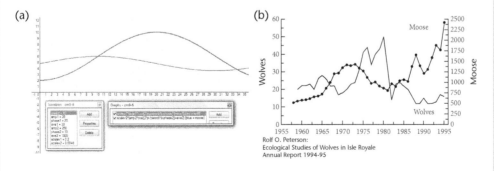

Figure 9.15 Variation of wolf and moose populations (a) FNGRAPH 9.6, (b) Peterson (2007)

[4] Cited in *Chance News*: http://www.dartmouth.edu/~chance/chance_news/recent_news/chance_news_5.04.html#wolf-moose. Graph at http://www.dartmouth.edu/~chance/gif/wolves-moose.gif

9.5.2 Application: modelling tidal data

There is a website called *Admiralty EasyTide* (http://easytide.ukho.gov.uk/), which provides free 7-day predictions of tidal levels at over 7000 ports worldwide. Figure 9.16 shows the output from one such prediction, for the port of Inishgort in Ireland in March 2009, when this chapter was being written. There is a graph of tidal height h in metres over a 7-day (168-hour) period, and the times and heights of high tide and low tide (i.e. maximum and minimum values) each day. Our task is to produce trigonometric functions that will give a good model for this data.

Looking at the graph, it is clear that one single sine or cosine function will not fit this data. In the first part of the graph, from midnight on Friday until the high tide at 08:00 on Monday, the curve is very close to a standard sine curve. But after that time, the amplitude of the oscillations decreases linearly with time, so we will need to modify the function to model this. We will treat these two parts of the data separately, and find functions $h_1(t)$ and $h_2(t)$ that produce close approximations to each part. We will plot these functions in FNGraph and compare the resulting graph with that provided by EasyTide.

We will measure time t in days (it might be more sensible to measure t in hours, but FNGraph cannot cope with the a range of 0 to 168 along the horizontal axis). We need to convert time expressed in days/hours/minutes into days as a decimal, measured from $t = 0.0$

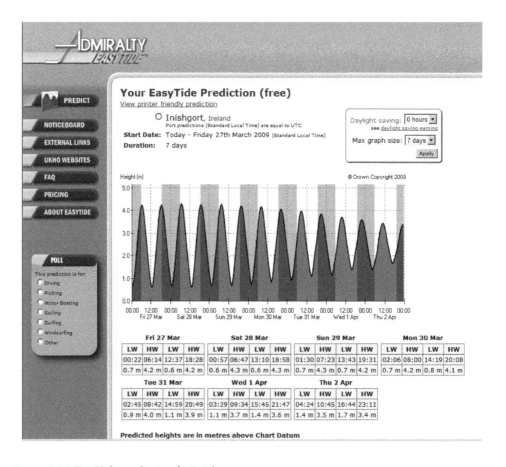

Figure 9.16 EasyTide prediction for Inishgort

at 00:00 on Friday. Thus the end of the first part of the graph, at 08:00 on Monday, i.e. after 3 days 8 hours, will be at time $t = 3.33$ days.

Although each cycle is slightly different, we are justified in approximating the graph as follows:

Part (i) of the graph ($t = 0.0$ to $t = 3.33$):

- The maximum value (high tide) is 4.2 metres.
- The minimum value (low tide) is 0.7 metres.
- Each cycle has equal period.

Part (ii) of the graph ($t = 3.33$ to $t = 7.0$):

- The amplitude decreases linearly with time.
- The average value remains constant, and equal to that in part (i).
- Each cycle has equal period.

These assumptions are illustrated in Figure 9.17.

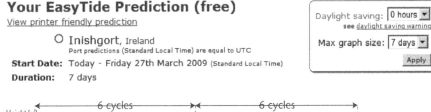

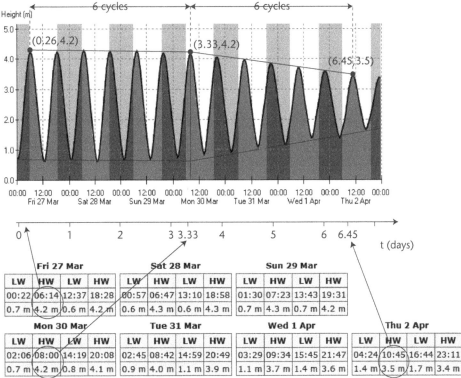

Figure 9.17 Analysing the EasyTide prediction

To produce the function modelling part (i) of the graph is straightforward. We will use the cosine function, so that the start of the first cycle is at the first high tide, at 06:14 on Friday. Converting to a decimal, this is when $t = 0.26$ days. There are six complete cycles from then until $t = 3.33$, so the mean duration of one cycle (i.e. the period) is $T_1 = \dfrac{3.33 - 0.26}{6} = 0.51$ days. We are taking the heights at high and low tide to be 4.2 and 0.7 metres, so the average height is $d = \frac{1}{2}(4.2 + 0.7) = 2.45$ metres. The amplitude is $A_1 = 4.2 - 2.45 = 1.75$ metres. Finally, the phase is the time at the start of the first cycle, namely $t_0 = 0.26$ days. Plugging these values into (9.19) we obtain our model of the tidal variation $h_1(t)$ in part (i) of the data:

$$h_1(t) = 1.75\cos\left(\frac{2\pi}{0.51}(t - 0.26)\right) + 2.45. \tag{9.23}$$

In part (ii) of the graph we assume that the average height d is 2.45 m, the same as in part (i). There are six complete cycles from the start of part (ii), when $t = 3.33$ days, up to the start of the final cycle, at 10:45 on Thursday; this time converts to $t = 6.45$ days. So we can work out the average part (ii) period: $T_2 = \dfrac{6.45 - 3.33}{6} = 0.52$ days. The first cycle starts when $t = 3.33$ days, so we use this as the phase t_0. But the maximum height H (and hence the amplitude A) are not constant, but a linear function of time.

Here, we find an expression for the amplitude, which in part (ii) is not constant but decreases linearly with time. At the start of part (ii), when $t = 3.33$ days, the high tide is $H = 4.2$ metres. But at the start of the final cycle, when $t = 6.45$ days, the high tide is only $H = 3.5$ metres. Thus, the amplitude A is 1.75 metres when t = 3.33 (as in part (i)), but when $t = 6.45$ the amplitude is $A = H - d = 3.5 - 2.45 = 1.05$ metres. To find the part (ii) amplitude A_2 as a function of t, means finding the equation of the straight line through the points (3.33, 1.75) and (6.45, 1.05). We covered this is Section 6.3.3; if we write A_2 as a linear function of t, $A_2 = mt + c$, we can find m and c by solving two simultaneous equations:

$$3.33m + c = 1.75 \quad \text{and} \quad 6.45m + c = 1.05.$$

Subtracting the two equations (so that the cs cancel out) gives $-3.12m = 0.7$, so that $m = -0.224$. Substituting back into either equation and solving for c gives $c = 2.497$. Our expression for amplitude as a function of time is therefore $A_2(t) = 2.497 - 0.224t$. We use this in the cosine function (9.19) to produce our model of the tidal variation $h_2(t)$ in part (ii) of the data:

$$h_2(t) = (2.497 - 0.224t)\cos\left(\frac{2\pi}{0.52}(t - 3.33)\right) + 2.45 \tag{9.24}$$

The two functions in (9.23) and (9.24) are plotted in Figure 9.18 and **FNGRAPH 9.7**. The blue curve is $h_1(t)$, plotted from $t = 0.0$ to $t = 3.33$, and the green curve is $h_2(t)$, plotted from $t = 3.33$ to $t = 7.0$ days. The composite curve is a good model of the EasyTide prediction.[5]

---

[5] Note: FNGraph has a bug which means it does not save fractional values in the 'Min' and 'Max' boxes of the graph properties window. You will need to change the value 3 to 3.33 in the properties for the two functions, in order to see the correct point at which the blue curve ends and the green curve starts.

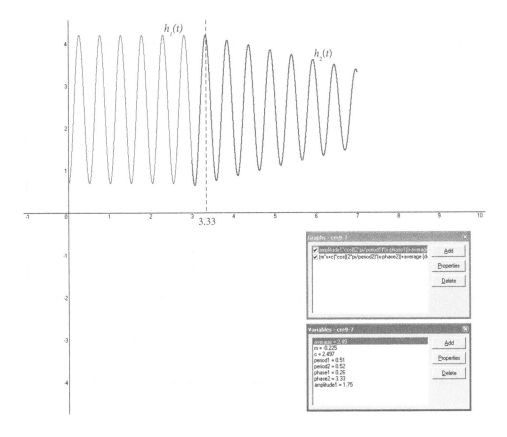

Figure 9.18 Model of the EasyTide prediction (FNGRAPH 9.7)

9.5.3 Application: modelling temperature variations

We saw in (9.24) a function that combined a cosine function with a linear function describing the amplitude. There are many other ways of combining trigonometric functions with linear, polynomial, or other trigonometric functions. For example, in situations in Nature where temperature varies with time, this is often governed by a circadian (daily) cycle, as well as a longer cycle over a month or a year. If we have trigonometric functions describing these two cycles, we can add them to obtain the required model, as in the following example.

In Figure 9.19 and FNGRAPH 9.8, we plot temperature y in degrees Celsius, against time t in days. The blue curve is of the function

$$y = \cos(2\pi t) - 4\cos\left(\frac{2\pi}{30}t\right) + 8. \tag{9.25}$$

This is made up from the addition of

(i) a function $-4\cos\left(\dfrac{2\pi}{30}t\right) + 8$ which has a period of 30 days, over which it rises from a minimum temperature of $-4 + 8 = 4$ degrees, to a maximum of $-(-4) + 8 = 12$ degrees and back down again; and

(ii) a function $\cos(2\pi t)$ with a daily cycle and an amplitude of 1°C.

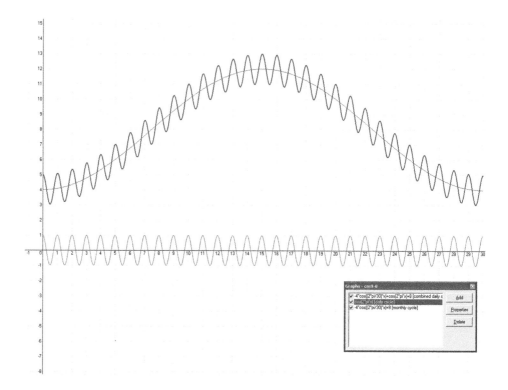

Figure 9.19 Temperature variation over daily and monthly cycles (FNGRAPH 9.8)

9.6 Reciprocal and inverse trigonometric functions

In Chapter 8 we introduced the idea of reciprocal and inverse functions. There are special functions that are the reciprocals and inverses of the sine, cosine, and tangent. We'll need these in Part II when we apply the theory of the calculus to trigonometric functions. Beyond that, these functions have limited use in the life sciences. But if you encounter one of them in a mathematical model you should be able to recognize what it means – or have a book in which you can look it up. We will therefore define these functions in this section.

9.6.1 Reciprocal trigonometric functions

Recall from Section 8.1 that the reciprocal of a function $f(x)$ is the function $\dfrac{1}{f(x)}$. The reciprocal of the cosine function is called the **secant**, and written as sec θ. So $\sec\theta = \dfrac{1}{\cos\theta}$. The reciprocal of the sine function is called the **cosecant** (written as cosec θ), and the reciprocal of the tangent function is called the **cotangent** (written as cot θ).

The definitions of these three new functions are:

$$\sec\theta = \frac{1}{\cos\theta}, \quad \mathrm{cosec}\,\theta = \frac{1}{\sin\theta}, \quad \cot\theta = \frac{1}{\tan\theta} = \frac{\cos\theta}{\sin\theta}. \tag{9.26}$$

(In the definition of cotangent, we have used the fact that $\frac{1}{\tan} = \frac{1}{\sin/\cos} = \frac{\cos}{\sin}$.)

As with the rational functions in Chapter 8, these also have vertical asymptotes at values of θ that make the denominator zero (because it's impossible to divide by zero). So the secant function will have a vertical asymptote at those values of θ for which $\cos \theta = 0$. This is seen in Figure 9.20, taken from FNGRAPH 9.9, where the graph of $\cos \theta$ is drawn in black, and that of $\sec \theta$ in blue. The two curves intersect at values of θ for which $\cos \theta = \sec \theta = 1$. You probably won't have separate keys for these functions on your calculator, since you can obtain $\sec(x)$ by pressing the [cos] key and then the [1/x] key), but they are standard functions in Excel and in FNGraph: `sec(x)`, `cosec(x)` and `cot(x)`.

Remember the important trigonometric identity $\cos^2 \theta + \sin^2 \theta = 1$ in equation (9.18)? There is a similar identity (i.e. an equation that is true for any value of θ) involving the reciprocal functions:

$$\sec^2 \theta = 1 + \tan^2 \theta. \tag{9.27}$$

You can obtain this identity from (9.18) by dividing both sides by $\cos^2 \theta$:

$$\frac{\cos^2 \theta + \sin^2 \theta}{\cos^2 \theta} = \frac{1}{\cos^2 \theta}$$

$$\frac{\cos^2 \theta}{\cos^2 \theta} + \frac{\sin^2 \theta}{\cos^2 \theta} = \frac{1}{\cos^2 \theta}$$

$$1 + \tan^2 \theta = \sec^2 \theta.$$

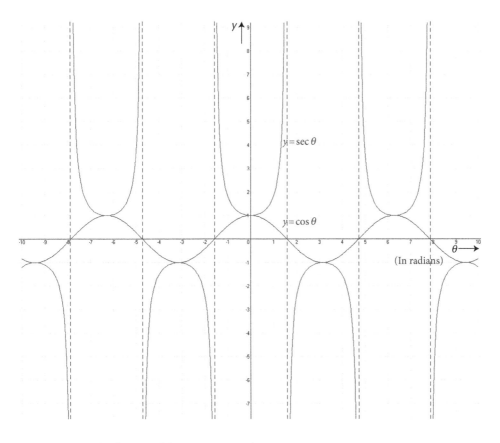

Figure 9.20 Graph of $y = \sec \theta$ (FNGRAPH 9.9)

9.6.2 Inverse trigonometric functions

Recall from Section 8.5 that, given a function $f(x)$, the inverse function $f^{-1}(x)$ is defined by

$$y = f^{-1}(x) \quad \text{if } y \text{ is a number satisfying} \quad f(y) = x.$$

The sine function $x = \sin\theta$ takes an angle θ as input, and produces a number x that lies between −1 and +1. So the inverse sine function, which is written as $\sin^{-1}x$, is a function that takes a number x between −1 and +1 as input, and produces an angle. Measuring angles in degrees for a moment, we know that $\sin 30° = \dfrac{1}{2}$, so $\sin^{-1}0.5$ will be 30° (or $\dfrac{\pi}{6}$ in radians). But there are other angles θ for which $\sin\theta = \dfrac{1}{2}$, for example 150° or 390° or −210°, and a function is only allowed to produce a single output value for each input value. We need to specify which value we are going to take; for the inverse sine function we take the angle that lies between −90° and +90° (we can write this as $\dfrac{-\pi}{2} < \theta < \dfrac{\pi}{2}$). This is called the **principal value** of $\sin^{-1}x$. To summarize:

The **inverse sine** of x (written $\sin^{-1}x$) is the angle between −90° and +90° whose sine is x.

$\theta = \sin^{-1}x$ means that $\sin\theta = x$.

Of course, x can only be a number between −1 and +1 (we can write −1 < x < +1). The expression '$\sin^{-1}2$' is meaningless; it says 'the angle whose sine is 2'. The domain of the function is the closed interval [−1, 1].

We saw in Figure 8.8 that the graph of the inverse function $y = f^{-1}(x)$ can be obtained by reflecting the graph of $y = f(x)$ in the diagonal line $y = x$. As the graph of $y = \sin x$ wiggles along the x-axis, reflecting this in $y = x$ will produce a sinusoidal curve wiggling up the y-axis. But because of the restriction to using just the principal value, we actually see only a small part of that curve. In Figure 9.21, reproducing FNGRAPH 9.10, you can see the graph of $\theta = \sin^{-1}x$ in blue, together with the sine curve (in red): the blue curve is a partial reflection of the red curve in the green 'mirror' line $y = x$.

Other inverse trigonometric functions – the inverse cosine and the inverse tangent – are defined similarly. Here are the formal definitions:

$\theta = \sin^{-1}x$ means that θ is the angle in the range $\dfrac{-\pi}{2} < \theta < \dfrac{\pi}{2}$ for which $\sin\theta = x$

$\theta = \cos^{-1}x$ means that θ is the angle in the range $0 < \theta < \pi$ for which $\cos\theta = x$ (9.28)

$\theta = \tan^{-1}x$ means that θ is the angle in the range $\dfrac{-\pi}{2} < \theta < \dfrac{\pi}{2}$ for which $\tan\theta = x$.

It's important to remember that the notation $\sin^{-1}x$ has nothing to do with reciprocals; it's a special one-off notation for the inverse. To avoid this confusion, and to have a name that can be used in computer programs such as Excel, there is an alternative name for $\sin^{-1}x$, which is arcsin (x). You can see this used in FNGRAPH 9.10. The inverse trigonometric functions are thus available in Excel and FNGraph as `arcsin(x)`, `arccos(x)`, and `arctan(x)`. Some calculators have $\boxed{\sin^{-1}}$ x etc., and some have $\boxed{\text{arcsin}}$ x etc. on the keypad.

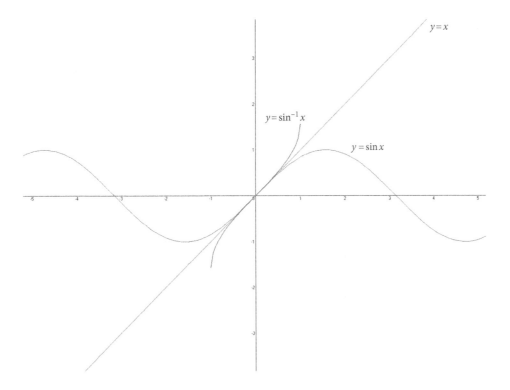

Figure 9.21 Graph of $y = \sin^{-1} x$ (**FNGRAPH 9.10**)

9.7 More trigonometric identities

If you studied Mathematics A-Level, you probably had to learn the following sets of trigonometric identities in addition to (9.18) and (9.27). We list them here for reference. They are usually written in terms of angles A and B; as identities, they are true for all values of the angles.

$$\sin(A + B) = \sin A . \cos B + \cos A . \sin B$$
$$\cos(A + B) = \cos A . \cos B - \sin A . \sin B$$
$$\tan(A + B) = \frac{\tan A + \tan B}{1 - \tan A . \tan B}.$$

(9.29)

By replacing B with $-B$ and using the fact that:

$$\sin(-\theta) = -\sin\theta$$
$$\cos(-\theta) = \cos\theta$$
$$\tan(-\theta) = -\tan\theta$$

(9.30)

we obtain

$$\sin(A - B) = \sin A . \cos B - \cos A . \sin B$$
$$\cos(A - B) = \cos A . \cos B + \sin A . \sin B$$
$$\tan(A - B) = \frac{\tan A - \tan B}{1 + \tan A . \tan B}.$$

(9.31)

By setting $B = A$ in (9.29) we obtain the **double-angle formulae**:

$$\sin(2A) = 2\sin A.\cos A$$
$$\cos(2A) = \cos^2 A - \sin^2 A$$
$$\tan(2A) = \frac{2\tan A}{1 - \tan^2 A}. \tag{9.32}$$

Finally, combining the above formula for $\cos(2A)$ with the identity (9.18):

$$\cos(2A) = \cos^2 A - \sin^2 A = 1 - 2\sin^2 A = 2\cos^2 A - 1. \tag{9.33}$$

9.8 The tangent function and the gradient of a curve

In addition to their use in modelling sinusoidal oscillations, there is an important link between trigonometry and the concept of 'instantaneous rate of change', which is the central idea of Part II of this book. This link is provided by the tangent function.

9.8.1 Definition of the tangent function

Following (9.9) in our revision of school trigonometry, we can define the tangent function as

$$\tan x = \frac{\sin x}{\cos x}. \tag{9.34}$$

Figure 9.22 shows the graph of this function between $x = -2\pi$ and $x = 2\pi$, drawn using FNGraph; we entered the function as `tan(x)`. Notice that there is no value of $\tan x$ when $x = \pm\frac{\pi}{2}, \pm\frac{3\pi}{2}, \pm\frac{5\pi}{2}, \ldots$ since for these values of x we have $\cos x = 0$, and we can't divide by zero. So the graph of $\tan x$ never crosses the dashed vertical lines drawn at those values of x; these are its vertical asymptotes.

9.8.2 The tangent function and the slope of a line

Looking back at the right-angled triangle ABC in Figure 9.5, we saw that

$$\tan B = \frac{\sin B}{\cos B} = \frac{b/c}{a/c} = \frac{b}{a} = \frac{\text{opposite}}{\text{adjacent}} = \frac{\text{length of vertical side}}{\text{length of horizontal side}}.$$

This will remind you of the right-angled triangles in Figure 6.6, which we used to define the slope of the straight-line graph $y = mx + c$ in equation (6.6) as

$$\text{slope} = m = \frac{\text{change in } y}{\text{change in } x} = \frac{\Delta y}{\Delta x} = \frac{y_1 - y_0}{x_1 - x_0}. \tag{9.35}$$

We can now say that the slope of the straight-line graph $y = mx + c$ in Figure 9.23 is:

$$m = \frac{\Delta y}{\Delta x} = \frac{y_1 - y_0}{x_1 - x_0} = \tan\theta, \tag{9.36}$$

where θ is the angle the line makes with the horizontal.

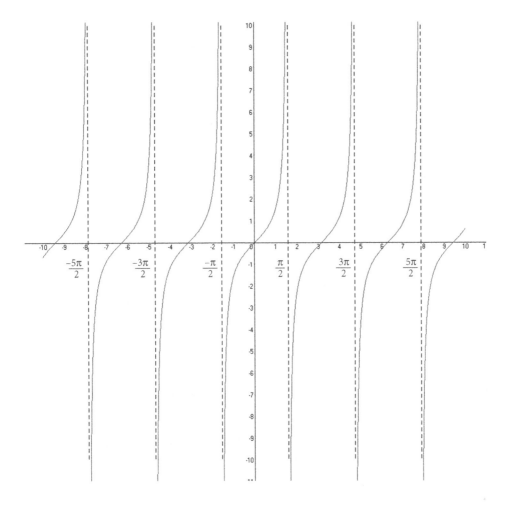

Figure 9.22 Graph of $y = \tan x$

9.8.3 The geometric tangent

The word 'tangent' has another, related meaning in geometry. Consider a curve in Cartesian coordinate space. A straight line which does not cross the curve, but which just touches it at one point, is called the **tangent to the curve** at that point.

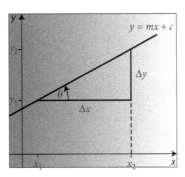

Figure 9.23 Relationship between slope m and the tangent

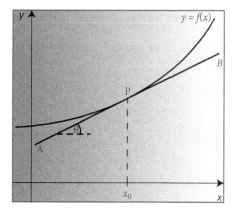

Figure 9.24 The tangent line

This geometric tangent line is shown in Figure 9.24. The line AB is the tangent to the curve $y = f(x)$ at the point P, where $x = x_0$.

We have defined the slope of a straight line, so the slope of the line AB is $\tan \theta$, and if the equation of the line AB is $y = mx + c$, then $m = \tan \theta$. Can we define the 'slope' of the curve, which is the graph of a function $y = f(x)$?

In Figure 9.25 the curve represents the track of a rollercoaster ride, with a rollercoaster car being winched up the initial ramp from the starting-point O. The car is tilted upward at an angle θ, where the slope of the ramp OA is $m = \tan \theta$. After reaching A the car levels off, and between B and C it is travelling horizontally (slope = 0). After point C it tilts downward, and when it reaches point D the car is sloping downwards (negative slope) along the dashed line drawn there, which is the tangent to the curve at D. The dashed lines at E, F, and G show the tangents to the curve at those points. At E the car is horizontal again momentarily, before it rolls up the slope. When it reaches point F it is tilted upwards at the same angle θ as when it was climbing the initial ramp; the tangent line at F is parallel to OA. When it reaches G it is horizontal for one stomach-dropping instant before it hurtles down the final run into the watersplash at H.

The rollercoaster is animated in SPREADSHEET 9.3. Move the car by dragging the slider. The orange line shows the tangent to the curve, and its slope at each instant appears in cell K32.

So the 'slope' of the curve OABCDEFGH, as experienced by the rollercoaster car, varies continuously as we move along the curve. Between O and A it is constant, between B and C it

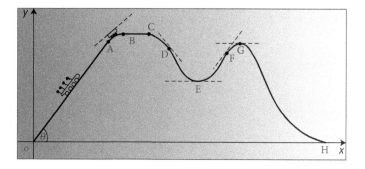

Figure 9.25 The rollercoaster ride

is zero, from C to E it is negative, at E it is zero, and so on. We call this 'variable slope' the **gradient** of the curve, and define it as:

> **The gradient of the curve y = $f(x)$ at the point P (Figure 9.24) is the slope of the tangent AB to the curve at P.**

We can also express this as:

> The gradient of the function $y = f(x)$ at $x = x_0$ is the slope of the tangent
> line to the curve at the point $(x_0, f(x_0))$.   (9.37)

Notice, from looking at points E and G, that:

> When the curve reaches a maximum or minimum, the tangent at that
> point is horizontal, i.e. the gradient at that point is zero.   (9.38)

The gradient of a function $y = f(x)$ has a physical significance: it is the **instantaneous rate of change** of the function at that point.

Consider again the quadratic function

$$s = ut + \frac{1}{2}at^2$$

(9.39)

which we saw in Section 5.2.3, equation (5.7), describing the motion of an animal that is accelerating. After t seconds, the animal has travelled $s(t)$ metres. The graph was plotted in Spreadsheet 5.2. The slope of the tangent to the curve at time $t = t_0$ will tell us the instantaneous rate of change of distance with time; that is, the instantaneous velocity of the animal. This is illustrated in SPREADSHEET 9.4, reproduced in Figure 9.26. Choose a time between 0 and 10 seconds in cell C8. The green line on the chart is the tangent to the curve at that point, and its slope is the instantaneous velocity, given in cell G12. But how is this calculated?

9.8.4 An approximation to the gradient

If we choose another time $t = t_1 > t_0$, we can work out the average velocity between t_0 and t_1, as

$$\text{average velocity} = \frac{\Delta s}{\Delta t} = \frac{s(t_1) - s(t_0)}{t_1 - t_0}.$$

(9.40)

You can set t_1 using the slider control on SPREADSHEET 9.4 (Figure 9.26); its value appears in cell G20, and the average velocity over this time interval is calculated in cell G22. It is the slope of the line joining the points $(t_0, s(t_0))$ and $(t_1, s(t_1))$, which is the red line in the chart. A line passing through two points on a curve is called a **secant line**. The chart initially shows $t_0 = 3$ and $t_1 = 8$ seconds, with the slider control pushed to the right. Now move the slider control gradually towards the left; this moves t_1 to the left on the chart. As t_1 gets closer to t_0, the slope of the secant line gets closer and closer to the slope of the tangent.

In the first chapter of Part II of this book, we will use this technique to work out a formula for the gradient of the curve. This will be the starting-point for a study of **calculus**, which is the gateway to a whole new world of mathematics that can be applied to the biosciences.

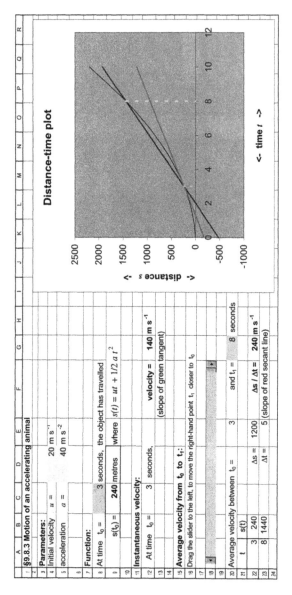

Figure 9.26 An approximation to the gradient (SPREADSHEET 9.4)

PROBLEMS

9.1–9.4: Convert the following angles in degrees, to radians. Express as a multiple of π, and also as a decimal correct to 3 decimal places:

9.1 225°   9.2 75°   9.3 $-72°$ 9.4 27°.

9.5–9.8: Convert the following angles in radians, to degrees:

9.5 $\pi/5$ 9.6 $-\dfrac{7\pi}{36}$ 9.7 -0.75 9.8 1.25.

9.9–9.12: Use FNGraph to draw the graph of $y = \sin x$ between $x = -2\pi$ and $x = 4\pi$. Print out four copies. On these, sketch also the graphs of the following functions, and explain how they differ from $y = \sin x$:

9.9 $y = 3\sin 2x$ 9.10 $y = \sin^2 x$ 9.11 $y = 2\sin\left(x - \dfrac{\pi}{4}\right)$ 9.12 $y = 1 - 2\sin x$.

9:13: Use a bracketing root-finding algorithm (see Section 8.6.1) in Excel to find a value of x between 1 and 6 for which $\sin x = \dfrac{1}{2}x$. Check your answer by drawing the graphs of $y = \sin x$ and $y = \dfrac{1}{2}x$ in FNGraph.

9.14: Use a bracketing minimization algorithm (see Section 8.6.2) in Excel to find a value of x between 3 and 6 for which $g(x) = 2\sin x - \sin 2x$ reaches its minimum. Check your answer by drawing the graph of $g(x)$ in FNGraph.

9.15–9.20: These problems are based on graphs of annual population censuses for game species in the Kruger National Park, given in Owen-Smith et al (2005). Herbivore populations tend to vary with the cycles of drought and heavier rainfall years.[6]

Tsessebe and roan are two of the rarer species of antelope in the Park. In the following two problems, use the information given to write a general cosine function (9.19) for the population, assuming it varies sinusoidally. Measure time t in years from 1985.

9.15 The population of tsessebe were at a maximum of 1200 animals in 1990, and at a minimum of 900 animals in 1992.

9.16 The population of roan varied around a mean population of 340 animals. They reached their maximum of 410 animals in 1986, and again in 1989.

9.17 Use a bracketing root-finding algorithm to find a time at which the combined tsessebe and roan populations were 1500 animals.

9.18 Use a bracketing root-finding algorithm to find a time at which the tsessebe population outnumbered that of roan antelope by three to one.

9.19 Use a bracketing minimization algorithm to find when the combined tsessebe and roan population reached a minimum.

9.20 Use a bracketing minimization algorithm to find when the combined tsessebe and roan population reached a maximum.

---

[6] You can read more about this at http://www.krugerpark.co.za/krugerpark-times-21-fewer-animals-17947.html.

10

Exponential and logarithmic functions

This final category of function is perhaps the most important for bioscientists, since so many nonlinear natural processes involve some form of exponential growth or exponential decay.

. . . .

10.1 Exponential functions to the base a

Some of the most important natural processes exhibit **exponential growth**. A population of cells on a Petri dish, or of rabbits in a field, will grow exponentially until the supply of food or nutrients runs low. Epidemics spreading through populations show similar behaviour – as does the growth of the debt on your student loan! In the next section we define what we mean by exponential growth, and relate it to the geometric growth model we have seen in Case Study A.

10.1.1 Discrete and continuous models

When we introduced our Case Study A of models of population growth, back in the Introduction and Chapter 1, we started with the example of a population of 100 cells on a Petri dish; the cells divide once every hour, so after one hour we will have 200 cells, after two hours 400 cells, after three hours 800 cells, and so on. We referred to this as **geometric growth**, defined by the update equation

$$N_{p+1} = \lambda N_p \tag{10.1}$$

where N_p is the population size after p generations, N_0 is the initial population size, and λ is the multiplication rate ($N_0 = 100$ and $\lambda = 2$ in the example). Equation (10.1) is called an **update equation**, because it uses information about the current population size to predict the size at some point in the future. Starting with the initial population size and using (10.1) repeatedly, gives us a series of 'snapshots' of the population at separated instants in time: $N_0, N_1, N_2, \ldots$ Such a model is called a **discrete model** ('discrete' means separate points, as distinct from a continuous line. We saw in Chapter 1 that the integers are a set of discrete points on the continuous real line).

When a population is large, it is more realistic to think of it as increasing continuously over time, rather than in sudden jumps. We would therefore need a **continuous model**, which gives the population $N(t)$ as a function of time t, which we can evaluate at any point in time. A continuous model of population growth, corresponding to the discrete geometric growth model, is called **exponential growth**.

To illustrate the difference, consider a bank savings account where the interest is paid at the end of each calendar year. If you allow the interest to be added to the capital and invested for the following year, this would be an example of geometric growth; the value of the account jumps suddenly on 1st January each year. Compare this with a bank overdraft, where the interest is calculated every day and added to the loan, so you see the amount owed growing at an ever-increasing rate. This is more what exponential growth looks like. Exponential growth starts slowly, but if unrestricted it develops surprisingly rapidly, and eventually outstrips the growth of even the highest-degree power function x^n. An initial population of 100 cells, doubling each hour, would reach 410,000 cells after 12 hours, 1.7 trillion cells after one day, and 3.7×10^{52} cells by the end of a week!

You can also think of exponential growth as a type of geometric growth where the 'snapshots' of population size are so close together as to make the growth look continuous. In both cases, the rate of growth is proportional to the current population size; a population of twice the size will be growing twice as fast.

If we consider our population of cells on a Petri dish as growing continuously, so that it has size $N(t)$ at time t (measured in hours), then we want to find a function that matches the 'snapshot' sizes of the discrete model, i.e.

$$N(0.0) = N_0 = 100, \quad N(1.0) = N_1 = 200, \quad N(2.0) = N_2 = 400, \text{ etc.}$$

Is there such a function?

10.1.2 Exponential function to the base a: $y = a^x$

Yes there is. We can take

$$N(t) = 100 \times 2^t \tag{10.2}$$

Then

$$N(0.0) = 100 \times 2^0 = 100 \times 1 - 100$$
$$N(1.0) = 100 \times 2^1 = 100 \times 2 = 200$$
$$N(2.0) = 100 \times 2^2 = 100 \times 4 = 400$$
$$N(3.0) = 100 \times 2^3 = 100 \times 8 = 800$$

and so on. But we can also evaluate this function at intermediate times, e.g. after 90 minutes the population will be $N(1.5) = 100 \times 2^{1.5} = 100 \times 2.83 = 2830$ cells.

Use the $\boxed{x^y}$ key on your calculator to evaluate $2^{1.5}$.

Figure 10.1 shows how this function 'joins up the dots' on the geometric growth model. The function

$$y = a^x \tag{10.3}$$

is called the **exponential function with base a**. Here, a is a fixed positive parameter; we were using $a = 2$ in equation (10.2).

In FNGRAPH 10.1 you can see the graphs of $y = 2^x$ in blue, and $y = 3^x$ in red. These are particular examples of (10.3), with $a = 2$ and $a = 3$ respectively. The curves appear in Figure 10.2.

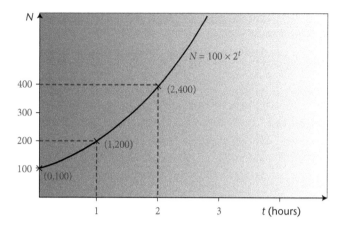

Figure 10.1 Exponential and geometric growth

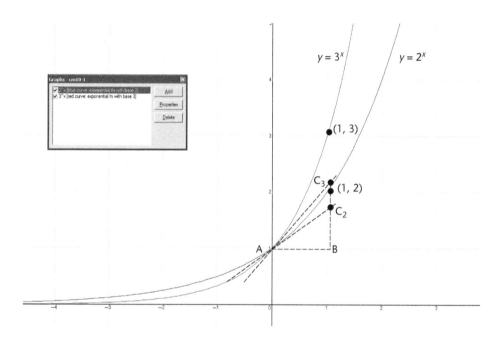

Figure 10.2 Gradients of $y = 2^x$ and $y = 3^x$ at $x = 0$ (FNGRAPH 10.1)

Rules of indices:
$a^0 = 1$ for any a

The graph of $y = 3^x$ climbs more steeply for positive x. Both curves intersect the y-axis at $y = 1$, since for any a, when $x = 0$, $y = a^0 = 1$.

So the graph of $y = a^x$ passes through the point $(0, 1)$, for any a.

Rules of indices:
$a^1 = a$ for any a

Another point on the graph of (10.3) is $(1, a)$, since $a^1 = a$.

Thus, the curve of $y = 2^x$ passes through the point $(1, 2)$, and the curve of $y = 3^x$ passes through the point $(1, 3)$. The graph of (10.3) is shown in Figure 10.3.

There is no value of x for which $a^x = 0$, so the curves never intersect the x-axis. When x is large and negative, say $x = -100$, then $a^x = a^{-100} = \dfrac{1}{a^{100}}$ which will be a very small number (provided $a > 1$). The negative x-axis is a horizontal asymptote for the curve $y = a^x$ as x tends to $-\infty$; using limit notation, we can write

$$\lim_{x \to -\infty} a^x = 0 \quad \text{(provided } a > 1\text{)}.$$

We will look at the situation for values of a less than 1, in Section 10.5. This would be a good point at which to revise the laws for exponents in Chapter 1. They were summarized in (1.19) in the following box:

$$a^m \times a^n = a^{m+n}$$
$$a^m \div a^n = a^{m-n}$$
$$(a^m)^n = a^{m.n}$$
$$(a \times b)^n = a^n \times b^n$$
$$\left(\frac{a}{b}\right)^n = \frac{a^n}{b^n}$$
$$a^1 = a$$
$$a^0 = 1 \tag{10.4}$$
$$a^{-n} = \frac{1}{a^n}$$
$$0^0 = 1.$$

Going back to the curves in Figure 10.2 (**FNGRAPH 10.1**) or Figure 10.3, as you move along the curve, the slope of the tangent to the curve, or **gradient** (see Section 9.8) becomes greater. A function exhibits **exponential growth** when its **rate of increase is proportional to its current value**. In the example above of the population of cells, the larger the population becomes, the faster it grows. The functions $y = a^x$, with $a > 1$, have this property; this claim will be justified in Chapter 11. We will now explore this property of exponential growth.

The example of an initial population of 100 cells, which doubles in size every hour, would be modelled by the function $N(t) = 100.2^t$ where $N(t)$ is the population size after t days. If we use the parameter N_0 to represent the initial population size, the function becomes $N(t) = N_0.2^t$.

More generally, the function $y = Ca^x$ will pass through the points $(0, C)$, $(1, Ca)$, $(2, Ca^2)$, and so on.

Problems: Now try the end-of-chapter problems 10.1–10.12.

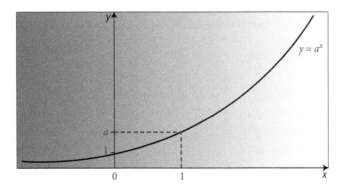

Figure 10.3 Exponential function to the base a ($a > 1$)

10.2 Exponential growth function $y = Ae^{kx}$

We have said that the functions $y = 2^x$, $y = 3^x$ and so on – in general $y = a^x$ – all demonstrate exponential growth. Which one should we use for modelling a particular experiment? In each of them, the instantaneous rate of change (or gradient) at any point is proportional to the value of y at that point; it's the constants of proportionality which differ.

There is one special value of a, for which the gradient of $y = a^x$ at any point is **equal to** the value of y at that point. This means that at the y-intercept, when $x = 0$ and $y = 1$, the slope of the tangent to the curve (see Section 9.8) will be equal to 1.

In Figure 10.2, the graphs of $y = 2^x$ and $y = 3^x$ are shown, together with the tangent lines to each curve at (0, 1). To find the slope of the tangent to $y = 2^x$, look at the right-angled triangle ABC_2 (as we did in Figure 6.6); the gradient is the slope of the hypotenuse AC_2. The points A and B are (0, 1) and (1, 1), so $AB = 1$ and the slope is $m_2 = \dfrac{BC_2}{AB} = BC_2$. From the graph you can see that this is less than 1; in fact it is 0.6932 correct to 4 significant figures. Similarly for $y = 3^x$ the slope of the tangent is $m_3 = \dfrac{BC_3}{AB} = BC_3$, which is a bit greater than 1 (actually 1.099). I claim that there is a value of a between 2 and 3 for which the gradient of the tangent at $x = 0$ will be exactly equal to 1.

You can find this special value of the base, using SPREADSHEET 10.1 (Figure 10.4). Initially, the spreadsheet shows a data table and a graph for the case when $a = 3$; this is set in the yellow cell C6. The chart also shows the tangent line to the curve at the point when $x = 0$. The slope of this line, which is 1.0986, appears in the green cell C21. Now try replacing the value in

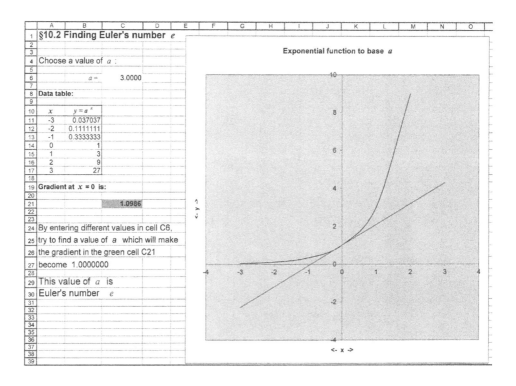

Figure 10.4 Finding the Euler's number (SPREADSHEET 10.1)

cell C6 by 2.9; you will see the data table and graph change, and the slope of the tangent drops to 1.0647. Try to find a value to enter in cell C6, which will produce 1.0000 in cell C21.

The value you have found, is called the **exponential constant or Euler's number,**[1] denoted by the symbol e. This is an irrational number; a more precise value than the one you have found, is

$$e = 2.718281828459045235360287471352 \ldots$$

It is an irrational number, like $\sqrt{2}$ and π. In your spreadsheet, you will now have the graph of $y = e^x$, which passes through the points $(0,1)$ and $(1, e)$.

The exponential constant is important because when scientists want to model exponential growth, they don't look for a value of a to use in the function $y = a^x$. Rather, they look for a value of the parameter k to use in the equation $y = e^{kx}$.

Euler's number e may be a special number, but it's still a number, and so the laws of exponents in (10.4) apply. In particular:

$$e^m \times e^n = e^{m+n}$$
$$e^m \div e^n = e^{m-n}$$
$$(e^m)^n = e^{m.n} \qquad\qquad (10.5)$$
$$e^{-n} = \frac{1}{e^n}$$
$$e^0 = 1.$$

So if you wanted to find a value of k that would make $y = e^{kx}$ behave like $y = 2^x$, we would use the fact that $e^{kx} = (e^k)^x$, so we want the value of k for which $e^k = 2$. We'll see how to find this (it's approximately 0.7), in the next section.

The function e^x is so important that it's hard-wired into your calculator. It may be a 2nd function key. To see the value of e, enter 1 into the calculator and press the e^x key (since $e^1 = e$). In Excel and FNGraph, the exponential function e^x is EXP(). In FNGRAPH 10.2 you can see the graph of $y = e^{kx}$, coded as exp(k*x). Try different values of the parameter k to see how this produces the range of graphs we saw in Section 10.1.

The general equation of exponential growth will therefore be

$$y = Ae^{kx} \qquad\qquad (10.6)$$

where A is the value of y when $x = 0$. In the case of the population $N(t)$ of cells, as described at the start of this chapter, we would model it as

$$N(t) = N_0.e^{kt}. \qquad\qquad (10.7)$$

If you take $N_0 = 100$ and $k = 0.7$, you will find that the population does approximately double each day. The coefficient k is called the **growth constant**: the larger the value of k, the faster the population increases.

EXAMPLE

A population $N(t)$ is given by $N(t) = 100e^{0.7t}$.
Initial population is $N(0) = 100$.
After 1 day, $N(1) = 100e^{0.7} = 100 \times 2.01375 = 201$ (correct to nearest whole number).
After 2 days, $N(2) = 100e^{1.4} = 100 \times 4.0552 = 406$.

Problems: Now try the end-of-chapter problems 10.13–10.14.

---

[1] The constant was first published by Leonhard Euler in 1736.

> ## CASE STUDY A9: Exponential growth of populations
>
> Our general model for geometric growth, from the update equation $N_{p+1} = \lambda N_p$, is
>
> $$N_p = \lambda^p N_0. \tag{10.8}$$
>
> Here, λ is the *multiplication rate*. In Part A3 we saw that when ecologists talk about a population of animals, they prefer to use the parameter R, the net growth rate per head of population. It is related to λ by $\quad \lambda = 1 + R$.
>
> This is a discrete model of population growth. The corresponding continuous model of exponential growth of populations is described by
>
> $$N(t) = N_0 e^{rt} \tag{10.9}$$
>
> where $N(t)$ is the population size at time t. In this equation the parameter r is the instantaneous net growth rate. As with R in the discrete model, it is the birth rate minus the death rate, but using the instantaneous rates, not the changes from one generation to the next. Comparing (10.9) with (10.8), the relation between λ and r is
>
> $$e^r = \lambda. \tag{10.10}$$
>
> The parameter r is sometimes called the Malthusian[2] parameter. Thomas Malthus proposed the **Malthusian Growth Model**, which foresaw the disastrous consequences of unchecked population growth, as follows:
>
> > *'The power of population is indefinitely greater than the power in the earth to produce subsistence for man. Population, when unchecked, increases in a geometrical ratio. Subsistence increases only in an arithmetical ratio. A slight acquaintance with numbers will show the immensity of the first power in comparison with the second.'*
>
> By 'arithmetical ratio' he means that subsistence grows linearly with time.

10.3 Logarithms

The logarithm has a lot to answer for; it is the topic which caused generations of schoolchildren to give up on maths. This is because they were introduced to it at an early age – before the days of pocket calculators, all long multiplication and division had to be done using tables of logarithms. Luckily those days are past, and we can now see the logarithm as an example of an inverse function; in fact, it's simply the inverse exponential function.

10.3.1 Definition of logarithms to base *a*

We said in Section 10.1 that the exponential function to the base a is $y = a^x$. Its inverse function is defined in just the same way as we defined the inverse function $f^{-1}(x)$ in Section 8.5:

$$\text{'}y = f^{-1}(x) \text{ means that } x = f(y).\text{'} \tag{8.14}$$

---

[2] Thomas Robert Malthus (1766–1834), English scholar.

We applied this to the trigonometric functions in Section 9.6.2, for example: '$\theta = \sin^{-1} x$ means that θ is the angle whose sine is x, i.e. $\sin \theta = x$.'

But we can't use the '−1' notation for the inverse exponential function, because it would be confused with the exponent. Instead, we call the inverse exponential function the **logarithm**.

Let's start with the exponential function to the base a; $f(x) = a^x$. Then the notation for its inverse function is $f^{-1}(x) = \log_a x$. It is called the **logarithm to the base a**. Following the definitions above,

$$y = \log_a (x) \quad \text{means that} \quad x = a^y. \tag{10.11}$$

For example, if we use the base $a = 2$, and the fact that $2^3 = 8$, we can take $x = 8$, $y = 3$ in (10.11) and deduce that $\log_2 8 = 3$. Similarly, $\log_{10} 1000 = 3$, since this is equivalent to saying that $1000 = 10^3$.

When we sketched the graph of $y = a^x$ in Section 10.1, we used our knowledge of exponents from Chapter 1, in particular that $a^0 = 1$ and $a^1 = a$, to deduce that the curve will pass through the points $(0, 1)$ and $(1, a)$. But now, from our definition of logarithms,

$$a^0 = 1 \text{ means that } \log_a 1 = 0 \quad \text{and} \quad a^1 = a \text{ means that } \log_a a = 1. \tag{10.12}$$

So the graph of $y = \log_a x$ will pass through the points $(1, 0)$ and $(a, 1)$. FNGRAPH 10.3 (Figure 10.5) shows the graph of $y = a^x$ (in red) and $y = \log_a x$ (in blue). The base a is set as a parameter at $a = 2$ initially. The exponential curve is reflected in the green line $y = x$ to produce its inverse, the logarithm curve. Change the value of a to 3 and see the effect on the graphs, which will now pass through $(1, 3)$ and $(3, 1)$ respectively. Try changing the parameter to $a = 10$; you'll need to zoom out using the ' − ' key. We'll discuss later what values of the base a are used in practice.

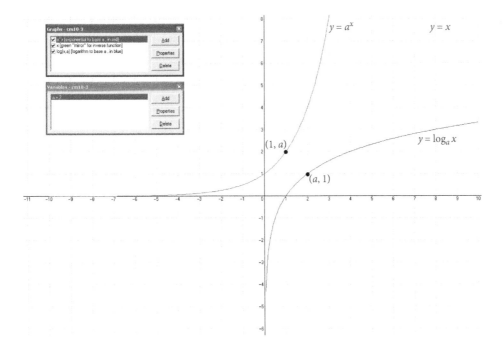

Figure 10.5 Graphs of $y = \log_a x$ and $y = a^x$ (FNGRAPH 10.3)

In FNGraph, $y = \log_a x$ is expressed by the function `log(x, a)`, while in Excel the function is `LOG(number, base)`, i.e. the function takes two arguments.

10.3.2 Laws of logarithms

One reason that logarithms are feared by students, is that they have their own rules of behaviour which need to be learned. Here they are:

$$\log_a(x.y) = \log_a x + \log_a y$$

$$\log_a\left(\frac{x}{y}\right) = \log_a x - \log_a y$$

$$\log_a(x^n) = n\log_a x$$

$$\log_a\left(\frac{1}{x}\right) = -\log_a x$$

$$\log_a 1 = 0, \quad \log_a a = 1. \tag{10.13}$$

But there's nothing mysterious: these come directly from the laws of exponents:

$$a^u \times a^v = a^{u+v} \quad a^u \div a^v = a^{u-v} \quad (a^u)^v = a^{u.v}.$$

Here's how to derive the first law of logarithms in (10.13). If you're happy with learning these rules without seeing why they are true, you can skip the rest of this section![3]

Suppose we take two numbers x and y, and find their logarithms to base a. Write these as u and v:

$$\log_a x = u \quad \text{and} \quad \log_a y = v.$$

From our definition of logarithms (10.11), this means that

$$x = a^u \quad \text{and} \quad y = a^v.$$

Using the laws of exponents,

$$x.y = a^u \times a^v = a^{u+v},$$

from which, using the definition (10.11) again, we deduce that $u + v = \log_a(x.y)$. Now substitute for u and v, and we have obtained the **first law of logarithms**:

$$\log_a(x.y) = \log_a x + \log_a y.$$

You should now be able to use the second law of exponents $a^u \div a^v = a^{u-v}$ to obtain the **second law of logarithms**:

$$\log_a\left(\frac{x}{y}\right) = \log_a x - \log_a y.$$

The third law can be derived similarly, and by putting $n = -1$ in this equation, you obtain the fourth law.

[3] But if you're in an exam, have forgotten the rule, and you're unable to work it out, don't blame me …

EXAMPLES

$$\log_a(72) = \log_a(8 \times 9)$$
$$= \log_a(8) + \log_a(9)$$
$$= \log_a(2^3) + \log_a(3^2)$$
$$= 3\log_a(2) + 2\log_a(3).$$

$$\log_a(0.4) = \log_a\left(\frac{2}{5}\right) = \log_a 2 - \log_a 5$$

$$\log_a(0.25) = \log_a\left(\frac{1}{4}\right) = -\log_a 4 = -2\log_a(2).$$

Problems: Now try the end-of-chapter problems 10.15–10.18.

10.3.3 Logarithms to base 2

We could take any positive number to be the base a. Perhaps the most obvious base to use is $a = 2$. This base is indeed sometimes used in models of cell culture growth and radioactive decay, and it is also used in computer science.

EXAMPLE

At the start of this chapter we gave the example of a cell culture with 100 cells initially, which doubles in size each hour. This can be modelled by the function

$$N(t) = 100.2^t$$

so that $N(0) = 100$, $N(1) = 200$, $N(2) = 400$, and so on (t is measured in hours). After how many hours will the population have reached one million cells? To solve this, we take the logarithms of both sides of the equation. Using logarithms to base 2, and the three laws of logarithms, we get

$$\log_2 N = \log_2(100.2^t)$$
$$= \log_2(100) + \log_2(2^t)$$
$$= \log_2(100) + t.\log_2(2)$$
$$= \log_2(100) + t$$
$$= \log_2(10^2) + t$$
$$= 2\log_2 10 + t.$$

$$\log_a(x.y) = \log_a x + \log_a y$$

$$\log_a(x^n) = n\log_a x$$

Solving for t,

$$t = \log_2 N - 2\log_2 10.$$

So setting $N = 10^6$,

$$t = \log_2 10^6 - 2\log_2 10$$
$$= 6\log_2 10 - 2\log_2 10 = 4\log_2 10.$$

Now all we have to do is to find the logarithm of 10 to base 2! Unfortunately your calculator probably doesn't have a $\log_2$ key. But we can get an idea of the value. If $x = \log_2 10$, it means that $2^x = 10$. We know that $2^3 = 8$ (less than 10) and $2^4 = 16$ (more than 10), so x will lie between 3 and 4. If $\log_2 10$ is between 3 and 4, then $t = 4\log_2 10$ will be between 12 and 16; there will be 1 million cells sometime between 12 and 16 hours after starting the experiment.

10.3.4 Logarithms to base 10 (common logarithms)

The most common choice for science applications is to take $a = 10$. Logarithms to base 10 are called **common logarithms**, and in this case the base is omitted, so if you see '$y = \log x$' you can assume it means $y = \log_{10} x$. The notation '$y = \lg x$' is also sometimes used.

Using base 10 has one very important advantage. The logarithm to base 10 of 5.678 is 0.7542; the logarithm to base 10 of 567.8 is 2.7542. In fact, if a number x is written in scientific notation as $x = s \times 10^n$, where s is a number between 1 and 9.9999 ..., then its logarithm to base 10 is

$$\log_{10} x = \log_{10}(s \times 10^n) = \log_{10}(s) + \log_{10}(10^n) = \log_{10}(s) + n\log_{10}(10) = n + \log_{10}(s)$$

using the final rule in (10.19).

So $\log_{10}(567.8) = \log_{10}(5.678 \times 10^2) = 2 + \log_{10}(5.678) = 2 + 0.7542 = 2.7542$.

This property can be useful in calculations involving logarithms and numbers in scientific notation. You should have a button marked $\log$ on your calculator; this produces the logarithm to base 10. In FNGraph the common logarithm function is $\lg(x)$, while in Excel the function is $LOG10()$, or $LOG()$ taking just one argument.

In (10.6) we wrote the general function for exponential growth as:

$$y = Ae^{kx}$$

where e is Euler's number $e = 2.71828 \ldots$ Taking common logarithms of both sides, we get

$$\begin{aligned}
\log y &= \log(Ae^{kx}) \\
&= \log(A) + \log(e^{kx}) \\
&= \log A + k\,x\log e \\
&= (k\log e)x + \log A.
\end{aligned} \qquad (10.14)$$

That is, the relationship between $\log y$ and x is a straight line, with y-intercept $c = \log A$, and slope $m = k \log e$. You can find the value of $\log e$ on your calculator; first get e by pressing 1 and then the $\exp$ or e^x key. Then press $\log$. The value is 0.4342944 ...

We can now answer the question of the previous section using common logarithms. We need to solve

$$N(t) = 100.2^t.$$

Taking common logarithms of both sides:

$$\begin{aligned}
\log_{10} N &= \log_{10}(100.2^t) \\
&= \log_{10}(100) + \log_{10}(2^t) \\
&= \log_{10}(100) + t.\log_{10}(2) \\
&= \log_{10}(100) + t\log_{10} 2 \\
&= \log_{10}(10^2) + t\log_{10} 2
\end{aligned}$$

$$= 2\log_{10} 10 + t\log_{10} 2$$
$$= 2 + t\log_{10} 2.$$

Now setting $N = 10^6$,

$$\log_{10} 10^6 = 2 + t\log_{10} 2$$
$$6\log_{10} 10 = 2 + t\log_{10} 2$$
$$6 = 2 + t\log_{10} 2$$
$$4 = t\log_{10} 2$$
$$t = \frac{4}{\log_{10} 2} = \frac{4}{0.3010} = 13.288.$$

Use the `log` key on your calculator.

So there should be 1 million cells in the population after about 13 hours 17 minutes.

10.3.5 Logarithms to base e (natural logarithms)

Apart from 10, the other frequently used base for logarithms is Euler's number e. Logarithms to base e are called **natural logarithms** (sometimes Napierian logarithms[4]). The natural logarithm $\log_e x$ is often written as $\ln x$, and in FNGraph and Excel, the notation is `ln(x)`. You will definitely have an `ln` or `log_e` key on your calculator.

If we now transform the exponential growth equation (10.6) by taking natural logarithms of both sides,

$$\ln y = \ln(Ae^{kx})$$
$$= \ln(A) + \ln(e^{kx})$$
$$= \ln A + kx\ln e \qquad (10.15)$$
$$= kx + \ln A$$

since $\ln e = \log_e e = 1$ by (10.12). So a plot of $\ln y$ against x should produce points lying on a straight line, whose slope is the growth constant k, avoiding any involvement of a constant such as $\log e$. We'll do this in Section 10.4.

This time we use natural logarithms to solve the example posed in Section 10.3.3: if an initial population of 100 cells grows exponentially, doubling every hour, how long will it take to reach 1 million cells? The equation to solve is

$$N(t) = 100.2^t.$$

We have to find t when $N = 10^6$. Taking natural logarithms of each side:

$$\ln N = \ln(100.2^t) = \ln 100 + t\ln 2 = \ln(10^2) + t\ln 2 = 2\ln 10 + t\ln 2.$$

The left-hand side is $\ln N = \ln(10^6) = 6\ln 10$, so $6\ln 10 = 2\ln 10 + t\ln 2$. Solving for t gives:

$$t = \frac{4\ln 10}{\ln 2} = \frac{4 \times 2.30259}{0.69315} = 13.288$$

as we found in the previous section.

---

[4] John Napier (1550–1617), 8th Laird of Merchistoun. Scottish mathematician, physicist, astronomer, and astrologer, and primary discoverer of the logarithm.

We claimed that a general exponential function $y = a^x$ can be written as $y = e^{kx}$. We can now answer the question: Given a value of a, what is the value of k? Taking natural logarithms:

$$e^{kx} = a^x$$
$$\ln(e^{kx}) = \ln(a^x)$$
$$kx \ln e = x \ln a$$
$$kx = x \ln a$$
$$k = \ln a.$$

Thus, the equation $N(t) = N_0 . 2^t$ can be written as $N(t) = N_0 . e^{kt}$ where $k = \ln 2$.

An aside: the logarithm function $\ln x$ is the inverse function of the exponential function e^x, just as the square root function $\sqrt{x}$ is the inverse of the squaring function x^2. That is, starting with x and applying one function and then its inverse takes you back to x. In algebra we can write

$$e^{\ln x} = x \quad \text{and} \quad \ln(e^x) = x. \tag{10.16}$$

Going back to (10.9) in Case Study A9, the growth rate or growth constant r tells us how fast a population of cells is growing. Another quantity used to describe this is the **cell doubling time** T_2. This is the time it takes for a population to double in size (for exponential growth this is a constant, independent of t). The initial population N_0 will have doubled to $2 N_0$ at time $t = T_2$, if

$$N(T_2) = N_0 e^{rT_2} = 2N_0.$$

Cancelling the common factor of N_0, we get

$$2 = e^{rT_2},$$

and by taking natural logarithms and solving for T_2, the cell doubling time T_2 is related to the growth constant r by

$$T_2 = \frac{\ln 2}{r}.$$

In the notation of this section, $T_2 = \dfrac{\ln 2}{k}$. \hfill (10.17)

Problems: Now try the end-of-chapter problems 10.19–10.34.

10.4 Fitting exponential curves to data

In Section 8.3 we developed a technique for fitting nonlinear equations such as the Michaelis–Menten equation to sets of experimental or observed data. By making a suitable transformation of the variables, we obtained transformed variables X, Y that were linearly related: $Y = mX + c$. The transformation was applied to the data-points, and the transformed points plotted and a best-fit least-squares straight line inserted using graph paper or Excel. From the calculated or measured values of the slope m and Y-intercept c, we could deduce the parameters of the original equation. The process is illustrated in Figure 8.6 (from SPREADSHEET 8.2). We will now apply this technique to equations involving exponential functions.

10.4.1 Fitting an exponential growth model

The technique is best illustrated by an example:

The following table (cited in Hoppensteadt and Peshkin 1992) gives data on the population of Sweden, as measured from censuses taken at 20-year intervals between 1780 and 1860.

Year t	Population N (1000s)
1780	2104
1800	2352
1820	2573
1840	3123
1860	3824

We will attempt to fit the exponential growth equation (10.9), namely $N = N_0 e^{rt}$; our aim is to estimate the values of the parameters N_0 and r. Taking natural logarithms of both sides, as in (10.15), produces

$$\ln N = rt + \ln N_0. \tag{10.18}$$

If we plot the points using the transformed variables X, Y defined by $X = t$ and $Y = \ln N$ (called a **semi-log** plot), then the transformed equation is the linear equation:

$$Y = mX + c$$

which describes a straight line where the slope $r = m$, and Y-intercept $c = \ln N_0$. We thus know the growth rate r immediately from m, and we find N_0 by using the definition of logarithms (10.11): if $c = \ln N_0$, then $N_0 = e^c$.

SPREADSHEET 8.2 (which used this technique to fit the Michaelis–Menten equation to experimental data) has been adapted to solve the current problem, in SPREADSHEET 10.2 (Figure 10.6). We find that $r = m = 0.0074$, and $c = -5.5412$, so $N_0 = e^c = 0.0039$. The exponential growth model for Sweden's population is thus

$$N = 0.0039 e^{0.0074t}. \tag{10.19}$$

This equation could be used to predict population sizes at times beyond the data range, i.e. to perform **extrapolation**. Inserting $t = 1890$ in (10.19) gives an estimated population for that year of 4,583,000. As you go further away from the end of the data, such predictions become less accurate. The equation would predict the population in 1970 as 8,279,000, which is surprisingly close to the published value of 8,042,706. But for the year 2000 the model predicts a population of 10,334,000, whereas the population in fact only rose to 8,872,110. It appears that in recent years Sweden's population would be better modelled by the logistic growth equation we saw in Section 7.4.2. Compare the above results with the predictions made by quadratic interpolation in problems 7.44–7.46.

10.4.2 Application: allometry

In Section 3.5 we introduced allometry: the study of power-law relationships between physical characteristics of an organism. These usually take the mass of the organism as the basic

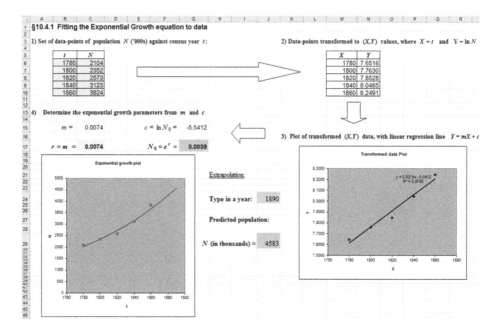

Figure 10.6 Fitting a curve using a semi-log plot (SPREADSHEET 10.2)

characteristic. Then for example the basal metabolic rate P (in kJ/day) for a mammal is related to its mass M (in kg) by

$$P = 2.20M^{0.76}.$$

Of course, this is not an exact relationship: it is a function that has been fitted, using the least-squares technique, to a large number of data-points collected from different species of mammal. To transform a general power-law relationship

$$P = aM^b \qquad\qquad\qquad (10.20)$$

to a linear form, take natural logarithms of both sides, as we did with (10.9) to derive (10.18) in the previous section. We obtain

$$\ln P = b.\ln M + \ln a. \qquad\qquad\qquad (10.21)$$

Our transformed variables are now $X = \ln M$ and $Y = \ln P$; the plot of $\ln P$ against $\ln M$ is called a **log–log plot**. The slope of the best-fit straight line will give us the parameter b, and the Y-intercept will allow us to calculate a.

We can also use logarithms to find the relationship between two characteristics. In the wild, e.g. on the plains of East Africa, the population density ρ of a herbivorous mammal such as antelope or zebra (measured in numbers per square kilometre) is related to its average mass M by

$$\rho = 91.2M^{-0.73}.$$

As the power of M is negative, this is an inverse relationship: the larger the animal, the fewer of them there will be in a given area.

The feeding and breeding area A that each animal requires, is related to its mass by

$$A = 2.3M^{1.02}.$$

How then is population density ρ related to feeding area A? Take natural logarithms of each equation:

$$\ln A = \ln 2.3 + 1.02\ln M$$
$$\ln \rho = \ln 91.2 - 0.73\ln M.$$

Solve for $\ln M$ in each of these equations:

$$\ln M = \frac{1}{1.02}(\ln A - \ln 2.3) = 0.98(\ln A - 0.8330) = 0.98\ln A - 0.816$$

$$\ln M = \frac{1}{0.73}(\ln 91.2 - \ln \rho) = 1.37(4.5131 - \ln \rho) = 6.183 - 1.37\ln \rho.$$

The left-hand side of each equation is the same, so the right-hand sides must be equal:

$$0.98\ln A - 0.816 = 6.183 - 1.37\ln \rho$$
$$0.98\ln A = 6.999 - 1.37\ln \rho$$
$$\ln A = 7.142 - 1.398\ln \rho.$$

We can now take the exponential of each side (the inverse process to taking logarithms):

$$e^{\ln A} = e^{7.142 - 1.398\ln \rho} = e^{7.142}.e^{-1.398\ln \rho} = 1264\left(e^{\ln \rho}\right)^{-1.398}$$
$$A = 1264\rho^{-1.398}.$$

Using the laws of exponents and of logarithms, we have obtained the power-law relation between feeding area and population density.

10.4.3 Application: allosteric regulation

In Section 8.4.4 the Michaelis–Menten equation (8.10) was modified to

$$v = \frac{v_{\max}.s^{\eta}}{K_m + s^{\eta}} \tag{10.22}$$

to model allosteric enzyme reactions, see equation (8.13). In addition to the Michaelis–Menten parameters $v_{\max}$ and K_m, s is raised to an exponent which is a third parameter η, called the Hill coefficient. Here is one possible transformation of (10.22) to a linear form, which uses logarithms:

$$v = \frac{v_{\max}.s^{\eta}}{K_m + s^{\eta}}$$

(multiply both sides by the denominator, to remove the fraction)

$$v_{\max}.s^{\eta} = K_m.v + v.s^{\eta}$$

$$v_{\max}.\frac{s^{\eta}}{K_m} = v + v.\frac{s^{\eta}}{K_m}$$

(divide both sides by K_m)

$$v = (v_{\max} - v)\frac{s^{\eta}}{K_m}$$

$$\frac{v}{v_{\max} - v} = \frac{s^{\eta}}{K_m}.$$

We can now take logarithms of both sides, and use the rules of logarithms in (10.13):

$$\ln\left(\frac{v}{v_{\max} - v}\right) = \eta.\ln s - \ln K_m. \tag{10.23}$$

If we had an estimate of the maximum reaction rate $v_{\max}$, we could use the transformed variables $X = \ln s$ and $Y = \ln\left(\frac{v}{v_{\max} - v}\right)$, which would be linearly related: $Y = \eta X - \ln K_m$. Hence we could fit the least-squares line and deduce η and K_m. The problem lies in estimating $v_{\max}$. One possible approach is to make an initial estimate of $v_{\max}$, fit the best-fit line (10.23) using this, and hence obtain first estimates for η and K_m. As well as displaying the line's equation in Excel, look also at the R^2 value. We now try different values of $v_{\max}$ with the aim of finding the value that results in a least-squares line that best fits the transformed data-points, i.e. which maximizes the R^2 value. The values to try could be found from a numerical algorithm such as golden section search (Section 8.6.2). A similar iterative approach to this problem, using Fibonacci search, is suggested by Atkins (1973).

Problems: Now try the end-of-chapter problems 10.35–10.40.

10.5 Exponential decay

At the start of this chapter we looked at the exponential function $y = a^x$, and drew its graph in Figure 10.3, for the case where $a > 1$. Certainly we cannot take the base a to be negative (why not? See Problem 10.6), but we could have $0 < a < 1$. What, for example, does the graph of $y = \left(\frac{1}{2}\right)^x$ look like?

Using the rules for exponents, $\left(\frac{1}{2}\right)^x = \left(2^{-1}\right)^x = 2^{-x}$, so the graph of $y = \left(\frac{1}{2}\right)^x$ can be obtained from that of $y = 2^x$ by replacing x by $-x$. As we saw in Section 6.5.3, the effect of this is to flip the curve horizontally about the y-axis. You can see this in Figure 10.7, where the graphs of $y = a^x$ and $y = a^{-x}$ are sketched. Looking at the graphs for positive values of x, the graph of $y = a^x$ shows exponential growth, while the graph of $y = a^{-x}$ shows **exponential**

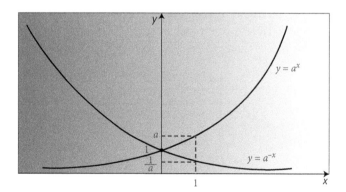

Figure 10.7 Graphs of $y = a^x$ and $y = a^{-x}$ ($a > 1$)

decay. In exponential decay, the function value decreases towards zero (the positive x-axis is a horizontal asymptote); in terms of populations, **the rate of decrease at a given time is proportional to the population size at that time**. The classic example of such behaviour is radioactive decay, but this behaviour is also present in chemical reactions, where the rate of reaction is proportional to the amount of reactant remaining.

10.5.1 Exponential decay function: $y = Ae^{-kx}$

We can move straight on to using the exponential constant e as the base for the decay function, writing it as

$$y = Ae^{-kx}. \tag{10.24}$$

Now the positive parameter k is the **decay constant**, governing how rapidly the decay occurs. The function is sketched in FNGRAPH 10.4 and programmed in SPREADSHEET 10.3 (see Figure 10.8).

Radioactive materials such as radium and uranium have unstable atoms, each of which has the potential to decay, a process in which it loses energy and becomes an atom of a different element or isotope. This is a random process, but if we have a very large number of atoms, we can say that a certain fixed proportion of them will decay at any given time. The number of atomic decays will be proportional to the total amount of radioactive material present, and so the process can be described by an exponential decay function:

$$N(t) = N_0 e^{-kt} \tag{10.25}$$

where $N(t)$ is the number of radioactive atoms remaining at time t.

Corresponding to the cell doubling time T_2 in Section 10.3.5 is the concept of the **half-life**. The half-life $T_{0.5}$ of a radioactive material is the time required for 50% of the atoms initially present in a sample to have decayed. Half-lives can vary from a matter of a few microseconds

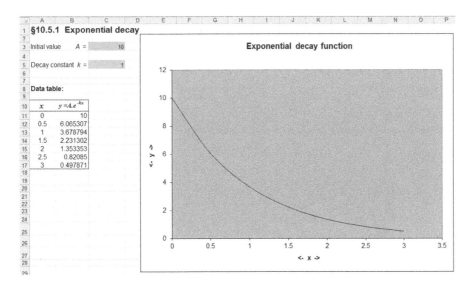

Figure 10.8 Exponential decay function $y = Ae^{-kx}$ (SPREADSHEET 10.3)

(polonium-214) to billions of years (uranium-238). The half-life is related to the decay constant in the same way as we saw with the cell doubling time:

When $t = T_{0.5}$ we have $N = \frac{1}{2}N_0$. So $\frac{1}{2}N_0 = N_0 e^{-kT_{0.5}}$. Cancelling the N_0 and taking natural logarithms of both sides we get

$$\ln\left(\frac{1}{2}\right) = -kT_{0.5}.$$

As $\ln\left(\frac{1}{2}\right) = \ln(2^{-1}) = -\ln 2$, we can relate the half-life $T_{0.5}$ to the decay constant k by

$$T_{0.5} = \frac{\ln 2}{k}. \tag{10.26}$$

To fit an exponential decay curve (10.24) or (10.25) to a set of data involves exactly the same process as we performed for exponential growth models in Section 10.4.1. The equation corresponding to (10.15) will be

$$\ln y = -kx + \ln A. \tag{10.27}$$

We again take $X = x$ and $Y = \ln y$, producing a semi-log plot, but now the best-fit straight line will have a negative slope.

The function e^{-x} can be used to construct models of physical processes other than exponential decay. A curve similar to that in Figure 8.1(c) is given by the function

$$y = A(1 - e^{-kx}). \tag{10.28}$$

This is sketched in Figure 10.9. It passes through the origin, and has a horizontal asymptote at $y = A$, since $e^{-kt} \to 0$ as $t \to \infty$. We will use it in Case Study C11 to produce an exponential model of animal speed.

As well as taking logarithms of both sides of an equation, we can take logarithms of inequalities (see Section 1.8.2). For example, if we want to know the range of values of x for which

$$10^6 < 3^x < 10^7,$$

we take logarithms to base 10, and solve for x:

$$\log\left(10^6\right) < \log\left(3^x\right) < \log\left(10^7\right)$$
$$6\log 10 < x\log 3 < 7\log 10$$

> $\log_{10} 10 = 1$

$$6 < 0.47712x < 7$$
$$\frac{6}{0.47712} < x < \frac{6}{0.47712}$$
$$12.575 < x < 14.671.$$

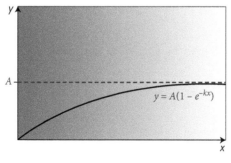

Figure 10.9 Graph of $y = A\left(1 - e^{-kx}\right)$

CASE STUDY C11: An exponential model of animal speed

In instalment C8 in Chapter 8, we modelled the speed of a lion or gazelle over time by a smooth curve producing part of a rectangular hyperbola:

$$v = \frac{v_{max}.t}{t + b} \tag{8.5}$$

where $b = \frac{v_{max}}{a_0}$. Here, v_{max} is the animal's maximum speed, and a_0 is its initial acceleration.

A very similar curve can be produced using a function of the form (10.28). We will show using calculus in Chapter 11 that the slope at the origin of the curve in Figure 10.9 is $A.k$. Thus, a curve with initial slope a_0 and horizontal asymptote $v = v_{max}$ is given by

$$v = v_{max}\left(1 - e^{-kt}\right). \tag{10.29}$$

We take $A = v_{max}$ and we want $v_{max}.k = a_0$, so set $k = \frac{a_0}{v_{max}}$. Using the notation for the hyperbolic model, we could write this as

$$v = v_{max}\left(1 - e^{-\frac{t}{b}}\right).$$

The lion which we encountered in Part C6 could accelerate at $a_0 = 6.25$ m s^{-2} and had a maximum speed of $v_{max} = 14$ m s^{-1}. The two velocity–time curves (8.5) and (10.29) for this data are drawn in **FNGRAPH 10.5** (Figure 10.10).

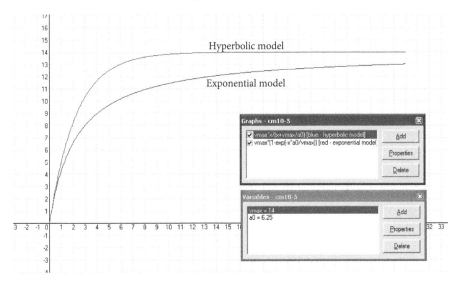

Figure 10.10 Hyperbolic and exponential models of animal speed (**FNGRAPH 10.5**)

10.5.2 Application: sensitization and habituation

If a simple cellular organism is given a stimulus such as a mild electric shock, it will respond by retreating from it. What happens if the stimulus is repeated over successive occasions? There are two basic behaviours. The first is **habituation**, in which the response becomes less and less intense, as the organism becomes accustomed to the sensation and tends to ignore it. The second, contrary response is **sensitization**; if the organism perceives the stimulus as noxious or life-threatening, it will become more alert to it, and will show an increased response when it happens again. Figure 10.11 sketches the habituation and sensitization curves, plotting response intensity I against stimulus repetition number m.

We may model these response curves using exponential decay functions:

$$I_h = I_{min} + \left(I_0 - I_{min}\right)e^{-kt}$$
$$I_s = I_{max} - \left(I_{max} - I_0\right)e^{-kt}.$$

(10.30)

10.5.3 Application: drug administration

This example is adapted from Jones and Sleeman (2003). When a patient swallows a drug capsule, it rapidly delivers a dose of the active ingredient into the bloodstream, at an initial concentration C_0. This concentration then decays exponentially over time as the drug is absorbed. So after t hours the blood concentration $C(t)$ will be $C_0 e^{-kt}$, where k is the decay constant. Now suppose the patient takes further capsules, one every T hours. Typically, a course of antibiotics is taken four times per day, so in this case $T = 6$. What happens to the drug concentration $C(t)$ in the patient's bloodstream, over a long course of treatment? Does it reach an equilibrium level, or cycle up and down, or increase without limit?

We will measure time t from the point at which the first capsule is taken. Let C_n be the drug concentration immediately after taking the nth capsule. So $C_1 = C_0$. Over the next T hours the concentration drops exponentially:

$$C(t) = C_1 e^{-kt} \quad 0 < t \le T.$$

(10.31)

After T hours, the concentration has dropped to $C(T) = C_1 e^{-kT}$, at which point the patient takes the second capsule, which provides an additional concentration of C_0. So

$$C_2 = C_1 e^{-kT} + C_0.$$

Over the following time interval, this new concentration decays exponentially. We saw in the example of drug administration in Section 5.6.1 how to use a horizontal shift so that the decay starts at $t = T$ instead of $t = 0$:

$$C(t) = C_2 e^{-k(t-T)} \quad T < t \le 2T.$$

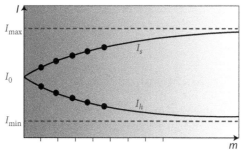

Figure 10.11 Habituation and sensitization curves

When $t = 2T$, and the concentration has dropped to $C(2T) = C_2 e^{-k(2T-T)} = C_2 e^{-kT}$, the third capsule is taken. The new concentration is then

$$C_3 = C_2 e^{-kT} + C_0.$$

In general, the drug concentration immediately after taking the $(n + 1)$th capsule, which is at time $t = nT$, is C_{n+1}, where

$$C_{n+1} = C_n e^{-kT} + C_0, \qquad (10.32)$$

and the variation of concentration with time over the subsequent T hours is

$$C(t) = C_{n+1} e^{-k(t-nT)}, \quad nT < t \le (n+1)T. \qquad (10.33)$$

Let $r = e^{-kT}$, so that (10.32) can be written as the update equation $C_{n+1} = rC_n + C_0$. Then

$$C_1 = C_0$$
$$C_2 = rC_1 + C_0 = (1 + r)C_0$$
$$C_3 = rC_2 + C_0 = r(1 + r)C_0 + C_0 = (1 + r + r^2)C_0$$

and so on. In general

$$C_{n+1} = (1 + r + r^2 + \cdots + r^n)C_0. \qquad (10.34)$$

The expression in brackets is a geometric series, whose sum we saw in (7.26) in Section 7.3.1:

$$C_{n+1} = \frac{1 - r^n}{1 - r} C_0. \qquad (10.35)$$

For the long-term concentration C_∞, we let $n \to \infty$, in which case $r^n \to 0$ provided $|r| < 1$, and so, as we saw in (7.27):

$$C_\infty = \frac{1}{1 - r} C_0. \qquad (10.36)$$

Since $r = e^{-kT}$, and k and T are positive, we know that $0 < r < 1$.

This model is programmed into SPREADSHEET 10.4 (see Figure 10.12). Taking $C_0 = 4$, $T = 6$, $k = 0.1$, so that $r = 0.5488$, equation (10.36) predicts that the long-term bloodstream concentration will be $C_\infty = 8.8655$ immediately after taking each capsule, dropping to $C_\infty - C_0 = 4.8655$ when the next capsule is due. You can see that this is borne out in the table in columns A, B, using (10.32). You can change the values of C_0, T, k on the spreadsheet and see the effect. If you unprotect the sheet and move the chart out of the way, you can see the data used to produce the curves.

We will meet this example again in Chapter 15, where we will use calculus to work out the total exposure of the patient to the drug over time.

Problems: Now try the end-of-chapter problems 10.41–10.44.

10.5.4 Example: radiocarbon dating[5]

The Earth's atmosphere contains carbon in the form of carbon dioxide (about 0.04% by volume). Almost all (98.9%) of the carbon in atmospheric CO_2 is the common isotope ^{12}C, which

[5] The radiocarbon dating technique was devised by Willard Libby in the late 1940s. He was awarded a Nobel Prize for his discovery in 1960.

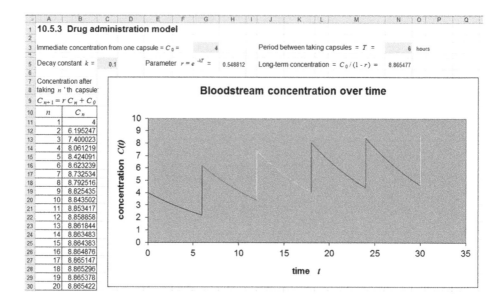

Figure 10.12 Drug administration model (SPREADSHEET 10.4)

CASE STUDY A10: An equation for logistic growth

In Part A8 of this Case Study we developed an update equation for the logistic growth model. In this model a population may grow exponentially at first, but the growth rate slows as the population size approaches the carrying capacity K (the maximum population size that can be supported by the environmental resources). Using this update equation (7.33) in Spreadsheet 7.4 (see Figure 7.19), we found that for small values of the maximum growth rate r_m, the population values $N_0, N_1, N_2, N_3, \ldots, N_p, \ldots$, plotted against generation p, lay on a smooth curve (which we sketched in Figure 7.17).

In Chapter 16 we will see how to solve a differential equation to derive the equation of this smooth curve, which is the continuous model for logistic growth. The equation is:

$$N(t) = \frac{K}{1 + \left(\dfrac{K}{N_0} - 1\right)e^{-r_m t}}. \tag{10.37}$$

The function is programmed into FNGRAPH 10.6 (see Figure 10.13). The curve is S-shaped or **sigmoid**. It is important to get the brackets in the correct places:

```
K/(1 + ((K/N0)−1)*exp(−rm*x))
```

The equation can be rearranged to:

$$\ln\left(\frac{K - N}{N}\right) = -r_m t + c \tag{10.38}$$

where $c = \ln\left(\dfrac{K}{N_0} - 1\right)$. If K can be estimated then the least-squares technique can be used to fit the curve to a set of data, as in Section 10.4.2, to obtain values for N_0 and r_m.

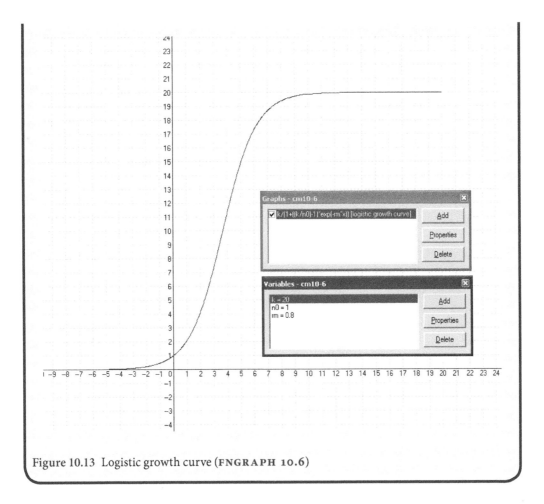

Figure 10.13 Logistic growth curve (**FNGRAPH 10.6**)

has an atomic mass of 12 daltons (see Section 4.1.2), and 1.1% is the stable isotope ^{13}C, which has an atomic mass of 13 Da. But there is a very small proportion (one atom in 8.33×10^{11}) of the unstable isotope ^{14}C. This is created from nitrogen (^{14}N) by cosmic radiation. Over time, the ^{14}C atoms change back to ^{14}N, releasing electrons in a process of radioactive beta decay. There is a balance between this decay and the generation of new ^{14}C atoms, which maintains the ^{14}C:^{12}C isotope ratio at a constant level in the atmosphere.

Atmospheric carbon dioxide enters the biosphere through photosynthesis, so the carbon in the bodies of living animals and plants reflects this ^{14}C:^{12}C ratio. The body of a person weighing 70 kg contains about 16 kg of carbon (Kutschera 2001). When the organism dies, the supply of fresh carbon from the atmosphere stops. The stable ^{12}C and ^{13}C isotopes in the dead body remain constant over time, but the number of ^{14}C atoms gradually reduces in a process of exponential decay:

$$^{14}C(t) = {}^{14}C_0\, e^{-kt} \tag{10.39}$$

where $^{14}C_0$ is the number of ^{14}C atoms at death and $^{14}C(t)$ is the number t years after death. The decay constant k can be found using (10.26), from the half-life $T_{0.5}$ of ^{14}C, which has been measured as 5730 ± 40 years.

Since the ^{12}C content remains constant, a similar equation applies to the ^{14}C:^{12}C ratio $P(t)$:

$$P(t) = P_0\, e^{-kt} \tag{10.40}$$

where $P(t) = \dfrac{^{14}C_t}{^{12}C_t}$, the ratio of the ^{14}C and ^{12}C content in the sample. Then (10.40) can be solved for t:

$$t = \frac{-1}{k}\ln\left(\frac{P(t)}{P_0}\right). \tag{10.41}$$

The ^{14}C content of a sample is measured by decay counting, or accelerator mass spectroscopy (AMS). In the Revision problems at the end of this chapter, you can apply this theory to the radiocarbon dating of 'Ötzi the Iceman', the mummified body of a Neolithic man, discovered high up in a remote mountain range on the Austrian–Italian border in 1991.[6]

The radiocarbon dating process is not quite as straightforward as this. The proportion of ^{14}C in the atmosphere does vary somewhat with time, especially over timescales of thousands of years. This variation is caused by climatic changes, and more recently human activity has interfered with this level through the burning of fossil fuels and nuclear emissions. Also, the amount of ^{14}C produced by cosmic radiation is dependent on the amount of cosmic rays entering the atmosphere (itself dependent on the strength of the Earth's magnetic field).[7] For samples dating back up to 12,000 years, it is possible to calibrate the age obtained from radiocarbon dating with ages given by dendrochronology (tree-ring dating), since fossilized tree-trunks can be dated by both methods.[8] Figure 10.14, from Kutschera (2001), shows the calibration curve by which a radiocarbon age on the vertical axis, in years BP (Before Present, taken as AD 1950) is mapped to a calibrated date in years BC on the horizontal axis. Unfortunately, in the case of Ötzi a radiocarbon age of 4550 ± 19 BP maps to three possible date ranges within the period between 3370 BC and 3100 BC, due to the 'wiggles' in the calibration curve.

. . . .

10.6 Example: reduction of cholesterol level

We will end the examples in Part I of this book on an optimistic note, with an example that brings in several pieces of maths from earlier chapters.

Recall the example in Section 7.2.4, where we fitted a quadratic curve to the following table of data on my cholesterol level y (in mmol l^{-1}) measured in May, July, and October 2008:

t (months from Jan 2008)	y (mmol l^{-1})
4	8.1
6	7.3
9	6.4

---

[6] Ötzi has his own website at http://www.iceman.it/, and can be visited in Bolzano, N. Italy.

[7] Creationists have seized on these factors to reconcile the Biblical account with scientific findings on the age of the Earth. See for example http://www.christiananswers.net, which adds the argument that 'Also, the Genesis flood would have greatly upset the carbon balance'.

[8] This calibration curve is part of the widely used INTCAL98 calibration, which allows archaeologists to adjust the radiocarbon ages for samples going back to 26000 BP.

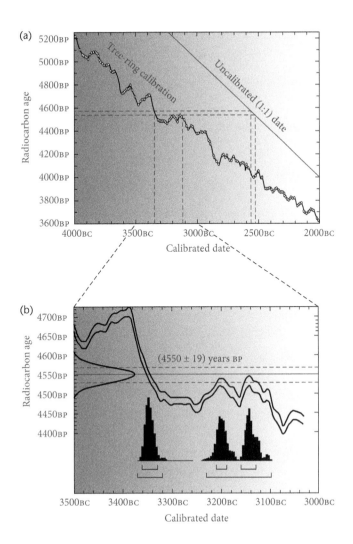

Figure 10.14 Radiocarbon dating calibration for Ötzi the Iceman (from Kutschera 2001)

That curve predicted that my cholesterol level would fall to a minimum of 5.68 mmol l^{-1} after 15 months, but then start rising again. I was hopeful that a more realistic model would produce the dashed curve shown in Figure 7.9, with the level gradually decaying down towards an asymptote at $y = c$. We now have such a function:

$$y = Ae^{-kt} + c \qquad (10.42)$$

and we can try to fit this data to (10.42) and obtain values for A, k and c. As there are three unknown parameters and three data-points, we hope to be able to calculate the parameters directly, without using a least-squares best-fit approach.

Inserting the three data-points in (10.42) gives us three simultaneous equations:

$$Ae^{-4k} + c = 8.1$$
$$Ae^{-6k} + c = 7.3 \qquad (10.43)$$
$$Ae^{-9k} + c = 6.4.$$

We can eliminate c by subtracting the first equation from the second, and the second from the third, leaving two equations involving A and k:

$$Ae^{-6k} - Ae^{-4k} = -0.8$$
$$Ae^{-9k} - Ae^{-6k} = -0.9. \tag{10.44}$$

Now we could divide the first equation by the second, and cancel the common factor of A on top and bottom:

$$\frac{e^{-6k} - e^{-4k}}{e^{-9k} - e^{-6k}} = \frac{8}{9}.$$

This looks like it cannot be simplified further, and we won't get very far if we take logarithms. But suppose we write $e^{-k} = u$. Then $e^{-6k} = (e^{-k})^6 = u^6$, etc. So we can rewrite our equation as:

$$\frac{u^6 - u^4}{u^9 - u^6} = \frac{8}{9}.$$

Now we can see that there is a common factor of u^4 which can be cancelled from top and bottom, leaving

$$\frac{u^2 - 1}{u^2(u^3 - 1)} = \frac{8}{9}.$$

There is one more common factor which can be cancelled. Recall from (1.20) that

$$u^2 - 1 = (u - 1)(u + 1)$$
$$u^3 - 1 = (u - 1)(u^2 + u + 1).$$

So we have another common factor of $(u - 1)$ to cancel, leaving

$$\frac{u + 1}{u^2(u^2 + u + 1)} = \frac{8}{9}.$$

Cross-multiplying, and collecting all the terms on the left-hand side:

$$F(u) = 8u^2(u^2 + u + 1) - 9(u + 1) = 0. \tag{10.45}$$

$F(u)$ is a quartic function, and we seek a root u lying between 0 and 1. Notice that $F(0) = -9$ and $F(1) = 6$, so there must be a root lying between them, at which $F(u) = 0$. To find it we use the method of *regula falsi* (Section 8.6.1). It is programmed in SPREADSHEET 10.5.

We find that the root occurs when $u = 0.8903$. Sketch $F(x)$ in FNGraph to check there are no other roots.

If $e^{-k} = u = 0.8903$, then $k = -\ln u = 0.1162$.

Now we have to find the other two parameters. Using one of the equations in (10.44), we can find A:

$$Au^6 - Au^4 = -0.8$$

$$A = \frac{-0.8}{u^6 - u^4} = \frac{-0.8}{0.8903^6 - 0.8903^4} = \frac{-0.8}{0.4980 - 0.6283} = \frac{-0.8}{-0.1303} = 6.141.$$

Finally, and most importantly from my point of view, we can put A and u into one of the equations in (10.43) to find the asymptote c:

$$c = 8.1 - Au^4 = 8.1 - 6.141 \times 0.6283 = 8.1 - 3.8585 = 4.2415.$$

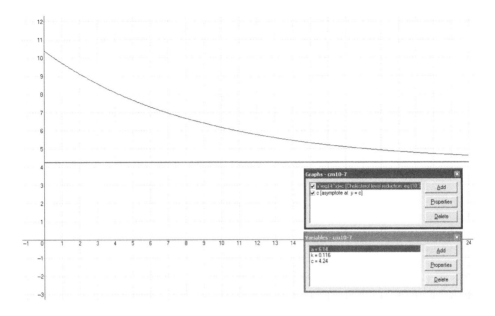

Figure 10.15 Cholesterol reduction model (FNGRAPH 10.7)

The model suggests that my cholesterol level could keep dropping, and I should aim towards a level of 4.25 mmol l^{-1}, which is a very healthy level.

The equation we have fitted is

$$y = 6.14e^{-0.116t} + 4.24. \tag{10.46}$$

The curve is sketched in FNGRAPH 10.7 (Figure 10.15), together with the asymptote at $y = 4.24$.

In Chapter 7 we asked: (i) when would my level fall to 6.0 mmol l^{-1}, and (ii) when would it fall to 5.5 mmol l^{-1}?

When $y = 6.0$ we need to solve $6.0 = 6.14e^{-0.116t} + 4.24$ for t. Simplifying and taking logarithms:

$$6.14e^{-0.116t} = 6.0 - 4.24 = 1.76$$

$$e^{-0.116t} = \frac{1.76}{6.14} = 0.2866$$

$$-0.116t = \ln(0.2866) = -1.2495$$

$$t = \frac{1.2495}{0.116} = 10.77.$$

So the level should fall to 6.0 after 10.8 months, i.e. in late October. A similar calculation says that it should fall to 5.5 mmol l^{-1} when $t = 13.65$, i.e. in the second half of February 2009.

. . . .

10.7 Extension: a stochastic model of exponential decay

Radioactive decay occurs when the nuclei of atoms of a radioactive isotope lose energy by emitting radiation and turning into stable isotopes. As we have seen in Section 10.5.1, radioactive carbon isotope ^{14}C is unstable, and atoms of this material will tend to transform into stable nitrogen ^{14}N atoms. The radiation thereby emitted can be detected by a Geiger counter.

This is a random process: we cannot predict when a particular atom will decay. But if there are a large number N_0 of radioactive atoms present initially, the number $N(t)$ still present at time t, will follow the exponential decay curve:

$$N(t) = N_0.e^{-kt}. \tag{10.25}$$

In SPREADSHEET 10.6 (Figure 10.16) you can see a demonstration of this. Cells B11 to B110 represent 100 atoms at time $t = 0$, and the value 1 in each cell indicates it is radioactive. The state of the atoms after one year, i.e. at time $t = 1$, is represented in cells C11 to C110, and so on. The time value in years for each column is given in row 8.

The likelihood that an atom will decay over one year is specified using a 'decay probability' p in cell C4. This is initially set to $p = 0.2$, meaning that each atom has a 20% likelihood of decaying during one year.

We can model the random nature of the radioactive decay process, using a random number function RAND(). When we use it in a cell formula, it is evaluated as a random number between 0.0 and 1.0. Note that this RAND function has no arguments, but we must write the empty brackets (). Then a logical test condition 'RAND() < $C$4' will have a 20% likelihood of being true, since there is a 20% likelihood that the random number will lie in the interval between 0 and 0.2. It is rather like tossing a die with five faces,[9] and saying that if the number is a 1 the atom decays, but if it is 2, 3, 4, or 5 the atom remains radioactive.

Cell C11, representing the first atom at time $t = 1$, has the formula

    =IF(B11=0, 0, IF(RAND()<$C$4, 0, 1))

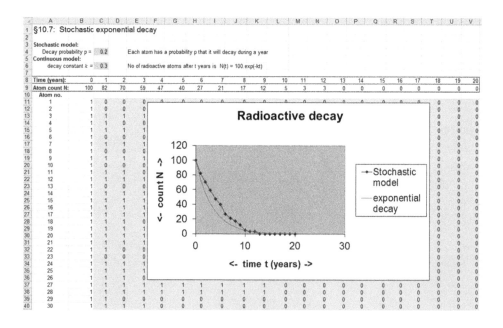

Figure 10.16 Stochastic model of radioactive decay (SPREADSHEET 10.6)

[9] Is it possible to make a die with five faces? For the answer, do a search for 'platonic solids'.

Here we are using two IF functions, nested. The outer one asks if the atom had already decayed at the previous time (B11=0): if this is true, then it remains decayed so we set the cell value to 0. If not (i.e. it is still radioactive), we call the random number function and if the condition RAND()<$C$4 is satisfied, we put 0 in the cell, to indicate it has decayed. If not, we put a 1 in the cell, to indicate it is still radioactive. So with C4 containing 0.2 and B11 containing 1, there is a 20% chance that cell C11 will contain 1.

The formula for cell C11 has been copied down the column to C110, then this column is copied across up to column V, representing the state at times $t = 1, 2, 3, \ldots, 20$.

The number of cells still radioactive at each time, is counted in row 9. The formula in the first cell, B9, is

=COUNTIF(B11:B110,'1')

We use the COUNTIF function, which counts the number of cells in the range B11 to B110 that have the value 1. This has been copied along the row for each time value. Finally, the Chart plots this data against time, in the dark line. Notice that this is not a smooth curve, because the data are coming from random events. If you change the value of the decay probability in C4, or even if you just click the Save icon, all the RAND() functions get re-evaluated and you get a different curve.

These data can be modelled by the continuous exponential decay function (10.25), choosing an appropriate value of the decay constant k. You can adjust k in cell C6, until the smooth curve most closely matches the experimental data plot. For a fixed value of p, different re-evaluations of the random-number functions will produce different data-sets, but they should all be approximated best by (10.25) with roughly the same value of k.

CASE STUDY A11: Gompertz curve for population mortality

A 25-year-old man goes to a insurance broker to buy life insurance, which will pay out a benefit to his family if he dies before the age of 65. For the insurance company to decide how much to charge, it must estimate the risk of an early death: from a population of 1,000 25-year-olds, how many will still be alive after 40 years? Benjamin Gompertz[10] produced the mathematical model still used by actuaries today. Starting from the assumption that the mortality rate increases exponentially with age, the Gompertz law he derived in 1825 says that, for such a population, the number still alive after t years is given by

$$N(t) = ae^{\left(-be^{kt}\right)}. \tag{10.47}$$

This fits remarkably well to real population data, with suitable choices of the parameters a, b, k. Notice that a is not simply equal to N_0, the initial population size.

The Gompertz curve is drawn in FNGRAPH 10.8 (Figure 10.17). It is programmed as a*exp(-b*exp(k*x)). The downward slope of the curve is small initially, but is then roughly constant throughout most of the time period. However, it reduces again at old age, creating a 'mortality plateau' which is indeed observed in real life. Some biologists have argued that this is because of genetic and environmental differences among the population, where those with 'healthy genes' and a healthy lifestyle survive into a ripe old age.

[10] Benjamin Gompertz FRS (1779–1865), English self-educated mathematician and actuary.

To produce a linear relationship to which we can fit a trend line, rewrite (10.47) as:

$$\frac{a}{N} = e^{\left(be^{kt}\right)}.$$

Now take natural logarithms of both sides, twice (see problem 10.48).

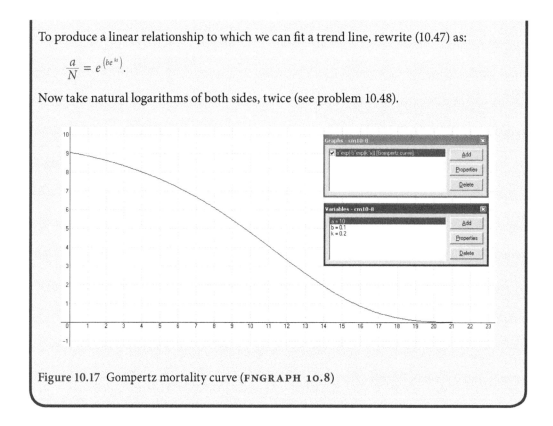

Figure 10.17 Gompertz mortality curve (**FNGRAPH 10.8**)

Problems: Now try the end-of-chapter problems 10.45–10.50.

 **PROBLEMS**

10.1–10.2: In a laboratory experiment recording the growth of cells in several Petri dishes, the student noted the initial population N_0 and then $N(t)$ at one-hour intervals. Due to a procedural irregularity (he lost the scrap of paper where he'd jotted down the earlier values), only the populations at $t = 5$ and $t = 6$ hours are known. Using your calculator, reconstruct the earlier values by fitting each set of results to the exponential growth function $N(t) = N_0\, a^t$, finding the positive numbers N_0 and a.

10.1

t (hours)	$N(t)$
0	
1	
2	
3	
4	
5	4 131
6	12 393

10.2

t (hours)	$N(t)$
0	
1	
2	
3	
4	
5	2 637
6	6 592

10.3–10.4: A second student performing the same experiment also lost his earlier results before recording the count at t = 5 hours. In a further procedural irregularity, he went for a coffee, got chatting, and only returned in time to record the count at $t = 7$ hours. Can you reconstruct the missing values by fitting the exponential growth function $N(t) = N_0\, a^t$?

10.3

t (hours)	$N(t)$
0	
1	
2	
3	
4	
5	873 964
6	
7	42 824 236

10.4

t (hours)	$N(t)$
0	
1	
2	
3	
4	
5	3 429
6	
7	8 778

10.5: Use FNGraph to sketch the graphs of the following functions, on the same plot:

$$y = 1.25^x, \qquad y = 0.8^x, \qquad y = 1.25^{-x}, \qquad y = 0.8^{-x}.$$

Explain why some of the curves coincide.

10.6: Now try to sketch the curves of:

$$y = -1.25^x, \qquad y = (-1.25)^x, \qquad y = (-1.25)^{-x}.$$

Explain why some of the curves don't appear.

10.7: Recall problem 1.45, where an Indian mathematician was paid 1 grain of wheat for the first square of the chessboard, 2 grains for the second square, 4 grains for the third, and so on. In problem 7.33 you found an expression for the total number of grains paid for the whole board. Let $P(n)$ be the number of grains paid for the nth square, and $S(n)$ be the total number of grains paid for the first n squares (so, for example, $P(3) = 4$ and $S(3) = 7$). Find the functions $P(n)$ and $S(n)$, and turn these into continuous models by replacing n by a continuous variable x. Draw the graphs of $P(x)$ and $S(x)$ in FNGraph. You may need to scale, e.g. by counting in millions of grains.

10.8: Repeat problem 10.7 with the modification that he is paid 1 grain of wheat for the first square of the chessboard, 3 grains for the second square, 9 grains for the third, and so on, tripling the amount paid each time.

10.9–10.12: Simplify the following expressions, where possible:

10.9 $\dfrac{1}{\left(\dfrac{1}{3}\right)^x}$ 10.10 $\dfrac{1}{2^x} - \dfrac{1}{2^{-x}}$ 10.11 $\dfrac{2^{3x}.3^{2x}}{2^{5x}.5^{2x}}$ 10.12 $\dfrac{2^x + 3^x}{2^x + 2^{2x}}.$

10.13: Using FNGraph, draw the curve $y = e^x$. Find the equation of the straight line that passes through $(0, 1)$ and has slope $m = 1$, and draw it in the same graph. Enlarge the graph around $(0, 1)$ to see if the line really is the tangent to the curve at this point.

10.14: Modify the previous graph by entering a parameter $k = 2$, and drawing the curve $y = e^{kx}$, and the line through (0, 1) with slope $m = k$. Try different values of k, positive and negative. Is the line tangent to the curve?

10.15–10.18: Without using your calculator, simplify:

10.15 $3 \log_7 243$

10.16 $2 \log_3 \left(\sqrt{27} \right)$

10.17 $\log_{10} 8 + 3 \log_{10} 5$

10.18 $2 \log_{10} 2 + \log_{10} 15 - \log_{10} 6$.

10.19–10.22: Given that $\ln 2 = 0.6931$ and $\ln 3 = 1.0986$, evaluate the following without using your calculator:

10.19 $\ln 4$ 10.20 $\ln 0.75$ 10.21 $\ln 12$ 10.22 $\ln 15 - \ln 10$.

Check your answers on your calculator.

10.23–10.24: In problem 10.7 you found the functions $P(n)$ and $S(n)$ expressing:

- the number of grains of wheat paid for the nth square;
- the total number of grains paid up to and including the nth square.

Write inequalities involving these functions, then take logarithms, to answer the following:

10.23 For which squares is the mathematician paid between one million and one hundred million grains?

10.24 At which square does the total amount paid start to exceed one billion grains?

10.25: In the bisection algorithm for root-finding (Section 8.6.1), the length of the interval bracketing the root is halved at each iteration. Suppose the length of the initial interval is d_0. Write down an expression for d_n, the length of the interval after n iterations.

Suppose $d_0 = 1.5$. How many iterations will be needed before we can quote the root correct to 3 decimal places? Use logarithms. For general d_0, show that $n > 3.32 \log_{10} (d_0) + 9.97$.

How many iterations would be needed before the interval length is reduced by a factor of 10^5?

10.26: Repeat problem 10.25 for the minimization algorithm golden section search (Section 8.6.2).

10.27: An initial population of 5000 cells exhibits exponential growth, with growth constant $k = 0.20$ hour^{-1}. After how many hours does the population:

(i)   reach 8000 cells?

(ii) reach double its initial size?

(iii) quadruple in size?

10.28: An initial population of 5000 cells grows to 5500 cells after 1 hour. After how many hours does the population:

(i)   reach 8000 cells?

(ii) reach double its initial size?

(iii) quadruple in size?

10.29–10.30: In the following problems, don't use your calculator, but you can use the values of ln 2 and ln 3 given in problems 10.19–10.22.

10.29 I have two Petri dishes. The cells in the first dish grow exponentially with growth constant $k = 0.25$ hour^{-1}. The second dish initially has only half as many cells, but they grow exponentially with growth constant $k = 0.35$ hour^{-1}. At what time will the populations of the two dishes be equal?

10.30 I have two Petri dishes. In the first dish I put 3000 cells, which grow exponentially with growth constant $k = 0.45$ hour^{-1}. In the second dish I put 2000 cells, which grow exponentially with growth constant $k = 0.65$ hour^{-1}. At what time will the populations of the two dishes be equal?

10.31–10.34: Repeat problems 10.1–10.4, but fit the exponential growth model $N(t) = N_0\, e^{kt}$. Also calculate the cell doubling time, in each case.

10.35: In SPREADSHEET 10.2 (Section 10.4.1) we fitted an exponential curve to the population data from Swedish censuses from 1780 to 1860, using a semi-log plot. Repeat the fitting but using the following population data from 1880 to 1960 (taken from problems 7.44–7.46). Now modify the spreadsheet to use more data-points, and use the data from 1780 to 1940 (omitting the 1960 data). In each case, use the equation to estimate the populations in 1970 and 2000.

Year t	Population N (1000s)
1880	4572
1900	5117
1920	5876
1940	6356
1960	7480

Hint: to insert a new row in the spreadsheet, highlight the location and choose Insert > Rows .... Do this inside the data table, not at the bottom, and Excel will automatically increase the set of points it uses in the graphs.

10.36: Use SPREADSHEET 10.2 to fit an exponential curve to the data for CO_2 emissions, from problems 3.13–3.20:

Year t	Atmospheric CO_2 N (ppm)
1959	315.98
1969	324.62
1979	336.78
1989	352.90
1999	368.14
2009	387.35

Source: National Oceanic and Atmospheric Administration, quoted on CO2now.org website.

Note: in the final graph, you may want the vertical axis to start at 300 instead of at 0. Put the cursor on the axis, right-click and choose Format Axis ... Then for Minimum, uncheck the Auto box and type in the minimum you want to use.

Hence answer problems 3.13–3.20 using your best-fit curve. You can answer problems 3.13–3.17 using the 'interpolation machine' in column J; for the last three problems you'll have to solve the equation $N(t) = N_0 e^{kt}$ for t; then you can program an 'inverse interpolation machine' on the spreadsheet. Compare your answers with those obtained by piecewise linear interpolation in Chapter 3.

The best-fit exponential curve looks almost linear, in the time-range 1960–2020. But perhaps the CO_2 emissions could be better modelled as consisting of a constant natural level d, plus a man-made component that is increasing exponentially:

$$N(t) = N_0 e^{kt} + d.$$

Modify SPREADSHEET 10.2 to fit such a curve to the data. Use cell G5 to enter a value of d. We now define the transformed variable $Y = \ln(N - d)$ in the right-hand data table. The line-fitting and the calculations of k and N_0 are unchanged, but in the final graph you'll have to modify the data that produces the line. Move the graph out of the way and you'll see the data. Change the formula in the first row (on my sheet it's cell D21) by adding '+ $G$5', and copy it down. You'll also have to change the formula that calculates N in the 'interpolation machine', in cell J30.

With $d = 0$ in cell G5 you should obtain the previous result, and the linear regression has an R^2 value of 0.9912. Try changing the value of d to increase the goodness-of-fit. When you have found a value that maximizes the R^2 value, repeat the calculations for problems 3.13–3.20. If the value of N0 in cell G18 appears as 0.0000, change the display format to scientific notation.

10.37: Researchers have proposed an allometric relationship in primates between the volume of the brain V (found by pouring small pellets into the cranial cavity in skulls) and the area A of the *foramen magnum* (a circular opening at the base of the skull, through which the spinal cord and arteries pass). Investigate this by taking logarithms to base 10 of both sides of the equation:

$$V = bA^k$$

and then modify Spreadsheet 10.2 to use a log-log plot for the following data:

Foramen magnum area A (cm²)	Brain volume V (ml)
0.52	17.5
0.83	32.0
1.27	55.5
1.56	72.5
1.98	99.0
2.42	128.5

What are the values of b and k that you obtain? Does it matter whether you use common logarithms or natural logarithms?

10.38: Allainé et al. (1987) propose that there is an allometric relationship within species between fecundity F (defined as the mean number of offspring per year per adult breeding pair) and mean body mass M. Investigate this using the following data for different primates:

Primate	Mean body mass (kg)	Fecundity
Golden lion tamarin	0.7	Two litters per year, each with two offspring
Slender loris	1.2	One litter each year, with two offspring
Spider monkey	10	One offspring every 3 years
Baboon	25	One offspring every two years
Archbishop of Canterbury	70	Two children
Gorilla	100	one offspring every four years

You may choose to remove an 'outlier' from the data; why is this justified?

To find the equation of the best-fit line through the transformed data, don't use Excel: calculate the parameters using the formulae in Section 8.7 (equation 8.33).

Hence predict the natural fecundity of the Archbishop of Canterbury.

10.39: Use the technique described in Section 10.4.3, to estimate the values of the Hill coefficient η, v_{max} and K_m in the allosteric regulation equation (10.22), from the following set of experimental data:

s	v
20	17.7
40	30.5
60	35.8
80	39.0
100	40.8
120	41.8

Program this in Excel by modifying SPREADSHEET 8.2. Put the estimated value of v_{max} in cell G5. You need to re-program the transformed variables X and Y, and the calculation of η and K_m from the values of the linear regression slope and intercept. You will also have to re-program the y-values for the data-points that are underneath the final graph. You have to make an initial guess for v_{max}, and look at the R^2-value of the best-fit parameters. Then adjust v_{max} and repeat. Refine your guess of v_{max}, attempting to make R^2 as close to 1.0 as possible. Find the best value of v_{max}, correct to one decimal place.

10.40: Repeat problem 10.39 using the following set of data:

s	v
20	25.3
40	33.0
60	36.8
80	39.0
100	40.8
120	41.8

10.41–10.42: The following data-sets correspond to the experiments in problems 10.1–10.4, but this time there was a pathogen present in the dish nutrient, so that the cells were dying. Fit $N(t) = N_0\, a^{-t}$ and $N(t) = N_0\, e^{-kt}$ in each problem. Calculate the half-life in each case.

10.41

t (hours)	$N(t)$
0	
1	
2	
3	
4	
5	1626
6	1084

10.42

t (hours)	$N(t)$
0	
1	
2	
3	
4	
5	721
6	
7	461

10.43: Re-solve problem 8.13 using the exponential model of animal speed $v(t)$, in Case Study C11, i.e. using equation (10.29), given its initial acceleration a_0 and maximum velocity v_{max}. Consider the lion for which $a_0 = 6.25$ m s^{-2} and $v_{max} = 14$ m s^{-1}, and a gazelle for which $a_0 = 3.75$ m s^{-2} and $v_{max} = 25$ m s^{-1}. If the two animals accelerate from rest at the same time $t = 0$, at what time t^* will their velocities be equal? You will need to use a numerical method (root-finding algorithm); in my solution I have adapted SPREADSHEET 8.4 to use *regula falsi*.

10.44: In the previous problem we know that the two animals' speeds are equal when $t = 0$ (both are stationary) and when $t = t^*$. At what time between 0 and t^* is the difference in speeds a maximum? In my solution I have modified SPREADSHEET 8.5 to solve this by golden section search.

10.45: Fit a logistic growth curve to the Swedish population data 1780–1940 (from Problem 10.35).

10.46: Fit a logistic growth curve equation (10.37) to the following data for the growth of a yeast population in culture.[11] The units of biomass are not given.

Time t (hours)	Yeast biomass $N(t)$	Time t (hours)	Yeast biomass $N(t)$
0	9.6	10	513.3
1	18.3	11	559.7
2	29.0	12	594.8
3	47.2	13	629.4
4	71.1	14	640.8
5	119.1	15	651.1
6	174.6	16	655.9
7	257.3	17	659.6
8	350.7	18	661.8
9	441.0		

[11] Data cited by Neal (2004).

10.47: Show that the Gompertz function (10.47) can be rewritten as

$$N(t) = N_0 \, e^{b\left(1 - e^{kt}\right)}$$

where N_0 is the initial population size.

10.48: Manipulate the Gompertz function (10.47) to obtain a linear relationship between transformed variables X, Y. Describe an algorithm to use this to fit a Gompertz curve to a set of population-time data.

10.49: The stochastic model of exponential decay, in Section 10.7, uses a decay probability p; in moving from one 'generation' to the next, each atom has a probability p of decaying. If the initial number of radioactive atoms N_0 is large, roughly how many radioactive atoms do you expect there to be after one generation? Use this in the continuous model (10.25), where t is the number of generations, to find the continuous decay constant k as a function of p. Deduce that the decay constant k in the continuous exponential-decay model (10.25), which matches the decay probability p in Spreadsheet 10.6, is given by $k = -\ln(1 - p)$. Program SPREADSHEET 10.6 to take the value of p entered in cell C4, work out this value of k and enter it in cell C6. Try entering different values of p, and see if the continuous curve closely matches the stochastic graph each time.

And finally …

10.50: Do you recall the final problem 1.68 in Chapter 1?

You are working for MI5, and intercept a new series of emails:

> From: olga@gov.slk
> To: I.Panic@gurglemail.com
> Ivo,
> You need a new PIN number. Our mathematicians have invented a more secure way of transmitting it to you.

> Choose a new natural number b. Use your calculator to raise the following number to the power b, and send me the result. The number is 3124943128.
> Comradely greetings, Olga

> From: I.Panic@gurglemail.com
> To: olga@gov.slk
> Here is the result: $2.979961062 \times 10^{47}$. Ivo

> From: olga@gov.slk
> To: I.Panic@gurglemail.com
> Okay. Here is a new number: $6.679390939 \times 10^{15}$. Take the bth root of this number, and the result will be the PIN number you need. Olga.

You should now be able to find out what the PIN number is. For more information about asymmetric key encryption see Davenport (2008).

❓ REVISION PROBLEMS

The following questions about radiocarbon dating and Ötzi the Iceman (see Section 10.5.4) involve facts and results from this chapter and also from earlier chapters in Part I. I am not going to spoon-feed you this material within the questions – or even to tell you what information to search for. You will have to decide what information you need, as well as how to solve the question!

R10.1: What is the carbon content of the human body, as a percentage (to the nearest %)?

R10.2: Estimate how much carbon is contained in a 1 mg sample of human tissue.

How much human tissue would be required to provide 1 mg of carbon?

R10.3: Show that the atomic mass of atmospheric (naturally occurring) carbon is 12.011 (to 3 decimal places).

R10.4: In *A Short History of Nearly Everything*, Bryson (2004) describes radiocarbon dating on pages 197–199. He states that there is a problem in radiocarbon dating very ancient material, because 'after eight half-lives, only 0.39 per cent of the original radioactive carbon remains, which is too little to make a reliable measurement, so radiocarbon dating works only for objects up to forty thousand or so years old'. Check this claim. How does this percentage (0.39% for ^{14}C) depend on the length of the half-life?

R10.5: Calculate the ^{14}C decay constant k, when the half-life $T_{0.5}$ is given

(a) in years;

(b) in seconds.

R10.6: Find an uncertainty interval for k in each case above.

R10.7: How many atoms are in 1 mg of carbon? How many atoms of carbon are in 1 mg of human tissue?

R10.8: How many ^{14}C atoms are in 1 mg of atmospheric carbon?

How many ^{14}C atoms are in 1 mg of human tissue?

How many ^{14}C atoms are in a 70 kg human body?

R10.9: In exponential decay, the rate of decay is proportional to the amount of material present, where the constant of proportionality is the decay constant k. In radioactive decay, the rate of decay can be measured as the radioactivity level, in becquerels. Show that the average human body weighing 70 kg has a radioactivity level of about 3700 becquerels.

R10.10: What level of radioactivity would be observed in a 1 mg tissue sample from an animal that died 5730 years ago? Using a Geiger counter, how long would we need to test such a sample for, if we want to observe at least 100 nuclear decays?

R10.11: Suppose that we can measure the $\dfrac{P_t}{P_0}$ ratio in (10.41) to the nearest 0.5%. What is the age of the oldest object that can be carbon-dated?

R10.12: In the initial radiocarbon dating of bone and tissue samples from Ötzi, measured in 1995, the ^{14}C content was found to be 53% of the ^{14}C:^{12}C ratio observed in living tissue, i.e. $P_t = 0.53P_0$. Show that this predicts Ötzi to be about 5 200 years old.

R10.13: Rewrite (10.41) using the half-life $T_{0.5}$ rather than the decay constant k. Given the uncertainty interval for the ^{14}C half-life, what would be the uncertainty interval for Ötzi's age, assuming that the 53% figure is exact?

R10.14: Supposing that the $T_{0.5}$ value is exact, what is the uncertainty interval for Ötzi's age if the 53% figure is correct to the nearest 0.5%?

R10.15: What is the uncertainty interval for Ötzi's age, considering the precision both of the half-life and of the 53% measurement?

R10.16: Later measurements conducted in 2000 in Oxford and Zürich produced an uncalibrated radiocarbon age of 4550 ± 19 BP (Before Present, where 'Present' is 1950[12]). What range of $\frac{P_t}{P_0}$ ratios (in percent) does this correspond to?

R10.17: The radiocarbon dates assume that the ^{14}C:^{12}C has remained constant over time. Suppose that at the time of Ötzi's death there was 10% less ^{14}C in the atmosphere than at present (but the same amount of ^{12}C). How would this affect the radiocarbon dates? Repeat the calculation of problem R10.12 using a corrected value of P_0.

R10.18: Now look at the graph in Figure 10.14. On the left-hand axis is the radiocarbon age, expressed as Years BP. Along the horizontal axis is the calibrated date expressed in Years BC. To convert a raw age t calculated from experimental data using equation (10.41) and expressed in units of 'years ago', first convert it into an uncalibrated radiocarbon age T expressed in Years BP. Then locate this age on the vertical axis, follow across to the jagged tree-ring calibration curve, and thence down to the horizontal axis, producing a calibrated date τ (Greek letter t, pronounced tau) expressed in Years BC. The straight line on the graph maps ages Before Present to dates BC without calibration adjustment.

What is different about this graph, that prevents us from using the coordinate geometry we have studied in Chapters 5 and 6? Instead, in questions R10.23 onwards, we will find functions that express the relationships between t, T and τ.

R10.19: Suppose that a new radiocarbon dating of a grass sample from the Ötzi site, made in 2010, produced a ^{14}C content that was 59% of the ^{14}C:^{12}C ratio observed in living plants. Calculate the raw age t of the sample, and hence the uncalibrated radiocarbon age T and, using the graph, the calibrated age τ.

R10.20: Suppose that a second new radiocarbon dating of a grass sample produced a calibrated date τ of 3800 BC. What would have been the isotope ratio $\frac{P_t}{P_0}$ that produced such a date?

R10.21: Express T as a function of t. That is, find the function $g(t)$ where $T = g(t)$, and t is being calculated from radiocarbon dating tests made in 2010.

[12] The year 1950 is taken as the reference point in radiocarbon dating because it was in that year that the first calibration curves and calibrated radiocarbon dates of samples were published.

On a copy of the graph in Figure 10.14, use a ruler to draw a best-fit straight line through the calibration curve. Find the coordinates of two points on the line, and hence find the function $f(T)$ where $\tau = f(T)$, defining your calibration line.

Using this function, what calibrated date would correspond to Ötzi's radiocarbon age of 4550 BP?

What is the function $i(T)$ where $\tau = i(T)$ defines the uncalibrated (1:1) line on the graph?

By equating these functions, find the date at which the two lines would intersect, i.e. when the radiocarbon and tree-ring dates would coincide.

On paper, draw a graph of τ against T, showing the two lines $\tau = i(T)$ and $\tau = f(T)$, and their point of intersection.

R10.22: We could also express equation (10.41) using a function h, i.e. we write the equation as $t = h\left(\dfrac{P_t}{P_0}\right)$. The process of radiocarbon dating can now be seen as taking the isotope ratio $\dfrac{P_t}{P_0}$ and applying three functions in succession to obtain the calibrated date τ:

$$t = h\left(\frac{P_t}{P_0}\right), \text{ then } T = g(t) \text{ and then } \tau = f(T) \,.$$

Use the functions you have found to re-solve problem R10.19 algebraically.

R10.23: Find the inverse functions $h^{-1}(t)$, $g^{-1}(T)$, $f^{-1}(\tau)$. Use these to re-solve problem R10.20 by:

$$T = f^{-1}(\tau)$$
$$t = g^{-1}(T)$$
$$\frac{P_t}{P_0} = h^{-1}(t).$$

Notice the order in which we apply the inverse functions, compared with that in R10.22.

R10.24: We could write the dating process in R10.22 in a single function Φ:

$$\tau = \Phi\left(\frac{P_t}{P_0}\right), \text{ where } \Phi\left(\frac{P_t}{P_0}\right) = f\left(g\left(h\left(\frac{P_t}{P_0}\right)\right)\right).$$

That is, Φ is the composition of three functions (see Section 5.6.3):

$$\Phi = f \circ g \circ h.$$

Compose the three functions to obtain an expression giving $\tau = \Phi\left(\dfrac{P_t}{P_0}\right)$.

Invert this to find an expression for $\Phi^{-1}(\tau)$.

How could you write the inverse function Φ^{-1}, as a composition of the inverse functions h^{-1}, g^{-1}, f^{-1}? Compose these in the correct order, and obtain the same expression for $\Phi^{-1}(\tau)$.

Historical interlude: finding the roots of polynomials

In Section 7.2.2 we saw the formula (7.10) for the roots of a general quadratic function, i.e. for the solutions of $ax^2 + bx + c = 0$. You may wonder if there is a method equivalent to (7.10) for finding the roots of a general cubic, i.e. the solutions of $ax^3 + bx^2 + cx + d = 0$. The answer is 'Yes', although it's a bit more complicated than a single formula; you have to follow an algorithm or recipe involving linear transformations to get the polynomial into a special form. What about finding the roots of a quartic, or higher-degree polynomials? These higher-degree polynomials look as simple as the quadratic, just a bit longer (see Section 7.4). And roots certainly exist; you can see them when you sketch the graphs of the functions (the roots are at the points where they cross the x-axis).

The story of the quest for methods for solving polynomial equations is a fascinating saga, going back 4,000 years and involving some of the most colourful mathematicians in history. It involves poetry, hedonism, financial ruin, doomed love, politics and revolution, and tragic and violent deaths. I am including some of the highlights here, to provide a breather before Part II, and in the hope that seeing some of that maths in its historical context will make it more meaningful to you. If you want, just skip ahead to Part II. But if you believe that the phrase 'colourful mathematicians' is a contradiction in terms, read on!

The Babylonians knew an algorithm for finding the roots of a quadratic. Stewart (2007) cites a Babylonian clay tablet, some 4,000 years old, with cuneiform writing which was translated in the 1930s as 'Find the side of a square if the area minus the side is 14,30'. The Babylonians used a number system with base 60, so the number 14,30 means $14 \times 60 + 30$, i.e. 870. (It is from the Babylonian system that we have 60 seconds in a minute, 60 minutes in an hour – and 360 degrees in a circle, which we used in Chapter 8.) So the tablet is asking the reader to solve the quadratic equation $x^2 - x - 870 = 0$. It goes on to provide an algorithm for doing this, obtaining the answer: $x = 30$.

To find a mathematician who was able to solve cubic equations, we need to move forward 3,000 years, to the Persian poet and mathematician Omar Khayyam (1048–1122). When not writing hedonistic poetry (including the famous Rubaiyat) with lines such as 'Enjoy wine and women and don't be afraid, God has compassion', Khayyam classified cubic equations into 14 different types, and described geometrical constructions involving intersecting lines, circles, and parabolas which would locate their roots.

But Omar Khayyam could not convert these geometric constructions into algebraic formulas such as equation (7.10). Credit for producing algebraic 'recipes' for finding the roots of cubic and quartic equations, is usually given to Girolamo Cardano (1501–1576), although much of the work was done by his students and protegés. Cardano, the son of a lawyer, frittered away his inheritance through his addiction to gambling. He was also a violent man; he once slashed a man's face with a knife when he suspected him of cheating. He would prowl the city of Padua at night, wearing a hood and carrying his sword (illegal under the Offensive Weapons Act of the day), looking for trouble. When he had lost all his money and his family were in the alms house, he suffered the final humiliation: he became a university lecturer.

Cardano's formula (originally developed by his protégé del Ferro) for the roots of the cubic function $p(x) = x^3 + ax - b$, is

$$x = \sqrt[3]{\frac{b}{2} + \sqrt{\frac{a^3}{27} + \frac{b^2}{4}}} + \sqrt[3]{\frac{b}{2} - \sqrt{\frac{a^3}{27} + \frac{b^2}{4}}}.$$

We checked that this was a root, in the last question in problem 1.66 at the end of Chapter 1. But, you say, $p(x)$ isn't the general cubic – there isn't a term in x^2. Actually, any cubic can be reduced to this form using a linear transformation: a horizontal shift (see Section 6.5.4). We consider $p(x - k)$ where the constant k is chosen to make the new coefficient of x^2 zero when we multiply out. Find the roots of that polynomial, then add k to them to get the roots of the original one (we saw an example of this in problem 7.30). In a further development, Cardano's student Ferrari developed a similar algorithm for finding the roots of a quartic equation; everyone assumed that a solution to the quintic would soon follow. It is worth mentioning that these Italian Renaissance mathematicians were not developing their formulae for any practical reason; the point was to gain prestige by solving equations that no-one else was able to. Like modern athletes, they would compete in solving problems at public events, for prize money. They therefore guarded their methods jealously, and entered into bitter and protracted disputes about who first thought of a technique.

But efforts to find a solution to the general quintic equation proved fruitless. The first intimation that something was wrong came in 1770. The Italian-French mathematician Joseph-Louis Lagrange (1736–1813) developed a method for solving the cubic equation by transforming it into a so-called 'auxiliary equation' which was a quadratic, so could be solved by (7.10), and those roots transformed back into the roots of the cubic. This technique also worked for the general quartic equation, since its auxiliary equation turned out to be a cubic. But when the method of 'Lagrange resolvents' was applied to the quintic equation, the auxiliary equation turned out to be not a quartic but a sextic (a polynomial of degree 6). In 1790, the new French revolutionary government set up a commission to devise a rational scientific system of weights and measures, and Lagrange was appointed its president. He is therefore largely responsible for the adoption of the metric system we described in Chapter 2.[1]

After Lagrange's work, mathematicians started to suspect that the general quintic equation could not be 'solved by radicals', i.e. solved with a formula involving square roots and cube roots. This was finally proved in 1823 by Niels Abel (1802–1829), a university student in Oslo, Norway. In the same year he fell in love with Christine (Crelly) Kemp, and they were determined to marry. But this was impossible because neither of them had any money; Abel needed to find a job. Abel's friends tried to help him find a university position, and he made a research visit to Paris, but nothing came of it. When he returned to Norway to be with Crelly, he developed

[1] In a further burst of 'decimania', the government also introduced in 1793 a new system of measuring time, with 100 seconds in a minute, 100 minutes in an hour and 10 hours in a day. For some reason this didn't catch on, and it was abandoned two years later.

tuberculosis. Crelly nursed him, but he became ever weaker. On 8 April 1829 a letter was sent to him offering him a professorship at the University of Berlin: something which would solve all their problems. But Abel never saw it: he had died two days before, aged just 26.

The final chapter in this story is the most mathematically significant and the most dramatic. It involves another student who died young and tragically. Evariste Galois was born in France in 1811. By the time he finished his schooling, France had become a monarchy again. Galois was a lad with revolutionary leanings and a short temper. He failed his entrance exam to the Ecole Polytechnique after throwing a blackboard duster at the examiner. He did pass the exam for the less-prestigious Ecole Preparatoire, but was expelled after publishing a letter in the press denouncing the college's director (who had prohibited the students from taking part in popular demonstrations against the king). Galois joined an armed revolutionary group within the National Guard, while earning his living as a private tutor of algebra. In 1831 Galois was arrested for publicly threatening the life of King Louis-Philippe while waving a dagger at a drunken party. The jury acquitted him, but one month later he was arrested leading a revolutionary demonstration, armed with a knife, pistols, and a rifle, on Bastille Day. He and his comrade Ernest Duchâtelet were thrown in prison. He spent ten months in prison, after which the facts become very hazy. He fell in love with a young girl, but the affair did not go well. After the break-up, Galois was challenged to a duel. On 30 May 1832 the two young men met and fired at each other at close range with pistols – only one of which was loaded. There have been suggestions that the duel was orchestrated by Galois' enemies, or by the secret police, to eliminate a troublemaker. But it is now thought that Galois' opponent was his comrade Duchâtelet. In a biography, Rothman (1982) concluded: 'We arrive at a very consistent and believable picture of two old friends falling in love with the same girl and deciding the outcome by a gruesome version of Russian roulette'. Galois was hit in the stomach, and died in hospital of peritonitis the next day.[2]

On the eve of this duel, Galois sent a close friend of his a bundle of mathematical papers, with a letter: 'Later there will be, I hope, some people who will find it to their advantage to decipher all this mess.' The documents were an explanation of the theory involved in a paper Galois had written, titled 'On the conditions of solubility of equations by radicals'. He had been trying for years to get versions of this paper published, but they had either been lost or were rejected as 'incomprehensible'. In this paper Galois had in fact found the answer: it is impossible to find formulae for the roots of the general polynomial equation of degree 5, or any higher degree. He proved this by taking the problem to a new level of abstraction. A cubic polynomial can have three roots, and we can form different permutations of these roots. Two permutations can be multiplied to form a new permutation. But there is only a finite number (six) of such permutations, so we have an abstract mathematical structure, now known as a **group**, with six elements and a 'multiplication table'. A quintic polynomial has five roots, and its group will involve 120 permutations. By looking at the internal structure of the multiplication table (in particular, whether the group can be split into subgroups), the theory tells us whether it is possible for the roots to be written as algebraic expressions involving square roots and cube roots, i.e. whether the equation is 'solvable by radicals'. This concept of a group became the basis of **Group Theory**, a mathematical description of structure and symmetry which is now a cornerstone of pure mathematics. Group theory is also used in particle physics to classify the different types of sub-atomic particles and the results of their interactions. Chemists use it to classify crystal structures and the symmetries of molecules. And

---

[2] When he died in his brother's arms, his last words were 'Don't cry, Alfred, it takes all the courage I can muster to die at the age of twenty.'

group theory now provides the basis for modern algorithms of public key cryptography: ways of exchanging secret information without the need for codebooks or agreed keywords (see final problems of Chapter 1), which are used for secure transmission of information over the Internet. The 'mess' left behind by this disruptive student when he threw away his life at the age of 20, was perhaps the greatest single contribution to mathematics since the time of Newton.

If you want to read more of the history of mathematics, especially of equation-solving and symmetry, then Stewart (2007), du Sautoy (2009), and Singh (2002) are excellent starting-places.

Online, you can find biographies of the mathematicians involved on Wikipedia and at http://www-history.mcs.st-andrews.ac.uk/. You can listen to podcasts of talks about symmetry at Professor Stewart's website: http://www2.warwick.ac.uk/newsandevents/podcasts/media/more/symmetry, and an 'In our Time' programme on the subject with Melvyn Bragg at http://www.bbc.co.uk/programmes/b00776v8.

Links are on the CoreMaths website.

Problem: Given the general cubic polynomial $p(x) = ax^3 + bx^2 + cx + d$, find the value of k in terms of a, b, c and d, so that $p(x - k)$ has no term in x^2.

Answer: $k = \dfrac{b}{3a}$.

CALCULUS AND DIFFERENTIAL EQUATIONS

11 Instantaneous rate of change: the derivative   337

12 Rules of differentiation   366

13 Applications of differentiation   389

14 Techniques of integration   426

15 The definite integral   451

16 Differential equations I   481

17 Differential equations II   509

18 Extension: dynamical systems   537

Instantaneous rate of change: the derivative

In the Introduction to this book we suggested a contrast between traditional nineteenth century biology, which explored the **structure** of Nature (by collecting and classifying specimens, and detailing their anatomy) and twentieth century biology which concentrated on **processes** (population growth, spread of disease, growth and development of an organism). All processes involve change, and change happens at a finite rate, so the concept of the **rate of change** is central to modern biology. There is a powerful mathematical rate-of-change theory called the **calculus,** the foundations of which were laid by Newton[1] in Cambridge and Leibniz[2] in Hanover (the first publication was in 1684 by Leibniz[3]). It has been used for centuries by engineers and physicists to solve practical problems, and now it has become an essential tool of modern biology. There are two branches of the theory: **differential calculus** (Chapters 11, 12, 13) and **integral calculus** (Chapters 14, 15), and we shall need both branches in Chapter 16, where we see how to model natural processes by **differential equations**.

11.1 Introduction to the calculus

A rate of change involves two variables: an independent variable and a dependent variable. For example, the velocity of a moving object is the rate of change of its position (dependent variable) over time (independent variable). In most of the applications we will see, time t is the independent variable, but this is not always the case. In the Michaelis–Menten equation describing an enzyme-catalysed biochemical reaction (Section 8.3) the rate of reaction v (which is itself a rate of change with respect to time) depends on the substrate concentration s. When s is small, a small increase in s causes a relatively large increase in the reaction rate, but when there is plenty of substrate present the same increase in s will have little effect on the

---

[1] Sir Isaac Newton (1643–1727), English physicist, mathematician, astronomer, natural philosopher, alchemist, and theologian.

[2] Gottfried Leibniz (1646–1716), German mathematician and philosopher.

[3] Leibniz and Newton argued bitterly about who originated the calculus. Newton claimed to have been developing the idea since 1666, but did not publish his theory. Mathematicians have continued the argument ever since; see the Wikipedia article on the 'Leibniz and Newton calculus controversy'. The lesson for academics everywhere is 'If you don't publish your findings, you won't get the credit for them'.

reaction rate. The question is: can we describe mathematically how a change in substrate concentration will affect the reaction rate – that is, can we find a formula for the rate of change of v with respect to s? This is where the power of calculus comes to the fore.

In fact, whenever we have a dependent variable y that is a function of an independent variable x, we can think about the rate of change of y with respect to x. By this we mean: how much does y change when we make a small change to the current value of x? The answer is a ratio: the change in y compared to a small change in x. It will also depend on the current value of x, as we saw in the reaction example above. Thus, the rate of change will itself be a function of x. In Chapters 6–10 of Part I, we saw how the theory of functions can be used to model the processes of Nature; now we have an exciting new concept to apply to those functions.

11.1.1 Differential calculus

The definition of rate of change in terms of a dependent variable y and an independent variable x seems rather abstract, so let's develop it using a practical example. A car starts at 12 noon from rest at a motorway service station and drives north along the motorway. Let s miles be the distance travelled after t hours have elapsed. So s is a function of t; we can write this as $s = s(t)$. At a particular time t_0, the car has a position $s(t_0)$. (This notation means 's evaluated when $t = t_0$'. If you're unsure about function notation, look back to Section 5.1.) But the car is also travelling at a particular speed v at that time. The speed of the car is the rate of change of distance s with respect to time t, measured in units of miles per hour (as we're on a UK motorway).

This speed v varies with time, so it is also a function of t, just like $s(t)$. To indicate this, we will write it as $v(t)$. The starting-point of the differential calculus is the question: if I know a function giving the position of the car at every instant of time, can I work out what its speed is at any specified instant of time? That is, if I know the function $s(t)$, how can I work out the function $v(t)$? It is important to distinguish between $v(t)$, which is the speed of the car at a particular *instant of time t*, and the average speed of the car over an *interval of time* from t_0 to t_1. We say that $v(t)$ represents the **instantaneous rate of change** of s with respect to t.

We have already seen in equation (3.6) how to calculate average speed. Suppose that at 3 p.m. (i.e. when $t = 3$), our car driver enters a section of motorway roadworks with a 50 mph speed limit. The roadworks are located 200 miles from the service station where he started, and are 15 miles long. Our car reaches the end of the roadworks at 3:20 p.m. The roadworks are monitored by an 'average speed check' system, with cameras at the start and end of the roadworks section, which record car number plates and the times they pass. They are connected to a computer that calculates each car's average speed through the section. In our case, the first camera is located at $s_0 = 200$ miles from the service station, and records our car passing at time $t_0 = 3.0$ hours after noon; the second camera is located at $s_1 = 215$ miles, and records our car at time $t_1 = 3.3333$ hours (3:20 p.m. in decimals). The computer calculates the distance interval as $\Delta s = s_1 - s_0 = 215 - 200 = 15$ miles, and the time interval as $\Delta t = t_1 - t_0 = 3.3333 - 3.0 = 0.3333$ hours. The average speed over the section is then

$$\text{average speed} = \frac{\Delta s}{\Delta t} = \frac{s_1 - s_0}{t_1 - t_0} = \frac{215 - 200}{3.3333 - 3.0} = \frac{15}{0.3333} = 45 \text{ mph} \tag{11.1}$$

which is within the 50 mph speed limit. The calculation is shown graphically in Figure 11.1(a). The average speed is the slope of the line connecting the points (t_0, s_0) and (t_1, s_1). This is the

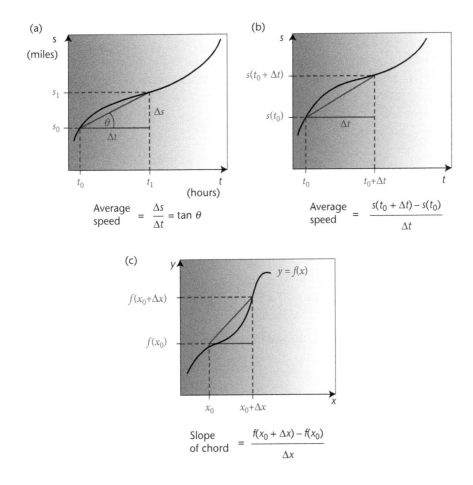

Figure 11.1 Average speed calculations

vertical distance divided by the horizontal distance, or 'rise over run', in the right-angled triangle ABC (see Section 6.3.3); it is also the tangent of the angle θ (see Section 9.8.2). Remember that 'Δs' means 'a difference between two s-values', or a finite-length interval in the s-direction on the graph; it is a single quantity, you can't split up the Δ from the s.

But this doesn't mean that the car was within the speed limit throughout its journey through the roadworks. In fact, there was also a GATSO roadside speed camera placed at the start of the roadworks. The driver sees a double-flash in his rear-view mirror: the camera takes a photograph as the rear of the car drives over the calibrations painted on the road and a second photograph 0.5 seconds later. The police work out the speed of the car by measuring the distance moved along the calibrations between the two photographs, and using exactly the same average speed formula $\frac{\Delta s}{\Delta t}$ as in (11.1). But now the time interval $\Delta t = 0.5$ seconds (or 0.14×10^{-3} hours) is much much shorter, so the average speed from this calculation is much much closer to the instantaneous speed $v(t)$ shown on the speedometer on the car dashboard at time $t = t_0$. Our driver was travelling at 60 mph when he entered the roadworks, and slowed gradually to 30 mph by the end, giving an average speed of 45 mph over the section, but the GATSO camera calculation gave a speed of 59.9 mph and he was prosecuted. The judge rejected my argument (you guessed I wasn't making all this up, didn't you?) that mathematically the concept of average speed was different from that of instantaneous speed; she said that if the average speed is measured over a small enough time interval, then it is virtually identical to the instantaneous speed at the start of the interval.[4]

In calculus, there's a special notation for the instantaneous rate of change. If we have variables x and y, where $y = y(x)$ (meaning that y is a function of x), then we can talk about the instantaneous rate of change of y with respect to x. The notation for this rate of change is $\frac{dy}{dx}$ (pronounced 'dee y dee x').

> The calculus notation for the instantaneous rate of change of y with respect to x, is $\frac{dy}{dx}$.

It's important always to remember that **this notation does not mean 'dy divided by dx'**. It's **a single notation, for the rate of change**. For this reason I prefer to read $\frac{dy}{dx}$ as 'dee y dee x' rather than 'dee y by dee x' or 'dee y over dee x', which you may also hear people say – and to say it quickly, without any pause between 'dee y' and 'dee x'!

This will be clearer if we go back to our speeding car example. There, we wrote $s = s(t)$ to mean that distance is a function of time. In calculus, the instantaneous rate of change of distance with respect to time, i.e. the instantaneous speed, will be written as $\frac{ds}{dt}$ (pronounced 'dee s dee t'). This is what we've been writing up till now as $v(t)$. So $\frac{ds}{dt} = v(t)$. Now you can see where the notation comes from: it's based on the average rate of change $\frac{\Delta s}{\Delta t}$, with the Greek letter 'delta' changing to 'd' to indicate this is a calculus quantity. In the next section we will see how to work out the velocity (speed) $\frac{ds}{dt}$ from knowing the function $s(t)$. We do it just like the judge said: we work out the average speed $\frac{\Delta s}{\Delta t}$ over an interval Δt, and make Δt smaller and smaller until, in the limit as Δt approaches zero, we get the instantaneous speed. We saw this process demonstrated graphically in Section 9.8.4. The process of finding $\frac{ds}{dt}$ from $s(t)$ is called **differentiation**; we say that we differentiate $s(t)$ to get $\frac{ds}{dt}$. The instantaneous rate of change $\frac{ds}{dt}$ is called the **derivative** of $s(t)$ with respect to t.

This is how $\frac{ds}{dt}$ is defined, but it would be impossibly laborious if we had to go through this process for every function. Instead, we find the derivatives of standard functions such as x^n, e^x, etc., together with rules for building up the derivatives of combinations of standard functions (such as the product $x^n.e^x$). For example, if a car makes a 96-mile journey and after t hours it has travelled s miles, where $s(t) = 6t^3(8 - 3t)$, then after reading this chapter you will be able to work out that its speed at any time t is given by $\frac{ds}{dt} = 72t^2(2 - t)$. In Chapter 12 we'll see the rules of differentiation, and the derivatives of trigonometric and logarithmic functions. This material forms the **differential calculus**.

11.1.2 Integral calculus

What if we know the car's speed at any time: can we work out $s(t)$? If a car travels at a constant speed of V miles per hour, then we can easily work out the distance travelled: after t hours it

---

[4] She could also have cited the **mean value theorem**, which in the context of distance/time graphs states that if you work out the average speed over a time interval, there will be at least one moment within that interval when the instantaneous speed is equal to the average speed. So, if a car is recorded as having an average speed of 60 mph, there must have been at least one moment when its instantaneous speed was 60 mph.

will have travelled $s = V.t$ miles. If its speed is not constant, but is given as a function of t, it is also possible to use calculus to find $s(t)$. This process is called **integration**; it is the reverse process to differentiation. We will study integration in Chapters 14 and 15. Again, we will see the integrals of standard functions, as well as techniques for integrating more complicated expressions. For example, the new VW Polo can accelerate from 0 to 60 mph in 16 seconds. Assuming constant acceleration over that period, the car's speed as a function of time is $\frac{ds}{dt} = 13500t$ (s in miles, t in hours). Once you have read Chapter 14, you will be able to prove this, and find its distance function as $s(t) = 6750t^2$, from which you can deduce how far it will have travelled in those 16 seconds (see Section 14.1.3). This material forms the **integral calculus**.

11.1.3 Differential equations

Along the way, we will see some important applications of calculus. But the most powerful use of calculus in the biosciences is in solving differential equations. A **differential equation** is simply an equation involving the derivative of a function as well as the function itself. We have already met such equations in Case Study A, as models of population growth. In Section 10.1.2 we defined **exponential growth** to mean that at any time t, the rate of increase of the population is proportional to its current size. Using our calculus notation, the rate of growth of a population $N(t)$ with respect to time t, can be written as $\frac{dN}{dt}$. The exponential growth model is then defined by the differential equation

$$\frac{dN}{dt} = rN. \tag{11.2}$$

Solving this differential equation, means finding a function $N(t)$ that satisfies the equation. We stated the solution in Chapter 10:

$$N(t) = N_0 e^{rt} \tag{10.9}$$

where N_0 is the initial population size, i.e. $N(0) = N_0$; this is called an **initial condition**.

Logistic growth (see Section 7.4.2) is also modelled by a differential equation. It is more complicated than (11.2) because it involves an extra factor $\left(1-\frac{N}{K}\right)$ that tends to zero as $N \to K$. This causes the rate of growth $\frac{dN}{dt}$ to slow down as the population N approaches its maximum sustainable size, the carrying capacity K. Here is the differential equation, which we'll see how to solve in Chapter 16:

$$\frac{dN}{dt} = r_m\left(1 - \frac{N}{k}\right).N \tag{11.3}$$

where r_m is the net growth rate. We met the solution in Case Study A10, namely equation (10.37).

Differential equations are central to modern biosciences, as well as to engineering, physics, etc. When epidemiologists come up with predictions for the likely course of a new strain of influenza through the population, they are not thinking up some complex mathematical formula for the number of cases of the disease at time t. Rather, they model the **instantaneous rate at which the disease is spreading**, as a function of the number of interactions between infected and susceptible individuals, and the mortality and recovery rates. This model, encapsulated in a differential equation, is of little use unless it can be **solved** to produce a graph plotting the extent of the disease over time. In Chapters 16 and 17 we will see some standard types of differential equation, and methods for solving them.

All this may sound very theoretical and too mathematical. But we will also continue our presentation of numerical methods alongside the theory, including simple but very powerful algorithms for solving differential equations numerically. We will program these algorithms in Excel spreadsheets.

Notation: We will sometimes write dy/dx and sometimes $\dfrac{dy}{dx}$. The former emphasises that this is not 'dee y divided by dee x', but they both mean exactly the same thing.

To end this introduction to calculus, here is another instalment of Case Study B, which we last met in Chapter 7. We can use our new concept of the derivative to build the set of differential equations modelling the growth of a tumour. Skip this for now, if you want to get started on how to work out derivatives.

CASE STUDY B8: Constructing the angiogenic tumour model

Let us use (11.2) and (11.3) to build the mathematical model for the growth of an angiogenic cancer tumour, which we have used in previous instalments of this Case Study. The model is proposed in Wodarz and Komarova (2005), Section 9.1. Starting from the exponential-growth model (11.2), and following the approach of Part 3 of Case Study A we split up the net growth rate r as the birth rate b minus the death rate d ($r = b - d$):

$$\frac{dN}{dt} = rN = (b - d).N = b.N - d.N.$$

(Note that in '$d.N$', d is the death rate – nothing to do with the derivative!) In real population growth the birth rate is not constant, but density-dependent, so that population size is limited by a carrying capacity k (even with a zero death rate). We therefore replace the term $b.N$ using (11.3):

$$\frac{dN}{dt} = r_m\left(1 - \frac{N}{k}\right).N - d.N.$$

This is the basic model for cell growth in the tumour. In the tumour there are healthy cells, non-angiogenic cancer cells and angiogenic cancer cells; we write the densities of these types of cell as $x_0(t)$, $x_1(t)$, $y(t)$ respectively at time t. Our model will consist of three equations, describing the rates of change of these three quantities. Solving our model would give us the graph of $y(t)$ over time.

The differential equation above is used to model the growth of healthy cells:

$$\frac{dx_0}{dt} = r_0\left(1 - \frac{x_0}{k_0}\right).x_0 - d_0 . x_0 \tag{A}$$

where r_0, k_0, d_0 are the growth rate, carrying capacity and death rate for healthy cells.

The differential equation for non-angiogenic cancer cells is similar, except that a certain proportion μ_0 of the replicating healthy cells may mutate into cancer cells. This gives an extra positive term on the right-hand side of our second equation:

$$\frac{dx_1}{dt} = r_1\left(1 - \frac{x_1}{k_1}\right).x_1 - d_1.x_1 + \mu_0.r_0\left(1 - \frac{x_0}{k_0}\right).x_0 \qquad (B)$$

where r_1, k_1, d_1 are the growth rate, carrying capacity, and death rate for the non-angiogenic cancer cells.

In turn, a certain proportion μ_1 of non-angiogenic cancer cells mutate into angiogenic cells. There is also an additional negative term on the right-hand side of the differential equation we will write for y. This reflects the balance between angiogenesis promoters and inhibitors produced by the cells. All cells (including cancer cells) produce inhibitors, but angiogenic cancer cells also produce promoters (see instalment B2 at the start of Chapter 1). Wodarz and Komarova model this situation using the term

$$-\frac{(p_0 x_0 + p_1 x_1 + p_2 y)}{(qy + 1)}\, y.$$

Increases in x_0 and x_1 will increase this death rate, while an increase in y contributes on both the numerator and denominator, modelling both an inhibiting and a promoting effect. Clearly a key question is the relative magnitudes of the coefficients p_0, p_1, p_2, q. The final equation for our model is thus

$$\frac{dy}{dt} = \mu_1.r_1\left(1 - \frac{x_1}{k_1}\right).x_1 + r_2\left(1 - \frac{y}{k_2}\right).y - d_2.y - \frac{(p_0 x_0 + p_1 x_1 + p_2 y)}{qy + 1}\, y. \qquad (C)$$

Equations (A)–(C) comprise our model of time-dependent tumour growth. The differential equations need to be solved to find $x_0(t)$, $x_1(t)$, $y(t)$: the densities of healthy, non-angiogenic and angiogenic cells in the tumour. For this we would need some **initial conditions**: the initial densities $x_0(0)$, $x_1(0)$, $y(0)$. These may be obtained by analysing the **equilibrium state**. This is a situation in which the cell densities are constant $\left(\text{so that } \dfrac{dx_0}{dt} = \dfrac{dx_1}{dt} = \dfrac{dy}{dt} = 0\right)$ and there are no mutations (i.e. $\mu_0 = \mu_1 = 0$). Setting the left-hand side of (A) to zero, we get

$$\left[r_0\left(1 - \frac{x_0}{k_0}\right) - d_0\right].x_0 = 0.$$

Then either $x_0 = 0$ (meaning there is no tumour!), or $r_0\left(1 - \dfrac{x_0}{k_0}\right) - d_0 = 0$, which can be solved to find the equilibrium density of healthy cells as

$$x_0 = \frac{k_0}{r_0}(r_0 - d_0).$$

Similarly, inserting $\dfrac{dx_1}{dt} = 0$ and $\mu_0 = 0$ in equation (B), the non-zero equilibrium density of non-angiogenic cancer cells is

$$x_1 = \frac{k_1}{r_1}\left(r_1 - d_1\right).$$

Finally, inserting $\dfrac{dy}{dt} = 0$ and $\mu_1 = 0$ in equation (C), we have at equilibrium that

$$r_2\left(1 - \frac{y}{k_2}\right).y - d_2.y - \frac{\left(p_0 x_0 + p_1 x_1 + p_2 y\right)}{q.y + 1}\,y = 0.$$

This is precisely the equation we wrote in Part B6 of this Case Study, in Chapter 7 (though we simplified it by assuming that the angiogenesis inhibition coefficients p are equal: $p_0 = p_1 = p_2$). To find the non-zero solution we divided through by y, then manipulated the equation to produce a quadratic equation in y, which we solved using the formula (7.10). As we are looking for positive roots, we take the $+$ sign in the formula. The root we obtained was

$$y = \frac{-Q + \sqrt{Q^2 - 4r_2.qk_2[(d_2 - r_2) + p(x_0 + x_1)]}}{2r_2.q}$$

where

$$Q = k_2.q(d_2 - r_2) + r_2 + k_2.p.$$

If $(d_2 - r_2) + p(x_0 + x_1) < 0$, i.e. if $p(x_0 + x_1) < r_2 - d_2$, then $y > 0$. Interpreting this, the angiogenesis inhibition is too weak, and there will be angiogenic cancer cells in the tumour.

11.2 Definition of the derivative

Let's stay with the example of the car travelling through roadworks. The average speed calculation is shown in (11.1). To find the instantaneous speed at time t_0 we use a **limiting process**: we gradually reduce the length of the time interval Δt, and as it approaches zero the average speed $\dfrac{\Delta s}{\Delta t}$ approaches the value of the instantaneous speed $\dfrac{ds}{dt}$. We presented this limiting process in Section 9.8; have a look back at this before proceeding.

This limiting process is written mathematically as:

$$\frac{ds}{dt} = \lim_{\Delta t \to 0}\left(\frac{\Delta s}{\Delta t}\right) \tag{11.4}$$

which we read as '$\frac{ds}{dt}$ is the limit of $\frac{\Delta s}{\Delta t}$ as Δt tends to zero'. We introduced the idea of limits in Section 6.4. A reminder: while $\frac{\Delta s}{\Delta t}$ means 'Δs divided by Δt', the variable $\frac{ds}{dt}$, read as 'ds (by) dt', is a single concept, which you shouldn't split up into ds and dt. The expression $\frac{ds}{dt}$ means '**the derivative of s with respect to t**', not 'ds divided by dt'.

The process is illustrated in SPREADSHEET 11.1, which we saw in Section 9.8.4. There's a screenshot in Figure 11.2. Drag the scrollbar button from right to left, to move the right-hand end of the time interval closer to t_0; the slope of the secant line (drawn in red) joining the two points on the curve, gets closer to that of the tangent line at $t = t_0$ (drawn in green).

At the start of the time interval, at time t_0, the value of s is $s(t_0)$. The time at the right-hand end of the interval is $t_0 + \Delta t$, so the value of s at that time is $s(t_0 + \Delta t)$. The difference in s-values is therefore $\Delta s = s(t_0 + \Delta t) - s(t_0)$. Using this expression in (11.4) we obtain a more useful definition of the derivative at time t_0:

$$\frac{ds}{dt} = \lim_{\Delta t \to 0} \left(\frac{s(t_0 + \Delta t) - s(t_0)}{\Delta t} \right). \tag{11.5}$$

This is illustrated in Figure 11.1(b).

To see this used in a numerical calculation, let's take $t_0 = 3$, and the simple function $s(t) = t^2$ (obtained by putting $u = 0$ and $a = 2$ in SPREADSHEET 11.1). If we take $\Delta t = 1$, the average speed over the interval $3 < t < 4$ is:

$$\frac{s(3 + \Delta t) - s(3)}{\Delta t} = \frac{s(3 + 1) - s(3)}{1} = \frac{4^2 - 3^2}{1} = 16 - 9 = 7 \text{ mph}.$$

Now take $\Delta t = 0.5$; the average speed over the interval $3 < t < 3.5$ is

$$\frac{s(3 + 0.5) - s(3)}{0.5} = \frac{3.5^2 + 3^2}{0.5} = \frac{12.25 - 9}{0.5} = \frac{3.25}{0.5} = 6.5 \text{ mph}.$$

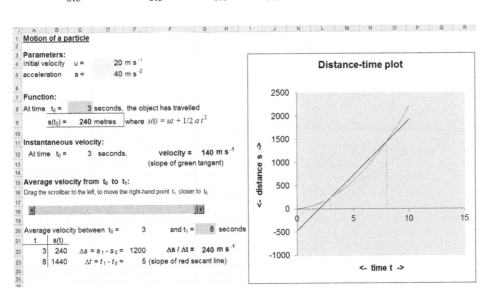

Figure 11.2 The derivative as a limiting process (SPREADSHEET 11.1)

Finally, take $\Delta t = 0.1$; the average speed over the interval $3 < t < 3.1$ is

$$\frac{s(3+0.1)-s(3)}{0.1}=\frac{3.1^2-3^2}{0.1}=\frac{9.61-9}{0.1}=\frac{0.61}{0.1}=6.1\,\text{mph.}$$

It looks as if the average speed is approaching 6 mph as $\Delta t \to 0$. But we can't check this by simply putting $\Delta t = 0$:

$$\frac{s(3+0.0)-s(3)}{0.0}=\frac{3^2-3^2}{0.0}=\frac{9-9}{0.0}=\frac{0}{0}=?$$

We can however confirm it by using an algebraic identity from Chapter 1. In (1.20) we saw that $(a+b)^2 = a^2 + 2ab + b^2$. So

$$s(3+\Delta t)=(3+\Delta t)^2=3^2+2(3)(\Delta t)+(\Delta t)^2=9+6.\Delta t+(\Delta t)^2.$$

The numerator of the fraction in (11.5) is then

$$s(3+\Delta t)-s(3)=(3+\Delta t)^2-3^2=\left(9+6.\Delta t+(\Delta t)^2\right)-9=6.\Delta t+(\Delta t)^2.$$

Now that the 9s have cancelled, there is a common factor of Δt in the remaining two terms. This will cancel with the Δt on the denominator of the fraction:

$$\frac{s(3+\Delta t)-s(3)}{\Delta t}=\frac{(3+\Delta t)^2-3^2}{\Delta t}=\frac{9+6.\Delta t+(\Delta t)^2-9}{\Delta t}$$

$$=\frac{6.\Delta t+(\Delta t)^2}{\Delta t}=\frac{\Delta t(6+\Delta t)}{\Delta t}=6+\Delta t.$$

Once we have cancelled the Δt and are left with the expression $6+\Delta t$, we are now able to substitute $\Delta t = 0$, giving us the value of 6 for the instantaneous rate of change at $t_0 = 3$.

The same algebraic trick will work for any value of t_0:

$$s(t_0+\Delta t)-s(t_0)=(t_0+\Delta t)^2-t_0^2=\left(t_0^2+2.t_0.\Delta t+(\Delta t)^2\right)-t_0^2=2.t_0.\Delta t+(\Delta t)^2.$$

Dividing by Δt gives

$$\frac{s(t_0+\Delta t)-s(t_0)}{\Delta t}=\frac{(t_0+\Delta t)^2-t_0^2}{\Delta t}=\frac{t_0^2+2t_0.\Delta t+(\Delta t)^2-t_0^2}{\Delta t}$$

$$=\frac{2t_0.\Delta t+(\Delta t)^2}{\Delta t}=\frac{\Delta t(2t_0+\Delta t)}{\Delta t}=2t_0+\Delta t$$

and substituting $\Delta t = 0$ we obtain $2t_0$ for the instantaneous rate of change $\frac{ds}{dt}$ when $t = t_0$. But t_0 could be any value of t. We can therefore write $\frac{ds}{dt}$ as a function of t, as $\frac{ds}{dt}=2t$. In fact, we have found our first derivative of a function:

$$\text{If}\quad s(t)=t^2,\quad\text{then its derivative is}\quad \frac{ds}{dt}=2t. \tag{11.6}$$

11.3 Differentiating polynomial functions

We'll now switch from using s and t, to the standard mathematical notation for functions, as used in Chapters 5–10, in which x is the independent variable, and y is the dependent variable. So y is a function of x, which we can write as $y = y(x)$. The instantaneous rate of change of y with respect to x, i.e. the **derivative of y with respect to x**, will then be denoted by $\frac{dy}{dx}$. The definition of the derivative which we gave in (11.4) can now be rewritten as

$$\frac{dy}{dx} = \lim_{\Delta x \to 0} \left(\frac{\Delta y}{\Delta x} \right). \tag{11.7}$$

If we draw a graph of the function as a smooth curve, then $\frac{dy}{dx}$ tells us the slope of the tangent line to the curve for any chosen value of x. We have seen this illustrated in Section 9.8.3, where we imagined the curve as a rollercoaster track, and the tangent line as the longitudinal axis of the car which travels along it. The process is animated in SPREADSHEET 11.2, shown in Figure 11.3.

The slope of the tangent to the curve $y = f(x)$ at a particular point is called the **gradient** of the curve at that point. So $\frac{dy}{dx}$ tells us the gradient of the function $f(x)$. The gradient is a generalization of the idea of slope (which applies to straight lines).

Look at the value of the gradient (in cell K32) as you move the car along the curve in SPREADSHEET 11.2. Notice that the gradient $\frac{dy}{dx}$ is positive when the car is travelling uphill and negative when the car runs downhill. It is zero when the car reaches the top or the bottom of a curve.

We have already found the derivative of one function. Writing (11.6) in our maths notation:

$$\text{If} \quad y(x) = x^2, \quad \text{then its derivative is} \quad \frac{dy}{dx} = 2x. \tag{11.8}$$

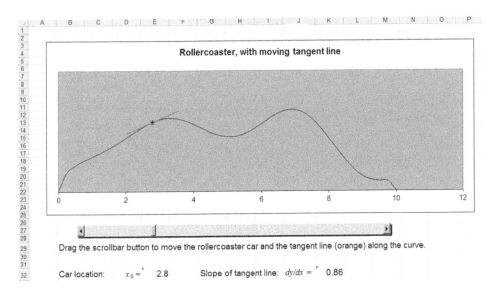

Figure 11.3 The gradient as the slope of the tangent (SPREADSHEET 11.2)

What does this mean in practice? Suppose you draw the graph of $y = x^2$. Choose any value of x, for example $x = 7$. Locate the point $(7, 0)$ on the x-axis, then move vertically up to meet the curve, which will be at the point $(7, 49)$. Draw the tangent to the curve at this point. The slope of the tangent line you have drawn will be $2 \times 7 = 14$. For any value of x, the gradient of the tangent line drawn at the point (x, x^2) on the curve will be $2x$. This is shown in Figure 11.4(a).

In the rollercoaster example in Section 9.8, we noted that at points where the function reaches a local maximum or minimum, the tangent to the curve is horizontal; we can now rephrase that by saying that the gradient is zero at such points. This gives us a way of solving optimization problems (see Section 8.6.2 for numerical algorithms): to find when $y(x)$ reaches a minimum or maximum, differentiate it and solve $\dfrac{dy}{dx} = 0$.

In the next section we'll obtain the derivatives of the power functions $y = x^3$, $y = x^4$, and so on.

11.3.1 The derivative of power functions $y = x^n$

In Chapter 5 we introduced $f(x)$ as the standard name for a function of x, which could be graphed by plotting or sketching the curve $y = f(x)$. We could rewrite the average-rate-of-change calculation (11.7) by considering an interval in the x-direction, starting at $x = x_0$, and of length Δx. When $x = x_0$, the value of y will be $f(x_0)$, and at the end of the interval the y-value will be $f(x_0 + \Delta x)$; see Figure 11.1(c). Then the basic definition of the derivative $\dfrac{dy}{dx}$ at the point $x = x_0$, from (11.5), becomes

$$\frac{dy}{dx} = \lim_{\Delta x \to 0} \left(\frac{f(x_0 + \Delta x) - f(x_0)}{\Delta x} \right). \tag{11.9}$$

We can use this definition to find the derivative of $y = x^3$ by using an algebraic identity, just as in Section 11.2. Now we need from (1.20) the identity for $(a + b)^3$:

$$(a + b)^3 = a^3 + 3a^2b + 3ab^2 + b^3.$$

If you use this identity to expand $(x_0 + \Delta x)^3$, and follow the method of Section 11.2, you should find that after the terms x_0^3 and $-x_0^3$ cancel out on the numerator, you are again able to cancel a common factor of Δx on numerator and denominator, leaving you with

$$\frac{dy}{dx} = \lim_{\Delta x \to 0} \left(\frac{(x_0 + \Delta x)^3 - x_0^3}{\Delta x} \right) = \lim_{\Delta x \to 0} \left(3x_0^2 + 3x_0 \cdot \Delta x + (\Delta x)^2 \right).$$

Setting $\Delta x = 0$, we find that when $y = x^3$, its derivative is $\dfrac{dy}{dx} = 3x^2$. We have obtained the derivative of $f(x) = x^3$ **from first principles**, i.e. using the definition (11.9). Don't worry, you will not have to do all differentiation from first principles!

In Figure 11.4 you can see the graphs of $y = x^2$ and $y = x^3$ with their tangent lines. They are also featured in SPREADSHEET 11.3, where you can move the tangent lines along the curves using the scrollbar. Notice that for $y = x^2$, the tangent slope is negative when x is negative and positive when x is positive. For $y = x^3$, the tangent is always sloping upwards (so $\dfrac{dy}{dx}$ is positive), except at the origin. Here the tangent is actually horizontal (so $\dfrac{dy}{dx}$ is zero), although the

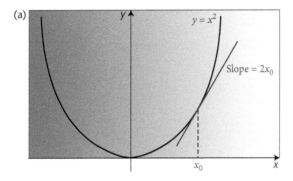

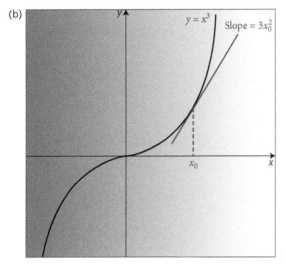

Figure 11.4 Derivatives of x^2 and x^3

point is not a maximum or minimum of the curve. It is called a **point of inflection**; we'll discuss this further in Section 13.1.4.

In fact, we could use the binomial theorem (which we are not covering in this book; it is summarized in Problem 11.4) to prove that in general

$$\text{If } y = x^n, \text{ then its derivative is } \frac{dy}{dx} = nx^{n-1}. \tag{11.10}$$

This is the first standard derivative which you must remember. So the derivative of x^4 is $4x^3$ and the derivative of x^{100} is $100x^{99}$ (using $n = 100$ in (11.10)). In FNGRAPH 11.1, you can see sketched the curves of $y = x^n$ (in blue) and its derivative function $\frac{dy}{dx} = nx^{n-1}$ (in green). Change the value of n in the 'variables' window. At the moment we're only considering when n is an integer; however the rule (11.10) applies also when n is a negative integer:

EXAMPLE

Differentiate $f(x) = \dfrac{1}{x^2}$ from first principles, and compare with the result from (11.10).

Solution:

$$f(x_0) = \frac{1}{x_0^2}, \text{ and so } f(x_0 + \Delta x) = \frac{1}{(x_0 + \Delta x)^2}. \text{ Then}$$

$$f(x_0 + \Delta x) - f(x_0) = \frac{1}{(x_0 + \Delta x)^2} - \frac{1}{x_0^2}$$

$$= \frac{x_0^2 - (x_0 + \Delta x)^2}{(x_0 + \Delta x)^2 x_0^2}$$

> Put over a common denominator

$$= \frac{x_0^2 - \left(x_0^2 + 2x_0.\Delta x + (\Delta x)^2\right)}{(x_0 + \Delta x)^2 x_0^2}$$

$$= \frac{x_0^2 - x_0^2 - 2x_0.\Delta x - (\Delta x)^2}{(x_0 + \Delta x)^2 x_0^2}$$

> The terms in x_0^2 cancel

$$= \frac{-2x_0.\Delta x - (\Delta x)^2}{(x_0 + \Delta x)^2 x_0^2} = \left(\frac{-2x_0 - \Delta x}{(x_0 + \Delta x)^2 x_0^2}\right)\Delta x.$$

> Now we can take out a factor of Δx

Now we can insert in (11.9):

$$\frac{dy}{dx} = \lim_{\Delta x \to 0}\left(\frac{f(x_0 + \Delta x) - f(x_0)}{\Delta x}\right) = \lim_{\Delta x \to 0}\left(\frac{-2x_0 - \Delta x}{(x_0 + \Delta x)^2 x_0^2}\right)$$

$$= \frac{-2x_0 - 0}{(x_0 + 0)^2 x_0^2} = \frac{-2x_0}{x_0^4} = \frac{-2}{x_0^3}.$$

So the derivative of $y = \frac{1}{x^2}$ is $\frac{dy}{dx} = -\frac{2}{x^3} = -2x^{-3}$.

You obtain the same result using rule (11.10) with $n = -2$.

The derivative of other basic functions can be found from first principles, i.e. using the definition (11.9). However, this usually requires some extra bits of algebra, which we haven't covered. As an example, consider the trigonometric function $y = \sin x$. Using (11.9), its derivative at x_0 is defined as

$$\frac{dy}{dx} = \lim_{\Delta x \to 0}\left(\frac{\sin(x_0 + \Delta x) - \sin(x_0)}{\Delta x}\right).$$

We could expand $\sin(x_0 + \Delta x)$ using the formula for $\sin(A + B)$ in (9.29):

$$\sin(x_0 + \Delta x) = \sin x_0 \cos \Delta x + \cos x_0 \sin \Delta x,$$

but to get much further we would need to know about how $\dfrac{\cos \Delta x - 1}{\Delta x}$ and $\dfrac{\sin \Delta x}{\Delta x}$ behave

for small angles Δx. As scientists, you need to be able to use calculus to solve problems, not to prove that Newton and Leibniz were right, so we will not worry too much about deriving all the standard derivatives and the rules you need to know, from first principles.

However, to get a proper feel for the subject, and to have confidence in what you are doing, you should see something of the structure and logic behind it, which is why you need to know where the definition (11.9) comes from, and how it can be used in some relatively simple cases.

11.3.2 Notation

Before going any further, we need to introduce some additional notation which will make it easier to talk about derivatives. We already did this with function notation: instead of writing a function of x by inventing a new dependent variable y, we can use a function $f(x)$. So instead of writing $y = x^4$ we can say $f(x) = x^4$. This enables us to talk about adding two functions together, as $f(x) + g(x)$. We can now talk about the derivative of f, rather than the derivative of y, writing it as $\dfrac{df}{dx}$ instead of $\dfrac{dy}{dx}$. Differentiation also has its own shorthand notation, using the 'prime' symbol (like an straight apostrophe) to indicate differentiation. We will sometimes write y' to mean $\dfrac{dy}{dx}$. Here is a summary of these notations:

'If y is the function of x given by $y = x^4$, then its derivative is $\dfrac{dy}{dx} = 4x^3$.' Here are three other ways in which we can express this fact:

- Instead of inventing the variable y, we use a function $f(x)$, and write $f(x) = x^4$. Then we can write the derivative as $\dfrac{df}{dx} = 4x^3$.

- We could combine the two equations $y = x^4$ and $\dfrac{dy}{dx} = 4x^3$ into one and dispense with y or f altogether, writing $\dfrac{d(x^4)}{dx} = 4x^3$, or alternatively $\dfrac{d}{dx}\left(x^4\right) = 4x^3$. You would read $\dfrac{d}{dx}(...)$ as 'the derivative with respect to x of ...', so we are saying 'the derivative of x^4 is $4x^3$'. This notation is neater if we have a long expression to differentiate. We will see shortly that you can differentiate long expressions term-by-term, for example

$$\frac{d}{dx}\left(x^4 + x^2 + x\right) = \frac{d}{dx}\left(x^4\right) + \frac{d}{dx}\left(x^2\right) + \frac{d}{dx}\left(x\right) = 4x^3 + 2x + 1.$$

- To avoid writing lots of 'dx's, there is a shorthand notation: we write $\dfrac{dy}{dx}$ as y' (which we read as 'y dash' or 'y prime'). This is okay provided it is clear what the independent variable is. So our statement would be written 'If $y = x^4$, then $y' = 4x^3$'. We could also use this shorthand with the function f: 'If $f(x) = x^4$, then $f'(x) = 4x^3$'. We have also written $y = y(x)$ to show that y depends on x, so we could write $y'(x)$ for its derivative.

These three notations mean the same thing, but are used in different circumstances, or according to personal preference. We said earlier that $\dfrac{dy}{dx}$ is a single concept and can't be split up; similarly, 'dx' is a single concept, like 'Δx', and you can't split up the d and the x. So the second notation above is taking a bit of a liberty!

11.3.3 The derivative of linear functions

In Chapter 6 we discussed constant functions $y = c$, the identity function $y = x$, proportionality functions $y = mx$, and the general linear function $y = mx + c$, together with their graphs.

The graph of the linear function $y = mx + c$ is a straight line with slope m, and the tangent at any point lies along the line itself, i.e. the slope of the tangent will also be m. If we perform the limiting process (11.7) we would find that $\dfrac{\Delta y}{\Delta x} = m$, no matter how large or how small Δx is. We saw this in equation (6.3). We can conclude that

$$\text{if } y = mx + c, \text{ then } \frac{dy}{dx} = m.$$

If $m = 0$, i.e. we have the constant function $y = c$, then the derivative is zero. The graph of a constant function is a line that is horizontal (i.e. of zero slope):

$$\text{if } y = c, \text{ then } \frac{dy}{dx} = 0.$$

These results agree with what we get by using the formula (11.10) for the derivative of power functions x^n. Remember, from Section 1.5 on exponents, that $a^1 = a$ and $a^0 = 1$ for any a. Then:

- When $n = 1$, $x^n = x^1 = x$, and by (11.10) its derivative is $nx^{n-1} = 1.x^0 = 1 \times 1 = 1$. So
$$\frac{d}{dx}(x) = 1.$$

- When $n = 0$, $x^n = x^0 = 1$, and by (11.10) its derivative is $nx^{n-1} = 0.x^{-1} = 0 \times x^{-1} = 0$. So
$$\frac{d}{dx}(1) = 0.$$

In fact, if $y = mx + c$, then

$$\frac{dy}{dx} = \frac{d}{dx}(mx + c) = m.\frac{d}{dx}(x) + \frac{d}{dx}(c) = m.1 + 0 = m.$$

Problems: Now try the end-of-chapter problems 11.1–11.6.

11.3.4 The derivative of polynomial functions

We saw polynomial functions in Chapter 7. The general quintic (degree 5) polynomial is

$$y = ax^5 + bx^4 + cx^3 + dx^2 + ex + k$$

where a, b, c, d, e, k are constant coefficients. How can we differentiate this? The good news is that we do it just like we did in the last line of the previous section: we differentiate term-by-term, using (11.10) for each power of x and keeping the coefficient in front of each term:

$$\frac{dy}{dx} = a.\left(5x^4\right) + b.\left(4x^3\right) + c.\left(3x^2\right) + d.(2x) + e.(1) + k.(0)$$
$$= 5ax^4 + 4bx^3 + 3cx^2 + 2dx + e.$$

We can encapsulate the principles we have used in two rules, which we'll call the **sum rule** and **coefficient rule**. If $f(x)$ and $g(x)$ are functions of x, and a is a constant coefficient, then

Sum rule:

$$\frac{d}{dx}\big(f(x) + g(x)\big) = \frac{df}{dx} + \frac{dg}{dx}$$

$$\frac{d}{dx}\big(f(x) - g(x)\big) = \frac{df}{dx} - \frac{dg}{dx} \qquad\qquad (11.11)$$

Coefficient rule:

$$\frac{d}{dx}\big(af(x)\big) = a \cdot \frac{df}{dx}.$$

So the derivative of $f(x) = x^5 + 7x^3 - 5x^2 + 2x - 13$ will be

$$\frac{df}{dx} = 5x^4 + 7(3x^2) - 5(2x) + 2(1) - 13(0)$$
$$= 5x^4 + 21x^2 - 10x + 2.$$

EXAMPLE

Back in the introduction Section 11.1.1, we cited the example of a car that had travelled s miles after t hours from its start-time, where $s(t) = 6t^3 (8 - 3t)$. We can now use calculus to work out its speed ds/dt at any time t. First we must multiply out the brackets to express the polynomial as a sum of terms:

$$s(t) = 6t^3(8 - 3t) = 48t^3 - 18t^4.$$

Now, differentiating term-by-term, and then taking out a common factor again:

$$\frac{ds}{dt} = 48(3t^2) - 18(4t^3) = 144t^2 - 72t^3 = 72t^2(2 - t).$$

As $(2 - t)$ is a factor of $\frac{ds}{dt}$, when $t = 2$ we will have $\frac{ds}{dt} = 0$. Thus, the car's speed will be zero after $t = 2$ hours.

PRACTICE

Differentiate with respect to x:

(i) $3x^7 - 7x^3 + 5$ (ii) $(2x - 1)(x + 1)$.

Answers: (i) $21x^6 - 21x^2$ (ii) $4x + 1$.

Problems: Now try the end-of-chapter problems 11.7–11.10.

> **CASE STUDY C12: Differentiating the animal motion model**
>
> We saw in instalment C6 the equation of motion for an animal running with constant acceleration:
>
> $$s(t) = d + ut + \tfrac{1}{2}at^2 \qquad\qquad (7.14)$$
>
> where d, u, a are constants. By differentiating this with respect to t, we immediately obtain the velocity equation:
>
> $$\frac{ds}{dt} = 0 + u.(1) + \tfrac{1}{2}a.(2t)$$
>
> or
>
> $$v(t) = u + at$$
>
> which was equation (7.13). If we differentiate again, we find that $\dfrac{dv}{dt} = 0 + a.(1) = a$, confirming that a is the rate of change of velocity with time, i.e. the acceleration.

11.4 Differentiating roots and reciprocals

More good news: the rule (11.10) for differentiating x^n applies not just when $n = 0, 1, 2, 3$, etc., but for any rational value of n. This will give us the derivatives of more standard functions.

Recall from Section 1.6 that $x^{1/2} = \sqrt{x}$, the square root of x. Using $n = \tfrac{1}{2}$ in (11.10) we find that the derivative of $\sqrt{x}$ is $\tfrac{1}{2}x^{-1/2}$, or $\dfrac{1}{2\sqrt{x}}$.

Writing the cube root of x as $x^{1/3}$, try to use (11.10), together with the rules of algebra in Chapter 1, to show that

$$\frac{d}{dx}\left(\sqrt[3]{x}\right) = \frac{1}{3\sqrt[3]{x^2}}.$$

If we take the exponent n to be negative, we obtain reciprocals of powers of x (Section 8.1.1), which we can now differentiate:

When $n = -1$, $x^{-1} = \dfrac{1}{x}$, and by (11.10) $\dfrac{d}{dx}\left(\dfrac{1}{x}\right) = \dfrac{d}{dx}\left(x^{-1}\right) = -1 . x^{-2} = -\dfrac{1}{x^2}$. You should have obtained this result in Problem 11.5, differentiating from first principles.

When $n = -\dfrac{1}{2}$, $x^{-\frac{1}{2}} = \dfrac{1}{\sqrt{x}}$, and by (11.10) $\dfrac{d}{dx}\left(\dfrac{1}{\sqrt{x}}\right) = \dfrac{d}{dx}\left(x^{-\frac{1}{2}}\right) = -\dfrac{1}{2} . x^{-\frac{3}{2}} = -\dfrac{1}{2x\sqrt{x}}$.

PRACTICE

Differentiate (i) $\sqrt[5]{x^3}$ (ii) $\dfrac{3}{2x^5}$.

Answers: (i) $-\dfrac{0.6}{\sqrt[5]{x^2}}$ (ii) $-\dfrac{15}{2x^6}$.

Problems: Now try the end-of-chapter problems 11.11–11.14.

11.5 Differentiating functions of linear functions

Suppose we want to differentiate $(5x + 1)^3$. To use (11.10), we must multiply out using the expansion of $(a + b)^3$ in (1.20), then differentiate term-by-term, and finally we may be able to take out a common factor or factorize the polynomial expression:

$$y = (5x + 1)^3$$
$$= (5x)^3 + 3(5x)^2 \cdot 1 + 3(5x) \cdot 1^2 + 1^3$$
$$= 125x^3 + 75x^2 + 15x + 1.$$

Differentiating term-by-term, then taking out a common factor of 15 and factorizing:

$$\frac{dy}{dx} = 125 \cdot (3x^2) + 75 \cdot (2x) + 15 \cdot (1) + 0$$
$$= 375x^2 + 150x + 15$$
$$= 15(25x^2 + 10x + 1)$$
$$= 15(5x + 1)^2.$$

It's not a coincidence that the derivative of $(5x + 1)^3$ involves a factor of $(5x + 1)^2$, in the same way as the derivative of x^3 is $3x^2$. In fact, the rule for differentiating a power of a linear function $(ax + b)$ is:

If $y = (ax + b)^n$,   then $\dfrac{dy}{dx} = a \cdot n(ax + b)^{n-1}$.   (11.12)

So when $y = (5x + 1)^3$, then $\dfrac{dy}{dx} = 5 \times 3(5x + 1)^2 = 15(5x + 1)^2$.

Another example: when $y = (7x - 6)^9$, then $\dfrac{dy}{dx} = 7 \times 9(7x - 6)^8 = 63(7x - 6)^8$. I would not want to check this by multiplying out the polynomial of degree 9!

What is the reasoning behind (11.12)? We have seen that

'nx^{n-1} is the derivative of x^n with respect to x'

so by substituting $(ax + b)$ for x in the above, we can say that

'$n(ax + b)^{n-1}$ is the derivative of $(ax + b)^n$ with respect to $(ax + b)$'.

Also, the derivative of $(ax + b)$ with respect to x is $\dfrac{d}{dx}(ax + b) = a \cdot 1 + 0 = a$.

We can rewrite these statements by saying that if $y = (ax + b)^n$, then

$$\frac{dy}{d(ax + b)} = n(ax + b)^{n-1}   \text{and}   \frac{d(ax + b)}{dx} = a.$$

Then the algebraic reasoning behind (11.12) is that

$$\frac{dy}{dx} = \frac{dy}{d(ax + b)} \times \frac{d(ax + b)}{dx} = n(ax + b)^{n-1} \times a = a.n(ax + b)^{n-1}.$$

This is an instance of the **chain rule** for differentiation of a 'function of a function', which we will meet at the start of Chapter 12. You can imagine that the $d(ax + b)$ terms in the equation above are cancelling out – although this isn't really what's happening, since dy/dx doesn't represent a fraction of dy and dx. The rule can be used with any function of a linear function $(ax + b)$, not just with powers. We can write the rule as:

$$\text{If } y = f(ax + b), \quad \text{then} \quad \frac{dy}{dx} = a \cdot f'(ax + b) \tag{11.13}$$

where $f'(ax + b)$ means taking $f'(x)$ and replacing each x in the expression with $(ax + b)$.

EXAMPLE

Differentiate $y = \dfrac{2}{5x - 3}$.

We write $y = f(5x - 3)$, where $f(x) = \dfrac{2}{x} = 2x^{-1}$.

By (11.10), $f'(x) = 2(-1)x^{-2} = \dfrac{-2}{x^2}$. So, replacing x by $(5x - 3)$, $f'(5x - 3) = \dfrac{-2}{(5x - 3)^2}$.

Then by (11.11) $\dfrac{dy}{dx} = 5 \cdot f'(5x - 3) = 5 \times \dfrac{-2}{(5x - 3)^2} = \dfrac{-10}{(5x - 3)^2}$.

This looks quite tricky; it's actually easier if you do the differentiation in your head. Write $f(x) = 2(5x - 3)^{-1}$. Think of the linear expression $5x - 3$ in brackets as a single variable Ξ, say, and differentiate with respect to it:

$$2(\Xi)^{-1} \rightarrow 2\left[-1(\Xi)^{-2}\right] = -2(\Xi)^{-2},$$

and finally multiply by the coefficient 5:

$$\frac{dy}{dx} = -2(5x - 3)^{-2} \cdot 5 = -10(5x - 3)^{-2}.$$

It is also possible to explain what's happening in (11.13) geometrically. We can consider $y = f(ax + b)$ as a **linear transformation** of the graph of $y = f(x)$. We saw the graphs of linear transformations in Section 6.5. We are interested in the effect of the transformation on the gradient of the curve, since this is the geometric meaning of the derivative. The effect of the transformation $y = f(x + b)$ is to cause a horizontal shift of the graph of $y = f(x)$ (Section 6.5.4). Such a shift has no effect on the shape of the curve itself. However, the transformation $y = f(ax)$ where $a > 1$, performs a horizontal squashing of the graph of $y = f(x)$ towards the y-axis (Section 6.5.3); this will make the gradient of the curve steeper by a factor of a. This is illustrated in FNGRAPH 11.2 (see Figure 11.5), where the original function $f(x) = x^3 + 5x^2 + 7x + 3$ is shown in blue. Tick the boxes to display the graphs of $y = f(x + b)$, $y = f(ax)$ and $y = f(ax + b)$, where the values of a and b are set in the Variables window.

PRACTICE

Differentiate (i) $(2x + 3)^5$ (ii) $\sqrt{3x - 2}$.

Answers: (i) $10(2x + 3)^4$ (ii) $\dfrac{3}{2\sqrt{3x - 2}}$.

Problems: Now try the end-of-chapter problems 11.15–11.18.

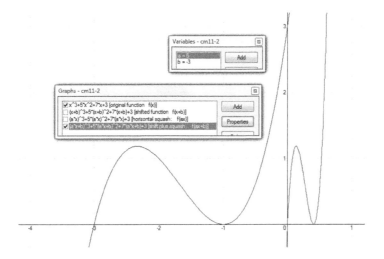

Figure 11.5 The graphs of $y = f(x)$ and $y = f(ax + b)$ (FNGRAPH 11.2)

11.6 Differentiating exponential functions

There is another type of function whose derivative we already know, because we defined it in terms of its instantaneous rate of change. In Chapter 10 we introduced the exponential function to base a. This is the function $f(x) = a^x$, where the base a is a positive constant. The graph of $y = a^x$ for $a > 1$ is sketched in Figure 11.6; the graph also shows the tangent to the curve at the point $(0, 1)$. You can see that as you move along the curve, the gradient will get steeper and steeper, and in Section 10.1 we claimed that the gradient at the point (x, y) was in fact proportional to the value of y. We can see this if we try to find the gradient of the function $f(x) = a^x$ at the point where $x = x_0$ using the definition of the derivative (11.9). I promised that I would only work through differentiation from first principles for simple functions like x^2, but in this case we will do it to practise the rules of algebra we met in Chapter 1. I also want to show how mathematics develops through logical reasoning, in addition to algebra and definitions. The most important things to remember are the derivatives we obtain, in equations (11.14–11.16).

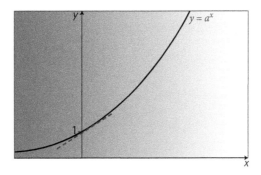

Figure 11.6 The graph of $y = a^x$

We follow the steps in the example in Section 11.3.1, but using the function $f(x) = a^x$:

$$\frac{dy}{dx} = \lim_{\Delta x \to 0}\left(\frac{f(x_0 + \Delta x) - f(x_0)}{\Delta x}\right)$$

> **Insert $f(x) = a^x$.**

$$= \lim_{\Delta x \to 0}\left(\frac{a^{x_0 + \Delta x} - a^{x_0}}{\Delta x}\right)$$

> Use the laws of exponents to write $a^{x_0 + \Delta x}$ as $a^{x_0}.a^{\Delta x}$.

$$= \lim_{\Delta x \to 0}\left(\frac{a^{x_0} \cdot a^{\Delta x} - a^{x_0}}{\Delta x}\right)$$

> Take out a common factor of a^{x_0} on the numerator.

$$= \lim_{\Delta x \to 0}\left(\frac{a^{x_0}\left(a^{\Delta x} - 1\right)}{\Delta x}\right)$$

$$= \lim_{\Delta x \to 0}\left(a^{x_0} \cdot \frac{a^{\Delta x} - 1}{\Delta x}\right)$$

$$= a^{x_0} \cdot \lim_{\Delta x \to 0}\left(\frac{a^{\Delta x} - 1}{\Delta x}\right).$$

In the last line we can take a^{x_0} outside the limit sign, since a^{x_0} does not involve Δx at all. So the derivative of $y = a^x$ is a^x multiplied by the constant of proportionality $K_a = \lim_{\Delta x \to 0}\left(\frac{a^{\Delta x} - 1}{\Delta x}\right)$. Thus, the gradient of $y = a^x$ is proportional to a^x, i.e. it is proportional to the value of y itself:

$$\frac{dy}{dx} = K_a.y.$$

Notice that the value of the proportionality constant K_a will depend on the choice of the base a, but not on the point x_0.

All curves $y = a^x$ pass through the point $(0, 1)$, since when $x = 0$ we have $y = a^0 = 1$. In Section 10.2 you used SPREADSHEET 10.1 to find a particular value of a for which the gradient at this point is equal to 1; that is, the gradient is equal to the value of y itself, or $K_a = 1$, meaning that $\frac{dy}{dx} = y$. This particular value of a is called the exponential or Euler constant e, which has the value 2.71828 ... As K_a is independent of the point x_0, this equality is true at all points of the curve: the gradient of the function $f(x) = e^x$ at any given point is equal to the function value e^x itself at that point. That is,

> If $y = e^x$ then $\dfrac{dy}{dx} = e^x$ also. $\hspace{4em}$ (11.14)

You can check this by evaluating the quantity $\left(\dfrac{e^{\Delta x} - 1}{\Delta x}\right)$ for very small values of Δx; you should see that $\lim_{\Delta x \to 0}\left(\dfrac{e^{\Delta x} - 1}{\Delta x}\right) = 1$.

If we have to differentiate $y = e^{kx}$, we can apply the 'function of a linear function' rule (11.13), using the fact that $\dfrac{d\left(e^{kx}\right)}{d(kx)} = e^{kx}$:

$$\frac{d}{dx}\left(e^{kx}\right) = \frac{d\left(e^{kx}\right)}{d(kx)} \times \frac{d(kx)}{dx} = e^{kx} \times k = ke^{kx},$$

and to generalize this further by including a coefficient A:

$$\text{If } y = Ae^{kx} \quad \text{then} \quad \frac{dy}{dx} = Ake^{kx}. \tag{11.15}$$

So if $y = 4e^{3x}$ then $\frac{dy}{dx} = 4(3e^{3x}) = 12e^{3x}$, and if $y = 4e^{-3x}$ then $\frac{dy}{dx} = 4(-3e^{-3x}) = -12e^{-3x}$.

We can use this result to find the derivative of the exponential function to base a: $y = a^x$. Recall from Section 10.3.5 that we can find a value of k such that $a^x = e^{kx}$. Taking natural logarithms of both sides gives

$$\begin{aligned} a^x &= e^{kx} \\ \ln(a^x) &= \ln(e^{kx}) \\ x.\ln a &= kx.\ln e = kx \\ k &= \ln a. \end{aligned}$$

So we can write a^x as $e^{(\ln a)x}$, whose derivative is $(\ln a).e^{(\ln a)x}$ by (11.15). But $e^{(\ln a)x}$ is a^x, so

$$\text{If } y = a^x \quad \text{then} \quad \frac{dy}{dx} = (\ln a).a^x \quad a > 0. \tag{11.16}$$

So if $y = 2^x$ then $\frac{dy}{dx} = (\ln 2).2^x = 0.6931(2^x)$, and if $y = 10^{-3x}$ then $\frac{dy}{dx} = -3(\ln 10)(10^{-3x})$.

Application: cellular learning

In Section 10.5.2 we defined two basic learning behaviours shown by simple organisms in response to repeated stimuli such as electric shocks: **habituation** (decreasing intensity of response as the organism becomes accustomed to the stimulus) and **sensitization** (increasing intensity of response to perceived threat). Both these time-dependent behaviours can be modelled by curves involving exponential decay functions e^{-kt}. If $I(t)$ denotes the response intensity at time t, and I_0 is the intensity in response to the initial stimulus, then:

- the **habituation** response behaviour can be modelled as

$$I(t) = I_{min} + (I_0 - I_{min})e^{-kt};$$

- the **sensitization** response behaviour can be modelled as

$$I(t) = I_{max} - (I_{max} - I_0)e^{-kt}.$$

For the habituation behaviour, the rate at which the response is decreasing, at time t, is given by

$$\frac{dI}{dt} = -k(I_0 - I_{min})e^{-kt} = -k(I(t) - I_{min}),$$

and in the sensitization behaviour, the rate at which the response is increasing, at time t, is given by

$$\frac{dI}{dt} = -\left[-k\left(I_{max} - I_0\right)e^{-kt}\right] = k\left(I_{max} - I_0\right)e^{-kt} = k\left(I_{max} - I(t)\right).$$

PRACTICE

Differentiate: (i) $6(2^{3x})$ (ii) $3e^{3x} - 2e^{-2x}$.

Answers: (i) $(18 \ln 2)2^{3x}$ (ii) $9e^{3x} + 4e^{-2x}$.

Problems: Now try the end-of-chapter problems 11.19–11.20.

11.7 Extension: small changes and errors

We said in (11.7) that $\frac{dy}{dx}$ is the value of the fraction $\frac{\Delta y}{\Delta x}$ as the interval Δx approaches zero. So if Δx is a relatively small change or increment in the variable x, which causes a change Δy in the dependent variable y, we would expect the value of $\frac{\Delta y}{\Delta x}$ to be very close to that of the derivative $\frac{dy}{dx}$ at that point. We use '$\approx$' to mean 'approximately equal to':

$$\frac{\Delta y}{\Delta x} \approx \frac{dy}{dx} \quad \text{when } \Delta x \text{ is small.}$$

CASE STUDY C13: Deriving the exponential model of animal speed

In instalment C11 in Chapter 10 we wrote down a model of animal speed $v(t)$ during acceleration from stationary until the animal reaches its maximum running speed v_{max}:

$$v = v_{max}\left(1 - e^{-kt}\right). \tag{10.29}$$

We can now show how we expressed the coefficient k using the physical parameter a_0, the initial acceleration. First multiply out the brackets:

$$v = v_{max} - v_{max}e^{-kt}.$$

Differentiating with respect to time t gives the acceleration $a(t)$ at time t:

$$a(t) = \frac{dv}{dt} = 0 - v_{max}\left(-ke^{-kt}\right) = v_{max}ke^{-kt}.$$

For the initial acceleration a_0, set $t = 0$:

$$a_0 = a(0) = v_{max}ke^{-0} = v_{max}k.1 = v_{max}k.$$

Solving $a_0 = v_{max}k$ for k, we obtain $k = \dfrac{a_0}{v_{max}}$, as we claimed in instalment C11.

This gives us a simple formula for estimating the change in y caused by a small change Δx in x:

$$\Delta y \approx \frac{dy}{dx} \cdot \Delta x. \qquad (11.17)$$

EXAMPLE

Over the first ten years of its life, the height h (in metres) of an oak sapling can be modelled as a function of time t (in years) by the equation

$$h = 1 + t - 0.05t^2.$$

Estimate how much the tree grows in one month, when the tree is 7 years old.
  Differentiating term-by-term, the derivative is

$$\frac{dh}{dt} = 0 + 1 - 0.05(2t) = 1 - 0.1t,$$

and when $t = 7$ the value of the derivative is $\dfrac{dh}{dt} = 1 - 0.1(7) = 1 - 0.7 = 0.3$ metres per year. We want the growth over one month, i.e. $\Delta t = \frac{1}{12}$. So, using (11.17), the change in height over one month will be

$$\Delta h \approx 0.3 \times \tfrac{1}{12} = 0.025 \text{ metres} = 25 \text{ mm}.$$

The same technique can be used to estimate small errors (see Section 2.6). In this case Δy and Δx represent the magnitude of errors, and we take the absolute value of the derivative:

$$\Delta y \approx \left| \frac{dy}{dx} \right| \cdot \Delta x. \qquad (11.18)$$

EXAMPLE

A weather balloon is inflated and its diameter is measured as $d = 36$ cm. Its volume V is then calculated from the formula for the volume of a sphere of radius r:

$$V = \tfrac{4}{3}\left(\pi r^3 \right) = \tfrac{4}{3}\left(\pi (0.18)^3 \right) = 0.02443 \,\text{m}^3.$$

If the measurement of d has a possible error of 0.5 mm, what is the possible error in this value of V?
If $\Delta d = 0.5 \times 10^{-3}$ metres, then the error in the radius will be half of that: $\Delta r = 0.25 \times 10^{-3}$. The derivative of V with respect to r is:

$$\frac{dV}{dr} = \frac{d}{dr}\left(\tfrac{4}{3}\pi r^3 \right) = \tfrac{4}{3}\pi \cdot \frac{d}{dr}\left(r^3 \right) = \tfrac{4}{3}\pi \cdot \left(3r^2 \right) = 4\pi r^2.$$

We can now use (11.18):

$$\Delta V \approx \left| \frac{dV}{dr} \right| . \Delta r = \left(4\pi r^2 \right) . \Delta r = 4\pi . (0.18)^2 . (0.25 \times 10^{-3}) = 0.102 \times 10^{-3} \,\text{m}^3.$$

We thus expect the volume of the sphere to be $0.02443 \pm 0.0001 \,\text{m}^3$.

We can also use (11.17) to answer **inverse problems**. If I know the value of the change Δy, I can estimate the value of Δx that caused that change. In the weather balloon example, suppose the balloon is released; as it rises, the gas expands. When the volume has increased by 1%, by how much will the diameter have changed?

Now $\dfrac{\Delta V}{V} = 0.01$, so $\Delta V = 0.01V = 0.2443 \times 10^{-3}$ m^3.

Then $\Delta V \approx \dfrac{dV}{dr} . \Delta r = \left(4\pi r^2\right). \Delta r$, so

$$\Delta r \approx \frac{1}{4\pi r^2} . \Delta V = \frac{1}{4\pi\left(0.18\right)^2}\left(0.2443 \times 10^{-3}\right) = 0.0006 \text{ m}.$$

The diameter will thus have increased by about 1.2 mm.

The calculations are easier if we are dealing with exponential growth or decay:

EXAMPLE

The population of a colony of bacteria has a population size $P(t) = 250e^{0.6t}$ after t hours. At a certain time the population is 3500; how long will it take to increase to 3600?

We don't know the time t_0 at which this occurs, but $P(t_0) = 3500$. Using (11.15) to differentiate $P(t)$:

$$\frac{dP}{dt} = \frac{d}{dt}\left(250e^{0.6t}\right) = 250 \times 0.6e^{0.6t}$$
$$= 0.6 \times 250e^{0.6t} = 0.6P = 0.6 \times 3500 = 2100.$$

So (11.17) gives $\Delta P \approx \dfrac{dP}{dt} . \Delta t = 2100 . \Delta t$, and if $\Delta P = 3600 - 3500 = 100$, we estimate $\Delta t \approx \dfrac{100}{2100} = 0.0476$ hours, or 2.86 minutes.

There was no need to work out any exponentials!

CASE STUDY A12: Differential equation for exponential growth

We saw in Part 9 of this case study, the continuous model for exponential growth of populations. In a population that is growing exponentially from an initial size N_0, the population size $N(t)$ at time t is given by

$$N(t) = N_0 e^{rt}. \tag{10.9}$$

We said that r is called the growth rate or growth constant. We can now see how it plays this role. Differentiating, we find that the instantaneous rate of growth of the population at time t is

$$\frac{dN}{dt} = N_0 re^{rt} = r(N_0\ e^{rt}).$$

Using the definition of $N(t)$ above, we arrive at the differential equation for exponential growth:

$$\frac{dN}{dt} = rN \tag{11.19}$$

(see Section 11.1.3) of which (10.9) is a solution.

So the instantaneous rate of change of the population $\frac{dN}{dt}$ is proportional to its current size N, and r is the constant of proportionality.

If a colony of bacteria is doubling in size every hour, we may write its mathematical model as

$$N(t) = N_0 2^t.$$

Using (11.16), its instantaneous rate of change at time t is then $\frac{dN}{dt} = (\ln 2).N$, so its growth rate is $\ln 2$.

We have claimed that the exponential growth model is the continuous-time model corresponding to the discrete-time model of geometric growth, defined by the update equation

$$N_{p+1} = N_p + R.N_p \tag{1.28}$$

which we introduced in this Case Study, back in Chapter 1. Here, N_p is the population size at the pth generation. We can now establish how this correspondence comes about.

Suppose that there is a fixed time-interval Δt from one generation to the next. The derivative $\frac{dN}{dt}$ at the start of the pth generation, can be approximated by the ratio of increments $\frac{\Delta N_p}{\Delta t}$ over that generation, where the population increment is $\Delta N_p = N_{p+1} - N_p$.

Thus

$$\frac{dN}{dt} \approx \frac{\Delta N}{\Delta t} = \frac{N_{p+1} - N_p}{\Delta t}.$$

Substituting into (11.19),

$$\frac{N_{p+1} - N_p}{\Delta t} = rN_p$$

which rearranges to give the update equation (1.28) where the discrete growth constant $R = r.\Delta t$.

PROBLEMS

11.1: Use the limiting process from Section 11.2 to obtain four increasingly accurate approximations to the gradient of the function $f(x) = \frac{1}{x}$ at $x_0 = 4.0$. Start with $\Delta x = 1.0$, so that $x_1 = x_0 + \Delta x = 5.0$. Find $f(x_1)$ and hence $\Delta y = f(x_1) - f(x_0)$. Then evaluate $\frac{\Delta y}{\Delta x}$. Repeat the evaluations for $\Delta x = 0.5, 0.25, 0.125$. Write the iterates correct to 4 decimal places.

11.2: Repeat problem 11.1 for $f(x) = \sin x$ at $x_0 = \dfrac{\pi}{4}$. Work in radians, to 4 decimal places. Start with $\Delta x = 0.2$ radians and make four evaluations, halving Δx each time.

11.3: Write an Excel spreadsheet to perform the limiting process from problems 11.1–11.2. Each row should calculate a new evaluation, halving the interval Δx from the row above. Use it to solve those problems, dragging the columns down until the ratio $\dfrac{\Delta y}{\Delta x}$ converges. Once you have learned how to differentiate the functions, check your answers against what you obtain from calculus.

11.4: Differentiate x^n, where n is a natural number, from first principles. Use the fact, from the binomial theorem, that the expansion of $(a + b)^n$, written with terms in ascending order of powers of b, is $a^n + na^{n-1} b + \cdots + b^n$.

11.5–11.6: Differentiate from first principles:

11.5 $y = \dfrac{1}{x}$

11.6 $y = \dfrac{cx}{x + b}$ b, c constants.

11.7–11.20: Differentiate with respect to x:

11.7 $3x^{10} + 5x^5 + 1$

11.8 $(x^2 - 1)(x + 1)$

11.9 $(2x^8 - 1)(3x^5 - 1)$

11.10 $(((x + 2)x - 3)x - 7)x + 2$

11.11 $\dfrac{5}{x} - \dfrac{6}{x^2}$

11.12 $\dfrac{3 - \dfrac{2}{x}}{x^2}$

11.13 $\sqrt[3]{x} + \dfrac{1}{\sqrt{x}}$

11.14 $\dfrac{3 - \dfrac{2}{x}}{\sqrt{x}}$

11.15 $(5x - 2)^7$

11.16 $(3 - 2x)^{17}$

11.17 $\dfrac{3}{(2x - 1)^2}$

11.18 $\dfrac{3}{\sqrt{2 - 3x}}$

11.19 $5e^{3x} - 2e^{2x} + e^{-x} + 100$

11.20 $(e^{2x} - e^{-2x})(2e^{3x} - e^{-x})$.

11.21: In Problem 8.21 we saw the equation $P = \dfrac{k}{V}$ relating the pressure P (in pascals) and volume V (in litres) of a fixed body of gas. Using the theory of Section 11.7, find an equation for the approximate change in pressure ΔP that results from a small change in volume ΔV. Given that $k = 37.1$, estimate the change in pressure resulting from a decrease in volume of 5 ml when (a) $V = 10$ litres, (b) $V = 1$ litre.

11.22: The equation $g = \dfrac{282.36}{f}$ can be used to convert a car's fuel consumption f expressed in miles per gallon, to its fuel efficiency g expressed in litres per 100 km (see problem 8.22). My car achieves 42.2 mpg on the motorway, but only 12.7 mpg in town. If, by driving more slowly, I could use 0.2 l less fuel per 100 km, estimate the effect on my fuel consumption on (a) motorway and (b) urban driving. Apply the theory of Section 11.7.

11.23: In the Gordon–Schaefer fisheries management model, the yield Y is related to the effort E by $Y = aE\left(1 - \dfrac{E}{k}\right)$ (problem 7.35). Taking $a = 0.7$ and $k = 80$, estimate the change in yield from an increase of 5 units of effort, when (a) $E = 25$, and (b) $E = 65$.

11.24: In the Revision problems at the end of Part I, the technique of radiocarbon dating uses the equation $P(t) = P_0 e^{-kt}$, where $P(t) = \dfrac{^{14}C_t}{^{12}C_t}$, the present-day ratio of the ^{14}C and ^{12}C content in the sample, and P_0 is the ratio in atmospheric carbon. The model gives t as the radiocarbon age of the sample. The decay constant $k = 1.21 \times 10^{-4}$ per year.

In 2006, researchers tested a fragment of olive tree found buried in lava on the island of Santorini in Greece, to estimate the date of the great volcanic eruption of Thera that devastated the Minoan civilization, in the late Bronze Age. They reported a radiocarbon age (before calibration) of approximately $t = 3386$ years ago. Deduce the ratio $\dfrac{P(t)}{P_0}$, and estimate the change in t if there were an increase of 0.1% in $P(t)$.

12

Rules of differentiation

We have now seen the derivatives of two basic functions (x^n and e^x), together with the rules (11.11) that allow us to differentiate an expression term-by-term, and to take constant coefficients outside the differentiation. Using these rules we can differentiate any polynomial function. But we can't yet differentiate $\sin x$, or a product such as $x.e^x$. In this chapter we'll see the derivatives of the remaining functions we studied in Part I, plus further rules of differentiation. With this armoury of tools, we'll be able to differentiate virtually any function no matter how complex it is – assuming that the derivative exists, of course. We discuss in Section 12.1 if this is always the case.

This chapter establishes all the results you'll need in order to differentiate, then Chapter 13 describes how differentiation can be used to solve problems. This seems to me (as a mathematician) the logical order in which to proceed. However, you (as a scientist) may be motivated to see some applications straight away. In this case, feel free to skip ahead to Chapter 13 first. Your knowledge of the basics from Chapter 11 should provide you with enough knowledge of differentiation to follow the arguments there. You may need the results from this chapter, as summarized in Section 12.9, in order to tackle the examples and problems, though.

12.1 Differentiable functions

We can find the derivative of a function $f(x)$ at any point where the tangent to the curve $y = f(x)$ exists and is unique. However, there may be isolated values of x where things go wrong. This can happen in three ways, illustrated in Figure 12.1(a)–(c):

(a) The function may not exist at that point, e.g. the function $f(x) = \dfrac{1}{x^2}$ does not exist for the value $x = 0$.

(b) There may be a sudden jump in function value. The 'staircase' function $f(x) = \text{int}(x)$ (where $\text{int}(x)$ means the integer part of x) and the 'sawtooth function' $f(x) = x - \text{int}(x)$, which we saw in Section 9.1.1, have sudden jumps in value when $x = 1, 2, 3, \ldots$ Such breaks in the graph of the function are called **discontinuities**. Functions that do not have breaks are called **continuous**; thus, polynomials are continuous functions, as are the sine and cosine (but not the tangent) functions.

(c) The graph of a function may be continuous, but have a sharp 'corner' where the gradient abruptly changes. This occurs in the absolute value function $y = |x|$

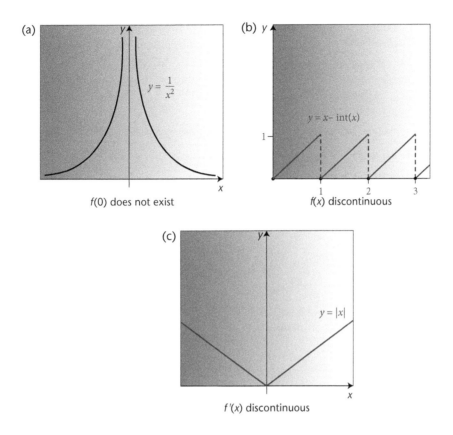

Figure 12.1 Examples of non-differentiability

(Section 1.3.4) at $x = 0$. Here, the gradient is -1 if you approach $x = 0$ from the left, but $+1$ if you approach from the right. There is not a unique gradient, so $\dfrac{df}{dx}$ does not exist at this point.

Mathematicians have put a lot of effort into studying these isolated points where the derivative of a function doesn't exist. Other scientists are content to work with the rest of the curve, where calculus can be used to solve practical problems, but you should be aware of what can go wrong. Functions that do not suffer from any of these problems are called **differentiable**.

12.2 The chain rule

In Section 11.5 we gave a rule for differentiating functions of a linear function, such as $(3x - 2)^7$. But we also frequently deal with a more general class of 'functions of a function'. For example, $f(x) = (3x^2 - 2)^7$ is a function of a function: it is the 7th power of the quadratic function $3x^2 - 2$. This becomes clearer if we invent an intermediate function $u(x)$: we can write $f(x) = u^7$, where $u(x) = 3x^2 - 2$.

As another example, $y = \sin(x^2)$ can be written as $y = \sin u$ where $u = x^2$. And $y = \sin^2 x$ can be written as $y = u^2$ where $u = \sin x$. The chain rule is a technique for differentiating a function of a function (assuming that we know the derivatives of the two functions involved).

In general, $y(x)$ can be written as a 'function of a function' if we can define a function $u(x)$, and then write y as a function of u only. It is essential that there should be no appearance of x in the expression of y as a function of u.

We now have two function definitions: $y = y(u)$ and $u = u(x)$. From the first one, we can differentiate y with respect to u, to obtain $\dfrac{dy}{du}$ (which will itself be a function of u). From the second, we can differentiate u with respect to x, to obtain $\dfrac{du}{dx}$. The **chain rule** tells us how to use these to obtain $\dfrac{dy}{dx}$:

> **Chain rule:**
>
> If $y = y(u)$ and $u = u(x)$, then $\quad \dfrac{dy}{dx} = \dfrac{dy}{du} \cdot \dfrac{du}{dx}.$ $\hfill$ (12.1)

In our first example above, we can write $y = (3x^2 - 2)^7$ as $y(u) = u^7$, where $u(x) = 3x^2 - 2$. Differentiating each of these functions:

$$\frac{dy}{du} = \frac{d}{du}(u^7) = 7u^6 \quad \text{and} \quad \frac{du}{dx} = \frac{d}{dx}(3x^2 - 2) = 3(2x) - 0 = 6x.$$

Then by the chain rule:

$$\frac{dy}{dx} = \frac{dy}{du} \cdot \frac{du}{dx} = (7u^6).(6x).$$

The final step is to substitute $u = 3x^2 - 2$ in every appearance of u, so that the derivative is a function of x only:

$$\frac{dy}{dx} = \frac{dy}{du} \cdot \frac{du}{dx} = (7u^6) \times (6x) = 7(3x^2 - 2)^6 \times (6x) = 42x(3x^2 - 2)^6.$$

Here's another example: differentiate $y = \sqrt{x^2 + 1}$.

We set $u = x^2 + 1$, so that $y = \sqrt{u} = u^{\frac{1}{2}}$.

Then $\dfrac{du}{dx} = 2x, \quad$ and $\quad \dfrac{dy}{du} = \dfrac{1}{2} u^{-\frac{1}{2}} = \dfrac{1}{2\sqrt{u}}.$

By the chain rule,

$$\frac{dy}{dx} = \frac{dy}{du} \cdot \frac{du}{dx} = \left(\frac{1}{2\sqrt{u}} \right) \times 2x = \frac{x}{\sqrt{u}}.$$

Substituting for u, we obtain

$$\frac{dy}{dx} = \frac{x}{\sqrt{x^2 + 1}}.$$

The chain rule can be extended to cover 'functions of a function of a function':

EXAMPLE

Differentiate $y = e^{\sqrt{x^2+1}}$.

Here, we need to invent two intermediate functions u and v:

Let $u = x^2 + 1$, $v = \sqrt{u}$, and then $y = e^v$.

Differentiating these three equations will give us $\dfrac{du}{dx}$, $\dfrac{dv}{du}$, and $\dfrac{dy}{dv}$:

$$\frac{du}{dx} = 2x, \quad \frac{dv}{du} = \frac{1}{2}u^{-\frac{1}{2}}, \quad \text{and} \quad \frac{dy}{dv} = e^v.$$

I performed the first two differentiations using the rule (11.10) for the derivative of x^n, and the third one using the rule (11.14) for differentiating e^x. The chain rule then says that

$$\frac{dy}{dx} = \frac{dy}{dv} \times \frac{dv}{du} \times \frac{du}{dx}$$

$$= e^v \times \left(\frac{1}{2\sqrt{u}}\right) \times 2x$$

$$= \frac{xe^v}{\sqrt{u}}$$

$$= \frac{xe^{\sqrt{x^2+1}}}{\sqrt{x^2+1}}.$$

In the last line we have substituted back for u and v as functions of x.

You can now see where the chain rule gets its name from. In

$$\frac{dy}{dx} = \frac{dy}{dv}.\frac{dv}{du}.\frac{du}{dx},$$ (12.2)

the right-hand side is like a chain with three links, where the denominator of one link matches the numerator of the next link. The chain always starts with 'dy' on top of the first link, and ends with 'dx' at the bottom of the last link; the names we use for the intermediate functions are immaterial.

PRACTICE

Differentiate $(x^2 + 3x - 1)^4$, $\sqrt{e^{3x} - 1}$, $(x + e^{2x})^5$.

Answers: $8x(x^2 + 3x - 1)^3$, $\dfrac{3e^{3x}}{2\sqrt{e^{3x} - 1}}$, $5(x + e^{2x})^4(1 + 2e^{2x})$.

EXAMPLE

The size of a colony of puffins is modelled by the function

$$P(t) = \frac{1000}{\sqrt{t^3 + 16}},$$

where $P(t)$ is the population of breeding pairs after t years. Use the chain rule to find the rate of decline of the population. What is the size and rate of decline of the population (a) initially, and (b) after four years?

Solution: We can write the equation as a function of a function: $P(t) = 1000u^{-\frac{1}{2}}$, where $u = t^3 + 16$. Differentiating each of these functions:

$$\frac{dP}{du} = 1000.\left(-\frac{1}{2}u^{-\frac{3}{2}}\right) = -\frac{500}{u^{\frac{3}{2}}}, \text{ and } \frac{du}{dt} = 3t^2. \text{ Use the chain rule to find the rate of}$$

change $\dfrac{dP}{dt}$:

$$\frac{dP}{dt} = \frac{dP}{du}.\frac{du}{dt} = -\frac{500}{u^{\frac{3}{2}}} \times 3t^2 = -\frac{1500t^2}{\left(t^3 + 16\right)^{\frac{3}{2}}}.$$

(a) When $t = 0$, $P(0) = \dfrac{1000}{\sqrt{16}} = 250$, and $\dfrac{dP}{dt} = 0$ because of the factor of t^2 on the numerator. So initially the population appears to be steady at 250 pairs.

(b) When t = 4, $P(4) = \dfrac{1000}{\sqrt{4^3 + 16}} = \dfrac{1000}{\sqrt{80}} = \dfrac{1000}{4\sqrt{5}} = 50\sqrt{5} = 111.8$. And the rate of change

is $\dfrac{dP}{dt} = -\dfrac{1500 \times 16}{\left(\sqrt{80}\right)^3} = -\dfrac{24000}{715.54} = -33.54$. So the population has reduced to 111 pairs

(rounding down to the nearest integer), and is declining at a rate of 33.5 pairs per year.

Problems: Now try the end-of-chapter problems 12.1–12.4.

. . . .

12.3 **The product and quotient rules**

The next two rules complete our toolbox of rules that will allow us to differentiate the most complicated functions, provided they are built up from standard basic functions whose derivatives we know.

12.3.1 **The product rule**

The chain rule may enable us to differentiate fiendishly complicated functions of functions of functions of …, but it can't help with differentiating a simple product such as $y = x.e^x$. For this we need a rule for finding $\dfrac{dy}{dx}$ when y is the product of two functions $u(x)$ and $v(x)$: i.e. when $y = u(x).v(x)$.

We go back to the definition of the derivative in (11.9). When x increases by a small amount Δx, this will cause small increases Δu in u, and Δv in v, and these cause the change Δy in y. If $y = f(x) = u(x).v(x)$, the increased value of y will be

$$y + \Delta y = f(x + \Delta x) = (u + \Delta u).(v + \Delta v).$$

The definition (11.9) of $\dfrac{dy}{dx}$ then becomes

$$\frac{dy}{dx} = \lim_{\Delta x \to 0} \left(\frac{f(x + \Delta x) - f(x)}{\Delta x} \right)$$

$$= \lim_{\Delta x \to 0} \left(\frac{(u + \Delta u)(v + \Delta v) - u.v}{\Delta x} \right)$$ **Multiply out ...**

$$= \lim_{\Delta x \to 0} \left(\frac{u.v + u.\Delta v + \Delta u.v + \Delta u.\Delta v - u.v}{\Delta x} \right)$$ **Cancel the first and last terms ...**

$$= \lim_{\Delta x \to 0} \left(\frac{u.\Delta v + \Delta u.v + \Delta u.\Delta v}{\Delta x} \right)$$ **Divide through by Δx ...**

$$= \lim_{\Delta x \to 0} \left(u.\frac{\Delta v}{\Delta x} + v.\frac{\Delta u}{\Delta x} + \Delta u.\frac{\Delta v}{\Delta x} \right).$$

Taking the limit as $\Delta x \to 0$, $\frac{\Delta u}{\Delta x}$ becomes $\frac{du}{dx}$, and $\frac{\Delta v}{\Delta x}$ becomes $\frac{dv}{dx}$. In the last term, $\frac{dv}{dx}$ is multiplied by Δu, which itself tends to zero, so the last term disappears. The result is the **product rule**:

Product rule:

If $y = u(x).v(x)$,   then $\dfrac{dy}{dx} = u.\left(\dfrac{dv}{dx}\right) + v.\left(\dfrac{du}{dx}\right).$ (12.3)

Here's how to use it, for the example we cited:

EXAMPLE

Differentiate $y = x.e^x$.

Let $u = x$ and $v = e^x$, so that $y = u.v$.

Differentiate u and v: $\dfrac{du}{dx} = 1$ and $\dfrac{dv}{dx} = e^x$.

Enter u, v, $\dfrac{du}{dx}$ and $\dfrac{dv}{dx}$ into the product rule:

$$\frac{dy}{dx} = u.\left(\frac{dv}{dx}\right) + v.\left(\frac{du}{dx}\right)$$

$$= (x).(e^x) + (e^x).(1)$$

$$= (x + 1)e^x.$$

Here's another example, which involves using the chain rule within the product rule:

EXAMPLE

Differentiate $y = x^2\sqrt{3x^2 + 1}$.

Let $u = x^2$ and $v = \sqrt{3x^2 + 1}$, so that $y = u.v$.

Differentiate u and v: $\dfrac{du}{dx} = 2x$ and $\dfrac{dv}{dx} = \dfrac{1}{2}\left(3x^2 + 1\right)^{-\frac{1}{2}}.6x = \dfrac{3x}{\sqrt{3x^2 + 1}}$ by the chain rule.

Enter u, v, $\dfrac{du}{dx}$ and $\dfrac{dv}{dx}$ into the product rule:

$$\frac{dy}{dx} = u.\left(\frac{dv}{dx}\right) + v.\left(\frac{du}{dx}\right)$$

$$= (x^2).\left(\frac{3x}{\sqrt{3x^2 + 1}}\right) + \left(\sqrt{3x^2 + 1}\right)(2x)$$

$$= \frac{3x^3}{\sqrt{3x^2 + 1}} + 2x\sqrt{3x^2 + 1}.$$

PRACTICE

Differentiate $x^4\sqrt{x+1}$, $x^5 e^{3x}$.

Answers: $4x^3\sqrt{x+1} + \dfrac{x^4}{2\sqrt{x+1}}$, $x^4(3x+5)e^{3x}$.

Here is an example of a more complicated product function to differentiate:

Application: cellular learning

We saw this example in Section 11.6, where we wrote functions $I(t)$ using exponential decay terms e^{-kt} to model the mechanisms of habituation and sensitization of an organism's behaviour in response to a stimulus. A more sophisticated two-phase response behaviour is called **habituating sensitization**: the response intensity initially increases while the stimulus is novel and its threat unknown, but when the organism learns that the stimulus does not result in harm, it becomes habituated to it (that is, it no longer reacts so strongly). The graph of such a two-phase learning behaviour is shown in the intermediate curve in Figure 12.2 (taken from Prescott and Chase 1999), where response intensity I is plotted against time t. The researchers observed this learning behaviour in the tentacle withdrawal response of the snail *Helix aspersa*.

A class of function whose graph can approximate this curve, is

$$I(t) = a.t^n.e^{-kt} + I_0 \tag{12.4}$$

for appropriate values of the parameters a, n, k and I_0. Provided that t is being raised to a positive power n, we can use our knowledge of the functions t^n and e^{-kt} from Part I to say that $I(0) = I_0$, and that the function approaches I_0 again for large values of t. Can we say anything further about the shape of this curve by looking at its derivative?

Differentiating (12.4) using the product rule to find the gradient of $I(t)$, we get

$$\frac{dI}{dt} = a\left[(t^n).\frac{d(e^{-kt})}{dt} + (e^{-kt}).\frac{d(t^n)}{dt}\right] + 0$$

> using the sum rule and coefficient rule (11.11), and the derivative of the constant I_0 is zero

$$= a\left[(t^n).(-ke^{-kt}) + (e^{-kt}).(nt^{n-1})\right]$$

> Find the derivatives using (11.10) and (11.15)

$$= a(-kt^n e^{-kt} + nt^{n-1}e^{-kt})$$

$$= a(nt - k)t^{n-1}e^{-kt}.$$

> Take out the common factors

When $t = k/n$, the factor $(nt - k)$ becomes zero, so that the gradient of $I(t)$ will be zero at this time; this is where the curve reaches its maximum. The gradient of the curve is also zero at $t = 0$, for $n = 2, 3, 4, \ldots$ Is this also true when $n = 1$? And what happens when n is not an integer, e.g. when $n = 0.5$ or $n = 1.5$? Try to work out $\dfrac{dI}{dt}$ in these cases, and use it to deduce the shape of the curve close to $t = 0$, and where the maximum will lie. Then check your conclusions by sketching the function (12.4) using FNGraph; this is programmed in FNGRAPH 12.1; the graph for the case of $n = 2$ is shown in Figure 12.3.

12.3.2 The quotient rule

The quotient rule is used when we need to differentiate an expression in which one function $u(x)$ is divided by another function $v(x)$. The rule is

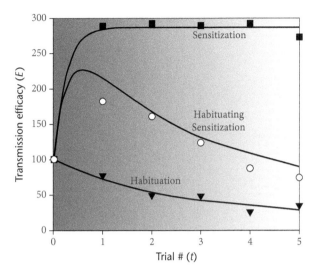

Figure 12.2 Habituating sensitization (from Prescott and Chase 1999)

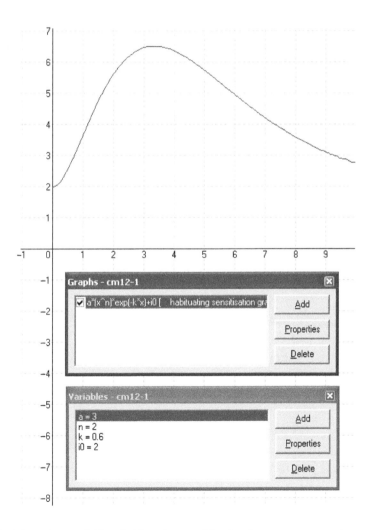

Figure 12.3 A function modelling habituating sensitization (FNGRAPH 12.1)

Quotient rule:

$$\text{If } y = \frac{u(x)}{v(x)}, \quad \text{then } \frac{dy}{dx} = \frac{v.\left(\dfrac{du}{dx}\right) - u.\left(\dfrac{dv}{dx}\right)}{v^2}. \tag{12.5}$$

EXAMPLE

Differentiate $y = \dfrac{e^{3x}}{2x + 7}$.

Here, we set $u(x) = e^{3x}$ and $v(x) = 2x + 7$.

Differentiating, $\dfrac{du}{dx} = 3e^{3x}$ and $\dfrac{dv}{dx} = 2$.

Inserting in (12.5):

$$
\begin{aligned}
\frac{dy}{dx} &= \frac{v.\left(\dfrac{du}{dx}\right) - u.\left(\dfrac{dv}{dx}\right)}{v^2} \\[2mm]
&= \frac{(2x + 7).(3e^{3x}) - \left(e^{3x}\right).(2)}{(2x + 7)^2} \\[2mm]
&= \frac{\left(6x + 21\right)e^{2x} - 2e^{2x}}{\left(2x + 7\right)^2} \\[2mm]
&= \frac{\left(6x + 19\right)e^{2x}}{\left(2x + 7\right)^2}.
\end{aligned}
$$

Here's a similar example, which requires more algebraic manipulation to produce the derivative in its simplest form:

EXAMPLE

Differentiate $y = \dfrac{x^2}{2x^3 + x - 5}$.

Here, we set $u(x) = x^2$ and $v(x) = 2x^3 + x - 5$.

Differentiating, $\dfrac{du}{dx} = 2x$ and $\dfrac{dv}{dx} = 6x^2 + 1$.

Inserting in (12.5):

$$
\begin{aligned}
\frac{dy}{dx} &= \frac{v.\left(\dfrac{du}{dx}\right) - u.\left(\dfrac{dv}{dx}\right)}{v^2} \\[2mm]
&= \frac{\left(2x^3 + x - 5\right).(2x) - \left(x^2\right).\left(6x^2 - 1\right)}{\left(2x^3 + x - 5\right)^2} \\[2mm]
&= \frac{\left(4x^4 + 2x^2 - 10x\right) - \left(6x^4 - x^2\right)}{\left(2x^3 + x - 5\right)^2} \\[2mm]
&= \frac{-2x^4 + 3x^2 - 10x}{\left(2x^3 + x - 5\right)^2} \\[2mm]
&= \frac{-x\left(2x^3 + x + 10\right)}{\left(2x^3 + x - 5\right)^2}.
\end{aligned}
$$

The quotient rule can be derived from first principles using (11.9), or from the product rule – see the Problems. While the product rule is easy to remember, and it doesn't matter which order you write the terms in, it is essential to get the terms on the numerator of the quotient rule in the correct order: in the numerator of (12.5), the first term must be $v.\left(\dfrac{du}{dx}\right)$, not $u.\left(\dfrac{dv}{dx}\right)$.

PRACTICE

Differentiate $\dfrac{2x+1}{x-3}$, $\dfrac{x}{1+e^x}$, $\dfrac{1+e^x}{1-e^x}$.

Answers: $\dfrac{-7}{(x-3)^2}$, $\dfrac{1+(1-x)e^x}{(1+e^x)^2}$, $\dfrac{-2e^{2x}}{(1-e^x)^2}$.

Application: Michaelis–Menten equation

In Section 8.4 we saw the Michaelis–Menten equation of enzyme kinetics, relating the reaction rate v to the substrate concentration s:

$$v = \frac{v_{\max}.s}{K_m + s}. \tag{8.10}$$

We can use the quotient rule to differentiate $v(s)$, giving us $\dfrac{dv}{dx}$. This could be used with equation (11.17):

$$\Delta v \approx \frac{dv}{ds}.\Delta s$$

to estimate the increase in reaction rate Δv resulting from a small increase Δs in substrate concentration. The differentiation is straightforward, since $v_{\max}$ and K_m are constants.

On the numerator we have $v_{\max}.s$, whose derivative with respect to s is $v_{\max}$. On the denominator we have $K_m + s$, whose derivative with respect to s is 1. So by the quotient rule

$$\frac{dv}{ds} = \frac{(K_m + s)(v_{\max}) - (v_{\max}.s)(1)}{(K_m + s)^2}$$
$$= \frac{(K_m.v_{\max} + v_{\max}.s) - v_{\max}.s}{(K_m + s)^2}$$
$$= \frac{K_m.v_{\max}}{(K_m + s)^2}.$$

(You should have obtained this result in Problem 11.6, differentiating from first principles.)

Then the increase in reaction rate when the substrate concentration increases from s to $s + \Delta s$, is

$$\Delta v \approx \frac{K_m.v_{\max}}{(K_m + s)^2}.\Delta s.$$

PRACTICE: Try to use the quotient rule to find $\dfrac{dv}{ds}$ in the case of allosteric regulation:

$$v = \frac{v_{\max}.s^\eta}{K_m + s^\eta}. \tag{8.13}$$

CASE STUDY C14: Deriving the hyperbolic model of animal speed

In instalment C8 in Chapter 8 we wrote down a model of animal speed $v(t)$ during acceleration from stationary until the animal reaches its maximum running speed v_{max}:

$$v = \frac{v_{max}t}{t + b}.$$

(8.5)

We can now show how we expressed the coefficient b using the physical parameter a_0, the initial acceleration. We use the quotient rule in exactly the same way as we did above, to differentiate (8.5). As in instalment C13, differentiating with respect to time t gives the acceleration $a(t)$ at time t:

$$a(t) = \frac{dv}{dt} = \frac{(t + b).v_{max} - v_{max}t.1}{(t + b)^2}$$

$$= \frac{t.v_{max} + b.v_{max} - v_{max}t}{(t + b)^2} = \frac{b.v_{max}}{(t + b)^2}$$

The first and last terms on the numerator cancel.

For the initial acceleration a_0, set $t = 0$:

$$a_0 = a(0) = \frac{b.v_{max}}{(0 + b)^2} = \frac{b.v_{max}}{b^2} = \frac{v_{max}}{b}.$$

Solving $a_0 = \frac{v_{max}}{b}$ for b, we obtain $b = \frac{v_{max}}{a_0}$, as we claimed in instalment C8.

Answer: $\frac{dv}{ds} = \frac{\eta.K_m.v_{max}.s^{\eta-1}}{(K_m + s^\eta)^2}.$

The product and quotient rules can be written more succinctly using the 'prime' notation we introduced in Section 11.3.2. Writing $\frac{du}{dx}$, $\frac{dv}{dx}$ as u', v', the rules become:

Product rule: $\frac{d}{dx}(u.v) = u.v' + v.u'$

Quotient rule: $\frac{d}{dx}\left(\frac{u}{v}\right) = \frac{v.u' - u.v'}{v^2}.$

(12.6)

Problems: Now try the end-of-chapter problems 12.5–12.16.

12.4 Differentiating trigonometric functions

The derivatives of $\sin x$ and $\cos x$ are refreshingly simple, although we won't attempt to differentiate them from first principles. As these functions are periodic (their graphs have a cycle of length 2π, which repeats), so their derivatives will also be periodic, with the same period 2π. We will guess the derivative of sin x by using this fact, and a description of what happens to the tangent as x goes from 0 to 2π along the curve $y = \sin x$. The curve is shown in the top graph of Figure 12.4.

At the point at which $x = 0$, the graph of $y = \sin x$ is sloping upwards. If you use FNGraph to draw the curve $y = \sin x$ and the line $y = x$, you will see that the line is the tangent to the curve at the origin. But the slope of the line $y = x$ is 1, so the gradient of the curve $y = \sin x$ at $x = 0$ is 1. By symmetry, at $x = \pi$, where the curve is sloping downwards, the tangent line will have a slope of -1, so the gradient of the curve at the point $x = \pi$ will be -1. These two tangents are shown in Figure 12.4(a).

We can now describe the behaviour of the curve and its gradient in the four quarters of one cycle, from $x = 0$ to $x = 2\pi$, as follows:

Angle x	Value of sin x	Slope of tangent
increases from 0 to $\pi/2$	increases from 0 to 1	decreases from 1 to 0
increases from $\pi/2$ to π	decreases from 1 to 0	decreases from 0 to -1
increases from π to $3\pi/2$	decreases from 0 to -1	increases from -1 to 0
increases from $3\pi/2$ to 2π	increases from -1 to 0	increases from 0 to 1

If you sketch a smooth curve with the behaviour described in the last column, over a cycle from 0 to 2π, you will arrive at the curve in Figure 12.4(b), which you should recognize as the cosine function. And this is what the derivative of sin x is:

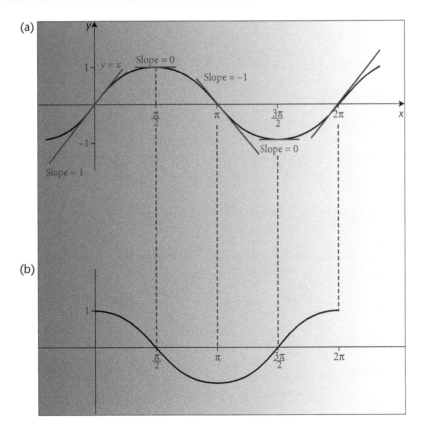

Figure 12.4 Graphs of (a) $y = \sin x$, and (b) the gradient of sin x

If $y = \sin x$,   its derivative is $\dfrac{dy}{dx} = \cos x.$ (12.7)

Compare the slope of the tangent to the curve $y = \sin x$ in Figure 12.4(a), with the value of $y = \cos x$ at the corresponding point on the curve in Figure 12.4(b). The process is animated in SPREADSHEET 12.1, with a tangent line moving along the upper curve, and a green dot moving along the lower curve, as you drag the scrollbar button. Notice how $\cos x$ is positive when the tangent slopes upwards, negative when the tangent slopes downwards, and zero when the tangent is horizontal (at $x = \pi$ and $x = 3\pi/2$).

Try applying the same argument to the graph of $y = \cos x$. You should see that the behaviour of the gradient is not that of $\sin x$ (as you might have been hoping), but of $-\sin x$:

If $y = \cos x$,   its derivative is $\dfrac{dy}{dx} = -\sin x.$ (12.8)

EXAMPLE

Differentiate (i) $y = \sin^2 x$ and (ii) $y = \cos^2 x$.
We use the chain rule:

(i) $u = \sin x$ and $y = u^2$

$$\frac{du}{dx} = \cos x, \quad \frac{dy}{du} = 2u$$

$$\frac{dy}{dx} = \frac{dy}{du}.\frac{du}{dx} = (2u)(\cos x) = 2\sin x \cos x.$$

(ii) $u = \cos x$ and $y = u^2$

$$\frac{du}{dx} = -\sin x, \quad \frac{dy}{du} = 2u$$

$$\frac{dy}{dx} = \frac{dy}{du}.\frac{du}{dx} = (2u)(-\sin x) = -2\sin x \cos x.$$

So by the sum rule, the derivative of $y = \sin^2 x + \cos^2 x$ would be $2\sin x \cos x - 2\sin x \cos x$, which is zero. You will not be surprised at this if you have remembered the trigonometric identity (9.18):

$$\cos^2\theta + \sin^2\theta = 1, \quad \text{for any angle } \theta.$$

To differentiate $\tan x$ we use the quotient rule:

Let $y = \tan x = \dfrac{\sin x}{\cos x} = \dfrac{u(x)}{v(x)}$. Then $u = \sin x$, $v = \cos x$, and differentiating, $\dfrac{du}{dx} = \cos x$, $\dfrac{dv}{dx} = -\sin x$, so by the quotient rule

$$\frac{dy}{dx} = \frac{v.u' - u.v'}{v^2} = \frac{(\cos x)(\cos x) - (\sin x)(-\sin x)}{(\cos x)^2}$$

$$= \frac{\cos^2 x + \sin^2 x}{\cos^2 x} = \frac{1}{\cos^2 x} = \sec^2 x.$$

So

$$\text{If } y = \tan x, \quad \text{its derivative is} \quad \frac{dy}{dx} = \sec^2 x \tag{12.9}$$

In the Problems you can differentiate the reciprocal functions: the secant, cosecant and cotangent. We will summarize the results at the end of this chapter.

Problems: Now try the end-of-chapter problems 12.17–12.24.

12.5 Implicit differentiation

There are occasions when, although we still have a variable y that depends on x, we are unable to write y explicitly as a function of x. For example, consider a function $y(x)$ that satisfies the equation

$$xy = e^{2y}. \tag{12.10}$$

It is impossible to manipulate this equation into an explicit definition of $y = f(x)$. But we can still differentiate it with respect to x, using the rules we have seen. We differentiate the left-hand side with respect to x using the product rule, where

$$u = x, \quad v = y, \quad \text{so that } \frac{du}{dx} = \frac{dx}{dx} = 1 \text{ and } \frac{dv}{dx} = \frac{dy}{dx} :$$

$$\frac{d}{dx}(xy) = x.\frac{dy}{dx} + y.\frac{dx}{dx} = x.\frac{dy}{dx} + y.1 = x.\frac{dy}{dx} + y.$$

To differentiate the right-hand side we use the chain rule, since e^{2y} is a function of y, and y is a function of x:

$$\frac{d}{dx}\left(e^{2y}\right) = \frac{d\left(e^{2y}\right)}{dy}.\frac{dy}{dx} = 2e^{2y}\frac{dy}{dx}.$$

Performing these differentiations to the left- and right-hand sides of (12.10) simultaneously:

$$\frac{d}{dx}(xy) = \frac{d}{dx}\left(e^{2y}\right)$$
$$x\frac{dy}{dx} + y = 2e^{2y}\frac{dy}{dx}.$$

Now rearrange this equation to solve for $\frac{dy}{dx}$:

$$2e^{2y}\frac{dy}{dx} - x\frac{dy}{dx} = y$$
$$\frac{dy}{dx} = \frac{y}{2e^{2y} - x}.$$

We'll need to use implicit differentiation in developing the theory of calculus, starting in the next section where we find the derivative of logarithmic functions. Rest assured that these will be more straightforward than in the following example which also involves the chain rule and trigonometric identities:

EXAMPLE

Find $\dfrac{dy}{dx}$; if $y = \tan(x + y)$.

Solution: The derivative of the left-hand side will be $\dfrac{dy}{dx}$; what about the right-hand side?

Using the chain rule, we set $u = x + y$, so the right-hand side is $\tan u$. Then

$$\frac{du}{dx} = \frac{d}{dx}(x + y) = \frac{d(x)}{dx} + \frac{d(y)}{dx} = 1 + \frac{dy}{dx},$$

and

$$\frac{d}{du}(\tan u) = \sec^2 u.$$

So

$$\frac{d}{dx}\big(\tan(x + y)\big) = \frac{d}{dx}(\tan u) = \frac{d(\tan u)}{du} \cdot \frac{du}{dx} = \sec^2 u \cdot \frac{du}{dx} = \sec^2(x + y) \cdot \left(1 + \frac{dy}{dx}\right).$$

We can now return to the original equation and differentiate both sides with respect to x simultaneously:

$$\frac{dy}{dx} = \big(\sec^2(x + y)\big)\left(1 + \frac{dy}{dx}\right).$$

Solving for $\dfrac{dy}{dx}$:

$$\frac{dy}{dx} = \frac{\sec^2(x + y)}{1 - \sec^2(x + y)}.$$

This can be simplified. Using the definition of $\sec x$ as $\dfrac{1}{\cos x}$, we multiply top and bottom by $\cos^2(x + y)$, to get

$$\frac{dy}{dx} = \frac{1}{\cos^2(x + y) - 1}.$$

But we have just recalled the trigonometric identity $\cos^2\theta + \sin^2\theta = 1$, so that $\cos^2\theta - 1 = -\sin^2\theta$. Using this,

$$\frac{dy}{dx} = \frac{1}{-\sin^2(x + y)} = -\mathrm{cosec}^2(x + y).$$

PRACTICE

Differentiate $x^2 + 5y = e^{3y}$, $y = x \sin y$.

Answers: $\dfrac{dy}{dx} = \dfrac{2x}{3e^{3y} - 5}$, $\dfrac{dy}{dx} = \dfrac{\sin y}{1 - x \cos y}$.

Problems: Now try the end-of-chapter problems 12.25–12.26.

. . . .

12.6 Differentiating logarithmic functions

We introduced logarithmic functions in Section 10.3. We can use implicit differentiation to find the derivative of the natural logarithm, $y = \ln x$. Recall how we defined the logarithm as the inverse of the exponential function:

$$y = \log_a (x) \quad \text{means that} \quad x = a^y. \tag{10.11}$$

So if $y = \ln x$ it means that $x = e^y$. We can use implicit differentiation to differentiate both sides with respect to x:

$$\frac{d(x)}{dx} = \frac{d(e^y)}{dx} = \frac{d(e^y)}{dy} \cdot \frac{dy}{dx} \quad \longleftarrow \boxed{\text{By the chain rule}}$$

$$1 = e^y \cdot \frac{dy}{dx}$$

$$\frac{dy}{dx} = \frac{1}{e^y}.$$

But $e^y = x$, so $\dfrac{dy}{dx} = \dfrac{1}{x}$. We have proved:

$$\text{If } y = \ln x, \quad \text{its derivative is} \quad \frac{dy}{dx} = \frac{1}{x}. \tag{12.11}$$

To find the derivatives of logarithms to base a:
$y = \log_a(x)$ means that $x = a^y$. Differentiating both sides with respect to x, and using the derivative of a^x which we found in equation (11.16):

$$\frac{d(x)}{dx} = \frac{d(a^y)}{dx} = \frac{d(a^y)}{dy} \cdot \frac{dy}{dx}$$

$$1 = (\ln a).a^y \cdot \frac{dy}{dx}$$

$$\frac{dy}{dx} = \frac{1}{(\ln a).a^y} = \frac{1}{x.\ln a}.$$

Thus,

$$\text{If } y = \log_a x, \quad \text{its derivative is} \quad \frac{dy}{dx} = \frac{1}{x.\ln a}. \tag{12.12}$$

This derivative can be used within the chain rule, product, and quotient rules:

EXAMPLE

Differentiate $y = \dfrac{\ln x}{x}$.

Solution: We need to use the quotient rule. Set $u(x) = \ln x$ and $v(x) = x.$

Differentiating, $\dfrac{du}{dx} = \dfrac{1}{x}$ and $\dfrac{dv}{dx} = 1$.

Substituting into the quotient rule:

$$\frac{dy}{dx} = \frac{x.\left(\frac{1}{x}\right) - (\ln x).1}{x^2} = \frac{1 - \ln x}{x^2}.$$

PRACTICE

Differentiate $x^2.\ln x$, $\ln^2 x$.

Answers: $x + 2x.\ln x$, $\dfrac{2\ln x}{x}$.

Problems: Now try the end-of-chapter problems 12.27–12.30.

12.7 Differentiating inverse trigonometric functions

For our final trick, we will differentiate the inverse trigonometric functions $\sin^{-1}x$ and $\tan^{-1} x$ which we introduced in Section 9.6.2. As a bioscientist you will probably never need to do this, but the results will provide us with standard integrals for the table at the end of Chapter 14, which you will need to use. As these are inverse functions, we proceed as we did for differentiating logarithms.

From the definition of $\sin^{-1} x$:

$$y = \sin^{-1}(x) \quad \text{means that} \quad x = \sin y.$$

Differentiating both sides with respect to x:

$$\frac{d(x)}{dx} = \frac{d(\sin y)}{dx} = \frac{d(\sin y)}{dy} \cdot \frac{dy}{dx}$$

$$1 = (\cos y) \cdot \frac{dy}{dx}$$

$$\frac{dy}{dx} = \frac{1}{\cos y}.$$

However, we need to write $\dfrac{dy}{dx}$ as a function of x. To do this we must express $\cos y$ in terms of $\sin y$. This can be done by solving the identity $\cos^2 \theta + \sin^2 \theta = 1$ for $\cos \theta$:

$$\cos \theta = \sqrt{1 - \sin^2 \theta}.$$

So

If $y = \sin^{-1} x$,   its derivative is $\dfrac{dy}{dx} = \dfrac{1}{\sqrt{1 - x^2}}.$ $\qquad$ (12.13)

In differentiating $\tan^{-1}x$ we will use the alternative form of the trigonometric identity:

$$\sec^2\theta = 1 + \tan^2\theta. \qquad (9.27)$$

Then

$$y = \tan^{-1}(x) \text{ means that } x = \tan y.$$

Differentiating both sides with respect to x:

$$\frac{d(x)}{dx} = \frac{d(\tan y)}{dx} = \frac{d(\tan y)}{dy} \cdot \frac{dy}{dx}$$

$$1 = (\sec^2 y) \cdot \frac{dy}{dx}$$

$$\frac{dy}{dx} = \frac{1}{\sec^2 y} = \frac{1}{1 + \tan^2 y} = \frac{1}{1 + x^2}.$$

So

> If $y = \tan^{-1} x$,   its derivative is $\dfrac{dy}{dx} = \dfrac{1}{1 + x^2}.$ (12.14)

We now know – or know how to find – the derivatives of all the basic functions we met in Part I, together with rules for differentiating combinations of these functions. We summarize these results in tables at the end of the chapter.

Problems: Now try the end-of-chapter problems 12.31–12.34.

. . . .

12.8 Higher-order derivatives

There is one more piece of the structure of the differential calculus which we need to put in place before we look at practical applications. We introduced the calculus through the example of the velocity $v(t)$ of a moving object, which is the instantaneous rate of change of the distance function $s(t)$ with respect to time t: $v(t) = \dfrac{ds}{dt}$. We can also talk about the rate of change of velocity with respect to time; this quantity is the instantaneous **acceleration** $a(t)$ of the object: $a(t) = \dfrac{dv}{dt}$. If s is measured in metres and t in seconds, the units of velocity are metres per second (m s^{-1}) and the units of acceleration are (metres per second) per second, or metres per second squared (m s^{-2}). To find the acceleration from $s(t)$ we need to differentiate twice.

> **Definition:** The function obtained by differentiating $s(t)$ twice is called the **second derivative** of $s(t)$ with respect to t, and is written as $\dfrac{d^2 s}{dt^2}$.   (12.15)

EXAMPLE

We can return to the example seen in sections 11.1 and 11.3. We saw that if a speeding car's distance function is $s(t) = 6t^3(8 - 3t) = 48t^3 - 18t^4$, then its velocity at time t is given by

$$\frac{ds}{dt} = 48(3t^2) - 18(4t^3) = 144t^2 - 72t^3 = 72t^2(2 - t).$$

We can now differentiate this velocity, and find that its acceleration at time t is given by

$$\frac{d^2 s}{dt^2} = 144(2t) - 72(3t^2) = 288t - 216t^2 = 72t(4 - 3t).$$

The acceleration will be zero when $t = 0$ and also when $t = \frac{4}{3}$ hours. At that instant in time, the car is travelling with a constant speed of $72\left(\frac{4}{3}\right)^2 \left(2 - \frac{4}{3}\right) = 72 \times \frac{16}{9} \times \frac{2}{3} = \frac{256}{3}$, or $85\frac{1}{3}$ miles per hour.

The **order** of a derivative is the number of times the function has been differentiated. So the full title of $\frac{dy}{dx}$ is 'the first-order derivative of y with respect to x' (or 'the first derivative of y'), and $\frac{d^2 y}{dx^2}$ is the second-order derivative. The second derivative plays an important role in finding maxima and minima of a function, as we'll see in the first section of Chapter 13.

Of course, we could carry on differentiating beyond $\frac{d^2 y}{dx^2}$, finding the third derivative, fourth derivative, and so on. You may suppose that these higher-order derivatives are fairly meaningless concepts, of interest only to mathematicians. but if you read a newspaper report that '*the latest data on the cost of living show that the rate of inflation is still rising, although not as steeply as last month*', they are talking about the third derivative of a price function!

In the next chapter we will write down higher derivatives of general polynomials of degree n. Here, we look at differentiating the power functions x^n, where n is a natural number. The first three derivatives of $f(x) = x^3$ are:

$$\frac{df}{dx} = 3x^2, \quad \frac{d^2 f}{dx^2} = 3(2x) = 6x, \quad \frac{d^3 f}{dx^3} = 6.$$

After three differentiations we have reached a constant; we could continue to differentiate, but all we get is zero:

$$\frac{d^4 f}{dx^4} = 0, \quad \frac{d^5 f}{dx^5} = 0, \quad \ldots$$

Repeating this process for $f(x) = x^6$, we reach a constant function after six differentiations. Fill in the gaps in the list below:

$$\frac{df}{dx} = 6x^5, \quad \frac{d^2 f}{dx^2} = 6(5x^4) = 30x^4, \quad \frac{d^3 f}{dx^3} = \ldots,$$

$$\frac{d^4 f}{dx^4} = \ldots, \quad \frac{d^5 f}{dx^5} = \ldots, \quad \frac{d^6 f}{dx^6} = 720.$$

You should see that the number 720 arises from $6 \times 5 \times 4 \times 3 \times 2$. There is a mathematical notation for this: we write it as 6!, pronounced 'six factorial'. You should have a factorial key on your calculator, labelled n! or x!. Here is the definition of what this key does:

If you take any natural number n, we define the quantity $n!$ (pronounced 'n factorial') as the product of all the integers from 1 up to n:

Definition: If n is a natural number, define

$$n! = n.(n - 1).(n - 2).(n - 3).\ldots.3.2.1. \tag{12.16}$$

Also define $0! = 1$.

So $6! = 6 \times 5 \times 4 \times 3 \times 2 \times 1 = 720,$ and $10! = 10 \times 9 \times \ldots \times 3 \times 2 \times 1 = 3\,628\,800.$

The factorial function $n!$ is not a continuous function: it is only defined at the discrete points $n = 0, 1, 2, 3, \ldots$ on the real line. It grows very quickly as n increases; the factorial of $n = 20$ is 2.4×10^{18}, for example.

We can now generalize what we have observed from differentiating $f(x) = x^3$ and $f(x) = x^6$, to repeatedly differentiating the general power function $f(x) = x^n$. After n differentiations we will obtain the nth derivative, written as $\dfrac{d^n f}{dx^n}$.

If $f(x) = x^n$ is differentiated n times, we obtain a constant function:

$$\frac{d^n}{dx^n}\left(x^n\right) = n!. \tag{12.17}$$

At the start of this chapter we described a curve as **continuous** if there are no sudden jumps in the function value, and **differentiable** if there are no sudden changes in the first derivative. If you can continue to differentiate the function, and all the derivatives are continuous, we say that the function is **smooth**. Polynomial functions, and $\sin x$ and $\cos x$ are examples of smooth functions.

EXAMPLE

If $y = \sin x$, then:

$$\frac{dy}{dx} = \cos x$$

$$\frac{d^2 y}{dx^2} = \frac{d}{dx}\left(\cos x\right) = -\sin x$$

$$\frac{d^3 y}{dx^3} = \frac{d}{dx}\left(-\sin x\right) = -\frac{d}{dx}\left(\sin x\right) = -\cos x$$

$$\frac{d^4 y}{dx^4} = \frac{d}{dx}\left(-\cos x\right) = -\frac{d}{dx}\left(\cos x\right) = -(-\sin x) = \sin x = y$$

and with further differentiation the cycle repeats.

In Section 11.3.2 we introduced various notations used for the derivative, and we can now expand this to cover notation for higher-order derivatives. Here is a list of notations that you might meet in books and research papers:

Type of notation	Function	1st derivative	2nd derivative	nth derivative
Standard notation	y	$\dfrac{dy}{dx}$	$\dfrac{d^2 y}{dx^2}$	$\dfrac{d^n y}{dx^n}$
Prime notation (favoured by mathematicians)	y	y'	y''	$y^{(n)}$
D-notation (used when solving some types of differential equation)	y	$\mathrm{D}y$	$\mathrm{D}^2 y$	$\mathrm{D}^n y$
Dot notation (used by engineers and scientists when the derivative is with respect to time t)	y	$\dot{y}$	$\ddot{y}$	

And if $y = f(x)$, writers may use f instead of y in any of the above examples!

You may see a rather similar notation in equations in research papers, using the symbol ∂, for example $\dfrac{\partial f}{\partial x}$. This is the **partial derivative** of f with respect to x; it is used when f is a function of more than one variable. If f is a function of three independent variables x, y, and z, which we write as $f = f(x, y, z)$, then $\dfrac{\partial f}{\partial x}$ means that we differentiate f with respect to x, treating all occurrences of y and z as if they were constants. We will introduce partial derivatives in Section 16.4.

Problems: Now try the end-of-chapter problems 12.35–12.38.

. . . .

12.9 Summary of standard derivatives, and rules of differentiation

y	$\dfrac{dy}{dx}$
c	0
x	1
x^n	nx^{n-1}
e^x	e^x
$a^x \ (a > 0)$	$(\ln a).a^x$
$\sin x$	$\cos x$
$\cos x$	$-\sin x$
$\tan x$	$\sec^2 x$
$\ln x$	$\dfrac{1}{x}$
$\log_a x$	$\dfrac{1}{x.\ln a}$
$\sin^{-1} x$	$\dfrac{1}{\sqrt{1-x^2}}$
$\tan^{-1} x$	$\dfrac{1}{1+x^2}$

Sum rule:

$$\frac{d}{dx}\big(f(x) + g(x)\big) = \frac{df}{dx} + \frac{dg}{dx}$$

$$\frac{d}{dx}\big(f(x) - g(x)\big) = \frac{df}{dx} - \frac{dg}{dx}$$

Coefficient rule:

$$\frac{d}{dx}\big(af(x)\big) = a.\frac{df}{dx}$$

Chain rule:

If $y = y(u)$ and $u = u(x)$, then

$$\frac{dy}{dx} = \frac{dy}{du} \times \frac{du}{dx}.$$

Product rule: $\dfrac{d}{dx}(u.v) = u.v' + v.u'$

Quotient rule: $\dfrac{d}{dx}\left(\dfrac{u}{v}\right) = \dfrac{v.u' - u.v'}{v^2}$

? PROBLEMS

12.1–12.11: Differentiate the following functions of x:

12.1 $\sqrt{3 + 4x}$ **12.2** $\sqrt{10 - x^2}$ **12.3** $(3x - 2)^5 - \dfrac{1}{(2x - 3)^2}$

12.4 $(6e^{2x} + x^2)^3$ **12.5** $x(1 - x)^2$ **12.6** $\sqrt{x}\left(2 + \dfrac{1}{x}\right)$

12.7 $x^2 e^x$ **12.8** $\left(x + \dfrac{1}{x}\right)e^{2x}$ **12.9** $\dfrac{x}{1 - x}$

12.10 $\dfrac{1 + x^2}{1 - x^2}$ **12.11** $\dfrac{1 + e^x}{1 - e^x}.$

12.12: Obtain the quotient rule for differentiating $y = \dfrac{u(x)}{v(x)}$ from the product rule. Write y as the product $y = u(x).[v(x)]^{-1}$, and differentiate the second factor $[v(x)]^{-1}$ using the chain rule.

12.13–12.16: Differentiate the following functions of x. You will need to use a combination of the chain rule, product rule, and/or quotient rule:

12.13 $(1 + x)\sqrt{1 - x}$ **12.14** $\dfrac{\sqrt{1 + e^x}}{x}$

12.15 $\sqrt{\dfrac{1 + x}{1 - x}}$ **12.16** $\dfrac{x}{\sqrt{1 + x^2}}.$

12.17–12.20: Differentiate the following trigonometric functions of x:

12.17 $x.\sin x$ **12.18** $\tan^2 x$

12.19 $\dfrac{\sin x}{x}$ **12.20** $\dfrac{\tan x}{x}.$

12.21: Differentiate $y = \sec x$ by writing it as $y = (\cos x)^{-1}$ and using the chain rule.

12.22: Differentiate $y = \operatorname{cosec} x$ and $y = \cot x$.

12.23: In Chapter 9 we saw the trigonometric identity $\sin 2x = 2\sin x \cos x$. Differentiate the left-hand side using the chain rule, and differentiate the right-hand side using the product rule. Are the two derivatives identical?

12.24: Repeat Problem 12.23 with the identity $\sec^2 x = 1 + \tan^2 x,$ using appropriate rules of differentiation.

12.25–12.26: Use implicit differentiation to find $\dfrac{dy}{dx}$ as a function of x and y:

  12.25 $x^2 + y^2 = r^2$ (r constant)   12.26 $\sin y = 1 - x^2$.

12.27–12.30: Differentiate the following functions involving logarithms:

  12.27 $\ln(1 + x^2)$ 12.28 $\dfrac{\ln x}{1 + x^2}$

  12.29 $\ln\left(\dfrac{1}{x}\right)$ 12.30 $\ln\left(\sqrt{1 + x^2}\right)$.

12.31–12.32: Use implicit differentiation to find the derivatives of:

  12.31 $\cos^{-1} x$ 12.32 $\sec^{-1} x$.

12.33–12.34: Use the rules and standard derivatives to differentiate:

  12.33 $x \sin^{-1} x$ 12.34 $\tan^{-1}(x^2)$.

12.35: Show that the function $y = x(x + 1)$ satisfies the differential equation

$$\frac{d^2 y}{dx^2} - 2\frac{dy}{dx} + 4y = 4x^2.$$

12.36: We start with $y = x^n$, and differentiate it k times, where n, k are natural numbers and $k < n$. Write down an expression for $\dfrac{d^k y}{dx^k}$, using factorial notation.

12.37: If $y = xe^x$, show that its nth order derivative is $\dfrac{d^n y}{dx^n} = (x + n)e^x$.

12.38: Find an expression for the nth order derivative of $y = xe^{ax}$, where a is a positive constant.

Applications of differentiation

The differential calculus gave a tremendous impetus to the development of both pure and applied mathematics. New fields of study grew up, such as optimization and polynomial approximation of functions, together with a powerful numerical method for finding roots – Newton's method – which is widely used on computers today. We will introduce all these topics in this chapter.

We start, however, by looking at functions, their graphs, and the significant features of those graphs. In Part I we introduced the basic functions of a single variable x, together with their graphs. Using logical argument and a lot of algebra, we were able to find roots (points where the curve cuts the x-axis) and turning-points (where the curve reaches a maximum or minimum). But these could only be found for simple functions. For example, we have a formula for the roots of a quadratic function, but formulae for the roots of cubic or higher-degree polynomials are either extremely complicated or non-existent. Similarly, we know where the trigonometric functions $\sin x$ and $\cos x$ reach their maxima, but we cannot determine the maximum of the function $y = \sin x + \cos x$. In the next section we will show how calculus can provide all the 'pre-calculus' results about roots and turning-points from Part I for the basic functions, and extend the analysis to more complicated differentiable functions. This is of more than theoretical interest; roots and turning-points have important practical significance. But we should perhaps have titled this chapter 'Mathematical applications of differentiation', which can in turn be applied to bioscience situations such as we've already seen.

13.1 Interpretation of graphs

Suppose you are a scientist trying to model a practical problem. Once you have formulated a mathematical model of your problem, solving it will produce a function expressing the variable you are interested in – usually as a function of time. You can try to visualize your function by sketching its graph. The final, crucial step is to **interpret** the function or graph, i.e. to extract significant information from which your practical conclusions can be drawn. The calculus can provide us with this information.

Here's a specific example. You are working in a hospital, and have formulated a model of the variation in blood glucose level in a patient with diabetes. Solving this will give us a function $g(t)$, expressing the blood glucose level at time t over a 24-hour period. You can plot or sketch the graph, but the patient's doctor will want you to explain the features that have clinical significance, such as:

- At what times will the glucose level fall outside the 'normal' range of 4–10 mmol per litre?
- During which time periods will the level be increasing, and when will it be decreasing?
- When will the level reach a maximum or minimum, and what will those extreme values be?
- In a period when the level is increasing, will it be levelling off to a stable plateau, or will it be rising ever more steeply to a dangerous 'spike' requiring medical attention?

All these questions can be answered by looking at the function and its first and second derivatives. To answer the question of when the function reaches a prescribed value, for example to find the time t at which $g(t) = 4$ mmol l^{-1}, we use techniques for finding the root of a function. The equation can be rewritten as $g(t) - 4 = 0$, which we solve by looking for a root of the function $f(t)$ defined by $f(t) = g(t) - 4$.

We already looked at the properties of graphs of linear and quadratic functions, in Chapters 6 and 7. To summarize:

§6.3 Linear function $f(x) = mx + c$

- The graph of $f(x)$ is a straight line, with slope m and y-intercept c.
- The function has one root, at $x = -\dfrac{c}{m}$.
- If the slope m is positive the graph slopes upwards, and the graph slopes downwards if $m < 0$.

§7.2 Quadratic function $f(x) = ax^2 + bx + c$

- The graph of $f(x)$ is a parabola.
- If $b^2 - 4ac > 0$ the graph has two roots, at $x = \dfrac{-b \pm \sqrt{b^2 - 4ac}}{2a}$.
- The graph has a single turning-point at $x = -\dfrac{b}{2a}$.
- If $b^2 - 4ac = 0$, the graph has just one root, at $x = -\dfrac{b}{2a}$ (which is also the turning-point). This is called a double root. If $b^2 - 4ac < 0$, there are no roots.
- If $a > 0$ the parabola is concave up, and the turning-point is a minimum. If $a < 0$ the parabola is concave down, and the turning-point is a maximum.

Recall that a curve is 'concave up' if it is like looking from above into a bowl, and 'concave down' if it is like an upturned bowl. We'll give a more mathematical definition in Section 13.1.4, when we discuss curvature.

We now compare the above results with what calculus can tell us about these functions. In the case of the linear function $f(x) = mx + c$, its derivative is $f'(x) = m.1 + 0 = m$, which is the gradient of the curve, i.e. the slope of the line. What about the quadratic function? In Chapter 7 we said that the turning-point of the parabola will be at $x = -\dfrac{b}{2a}$ because that is midway between the two roots. We then went on to claim (without any justification!) that this would be true even if the quadratic function had no roots. We can now prove this using calculus. Differentiate $f(x) = ax^2 + bx + c$ and we get $f'(x) = a(2x) + b(1) + c(0) = 2ax + b$. At a turning-point the slope of the tangent is zero, so setting $2ax + b = 0$ and solving for x gives $x = -\dfrac{b}{2a}$. This result is irrespective of whether the graph has two, one, or no roots. The second derivative is $f''(x) = a.2 + b.0 = 2a$, so when $f''(x) > 0$ it means that $a > 0$, and we have a parabola that is concave up, with a minimum; when $a < 0$ the parabola will be concave down, with a maximum.

We will now use the function $f(x)$ and its first and second derivatives $f'(x)$ and $f''(x)$ to identify significant features of the graph of a general smooth function $f(x)$. We will create a catalogue (a taxonomy, so to speak) of function properties, depending on whether $f(x), f'(x), f''(x)$ – and even the third derivative $f'''(x)$ – are positive, negative, or zero. Some properties apply to a single point (e.g. 'the function has a maximum at $x = 2$'), while others describe the behaviour of the function over an interval (e.g. 'the function is decreasing over the interval $2 < x < 5$'). With such a taxonomy we could describe algebraically the geometric features of a function curve without the need to plot or sketch it. This is important because any problem-solving computer algorithm is based on algebra, not geometry.

Recall from Chapter 1 that the expression $a < x < b$ means 'the interval from $x = a$ to $x = b$' on the real line, i.e. all the values of x between a and b. In writing $a < x < b$ we are conflating the two inequalities $a < x$ and $x < b$.

13.1.1 Gradients

We will consider the graph of a differentiable function $y = f(x)$ over an interval from $x = a$ to $x = b$ on the x-axis; there is an example in Figure 13.1.

If we just look at the sign of the function value over the interval, we can make the following rather obvious definition:

If $f(x) > 0$ for all values of x between a and b, we say that **the function is positive over the interval** $a < x < b$.

If $f(x) < 0$ for all values of x between a and b, we say that **the function is negative over the interval** $a < x < b$.

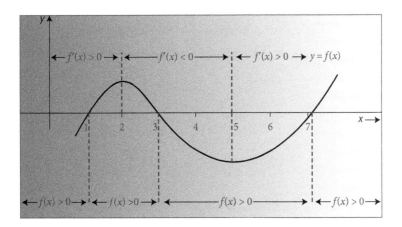

Figure 13.1 Positive and negative gradients

If the function is positive over the interval, it means that the graph of $y = f(x)$ will lie above the x-axis on that interval. If the function is negative, the graph will lie below the x-axis. In Figure 13.1, the function is positive on the interval $1 < x < 3$, and when $x > 7$. It is negative when $x < 1$ and on the interval $3 < x < 7$.

Now think about the sign of the first derivative $f'(x)$ on this interval. If $f'(x)$ is positive throughout the interval, then the gradient of the curve is positive, meaning that the tangent to the curve is always sloping upwards. As we noted in the introduction to Section 11.3 (using the analogy of a rollercoaster car travelling along the curve), this means that the function is increasing in value. We can state this in a definition:

> If $f'(x) > 0$ for all values of x between a and b, we say that **the function is increasing over the interval** $a < x < b$.
>
> If $f'(x) < 0$ for all values of x between a and b, we say that **the function is decreasing over the interval** $a < x < b$.

The function in Figure 13.1 is increasing when $x < 2$ and when $x > 5$, and it is decreasing on the interval $2 < x < 5$.

13.1.2 Roots

A **root** of $f(x)$ is a value of x at which $f(x) = 0$. So if $f(x)$ has a root at $x = a$, we can write this as $f(a) = 0$. The graph of $y = f(x)$ will touch the x-axis at $x = a$.

We can identify two different types of root, depending on whether the graph of $y = f(x)$ cuts through the x-axis, or does something more complicated. In the graph in Figure 13.2, there are roots of $y = f(x)$ at $x = a$, $x = b$, and $x = c$. That is, $f(a) = f(b) = f(c) = 0$. At the point $x = a$, the tangent is sloping downwards, so $f'(a) < 0$. At the point $x = b$, the tangent is sloping upwards, so $f'(b) > 0$. Roots such as these, where the gradient $f'(x)$ at the root is non-zero, are called **simple roots**. The curve is cutting through the x-axis at an angle.

The other possibility is that $f'(x)$ is zero at the root. An example of this is the root at $x = c$ in Figure 13.2. It is now possible for the curve to turn back, instead of crossing the x-axis, making the root either a minimum, as in the diagram, or a maximum (see the next section). There is another possibility, as seen in Figure 11.4(b) at the point $(0, 0)$ on the graph of $y = x^3$. Such a point is called a point of inflection, which we'll define in Section 13.1.4.

You could create a second root like that at $x = c$, by pulling up the left-hand part of the curve in Figure 13.2 (as indicated by the arrows and the dashed curve), making the two simple

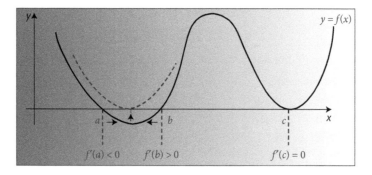

Figure 13.2 Formation of a double root

roots at *a* and *b* come closer together until they coalesce. This is why we refer to a quadratic function having a **double root**. In the case of the root of the function $y = x^3$, it is as if three separate roots have come together. This is called a triple root, or a root of multiplicity 3. We won't worry about the multiplicity, we will just refer to non-simple roots as multiple roots.

> The function $f(x)$ has a **root** at $x = a$ if $f(a) = 0$.
>
> This is a **simple root** if $f'(a) \neq 0$. Otherwise, if $f'(a) = 0$, it is a **multiple root**.

EXAMPLE

(i) The function $f(x) = \sin x$ has a root at $x = \pi$, since $f(\pi) = \sin \pi = 0$. Differentiating, we get $f'(x) = \cos x$, and at $x = \pi$, $f'(\pi) = \cos \pi = -1$, so this is a simple root of $\sin x$.

(ii) Now consider the function $g(x) = \sin^2 x$. This also has a root at $x = \pi$, since $g(\pi) = \sin^2 \pi = 0^2 = 0$. But now the function derivative is $g'(x) = \dfrac{d}{dx}(\sin^2 x) = (2\sin x)(\cos x)$ using the chain rule. And now $g'(\pi) = 2\sin \pi \cos \pi = 2(0)(-1) = 0$, so this is a multiple root. **FNGRAPH 13.1** shows the graphs of these two functions, so that you can see the difference in their curves at the roots. See Figure 13.3.

13.1.3 Critical points

Using our rollercoaster analogy, we have seen that when the curve reaches a maximum or minimum, the rollercoaster car is momentarily travelling horizontally, so the tangent to the curve at these points is horizontal, meaning that the gradient at these points is zero. A point on the curve where the gradient is zero is called a **critical point**.[1] A critical point may be a **turning-point** (a maximum or a minimum) or a point of inflection (which we'll discuss in the next section). To decide what type of critical point you are dealing with, you need to look at the sign of the second derivative at that point.

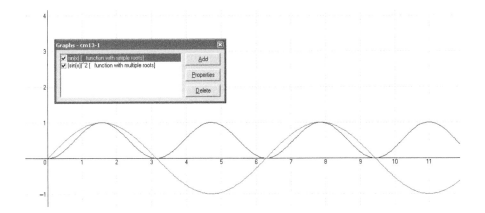

Figure 13.3 Graphs of $\sin x$ and $\sin^2 x$ (**FNGRAPH 13.1**)

---

[1] Points at which the derivative are not defined, are also classed as critical points.

In the case of the quadratic function $f(x) = ax^2 + bx + c$ we have seen that the second derivative is $f''(x) = 2a$, so that when $f''(x) > 0$ then the coefficient a is positive and the curve is a parabola that is concave up, with a minimum. When $f''(x) < 0$ then a is negative and the curve is a parabola that is concave down, with a maximum; see Figure 7.5. This is the general rule: if $f''(x) < 0$ at a turning-point then we have a maximum, and if $f''(x) > 0$ at a turning-point we have a minimum.

In the upper graph of Figure 13.4 we show a typical cubic function, with a maximum at $x = a$ and a minimum at $x = b$. The tangent is horizontal at these points, so $f'(a) = f'(b) = 0$. At point $x = a$ the curve is increasing (positive gradient) on the left of a, and decreasing (negative gradient) on the right, so the gradient is changing from positive to negative at $x = a$. At the point $x = b$ the situation is reversed. The graph of the gradient $y = f'(x)$ is sketched in the middle graph of Figure 13.4, and you can see that the gradient of that curve is negative at $x = a$ and positive at $x = b$. The gradient of the gradient is the derivative of the derivative, i.e. the second derivative $f''(x)$, sketched in the bottom graph. As predicted, $f''(x)$ is negative at $x = a$ and positive at $x = b$.

This gives us a way of distinguishing between maxima and minima algebraically, without needing to sketch a graph:

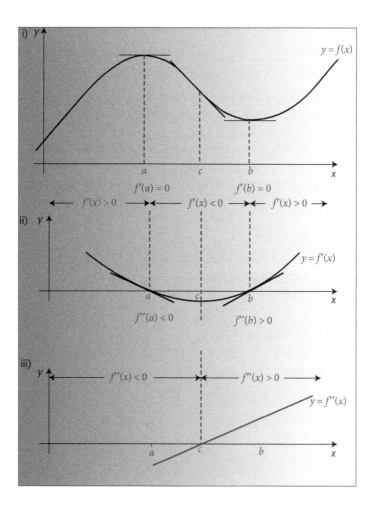

Figure 13.4 A cubic function and its derivatives

The function $f(x)$ has a **minimum** at $x = a$ if $f'(a) = 0$ and $f''(a) > 0$.

The function $f(x)$ has a **maximum** at $x = a$ if $f'(a) = 0$ and $f''(a) < 0$.

If $f'(a) = 0$ and $f''(a) = 0$, we are unable to determine what type of critical point it is. It may be a point of inflection (as seen in Figure 11.4(b) at the point $(0, 0)$ on the graph of $y = x^3$) but it could also be a minimum (as at the point $(0, 0)$ on the graph of $y = x^4$) or a maximum (e.g. on the curve $y = -x^4$).

EXAMPLE

Find a critical point of $y = \sin x + \cos x$.

Solution: Differentiate: $\dfrac{dy}{dx} = \cos x - \sin x$.

Setting $\dfrac{dy}{dx} = 0$ means that $\sin x = \cos x$. We could sketch graphs of $\sin x$ and $\cos x$ and see where they intersect. Alternatively, divide both sides by $\cos x$ to obtain

$\tan x = 1$.

Using your knowledge of trigonometric functions (or your calculator to find $\tan^{-1} 1$) you should see that the principal solution is $x = \dfrac{\pi}{4}$ or $45°$. At this point $y\left(\dfrac{\pi}{4}\right) = \dfrac{1}{\sqrt{2}} + \dfrac{1}{\sqrt{2}} = \dfrac{2}{\sqrt{2}} = \sqrt{2}$. To determine the nature of this critical point $\left(\dfrac{\pi}{4}, \sqrt{2}\right)$, differentiate again:

$y'' = -\sin x - \cos x = -y$ so $y''\left(\dfrac{\pi}{4}\right) = -\sqrt{2} < 0,$ so the point is a local maximum.

Check this by drawing the curve in FNGraph.

13.1.4 Curvature

The sign of the second derivative of $f(x)$ over an interval determines the curvature of the function. This may be either 'concave up' or 'concave down'. 'Concave up' means that when you look down at the curve from above, it looks concave like a bowl. A more mathematical definition is that a curve is **concave** if, joining any two points on the curve by a secant line, the line would lie in front of the curve. This is illustrated in Figure 13.5.

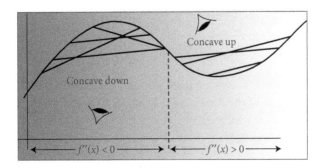

Figure 13.5 Curvature

We have seen from our analysis of quadratic functions, and also from Figure 13.4, that 'concave up' curvature on an interval corresponds to a negative second derivative, and 'concave down' curvature occurs when the second derivative is positive:

> If $f''(x) > 0$ for all values of x between a and b, we say that **the function is 'concave up' over the interval** $a < x < b$. This is also called **positive curvature.**
>
> If $f''(x) < 0$ for all values of x between a and b, we say that **the function is 'concave down' over the interval** $a < x < b$. This is also called **negative curvature.**

In the lower graph of Figure 13.4 we sketched the graph of the second derivative, $y = f''(x)$. The curve of $y = f(x)$ is 'concave down' (negative curvature) on the interval $x < c$, and is 'concave up' (positive curvature) on the interval $x > c$.

At the point $x = c$ on the graph, the curvature changes from negative to positive. This is called a **point of inflection** (or 'inflexion' in US English). To say that the curvature changes sign at $x = c$ means that this point should be a simple root of $y = f''(x)$, i.e. that the third derivative is non-zero at that point:

> The function $f(x)$ has a **point of inflection** at $x = c$ if $f''(c) = 0$ and $f'''(c) \neq 0$.

EXAMPLE

Find the turning points and points of inflection of the cubic function $f(x) = 2x^3 - 6x^2 + 5$.
Solution: Differentiating twice:

$$f'(x) = 6x^2 - 12x = 6x(x - 2), \quad \text{and} \quad f''(x) = 12x - 12 = 12(x - 1).$$

So there are critical points (points at which $f'(x) = 0$), at $x = 0$ and $x = 2$.
When $x = 0$, $f''(0) = -12 < 0$, so this point is a maximum.
When $x = 2$, $f''(2) = 12 > 0$, so this point is a minimum.
The second derivative is negative on the interval $x < 1$, and positive for $x > 1$. So the function will be concave down for $x < 1$, and concave up for $x > 1$. At $x = 1$, $f''(1) = 0$, and as the curvature changes from negative to positive at this point, and $f'''(x) = 12 \neq 0$, we have a point of inflection at $x = 1$.

Looking back to the top two graphs of Figure 13.4, the point of inflection at $x = c$ corresponds to a point where the gradient of the curve reaches a maximum or minimum. This has a physical significance in biosciences applications. In Section 11.7 we wrote the approximate equation

$$\Delta y \approx \frac{dy}{dx} . \Delta x \tag{11.17}$$

which predicts the change in y resulting from a small change Δx in x. If Δx is fixed, the size of Δy will depend on the size of the derivative. So the point at which the derivative is a maximum represents the point at which a given change in x will have the largest effect on y, i.e. it is where the dependent variable is most sensitive to a change in the independent variable. In the example of a population size plotted against time, a point of inflection corresponds to the time when the population is growing most rapidly. This example will be studied in the case study at the end of this section. You can see the point of inflection on the population curve, in Figure 13.8.

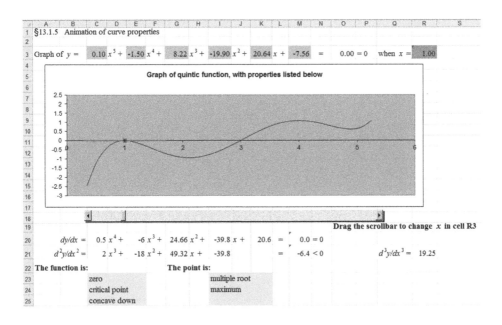

Figure 13.6 Animation of curve properties (SPREADSHEET 13.1)

CASE STUDY A13: Analysing the Ricker update equation

In instalment A8 of this Case Study (in Section 7.4.2) we wrote an update equation for logistic growth:

$$N_{p+1} = N_p + r.N_p\left(1 - \frac{N_p}{K}\right)$$ (7.33)

where N_p and N_{p+1} are the populations in the pth and $(p + 1)$th generations. The general form of such a population update equation is

$$N_{p+1} = F(N_p),$$

for some function $F(N)$. In fisheries management there is a model of populations of fish stocks called the Ricker model, developed for species such as Pacific salmon where the adult fish population feeds on its own eggs and larvae (Clark 2005, Section 7.4). A simplified form of this equation (taking $K = 1$) is

$$N_{p+1} = N_p\, e^{r(1-N_p)}.$$

We shall analyse this function F, writing $y = F(x)$ where $F(x) = xe^{r(1-x)}$.

Notice that $F(0) = 0$; this is the only root of the function.

Differentiating using the product rule: let $u = x$ and $v = e^{r(1-x)}$. Check that $\dfrac{dv}{dx} = -re^{r(1-x)}$ by the chain rule, since $r(1 - x) = r - rx$, whose derivative is $-r$. Then

$$\frac{dF}{dx} = x.\left(-re^{r(1-x)}\right) + e^{r(1-x)}.1$$
$$= (1 - rx)e^{r(1-x)}$$

so there is one critical point, when $x = \dfrac{1}{r}$. Differentiating again by the product rule, now with $u = 1 - rx$, you should find that

$$
\begin{aligned}
\frac{d^2 F}{dx^2} &= (1-rx).(-re^{r(1-x)}) + e^{r(1-x)}.(-r) \\
&= (-r + r^2 x - r)e^{r(1-x)} \\
&= -r(2 - rx)e^{r(1-x)}
\end{aligned}
$$

which is negative when $x = \dfrac{1}{r}$, so the critical point is a maximum.

There is a point of inflection when the second derivative is zero, which occurs if the factor $2 - rx = 0$. This occurs when $x = \dfrac{2}{r}$.

You should sketch the curve in FNGraph, trying different values of r, and identifying the maximum at $x = \dfrac{1}{r}$ and the point of inflection at $x = \dfrac{2}{r}$.

13.1.5 Summary

We will now summarize the properties of functions that we have identified. We divide these up into properties that apply over an interval, and properties that can occur at individual points.

In the table below, if the function or its derivatives have the condition specified at all points on an interval, then we say that the function has the corresponding property on that interval:

Condition on the interval	Property on the interval
$f(x) > 0$	$f(x)$ is positive
$f(x) < 0$	$f(x)$ is negative
$f'(x) > 0$	$f(x)$ is increasing
$f'(x) < 0$	$f(x)$ is decreasing
$f''(x) > 0$	$f(x)$ has positive curvature (concave up)
$f''(x) < 0$	$f(x)$ has negative curvature (concave down)

(13.1)

In the following table, if the condition shown is satisfied at a point $x = a$, then that point has the corresponding property:

Condition at $x = a$	Property at $x = a$
$f(a) = 0, f'(a) \neq 0$	simple root
$f(a) = 0, f'(a) = 0$	multiple root
$f'(a) = 0, f''(a) > 0$	minimum
$f'(a) = 0, f''(a) < 0$	maximum
$f''(a) = 0, f'''(a) \neq 0$	point of inflection

(13.2)

These tables of properties are animated in SPREADSHEET 13.1 (Figure 13.6). This shows the graph of a quintic polynomial. You can drag the point along the curve using the scrollbar; the

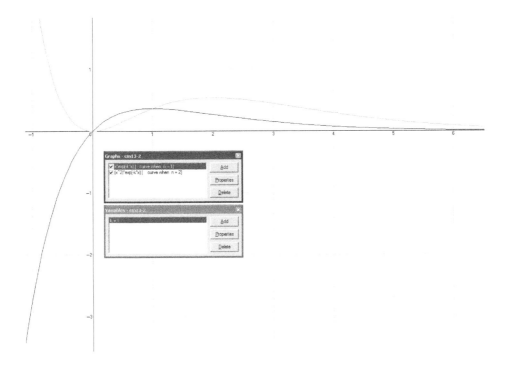

Figure 13.7 Graph of $y = xe^{-kx}$ and $y = x^2 e^{-kx}$ (FNGRAPH 13.2)

value of x appears in cell R3. Below the graph, the values of the first, second, and third derivatives are calculated. Check the formulas in the cells, which use the standard derivative $\dfrac{d(x^n)}{dx} = n.\,x^{n-1}$. Below this, the properties in table (13.1) are indicated, for the current value of x, under the heading 'The function is:'. They will change as the point moves and the function goes from increasing to decreasing, etc. Under the heading 'The point is:' the relevant text will appear when any of the conditions in table (13.2) are satisfied. Look at the formulas in the cells in columns C and I to see how these conditions are programmed using 'if' functions. You can change the values of the coefficients in the polynomial, in row 3, to analyse different curves.

We now give an example of using this list of properties to interpret the graphs of two functions. We want to show just how much information can be extracted systematically from the functions, so it will be quite a long example. In practice, the scientist will probably know the general shape of the function curve from other sources, or from using a curve-sketching package such as FNGraph. But calculus will be needed to find the exact coordinates of the turning-points, for example.

EXAMPLE

Discuss the properties of the functions $f_1(x)$ and $f_2(x)$, defined by

$$f_n(x) = Ax^n e^{-kx} \tag{13.3}$$

where A and k are positive constants. We saw such functions as possible models of the cellular learning behaviour of habituating sensitization in Section 12.3.1.

Discussion: The function $f_1(x)$ is obtained by taking $n = 1$ in (13.3), and for $f_2(x)$ we take $n = 2$. Notice first of all that when a function is multiplied by a positive constant A, then all its derivatives are also multiplied by A, and that this has no effect on the sign of the derivative, so all the conditions listed in the tables above will be unaffected. We can therefore ignore the A, and just analyse the two functions

(i) $f_1(x) = xe^{-kx}$ and   (ii) $f_2(x) = x^2e^{-kx}$.

Notice also that the factor e^{-kx} is positive for all values of x, and that its derivative is $-ke^{-kx}$, see equation (11.15). Finally, remember that $e^0 = 1$.

Solution: (i) First, we analyse $f_1(x) = xe^{-kx}$.

Although this is a totally different biological application, the function $f_1(x)$ is very similar to the Ricker function we just saw in Case Study A13. The following algebra should therefore look familiar to you:

The function is negative for $x < 0$, and positive for $x > 0$. It is zero when $x = 0$, so this is a root of the function: as $f_1(0) = 0$ the root is at the origin $(0, 0)$.

Differentiating $f_1(x)$ using the product rule:

$$f_1'(x) = x.(-ke^{-kx}) + (1).(e^{-kx}) = (1 - kx)e^{-kx}$$

When $x = 0$, $f_1'(0) = (1 - 0).e^0 = 1$, so the root at $x = 0$ is a simple root; the curve cuts through the x-axis with a slope of 1, i.e. at an angle of 45°.

The derivative will be positive when $1 - kx > 0$, i.e. when $x < \dfrac{1}{k}$, and negative when $x > \dfrac{1}{k}$, so the function $f_1(x)$ will be increasing when $x < \dfrac{1}{k}$, and decreasing when $x > \dfrac{1}{k}$,

The derivative is zero when $x = \dfrac{1}{k}$, so this is a critical point. To determine its type, differentiate again (another use of the product rule):

$$f_1''(x) = (1 - kx).(-ke^{-kx}) + (-k).(e^{-kx}) = -k(2 - kx)e^{-kx}.$$

When $x = \dfrac{1}{k}$, $f_1''\left(\dfrac{1}{k}\right) = -k(2 - 1)e^{-1} = \dfrac{-k}{e} < 0$, so the critical point is a maximum.

In general, $f_1''(x)$ will be negative when $2 - kx > 0$, i.e when $x < \dfrac{2}{k}$, and positive when $x > \dfrac{2}{k}$, so the function $f_1(x)$ will be concave up when $x < \dfrac{2}{k}$, and concave down when $x > \dfrac{2}{k}$.

The second derivative is zero when $x = \dfrac{2}{k}$, and the curvature changes from negative to positive, so this is a point of inflection.

Can you sketch the graph of $f_1(x)$ from this information? Take $k = 1$; then there is a maximum at $x = 1$, and a point of inflection at $x = 2$. There is no minimum. To see the graph, look at FNGRAPH 13.2 (Figure 13.7).

(ii) We now analyse $f_2(x) = x^2e^{-kx}$.

This time, the function is positive for all non-zero values of x. It is zero when $x = 0$, so this is a root of the function; thus, like $f_1(x)$ it also passes through the origin.

Differentiating $f_2(x)$ using the product rule:

$$f_2'(x) = x^2.(-ke^{-kx}) + (2x).(e^{-kx}) = (2x - kx^2)e^{-kx} = x(2 - kx)e^{-kx}.$$

When $x = 0$, $f_2'(0) = 0.(2 - 0).e^0 = 0$, so the root at $x = 0$ is a multiple root; the tangent to the curve is horizontal at this point.

The derivative is zero for two values of x: when $x = 0$ (the multiple root) and when $2 - kx = 0$, which is when $x = \dfrac{2}{k}$, To determine their types, differentiate again:

$$f_2''(x) = (2x - kx^2).(-ke^{-kx}) + (2 - 2kx).(e^{-kx})$$
$$= (-2kx + k^2x^2 + 2 - 2kx)e^{-kx}$$
$$= (k^2x^2 - 4kx + 2)e^{-kx}.$$

When $x = 0$, $f_2''(0) = (0 + 0 + 2).e^0 = 2 > 0$, so the multiple root at the origin is a minimum.

When $x = \dfrac{2}{k}$, $f_2''\left(\dfrac{2}{k}\right) = (4 - 8 + 2)e^{-1} = \dfrac{-2}{e^2} < 0$, so the critical point at $x = \dfrac{2}{k}$ is a maximum.

The second derivative is zero when the factor $(kx)^2 - 4(kx) + 2$ is zero. This is a quadratic in kx. Treating this as a single variable and using the formula for the roots of a quadratic, the roots are

$$kx = \frac{-(-4) \pm \sqrt{(-4)^2 - 4(1)(2)}}{2(1)} = \frac{4 \pm \sqrt{16 - 8}}{2} = \frac{4 \pm \sqrt{8}}{2} = \frac{4 \pm 2\sqrt{2}}{2} = 2 \pm \sqrt{2}.$$

So there are points of inflection at $x = c$ and $x = d$, where $c = \dfrac{2 - \sqrt{2}}{k}$ and $d = \dfrac{2 + \sqrt{2}}{k}$.

The function has positive curvature when $x < c$ (we have seen that $f''(0) = 2 > 0$) and negative curvature over the interval $c < x < d$ (e.g. at $x = \dfrac{2}{k}$ we saw that $f_2''\left(\dfrac{2}{k}\right) < 0$). It has positive curvature again when $x > d$.

To summarize: the function $f_2(x)$ is at a minimum at the origin (so its gradient is zero at that point), it goes through a point of inflection at $x = \dfrac{2 - \sqrt{2}}{k} = \dfrac{0.5858}{k}$, reaches a maximum at $x = \dfrac{2}{k}$, then decreases through another point of inflection at $x = \dfrac{2 + \sqrt{2}}{k} = \dfrac{3.4142}{k}$. In FNGRAPH 13.2 (Figure 13.7), where we take $k = 1$, you should see the maximum at $x = 2$, and the points of inflection at $x = 0.5858$ and $x = 3.4142$.

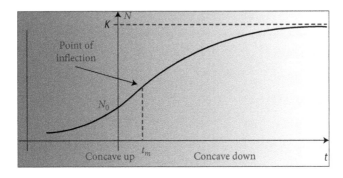

Figure 13.8 Logistic growth function

CASE STUDY A14: The point of inflection in the logistic growth curve

In instalment A10 of this Case Study (in Section 10.5) we wrote the function which solves the continuous logistic growth model, as

$$N(t) = \frac{K}{1 + \left(\dfrac{K}{N_0} - 1\right)e^{-r_m t}}. \qquad (10.37)$$

Recall that N_0 is the initial population size, and K is the carrying capacity (i.e. the maximum population limit). The curve is S-shaped (or sigmoid), as shown in Figure 13.8, from which you can see that it will have a point of inflection at some positive time t_m provided that N_0 is small enough. This is the time at which the population is growing most rapidly. We can work out t_m by finding when the second derivative of $N(t)$ is zero. Sounds simple enough, doesn't it?

First we can simplify the function we need to deal with. Write the coefficient $\left(\dfrac{K}{N_0} - 1\right)$ as a single parameter k, say. Then we could write the function in (10.37) as

$$N(t) = K.f(t), \text{ where } f(t) = \frac{1}{1 + ke^{-rt}}.$$

From now on we'll drop the subscript on r.

Notice first that the parameter K on the numerator of (10.37) just scales the function, so will have no effect on when the function or its derivatives become zero. That is because $N'(t) = K.f'(t)$ and $N''(t) = K.f''(t)$, which tells us that the time at which $N''(t) = 0$ will also be the time at which $f''(t) = 0$. So we can find the point of inflection by trying to solve $f''(t) = 0$.

To differentiate $f(t)$, we write it as $(1 + ke^{-rt})^{-1}$ and use the chain rule:

$$f'(t) = -(1 + ke^{-rt})^{-2}.(-kre^{-rt}) = \frac{kre^{-rt}}{(1 + ke^{-rt})^2}.$$

To differentiate again, we will use the quotient rule:
Let $u = kre^{-rt}$ and $v = (1 + ke^{-rt})^2$. Then

$$\frac{du}{dt} = kr(-re^{-rt}) = -kr^2 e^{-rt},$$

and

$$\frac{dv}{dt} = 2(1 + ke^{-rt}).k(-re^{-rt})$$
$$= -2kre^{-rt}(1 + ke^{-rt}).$$

So

$$f''(t) = \frac{(1 + ke^{-rt})^2.(-kr^2 e^{-rt}) - (kre^{-rt})(-2kre^{-rt}(1 + ke^{-rt}))}{(1 + ke^{-rt})^4}.$$

As we are only looking for values of t that will make the second derivative zero, we can forget about the denominator (which is always positive), multiply out the numerator, and set it to zero:

$$-kr^2 e^{-rt}(1 + 2ke^{-rt} + k^2 e^{-2rt}) + 2k^2 r^2 e^{-2rt}(1 + ke^{-rt}) = 0.$$

Remember from laws of exponents that $(e^{-rt})^2 = e^{-2rt}$. After more multiplying out and collecting terms you should find that

$$kr^2 e^{-rt}(k^2 e^{-2rt} - 1) = 0.$$

The expression in the brackets is the difference of two squares, which can be factorized:

$$kr^2 e^{-rt}(ke^{-rt} - 1)(ke^{-rt} + 1) = 0.$$

As k, r, and e^{-rt} are all positive, the only solution is that $ke^{-rt} = 1$. To solve this for t, we must take logarithms. You should find that the solution is $t = \dfrac{1}{r} \ln k$. The time at which the fastest growth occurs, is therefore

$$t_m = \left(\frac{1}{r_m}\right) . \ln\left(\frac{K}{N_0} - 1\right).$$

Now we can quantify our remark at the start, that the curve will have a point of inflection provided N_0 is small enough. To have $t_m > 0$, we need $\ln\left(\dfrac{K}{N_0} - 1\right) > 0$, so $\dfrac{K}{N_0} - 1 > 1$ (since $\ln x$ is an increasing function and $\ln 1 = 0$). This is satisfied if $N_0 > \dfrac{1}{2}K$. So the growth curve will be sigmoid in shape, with a point of inflection at $t_m > 0$, provided the initial population size is less than half the carrying capacity.

Test this conclusion by programming (10.37) in FNGraph and experimenting with different values of N_0 and k.

Problems: Now try the end-of-chapter problems 13.1–13.4.

13.2 Optimization

Optimization is the task of finding the conditions that produce the best result in a particular situation, subject perhaps to some constraints. In a mathematical model, we need to define the 'result' in terms of a variable that is a function of the conditions; this is called the **objective function**. And we define 'best' to mean that we make this objective function as great, or as small, as possible, depending on the application. If I am investing in the stock market, my objective function would be the total profit I make over the year, and I want to maximize it. The 'conditions' or variables are the companies I choose to invest in, and the number of shares in each that I buy. A constraint may be that I want to invest in at least five different companies, to avoid the risk of losing everything. The task of optimization is to find the choice of companies and the numbers of shares in each, which will **maximize** my profit.

13.2.1 Optimization in the biosciences

In a model of the spread of an epidemic, the objective function may be the total number of deaths over the course of the outbreak. This can be affected by the health education, treatment, and vaccination programmes that are implemented (and these variables may be subject to financial and logistical constraints). The aim here is to **minimize** the objective function. In some situations we may face a maximization and a minimization problem simultaneously. In planning a chemotherapy treatment regime for a patient, the oncologist aims to maximize the number of cancer cells killed, while trying also to minimize the damaging side-effects. One way of tackling this is to design an objective function that combines both these aims, in some sort of trade-off.

Nature has solved many complex optimization problems herself. For example:

- The spiral patterns seen in the arrangements of seeds on the flower head of a sunflower reflect an 'optimal packing' of the given number of seeds on the smallest circular disk, so that they are uniformly spread, and that this uniform density is maintained as the flower head grows and produces more seeds.

- As the sunflower stem grows upward and puts out new leaves, these are distributed in a pattern around the stem (as seen looking down on the plant from above). The pattern is 'chosen' by Nature to maximize the total amount of sunlight and rainwater that can reach the leaves and ground at all stages of development. That is, it minimizes the extent to which the upper leaves overhang the lower ones.

In both of these situations, the solutions found by Nature involve the Fibonacci sequence, which we met in Chapter 7. You can find more information, with many photographs, on Dr Ron Knott's Fibonacci website.[2]

An example of Nature performing optimization at the molecular level, is **protein folding**. Each protein is a particular sequence of amino acids, linked to form a polypeptide chain, with weak interactive forces (hydrogen bonds, electrostatic interactions, and hydrophobic interactions) acting between them. When it is first translated from mRNA, this chain has a random coil shape, but the protein then folds into its own characteristic structure comprising helices, sheets, and turns. It is the three-dimensional structure of the folded protein, rather than the amino acid sequence itself that determines important properties of the protein; folding errors (where the correct three-dimensional structure fails to be adopted) cause diseases such as cystic fibrosis, and recently they have been linked to prion diseases such as BSE.

The task of **protein structure prediction** (PSP) – that is, trying to predict the three-dimensional folded structure a protein will adopt, knowing only its amino acid sequence (the so-called 'primary structure') – is one of the greatest current challenges in molecular biology. A breakthrough was made in the early 1990s, with the recognition that the protein tends to fold into a structure that minimizes its energy (modelled by the sum of all the interactions between the amino acids), in the same way as a ball rolling on an uneven surface will stop at a position where it has minimized its potential energy.[3]

13.2.2 One-dimensional unconstrained optimization

Practical optimization problems may involve objective functions of many variables, with complicated constraints. In this section we will consider optimizing functions of only one

---

[2] www.maths.surrey.ac.uk/hosted-sites/R.Knott/Fibonacci/fib.html.

[3] There are PSP software packages that use this and other approaches to try to solve the PSP problem, but they require huge amounts of computing resources. You can participate in this research by allowing PSP software to use idle CPU time on your PC or PlayStation: an example is the Folding@home project.

variable, with no constraints: **one-dimensional unconstrained optimization**. In this context, the maximization problem can be defined as:

'Find the values of x in the domain of a given smooth function $f(x)$ at which the function reaches a maximum.'

Such values of x are called maximizers. It is often more important to know the maximizer, than to know the maximum function value at that point.

A minimization problem is defined by replacing 'max' by 'min' in the paragraph above.

We mentioned optimization in Section 8.6.2, where we used a class of numerical methods called bracketing methods to find a local minimum. We'll return to numerical methods at the end of this chapter. Here we will use the results of Section 13.1 to define the task of optimization. Recall from Section 8.6.2 the notation $[a, b]$ to mean the closed interval $a \leq x \leq b$ (that is, the interval includes the end-points $x = a$ and $x = b$).

We can distinguish between two types of optimum: **local optima** and **global optima**. These are illustrated in Figure 13.9, showing a function $f(x)$ defined on the interval $[a, b]$:

- Any point at which $f'(x) = 0$ and $f''(x) < 0$ will be a local maximum, and points at which $f'(x) = 0$ and $f''(x) > 0$ will be local minima.

- The global maximum is the point at which $f(x)$ achieves its greatest value, over the whole interval; this may be the highest local maximum, or it may occur at an end-point of the interval. As an analogy, in the mountain range of the Alps, the summits of the Eiger (3970 m) and Jungfrau (4158 m) are local maxima, but the summit of Mont Blanc (4810 m) is the global maximum.

- Similarly, the global minimum is the point at which $f(x)$ achieves its least value, over the whole interval. In Figure 13.9, the global minimum is at the end-point $x = a$.

We have already developed in the previous section the theory we need to solve the optimization problem. Here is an algorithm for finding the global minimum of the function $f(x)$ on the interval $[a, b]$:

1. Differentiate $f(x)$ twice, to obtain $f'(x)$ and $f''(x)$.

2. Solve $f'(x) = 0$ to find the critical points.

3. Evaluate $f''(x)$ at the critical points. If $f'(x) = 0$ and $f''(x) < 0$, then x is a local minimizer.

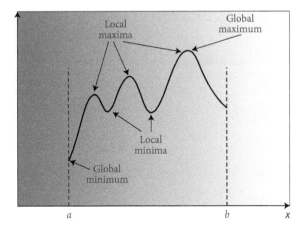

Figure 13.9 Local and global optima

CASE STUDY C15: Fisheries management: using calculus to find the Maximum Economic Yield

Recall the Gordon–Schaefer bioeconomic model of fisheries management described in instalment C7 in Chapter 7, and illustrated in Figure 7.15. The equation of the yield-effort curve is

$$Y(E) = aE\left(1 - \frac{E}{k}\right) \tag{7.30}$$

where $Y(E)$ is the annual yield (in dollars) resulting from a level of effort E. Differentiating with respect to E using the product rule:

$$
\begin{aligned}
Y'(E) &= a\left[E.\left(-\frac{1}{k}\right) + \left(1 - \frac{E}{K}\right)(1) \right] \\
&= a\left[-\frac{E}{K} + 1 - \frac{E}{K} \right] \\
&= a\left(1 - \frac{2E}{K}\right).
\end{aligned}
$$

So $Y'(E) = 0$ if $1 - \frac{2E}{K} = 0$, i.e. when $E = \frac{1}{2}K$. This is the value E_2 on the graph in Figure 7.15, called the Maximum Sustainable Yield point.

However, this level of effort does not necessarily produce the maximum profit. The straight line on the graph is the cost-effort line $C = mE$, where m is the dollar cost per unit of effort. Typical graphs are shown in FNGRAPH 13.3 (see Figure 13.10). For a particular effort E, we have:

profit = yield − cost,

or

$$
\begin{aligned}
P(E) &= Y(E) - C(E) \\
&= aE\left(1 - \frac{E}{K}\right) - mE.
\end{aligned}
$$

The Maximum Economic Yield point E_1 is the value of E that maximizes $P(E)$. Differentiating $P(E)$:

$$
\begin{aligned}
P'(E) &= a\left(1 - \frac{2E}{K}\right) - m \\
&= (a - m) - \frac{2aE}{K}.
\end{aligned}
$$

Setting $P'(E) = 0$ and solving for E:

$$E_1 = \frac{(a - m)K}{2a}. \tag{13.4}$$

$P''(E_1) = \frac{-2a}{K} < 0$, so this point will be a maximizer of profit P.

In Figure 13.10 we use $a = 5$, $m = 2$, $K = 10$, so in this case $E_1 = 3$.

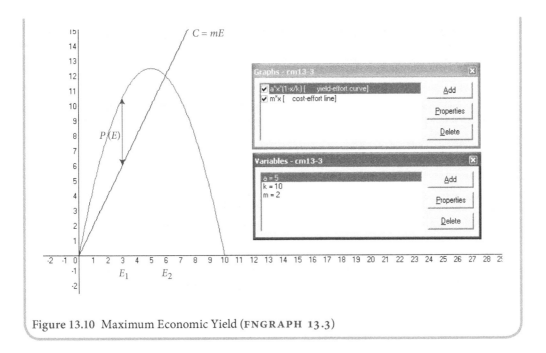

Figure 13.10 Maximum Economic Yield (FNGRAPH 13.3)

4. Calculate the value of $f(x)$ at each local minimizer. Also calculate $f(a)$ and $f(b)$, the function values at the end-points of the interval.

5. The least value found in step 4 is the global minimum. The value of x that produced it is the global minimizer.

We could easily adapt this algorithm to one for finding the local and global maxima of $f(x)$: replace 'min' by 'max', 'least' by 'greatest', and $f''(x) < 0$ by $f''(x) > 0$.

Alternatively, to find the maxima of $f(x)$, use the above algorithm to find the minima of the function $g(x)$, defined by $g(x) = -f(x)$. Recall that the graph of $-f(x)$ is obtained by flipping the graph of $f(x)$ around the x-axis; then the minima will become maxima, and vice versa!

There is one snag in the practical use of this algorithm for optimization. What if we are unable to solve the equation $f'(x) = 0$? In Section 8.6.1 we saw some numerical methods for locating the root of an equation, by finding a bracket around the root and then narrowing it down. Later in this chapter we'll see more powerful calculus-based methods for root-finding and optimization.

13.2.3 Application: tubular bones

Nature has proved to be very effective at solving optimization problems through the processes of evolution. As an example of natural optimization, consider bone structure. This application is adapted from Alexander (1982). The bones in an animal's limbs should ideally be both light and strong, but these are conflicting objectives. A solid bone would be strong, but much too heavy. Nature's solution is for the bone to have a hollow central cavity. We can model the bone as a thick tube. When a force is applied at right angles at one end of the bone (as happens when the animal is running), this causes a **bending moment** in the bone. Engineers have developed equations defining the maximum bending moment M that a solid rod can bear without breaking; it is proportional to the cube of the rod's cross-sectional radius r:

$$M = Kr^3,$$

where K is a measure of the strength of the material.

Now consider a bone which is a thick tube with a hollow centre. In cross section it is like a ring, with outer radius r and inner radius less than r; we'll write it as $k.r$, as in Figure 13.11(a). The parameter k is in the range $0 < k < 1$. When $k = 0$ the bone is solid, and when k is close to 1 the bone is a very thin wall enclosing a central cavity. From engineering theory, the bending moment that this tubular rod can support is

$$M = K(1 - k^4)r^3.$$

Put another way, to survive a bending moment of magnitude M the bone will need an outer radius r given by

$$r = \sqrt[3]{\frac{M}{K(1-k^4)}} \tag{13.5}$$

for a given value of k.

A bone with a larger value of k (i.e. a thinner wall) will have to have a larger radius; these will both affect the bone's volume and hence its mass. We ask: is there a value of k that will minimize the bone's mass?

The mass m_1 of the bone is given by

$$m_1 = \rho V = \rho Al \tag{13.6}$$

where ρ is the density of the bone material and V its volume. This volume is the cross-sectional area A multiplied by the bone length l. The cross-sectional area A is shown shaded in Figure 13.11(a). It is calculated as the difference in area between the outer and inner circles, of radius r and kr respectively:

$$A = \pi r^2 - \pi(kr)^2 = \pi r^2 - \pi k^2 r^2 = \pi(1 - k^2)r^2.$$

Substituting in (13.6):

$$m_1 = \pi \rho l(1 - k^2)r^2. \tag{13.7}$$

This is not the total mass of the bone, however. In mammals the central bone cavity is not empty but filled with bone marrow. If we know the density ρ_0 of the marrow, its mass will be

$$m_0 = \rho_0 A_0 l = \rho_0 l.\pi(kr)^2 = \pi \rho_0 l k^2 r^2 \tag{13.8}$$

since its cross-sectional area A_0 is the area of the inner circle, of radius $k.r$, in Figure 13.11(a). The total mass of the bone is then $m = m_1 + m_0$.

Bone marrow is about half as dense as the bone wall, so we will take $\rho_0 = \frac{1}{2}\rho$. Using (13.7) and (13.8) we obtain the total mass as

$$m = m_1 + m_0$$
$$= \pi \rho l(1 - k^2)r^2 + \pi.\frac{1}{2}\rho.l.k^2 r^2$$
$$= \pi \rho l\left(1 - k^2 + \frac{1}{2}k^2\right)r^2$$
$$= \pi \rho l\left(1 - \frac{1}{2}k^2\right)r^2.$$

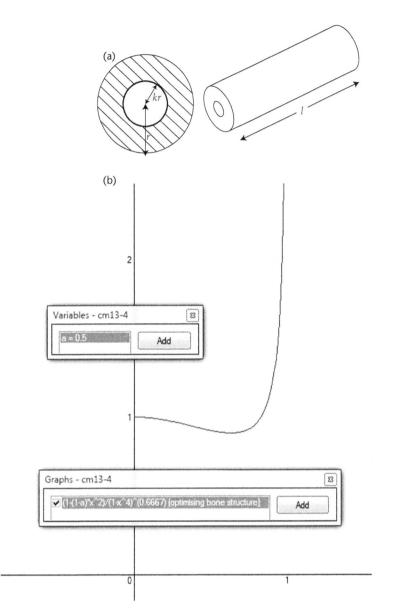

Figure 13.11
Optimal bone
structure coefficient k
(FNGRAPH 13.4)

Finally, by substituting for r from (13.5) we obtain m as a function of the variable k only:

$$m = \pi\rho l\left(1 - \frac{1}{2}k^2\right)\left(\sqrt[3]{\frac{M}{K(1-k^4)}}\right)^2$$

$$= \pi\rho l\left(1 - \frac{1}{2}k^2\right)\left(\frac{M}{K(1-k^4)}\right)^{\frac{2}{3}}.$$

Collecting all the constants together we can write this as

$$m = C\left(1 - \frac{1}{2}k^2\right)\left(1 - k^4\right)^{-\frac{2}{3}} \tag{13.9}$$

where $C = \pi \rho l \left(\dfrac{M}{K} \right)^{\frac{2}{3}}$. I have sketched this function (taking $C = 1$) in FNGRAPH 13.4, reproduced in Figure 13.11(b). The parameter k goes from 0 to 1 along the horizontal axis. I have introduced a parameter a; the bone marrow density is $a.\rho$, so in our example $a = 1/2$. You can see that there is indeed a value of k at which the mass is minimized. To find this value of k we need to differentiate (13.9) and solve $\dfrac{dm}{dk} = 0$.

Using the product rule, and the chain rule to differentiate the second factor with $u = 1 - k^4$:

$$\frac{dm}{dk} = C\left(1 - \frac{1}{2}k^2\right)\left(-\frac{2}{3}(1-k^4)^{-\frac{5}{3}}.(-4k^3)\right) + C\left(-\frac{1}{2}.2k\right)(1-k^4)^{-\frac{2}{3}}$$

$$= \frac{8}{3}Ck^3\left(1 - \frac{1}{2}k^2\right)(1-k^4)^{-\frac{5}{3}} - Ck\,(1-k^4)^{-\frac{2}{3}}.$$

By the laws of indices,

$$(1-k^4)^{-\frac{5}{3}} = (1-k^4)^{-1}(1-k^4)^{-\frac{2}{3}}$$

so we can take out a common factor of $Ck(1-k^4)^{-\frac{2}{3}}$:

$$\frac{dm}{dk} = Ck\left(1-k^4\right)^{-\frac{2}{3}}\left[\frac{8}{3}k^2\frac{\left(1-\frac{1}{2}k^2\right)}{1-k^4} - 1\right].$$

To solve $\dfrac{dm}{dk} = 0$ we need the factor in square brackets to be zero. This occurs when

$$\frac{8}{3}k^2\left(1 - \frac{1}{2}k^2\right) - (1-k^4) = 0$$

which simplifies to

$$-\frac{1}{3}(k^4 - 8k^2 + 3) = 0.$$

The polynomial in brackets is a quadratic in k^2, whose roots are given by the formula:

$$k^2 = \frac{8 \pm \sqrt{64 - 12}}{2} = 4 \pm \sqrt{13}.$$

We are interested in the smaller value of k, so we take

$$k = \sqrt{4 - \sqrt{13}} = \sqrt{4 - 3.6056} = \sqrt{0.3944} = 0.628.$$

Thus, the optimal bone structure combining strength with lightness will have a central cavity whose radius is about 63% of the outer bone radius.

This was a long calculation, and I'm certainly not going to try differentiating again to check that the critical point is a minimum, but Nature has got there before us. Alexander (1982) quotes measurements of k in the femur bones of various mammals, all in the range $0.54 < k < 0.63$, including a fox (0.63) and a camel (0.62). You can see in Figure 13.11(b) that the 'bowl' around the critical point is very shallow on the left side, so any value of k in the range found in Nature will produce equally efficient bones. If you experiment with FNGRAPH 13.4 for different values of a, you will see that with lighter bone marrow density the optimal value of k increases. This is borne out in the bones of birds, which are filled with air and which have very thin walls $(k \approx 0.9)$.

In fact, Nature is so good at performing optimization that new computational algorithms based on natural processes are being developed for otherwise intractable problems – a branch of mathematics called **evolutionary optimization.**

Problems:　Now try the end-of-chapter problems 13.5–13.10.

13.3 **Related rates**

The topic of related rates is an application of the chain rule. Suppose we have a variable y that depends on x: $y = y(x)$. And say that both x and y are changing with time t. How are the rates of change $\dfrac{dx}{dt}$ and $\dfrac{dy}{dt}$ related? By differentiating $y = y(x)$ with respect to x we obtain $\dfrac{dy}{dx}$. These three derivatives are linked by the chain rule:

$$\frac{dy}{dt} = \frac{dy}{dx} \cdot \frac{dx}{dt}. \tag{13.10}$$

If we know two of these derivatives, we can work out the third one.

EXAMPLE

There are muscles in the human eye that change the curvature of the lens, and hence its focal length f, so that we can focus on objects that are different distances away. The image distance v, from the lens to the retina at the back of the eye, is fixed at $v = 17$ mm. The focal length needed in order to focus on an object that is a distance u away from you, is given (approximately) by the thin lens formula:

$$\frac{1}{f} = \frac{1}{u} + \frac{1}{v}.$$

Solving this for f:

$$f = \frac{u.v}{u + v}. \tag{13.11}$$

If I am running to catch a bus, and approaching it at a speed of 0.5 m s^{-1}, how rapidly must the eye focal length change to keep the bus in focus, when it is (i) 3 metres and (ii) 1 metre away from me?

Solution: We know $v = 17$ mm, so working in millimetres we have from (13.11)

$$f = \frac{17u}{u + 17}.$$

Differentiating using the quotient rule:

$$\frac{df}{du} = \frac{(u + 17)(17) - (17u).1}{(u + 17)^2} = \frac{289 + 17u - 17u}{(u + 17)^2} = \frac{289}{(u + 17)^2}.$$

Note that $\dfrac{df}{du}$ is a dimensionless quantity, as f and u have the same dimensions.

We know that $\dfrac{du}{dt} = -500$ mm s^{-1} (negative, as I am approaching the bus, so the distance is decreasing with time).

By the chain rule:

$$\frac{df}{dt} = \frac{df}{du} \cdot \frac{du}{dt} = \frac{289}{(u+17)^2} \times (-500) = \frac{-144500}{(u+17)^2} \text{ mm per second.}$$

(i)   When $u = 3$ m $= 3000$ mm,

$$\frac{df}{dt} = \frac{-144500}{(3017)^2} = -0.0159 \text{ mm per second.}$$

(ii)   When $u = 1$ m $= 1000$ mm,

$$\frac{df}{dt} = \frac{-144500}{(1017)^2} = -0.140 \text{ mm per second.}$$

At the nearer distance, the muscles must be changing the focal length nearly ten times as rapidly.

In the following example, the unknown quantity is on the right-hand side of (13.10):

EXAMPLE

As a sunflower flower develops, it produces seeds at a constant rate. Assuming that the seed density on the circular flower head stays constant (as observed above), this means the head area $A(t)$ will increase at a constant rate. If each seed occupies 3 mm^2, and the head generates 20 new seeds per day, how rapidly is the diameter of the head increasing when it is (i) 2 cm and (ii) 5 cm in diameter?

Solution: Rather than use the diameter, we will work with the radius r. The area of the circular head is $A = \pi r^2$. Differentiating,

$$\frac{dA}{dr} = 2\pi r.$$

We know $\dfrac{dA}{dt}$:   20 seeds $\times$ 3 mm^2 = 60 mm^2 per day.

To find $\dfrac{dr}{dt}$ we use the chain rule, substitute in, and solve for $\dfrac{dr}{dt}$:

$$\frac{dA}{dt} = \frac{dA}{dr} \times \frac{dr}{dt}$$

$$60 = (2\pi r)\frac{dr}{dt}$$

$$\frac{dr}{dt} = \frac{60}{2\pi r}.$$

(i)   When the diameter is 2 cm, the radius will be 1 cm, or 10 mm. This gives $\dfrac{dr}{dt} = \dfrac{3}{\pi}$ = 0.955 mm per day. As the diameter is $2r$, the rate of increase of the diameter is

$$2\frac{dr}{dt} = 2 \times 0.955 = 1.91 \text{ mm per day.}$$

When the diameter is 5 cm, the radius will be 25 mm. This gives $\dfrac{dr}{dt} = \dfrac{6}{5\pi} = 0.38$ mm per day. Hence the diameter is increasing at 0.76 mm per day.

In the equation $A = \pi r^2$ relating A and r in this example, it is difficult to say which is the independent and which the dependent variable. Is a change in radius causing a change in area, or does the increasing area cause the radius to grow? That is, should we work out $\dfrac{dA}{dr}$ or $\dfrac{dr}{dA}$? We have treated A as a function of r and found $\dfrac{dA}{dr} = 2\pi r$, but we could also have rewritten the equation as:

$$r = \sqrt{\dfrac{A}{\pi}} = \dfrac{\sqrt{A}}{\sqrt{\pi}} = \dfrac{1}{\sqrt{\pi}} A^{\frac{1}{2}}, \text{ and then differentiated to get}$$

$$\dfrac{dr}{dA} = \dfrac{1}{\sqrt{\pi}} \left(\dfrac{1}{2} A^{-\frac{1}{2}} \right)$$

$$= \dfrac{1}{2\sqrt{\pi}} \left(\dfrac{1}{\sqrt{A}} \right)$$

$$= \dfrac{1}{2\sqrt{\pi}} \left(\dfrac{1}{\sqrt{\pi r^2}} \right) \quad \boxed{\text{substituting } A = \pi r^2}$$

$$= \dfrac{1}{2\sqrt{\pi}} \left(\dfrac{1}{\left(\sqrt{\pi} \right) r} \right)$$

$$= \dfrac{1}{2\pi r}.$$

We have found that $\dfrac{dr}{dA} = 1 \Big/ \dfrac{dA}{dr}$. This is a useful trick to remember:

If variables x and y are related, and we can work out $\dfrac{dy}{dx}$ and $\dfrac{dx}{dy}$, then

$$\dfrac{dx}{dy} = \dfrac{1}{\dfrac{dy}{dx}}. \tag{13.12}$$

We could have deduced this from the chain rule: $\dfrac{dx}{dy} \times \dfrac{dy}{dx} = \dfrac{dx}{dx} = 1.$

In the above example, it was less work to use $\dfrac{dA}{dt} = \dfrac{dA}{dr} \times \dfrac{dr}{dt}$, and solve for $\dfrac{dr}{dt}$, than to work out $\dfrac{dr}{dA}$ and find $\dfrac{dr}{dt}$ directly from $\dfrac{dr}{dt} = \dfrac{dr}{dA} \times \dfrac{dA}{dt}$, but had we done this we would have produced the same answer in the end.

Problems: Now try the end-of-chapter problems 13.11–13.12.

13.4 Polynomial approximation of functions

In our exploration of the different categories of mathematical function in Chapters 5–10 of this book, we started with linear functions, which we extended to polynomials. These are in many ways the simplest functions to deal with. But they have a surprising and very useful property: one might say that polynomials are the stem cells of the function world. They can be used to produce any of the other types of smooth function – or at least, to produce something that behaves just like another function, on at least part of the graph. In this section we'll see how to approximate complicated functions by polynomials. You may need to do this if solving your model of a biological problem has produced a very complicated function $f(x)$. While you may be able to differentiate $f(x)$ it may be impossible to find the roots or critical points as we have done in Section 13.1.

We take a smooth function $f(x)$, which we can differentiate as many times as we like over its domain. For example, $f(x)$ may be a trigonometric or logarithmic function, or the sigmoid function in Case Study A13 describing logistic growth. We are going to make polynomial approximations of $f(x)$ around the point $x = 0$. We already made polynomial approximations to a function in Part I, using the idea of interpolation. In Section 3.6 we found a linear approximation to $f(x)$ by fitting a straight line through two points on the graph of $y = f(x)$. In Section 7.5 we performed quadratic interpolation, fitting a quadratic curve through three points on the graph. This approach becomes quite complicated if we try to extend it to cubic and higher-degree interpolation. It also has another drawback: if you fit a polynomial of high degree through several points on a smooth curve, you can get a very 'wiggly' curve which looks nothing like the original function $f(x)$. This is illustrated in Figure 13.12.

Instead of using data from several different points on the curve $y = f(x)$, we will take our data from just one point, namely $x = 0$. We use the function value $f(0)$. We also differentiate $f(x)$ several times to get $f'(x), f''(x), f'''(x)$, etc, and evaluate these derivatives at $x = 0$. So the data we work with are the values of $f(0), f'(0), f''(0), f'''(0)$, and so on. Note that $f'(0)$ tells us the gradient of $f(x)$ at $x = 0$, and $f''(0)$ tells us the curvature at that point.

13.4.1 Linear approximation of $f(x)$ around $x = 0$

For the linear approximation we take the two pieces of data $f(0)$ and $f'(0)$, and construct the linear polynomial $p_1(x)$ defined by

$$p_1(x) = f(0) + f'(0).x. \tag{13.13}$$

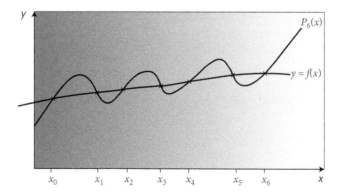

Figure 13.12 A high-degree interpolating polynomial

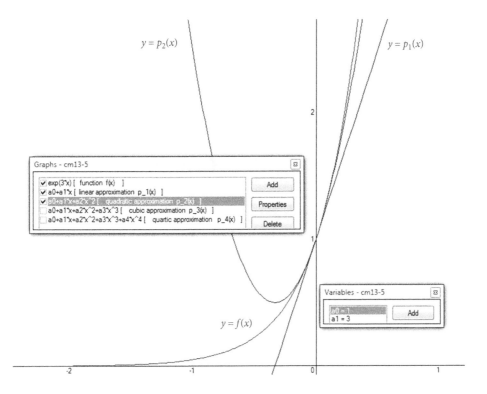

Figure 13.13 Polynomial approximations to $f(x) = e^{3x}$ around $x = 0$ (FNGRAPH 13.5)

For example, suppose we want a linear approximation to $f(x) = e^{3x}$. Then $f(0) = e^0 = 1$, and $f'(x) = 3e^{3x}$, so $f'(0) = 3e^0 = 3$. We construct the linear polynomial

$$p_1(x) = f(0) + f'(0).x$$
$$= 1 + 3x.$$

In FNGRAPH 13.5 (Figure 13.13) you can see the graphs of $f(x)$ and $p_1(x)$ around $x = 0$. The polynomial $p_1(x)$ is the best straight-line approximation to the function at the point where $x = 0$. The line and the curve have the same value at $x = 0$, and also have the same gradient at that point.

To explain why this is, we will write the polynomial as $p_1(x) = a_0 + a_1 x$, where $a_0 = f(0)$ and $a_1 = f'(0)$.

Then evaluating $p_1(x)$ at $x = 0$: $\quad p_1(0) = a_0 + a_1.0 = a_0 = f(0)$.

Differentiate $p_1(x)$:

$$p_1'(x) = a_0.0 + a_1.1 = a_1 = f'(0).$$

So the linear approximation satisfies $p_1(0) = f(0)$ and $p_1'(0) = f'(0)$. That is, at $x = 0$ the linear polynomial $p_1(x)$ and the function $f(x)$ agree in terms of function value and slope.

The linear approximation is not a very good approximation to the curve. For a better approximation we can try to fit a quadratic polynomial.

13.4.2 Quadratic approximation of $f(x)$ around $x = 0$

Extending the approach of the previous section, we define the quadratic polynomial

$$p_2(x) = a_0 + a_1 x + a_2 x^2. \tag{13.14}$$

We want $p_2(x)$ to agree with $f(x)$ at $x = 0$ in terms of function value, first derivative and also the second derivative. To achieve this we take

$$a_0 = f(0), \quad a_1 = f'(0) \quad \text{and} \quad a_2 = \frac{1}{2} f''(0).$$

Then $p_2(0) = a_0 + a_1.0 + a_2.0^2 = a_0 = f(0)$, as before. Differentiating $p_2(x)$, $p_2'(x) = a_0.0 + a_1.1 + a_2(2x) = a_1 + 2a_2 x$, so $p_2'(0) = a_1 + 2a_2.0 = a_1 = f'(0)$.

Differentiating $p_2(x)$ again:

$$p_2''(x) = a_1.0 + 2a_2.1 = 2a_2 = 2\left(\frac{1}{2} f''(0)\right) = f''(0).$$

Thus, $p_2(x)$ has the same function value, gradient and curvature as $f(x)$ at $x = 0$.

In our example of $f(x) = e^{3x}$, we get $a_0 = f(0) = e^0 = 1$ and $a_1 = f'(0) = 3.e^0 = 3$, as in the linear approximation. The second derivative of $f(x)$ is $f''(x) = 3(3e^{3x}) = 9e^{3x}$, so $a_2 = \frac{1}{2} f''(0) = \frac{1}{2}(9) = 4.5$. Our quadratic approximation is

$$p_2(x) = a_0 + a_1 x + a_2 x^2$$
$$= 1 + 3x + 4.5x^2.$$

You can see this quadratic curve in FNGRAPH 13.5, by clicking on the third function in the list in the Graphs window; the curve is the parabola in Figure 13.13. It is a good approximation to $f(x)$ for positive x, although it cannot model the exponential graph for negative x.

Notice that the coefficients of the first two terms in $p_2(x)$ are the same as in $p_1(x)$. To make a higher-degree approximation, we only need to work out the coefficients of the extra terms.

We might hope that a higher-degree polynomial approximation would provide an even better 'fit' to the function. In order to extend the quadratic approximation to a cubic approximation, we will add a term '$+a_3 x^3$' to the quadratic polynomial (13.14):

$$p_3(x) = a_0 + a_1 x + a_2 x^2 + a_3 x^3.$$

When we differentiate $p_3(x)$ three times, the first three terms will vanish, so let's just differentiate the third term:

$$y = a_3 x^3$$
$$\frac{dy}{dx} = a_3(3x^2) = 3a_3 x^2$$
$$\frac{d^2 y}{dx^2} = 3a_3(2x) = 6a_3 x$$
$$\frac{d^3 y}{dx^3} = 6a_3.$$

So if we want the third derivative of $p_3(x)$ to equal $f'''(0)$, we must take $a_3 = \frac{1}{6} f'''(0)$. You can check that for a quartic approximation (i.e. adding a '$+a_4 x^4$' term to $p_3(x)$) we must take $a_4 = \frac{1}{4.3.2} f^{(4)}(0) = \frac{1}{24} f^{(4)}(0)$. Recall our notation, that $f^{(n)}(x)$ means the nth derivative $\frac{d^n f}{dx^n}$.

You can see the graphs of $p_3(x)$ and $p_4(x)$ in FNGRAPH 13.5. The cubic approximation fits $f(x)$ closely for $x \geq 0$, but again cannot match the exponential curve for negative x. A quartic approximation does manage to match the function for $-0.5 \leq x \leq 0$, as shown in FNGRAPH 13.5. You should try calculating – and programming in FNGraph – quintic and sextic approximations.

13.4.3 Maclaurin series expansions of functions

In FNGRAPH 13.5 you saw that the graphs of the polynomial functions

$$p_1(x) = a_0 + a_1 x$$
$$p_2(x) = a_0 + a_1 x + a_2 x^2$$
$$p_3(x) = a_0 + a_1 x + a_2 x^2 + a_3 x^3$$
$$p_4(x) = a_0 + a_1 x + a_2 x^2 + a_3 x^3 + a_4 x^4$$

become closer and closer approximations to that of $f(x)$, as we increase the degree of the polynomial. In fact, provided the function and all its derivatives exist at $x = 0$, if we could take an 'approximating' polynomial with an infinite number of terms its graph would look exactly like the graph of $f(x)$; that is, it actually is $f(x)$.

A polynomial (where the terms are coefficients multiplying increasing powers of x) with an infinite number of terms is called a **power series**. If $f(x)$ can be written as

$$f(x) = a_0 + a_1 x + a_2 x^2 + a_3 x^3 + \cdots + a_n x^n + \cdots \tag{13.15}$$

we say that the right-hand side is the **power series expansion** of $f(x)$. All we need is a formula for calculating the coefficients $a_1, a_2, a_3, \ldots, a_n, \ldots$

We saw how to do this for the first few coefficients, in the previous two sections:

$$a_0 = f(0)$$
$$a_1 = f'(0)$$
$$a_2 = \frac{1}{2} f''(0)$$
$$a_3 = \frac{1}{3.2} f'''(0) = \frac{1}{6} f'''(0)$$
$$a_4 = \frac{1}{4.3.2} f^{(4)}(0) = \frac{1}{24} f^{(4)}(0).$$

You can see the pattern here. To write an expression for a_n, we use the factorial notation introduced in Section 12.7. Recall definition (12.16), that $n! = n.(n-1).(n-2).(n-3)\ldots3.2.1$. So $6! = 6 \times 5 \times 4 \times 3 \times 2 \times 1 = 720$.

We can now define the coefficients in our power series expansion. The result is called the **Maclaurin[4] series expansion** of $f(x)$:

If a function $f(x)$ and all its derivatives exist at $x = 0$, then we can write $f(x)$ as a Maclaurin series expansion

$$f(x) = a_0 + a_1 x + a_2 x^2 + a_3 x^3 + \cdots + a_n x^n + \cdots$$

where the coefficients are given by

$$a_n = \frac{1}{n!} f^{(n)}(0). \tag{13.16}$$

[4] Colin Maclaurin (1698–1746) was a Scottish mathematician, a protégé of Isaac Newton.

As an example, take the function $f(x) = e^x$. Then $f(0) = 1$, and we know that the derivative of e^x is e^x, so all the higher derivatives of $f(x)$ will also be e^x.

At $x = 0$ the values of $f(x)$ and all its derivatives are 1, so we have $a_n = \dfrac{1}{n!}(1) = \dfrac{1}{n!}$. The Maclaurin series expansion of e^x is thus

$$e^x = 1 + x + \frac{1}{2!}x^2 + \frac{1}{3!}x^3 + \cdots + \frac{1}{n!}x^n + \cdots$$

If we put $x = 1$ in this equation, we obtain the **series expansion for e**:

$$e = 1 + 1 + \frac{1}{2!} + \frac{1}{3!} + \cdots + \frac{1}{n!} + \cdots$$

As the factorial function grows rapidly as n increases, this series will converge rapidly. If you sum the first seven terms in the series, up to the term involving 6!, you obtain 2.718056, which is close to the true value of e, namely $2.71828\ldots$

An even more surprising series expansion is the following, for π:

$$\pi = 4\left(1 - \frac{1}{3} + \frac{1}{5} - \frac{1}{7} + \cdots\right).$$

This was derived by Leibniz, by finding the Maclaurin series expansion of $\tan^{-1} x$, and using the fact that $\tan^{-1} 1 = \dfrac{\pi}{4}$ (since $\tan 45° = 1$). You could try to obtain this result yourself. However, summing the first seven terms of this series I get $3.283738\ldots$, which is not a very good approximation to $\pi = 3.14159\ldots$ In this case the series converges only slowly.

You will see the Maclaurin series expansions of other functions in the Problems.

13.4.4 Taylor series expansions of functions

The Maclaurin series expansion of a function may be valid for all values of x, as in the case of the expansion of e^x given above, but for other functions it may only work in an interval around $x = 0$. In the latter case, we would want to have a power series expansion that works in the part of the graph we are interested in. Suppose we are interested in approximating $f(x)$ around the point where $x = a$. The power series expansion around this point is called a **Taylor** [5] **series**, and is given by:

If a function $f(x)$ and all its derivatives exist at $x = a$, then we can write $f(x)$ as a Taylor series expansion

$$f(x) = a_0 + a_1(x - a) + a_2(x - a)^2 + a_3(x - a)^3 + \cdots + a_n(x - a)^n + \cdots$$

where the coefficients are given by

$$a_n = \frac{1}{n!} f^{(n)}(a). \tag{13.17}$$

---

[5] Brook Taylor (1685–1731) was an English mathematician, also of Newton's time. He was a member of the committee of the Royal Society that adjudicated on whether Newton or Leibniz was the true father of the calculus.

This is a more general result than (13.16); if you set $a = 0$ you obtain the Maclaurin series.

Of course, you are not going to use these results by evaluating all the terms in the series, as there are infinitely many of them. But if the power series expansion is valid, successive terms will be getting smaller and smaller, and we can choose where to halt. A **truncated Taylor series**, stopping after the first $(n + 1)$ terms, would be

$$f(x) = a_0 + a_1(x - a) + a_2(x - a)^2 + a_3(x - a)^3 + \cdots + a_n(x - a)^n \qquad (13.18)$$

where the coefficients are still as given in (13.17).

We will use the linear and quadratic Taylor series expansions in the next section, to produce a numerical method for finding roots of functions.

Problems: Now try the end-of-chapter problems 13.13–13.14.

13.5 Extension: numerical methods for finding roots and critical points

We have seen that the roots of $f(x)$ can be found by solving $f(x) = 0$, and its critical points by solving $f'(x) = 0$. In real-life problems this is easier said than done. If $f(x)$ is a polynomial we can try to factorize it; if it is a trigonometric function we can use our knowledge of sine and cosine graphs from Chapter 9. But if we have $f(x) = 4\cos x - x$, for example, there is no way of obtaining an analytical solution of $f(x) = 0$. We can see that there is one positive and two negative roots, by using FNGraph to sketch the graphs of $y = 4\cos x$ and of $y = x$, and looking at where the two graphs intersect. This is done in **FNGRAPH 13.6**. But we can't work out these roots as expressions in x.

We discussed the root-finding problem in Section 8.6.1, where we developed **numerical methods** that can produce the root, not as an algebraic expression but as a number (to a prescribed precision). We saw the **method of bisection** for finding a root:

1. Find an interval $[a, b]$ that brackets the root, by checking if $f(a).f(b) < 0$.
2. Calculate the interval midpoint $c = \dfrac{1}{2}(a + b)$, and evaluate $f(c)$.
3. If $f(c) = 0$, then c is the root. Otherwise, depending on the sign of $f(c)$, the root will lie in either $[a, c]$ or $[c, b]$.
4. Go back to Step 2, using the new smaller interval.
5. Stop when the interval is small enough to provide the desired precision.

This is an example of an **iterative method**, where we perform a relatively simple calculation over and over again, each time getting closer to the answer.

Provided that the initial interval meets the requirement in Step 1, the method of bisection is guaranteed to find a root. It does have some drawbacks, though:

- Finding a and b for the initial interval can be a matter of trial and error.
- If the initial interval encloses several roots, the method will home in on one of them (we can't predict which), and will give no indication that other roots exist.
- The method is slow to converge (As a personal opinion, it is also rather mindless, like producing a copy of the *Mona Lisa* using painting-by-numbers).

We now have a powerful new mathematical tool, differential calculus, which will give us a much more impressive method, published by Raphson[6] in 1690. It is perhaps the best-known and most widely used numerical method today, because of its speed of convergence and its suitability for implementation in a computer program (or Excel spreadsheet). When applied to root-finding, it is usually called the Newton–Raphson method (Section 13.5.1), and when used to find critical points it is called Newton's method (Section 13.5.2).

The starting-point for deriving these algorithms is a truncated Taylor Series approximation to $f(x)$.

13.5.1 Newton–Raphson method for finding roots

We are given a differentiable function $f(x)$, and we want to find a root x^* where $f(x^*) = 0$. We need an initial guess for x^*, which we'll call x_0. We can usually obtain x_0 from experiment or experience of the physical problem we are solving; if not, use $x_0 = 0$ or make an educated guess.

If we write the Taylor series expansion of $f(x)$ around $x = x_0$, from (13.17), and truncate it after the first two terms, we get

$$f(x) \approx f(x_0) + f'(x_0).(x - x_0) \tag{13.19}$$

as a linear approximation to $f(x)$ in an interval around x_0. The graph of

$$y = f(x_0) + f'(x_0).(x - x_0)$$

is a straight line. We can find its root x_1 (where it crosses the x-axis) by setting $y = 0$ and solving for x:

$$f(x_0) + f'(x_0).(x - x_0) = 0$$
$$f'(x_0).(x - x_0) = -f(x_0)$$
$$x - x_0 = -\frac{f(x_0)}{f'(x_0)}$$
$$x_1 = x_0 - \frac{f(x_0)}{f'(x_0)}.$$

This root x_1 of the linear approximation to $f(x)$ should be a better approximation to the root x^* of $f(x)$ than x_0 was.

The situation is shown graphically in Figure 13.14(a). The linear approximation is a line with the same function value $f(x_0)$ and gradient $f'(x_0)$ as $f(x)$ at $x = x_0$, as shown. Looking at the triangle ABC,

$$\text{slope of AC} = f'(x_0) = \tan \theta = \frac{BC}{AB} = \frac{f(x_0) - 0}{x_0 - x_1},$$

which, if you solve for x_1, gives the same formula as obtained above.

We have performed the first iteration of our iterative method. For the next iteration, start with x_1, form a truncated Taylor series expansion around $x = x_1$, and obtain the formula for its root x_2:

$$x_2 = x_1 - \frac{f(x_1)}{f'(x_1)}.$$

[6] Joseph Raphson (c.1648 – c.1715), English mathematician. Supported Newton's claim to be the father of calculus.

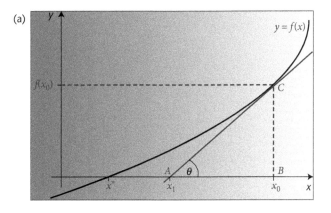

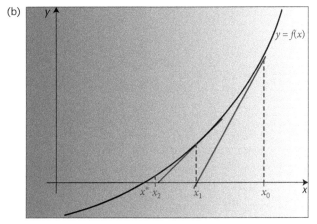

Figure 13.14 The Newton–Raphson method

Written as an iterative method:

> The **Newton–Raphson method** for finding a root x^* of a differentiable function $f(x)$,
> from an initial guess x_0, is given by the iterative formula
>
> $$x_{k+1} = x_k - \frac{f(x_k)}{f'(x_k)} \quad \text{for } k = 1, 2, 3,\ldots \tag{13.20}$$

The first two iterations we found are given by taking $k = 0$ and $k = 1$ in (13.20). Continuing
with $k = 2, 3, 4, \ldots$ we obtain a sequence of iterates $x_0, x_1, x_2, x_3, x_4, \ldots$ which (we hope) will
converge to the root x^*; see Figure 13.14(b).

We can prove that if the sequence converges, it will converge to a root of $f(x)$; this is called
the **limit** of the sequence – the same idea of limit as we used in Section 6.4.4.

Here is a simple example worked by hand:

EXAMPLE

Find the larger root of the function $f(x) = e^x - 5x$, correct to 4 decimal places.

We use FNGraph to draw the graphs of $y = e^x$ and $y = 5x$, see **FNGRAPH 13.7**. The roots
of $f(x)$ are the values of x where the two curves intersect. There is a smaller root close to
$x = 0$, and a larger root between 2 and 3. To find the latter, we will take $x_0 = 2.5$ as our
initial guess.

Differentiating, $f'(x) = e^x - 5$, so the iterative formula (13.20) becomes

$$x_{k+1} = x_k - \frac{e^{x_k} - 5x_k}{e^{x_k} - 5}.$$

Starting from $x_0 = 2.5$, we get

$$x_1 = x_0 - \frac{e^{x_0} - 5x_0}{e^{x_0} - 5} = 2.5 - \frac{e^{2.5} - 12.5}{e^{2.5} - 5} = 2.5 - \frac{-0.32}{7.18} = 2.5 + 0.044 = 2.544.$$

The next iteration is

$$x_2 = x_1 - \frac{e^{x_1} - 5x_1}{e^{x_1} - 5} = 2.544 - \frac{e^{2.544} - 12.72}{e^{2.544} - 5} = 2.544 - \frac{0.01}{7.73}$$
$$= 2.544 - 0.0014 = 2.5426.$$

One more iteration:

$$x_3 = x_2 - \frac{e^{x_2} - 5x_2}{e^{x_2} - 5} = 2.5426 - \frac{e^{2.5426} - 12.7132}{e^{2.5426} - 5} = 2.5426 - \frac{0.0005}{7.7127}$$
$$= 2.5426 - 0.000065 = 2.5425.$$

We should really do one more iteration, for x_4, but I am confident that the change from x_3 will be of size 10^{-8}. You can see how the size of the correction at each iteration is reducing: 0.044, 0.0014, 0.000065; the number of zeroes after the decimal point is doubling at each iteration.

I conclude that the larger root of $f(x)$ is 2.5425, correct to 4 decimal places.

You should try this iteration starting with $x_0 = 0$, to find the smaller root.

Notice that in the first two iterations I did not record many decimal places while performing the calculations. This is an attractive feature of iterative methods: the intermediate iterates can be calculated roughly, and the required degree of precision used only at the end. In fact, even if you make an error in the calculation of an iterate, you should still obtain the correct answer in the end, though it may take one or two extra iterations to get there.

There is a tutorial on the Newton–Raphson method, with programmed examples and ones for you to try, in SPREADSHEET 13.2.

The iteration can fail to converge if the curve $y = f(x)$ has a critical point between x_0 and x^*, as illustrated in Figure 13.15. However, it can be proved that, provided your initial guess is close enough to the root, the Newton–Raphson iteration will converge. Not only that, but it will converge incredibly rapidly, as we saw in the example. In the method of bisection, the size of the interval bracketing the root was halving at each iteration. This is called **linear convergence**, where the error at each iteration is a fixed proportion of the previous error. The Newton–Raphson method exhibits **quadratic convergence**: the error at each iteration is proportional to the **square** of the previous error. What this means in practice is that once quadratic convergence sets in, the number of correct decimal places doubles at each iteration. If x_1 is correct to 2 decimal places, then x_2 will be correct to 4 decimal places, and x_3 correct to 8 decimal places. One more iteration will give a solution correct to 16 decimal places – reaching the precision capability of most computers.

The phenomenon of quadratic convergence appears when x^* is a simple root of $f(x)$. If x^* is a multiple root, the Newton–Raphson iteration will still converge, although it will only be linear convergence.

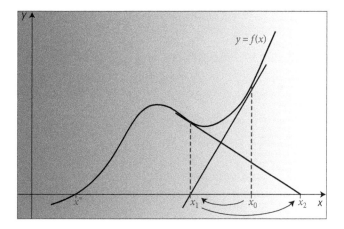

Figure 13.15 Failure of the Newton–Raphson method

Problems: Now try the end-of-chapter problems 13.15–13.18.

13.5.2 Newton's method for optimization

To find local maxima and minima of a differentiable function $f(x)$, we again use a truncated Taylor series expansion of $f(x)$ around an initial guess x_0, but this time we take the first three terms:

$$f(x) \approx f(x_0) + f'(x_0).(x - x_0) + \frac{1}{2}f''(x_0).(x - x_0)^2.$$

Differentiating:

$$f'(x) \approx f'(x_0) + f''(x_0).(x - x_0).$$

Setting this approximation equal to zero (since a critical point of $f(x)$ is where $f'(x) = 0$) and solving for x, as in the previous section, we obtain the iteration

$$x_{k+1} = x_k - \frac{f'(x_k)}{f''(x_k)} \quad \text{for } k = 0, 1, 2, 3, \ldots \tag{13.21}$$

This is Newton's method for optimization. You will see that this is not a new method: it is simply using the Newton–Raphson method to find a root of the function $f'(x)$.

In SPREADSHEET 13.3 we have programmed (13.21) to find the critical points of quintic polynomials, based on SPREADSHEET 13.1. Now, the value of x chosen via the scrollbar, is used as the starting-point x_0 for the iteration. The function entered into the spreadsheet has maxima at $x = 1$ and $x = 4$, and minima at $x = 2.11$ and $x = 4.89$. When x_0 is close to a maximizer or minimizer, the iteration will converge to it very rapidly. There are other starting-points, though, close to points of inflection, i.e. where $f''(x) \approx 0$, which send the iteration off on a wild goose chase. You can change the coefficients to try out other polynomials.

Problems: Now try the end-of-chapter problems 13.19–13.20.

? PROBLEMS

13.1–13.4: Analyse the following functions of x, and hence sketch their curves. You need to find the locations of the roots (if any), the locations and type of any critical points, and of any points of inflection. You may also need to use the theory of asymptotes from Part I.

13.1 $y = x^3 - 4x$ 13.2 $y = x^2 + \dfrac{16}{x}$

13.3 $y = \sin(x^2)$ 13.4 $y = (x^2 - 1)e^{-x}$.

13.5–13.8: Find the global minimum of the function, over the interval. You need to use calculus to look for local minima inside the interval, and also check the end-points of the interval.

13.5 $y = 2x^3 + 3x^2 - 12x + 5$ over the interval $[-4, 2]$

13.6 $y = x^4 + 3x^3 + x^2 + 1$ over the interval $[-3, 1]$

13.7 $y = \sin(x^2)$ over the interval $[-1, 3]$

13.8 $y = e^x - 5|x|$ over the interval $[-1, 3]$. You will need to consider the sub-intervals for $x < 0$ and $x \geq 0$ separately.

13.9: We want to fence off a rectangular field along a riverbank, as a nature reserve. We have a fixed length of fencing D, with which to fence the three sides of the field (the river forms the fourth side). The field will have length a (along the river) and width b. What are the values of a and b that will maximize the enclosed area A? Write down an equation expressing D in terms of a and b. The area is $A = a.b$, and by substituting for one of these variables using your equation you can write A as a function of just one variable. Then differentiate to find the critical point. You should find that the largest area is enclosed when $a = \dfrac{1}{2}D$ and $b = \dfrac{1}{4}D$.

13.10: We want to design a soft-drink can that contains a fixed volume V of drink and will require the least amount of metal to make. Suppose the can has radius r and height h. The volume of the can is $V = \pi r^2 h$. The area A of metal sheet needed comprises the rectangle which is curved to make the wall, plus the circular top and bottom. Write an equation expressing A in terms of r and h. Then follow the method of the previous problem to find the value of r that minimizes A. Show that for this value, the can height will be equal to its diameter.

13.11: The rate of photosynthesis depends on the light intensity $I(t)$ (in lumens per square metre) falling on the plant at time t. At a particular time of day this is given by $I(t) = I_0 \sin \theta$, where θ is the angle of elevation of the Sun above the horizon, and I_0 is the maximum intensity when the sun is directly overhead. At 10 a.m. the sun is at an elevation of 30° and is rising at a rate of 20 degrees per hour; show that the incident light intensity is increasing at a rate of $0.30\, I_0$ lumens m^{-2} hr^{-1}.

13.12: The skin area A (cm^2) of marsupial mammals is related to their weight M (kg) by the allometric relation:

$A = 1215\, M^{0.61}$.

How rapidly is the skin area increasing, for a young koala weighing 5 kg that is growing at a rate of 200 grams per month?

13.13–13.14: Find the first four non-zero terms of the Maclaurin series expansions of the following functions. Evaluate the series at $x = 1.0$, and compare with the value of the function at that point.

13.13 $f(x) = \sin x$

13.14 $f(x) = \ln (1 + x)$.

13.15: Re-solve problem 8.32 using the Newton–Raphson method. That is, perform four iterations to find a root of the function $f(x) = x^5 - 3$. Start at $x_0 = 1$.

13.16: Re-solve problem 9.13 using the Newton–Raphson method. That is, perform four iterations to find a value of x (in radians) for which $\sin x = \frac{1}{2}x$. Start at $x_0 = 3$.

13.17: Show that the Newton–Raphson iteration for finding a value of x for which $\ln x = ax$ (where a is a positive constant) can be simplified to

$$x_{k+1} = \frac{x_k(1 - \ln x_k)}{1 - ax_k}.$$

Perform four iterations by hand, to find the solution to $\ln x = 0.2x$, correct to 4 decimal places.

13.18: In Problem 8.13 we considered a lion with initial acceleration $a_0 = 6.25$ m s^{-2} and maximum velocity $v_{max} = 14$ m s^{-1}, and a gazelle with initial acceleration $a_0 = 3.75$ m s^{-2} and maximum velocity $v_{max} = 25$ m s^{-1}. Both animals accelerate from rest at time $t = 0$. You calculated the time t^* at which their velocities would be equal, using the hyperbolic model of animal speed. If we instead use the exponential model:

$$v(t) = v_{max}\left(1 - e^{-kt}\right) \quad \text{where } k = \frac{a_0}{v_{max}},$$

we need to solve the resulting equation using a numerical method. Program the Newton–Raphson method in Excel, to solve this problem.

13.19: Re-solve problem 8.34 using the Newton's method for optimization. That is, perform four iterations to find a minimum of the function $f(x) = x^5 - x - 3$. Start at $x_0 = 1$.

13.20: Re-solve problem 9.14 using the Newton's method. That is, perform four iterations to find a value of x (in radians) for which $g(x) = 2\sin x - \sin 2x$ reaches its minimum. Start at $x_0 = 3$.

14

Techniques of integration

We have now completed our study of the differential calculus. In this chapter and the next we describe the **integral calculus**. Integration is the inverse process to differentiation. We can define the integral of a function $f(x)$ as 'the most general function that, when you differentiate it, produces $f(x)$'. (We'll explain the 'most general' bit in a moment). Compare the definition in Chapter 1 of the square root of a number a as 'the positive number that, when you square it, produces a'. Taking the square root is the inverse operation to squaring a number. In the same way, integration is the inverse operation to differentiation. When integration is thought of in this sense, it is referred to as **anti-differentiation**, and the integral as the **anti-derivative**. For example, when you differentiate $\sin x$ you get $\cos x$, so $\sin x$ is an anti-derivative of $\cos x$.

The derivative has a physical meaning (the instantaneous rate of change), and a geometrical significance (the slope of the tangent) which gave us the definition of the derivative in (11.9). That definition can be used to work out the derivatives of functions from first principles. The integral also has a geometrical significance, which we'll see in Chapter 15, but this is of little use in finding the integrals of functions. Essentially, integration is an **inverse problem**: find a function that produces $f(x)$ when differentiated. Inverse problems are more difficult to solve than the corresponding direct problem: they are like saying 'Here is the answer, now work out what the question was.'[1] This involves an element of guesswork, and there are even fairly simple functions that cannot be integrated. However, it's not too difficult to write down the integrals of the standard functions whose derivatives we summarized in Section 12.9, and there are rules (analogous to the chain rule and product rule for differentiation) to help us integrate more complicated functions.

14.1 The integral as anti-derivative

In our example of the speeding car in Section 11.1, the integral does have a physical meaning. We now know that if the car were being tracked on traffic cameras, so that the function $s(t)$ defining its position at time t were known, then by differentiating $s(t)$ we can obtain the function $v(t)$ describing its speed at any time: $v(t) = \dfrac{ds}{dt}$. Now suppose that the driver monitored the

[1] In the book *The Hitchhiker's Guide to the Galaxy*, quoted in the preface, the task of finding the Ultimate Answer (which is 42) is trivial compared to the inverse problem: of finding the Great Question of Life, the Universe and Everything, to which 42 is the answer!

instantaneous speed $v(t)$ from inside the car, via the speedometer. From knowing $v(t)$ at every value of t, could he work out the distance travelled since $t = 0$, i.e. $s(t)$? This is a task of integration. Here, $s(t)$ will be the integral of $v(t)$.

In the case of motion with constant speed, you will remember from schooldays a simple formula:

$$\text{speed} = \frac{\text{distance}}{\text{time}}.$$

Rearranging, we get:

$$\text{distance} = \text{speed} \times \text{time}.$$

So if the car travels at a constant speed V, this equation gives us

$$s(t) = V.t.$$

We have performed our first integration.

When we explored the derivatives of standard mathematical functions in Chapter 12, we started with the power functions x^n (which, for $n = 1, 2, 3, \ldots$, are the functions x, x^2, x^3, etc.). We used the definition (11.9) to show that the derivative of x^n is nx^{n-1}. We now ask: what is the integral of x^n? First we need to state a definition of the integral, and introduce the mathematical notation for it.

14.1.1 Definition and notation

We now switch to using the mathematical variables x and y, and the function $f(x)$. Instead of differentiating $f(x)$ with respect to x, our task now is to integrate $f(x)$ with respect to x.

We wrote $\frac{df}{dx}$ to mean 'the derivative of $f(x)$ with respect to x'. We now write

$$\int f(x)dx$$

to mean 'the integral of $f(x)$ with respect to x'. The $\int$ sign is an elongated S, and comes from the interpretation of the integral as the limit of a sum (which is written using the symbol Σ, a Greek capital S) – we'll see this in Section 15.2 – in the same way as the derivative $\frac{dy}{dx}$ is the limit of the ratio $\frac{\Delta y}{\Delta x}$ (written using the Greek letter 'delta'). The function to be integrated, in this case $f(x)$, is called the **integrand**.

So our definition of the integral is:

> **Definition:**
>
> The **integral** of the function $f(x)$ with respect to x, written $\int f(x)\, dx$, is the most general function of x that, when differentiated with respect to x, produces $f(x)$. (14.1)

If we have to give this function a name, we'll use I, or $F(x)$. So

$$\text{if } F(x) = \int f(x)\, dx, \text{ it means that } \frac{dF}{dx} = f(x). \tag{14.2}$$

We said 'the most general function' because there will be a whole family of functions that, when differentiated, produce a given $f(x)$. This is because some information is lost through differentiation, namely, any constant term disappears. To give an example, we know that the derivative of x^2 is $2x$. But the derivatives of $x^2 + 1$, or $x^2 - 37$, are also $2x$. This is because we differentiate term-by-term, and the derivative of a constant is zero. So in fact the derivative of $x^2 + c$, where c is any constant, will be $2x$. This is the most general function whose derivative is $2x$, so this is the integral of $2x$;

$$\int 2x \, dx = x^2 + c. \tag{14.3}$$

This constant term that appears is called the **constant of integration**. Geometrically speaking, adding '+ c' to a function performs a vertical shift of the curve. The graph of $y = x^2 + c$ is a parabola that is concave up, with its minimum at $(0, c)$. This is sketched in FNGRAPH 14.1; you can change the value of c in the 'Variables' window. Notice that a vertical shift does not change the gradient of the curve at all. When $x = 3$, for example, the slope of the tangent to the curve will be 6, whatever the choice of c.

So satisfying the 'most general function' requirement means writing '+ c' at the end of the integral. This might seem rather pernickety, but is actually important when we need to find the value of c that solves our particular problem. Until a particular value of c has been determined, it is called an **arbitrary constant**.

14.1.2 The integrals of power functions, and the coefficient rule

Now we can answer the question posed in the introduction to this section: what is the integral of x^n?

We've just seen that the integral of $2x$ is $x^2 + c$, since $\frac{d}{dx}(x^2) = 2x$. Similarly, the integral of $3x^2$ is $x^3 + c$, since $\frac{d}{dx}(x^3) = 3x^2$. But $2x$ and $3x^2$ are not standard functions; we want to know the integrals of the power functions $x, x^2, x^3, \ldots$, and in general the integral of x^n.

Recall the coefficient rule for differentiation in (11.11): if a is a coefficient multiplying $f(x)$, we can take it outside the differentiation:

$$\frac{d}{dx}(a.f(x)) = a.\frac{df}{dx}.$$

So $\frac{d}{dx}\left(\frac{1}{2}x^2\right) = \frac{1}{2}\frac{d}{dx}(x^2) = \frac{1}{2}(2x) = x$. This means that $\int x \, dx = \frac{1}{2}x^2 + c$. There is thus a coefficient rule for integration, which says that we can take coefficients outside the integration sign:

Coefficient rule:

$$\int a.f(x) \, dx = a\int f(x) \, dx \tag{14.4}$$

where a is a constant.

We can use the coefficient rule to find the integral of x^2, using the fact that the derivative of x^3 is $3x^2$:

$$\int 3x^2 \, dx = x^3 + c$$
$$3\int x^2 \, dx = x^3 + c$$
$$\int x^2 \, dx = \frac{1}{3}x^3 + c.$$

Notice that we don't write '$+\frac{1}{3}c$' in the last line. As c is an arbitrary constant, $\frac{1}{3}c$ could be any constant too. As long as the constant of integration remains arbitrary, we just write it as '$+ c$'.

In Chapter 11 we saw the derivative of the power function x^n:

If $y = x^n$, then its derivative is $\dfrac{dy}{dx} = nx^{n-1}$. $\hspace{2cm}$ (11.10)

We could write this using $(n + 1)$ instead of n:

If $y = x^{n+1}$, then its derivative is $\dfrac{dy}{dx} = (n+1)x^n$.

From our definition of the integral as anti-derivative, this means that

$$\int (n+1)x^n \, dx = x^{n+1} + c.$$

Performing the algebra we used above:

$$\int (n+1)x^n \, dx = x^{n+1} + c$$
$$(n+1)\int x^n \, dx = x^{n+1} + c$$
$$\int x^n \, dx = \frac{1}{n+1}x^{n+1} + c.$$

We now have the integral of the power functions:

$$\int x^n \, dx = \frac{1}{n+1}x^{n+1} + c. \hspace{2cm} (14.5)$$

You need to remember this result: it's just as important as knowing that $\dfrac{d}{dx}(x^n) = nx^{n-1}$. You'll recall that this result (11.10) doesn't just work for $n = 1,2,3, \ldots$, it can be used when n is a rational number, or zero, or negative. The same thing applies to (14.5). Thus:

- $n = 0 : \int 1 \, dx = x + c$

  Using $\dfrac{1}{a/b} = \dfrac{b}{a}$

  and $x^{\frac{3}{2}} = x^1.x^{\frac{1}{2}} = x.\sqrt{x}$

- $n = \frac{1}{2} : \int \sqrt{x} \, dx = \int x^{\frac{1}{2}} \, dx = \dfrac{1}{\frac{1}{2}+1}x^{\frac{1}{2}+1} + c = \dfrac{1}{\frac{3}{2}}x^{\frac{3}{2}} + c = \frac{2}{3}x\sqrt{x} + c$

- $n = -2 : \int \dfrac{1}{x^2} \, dx = \int x^{-2} \, dx = \dfrac{1}{-2+1}x^{-2+1} + c = -x^{-1} + c = -\dfrac{1}{x} + c.$

You should check these results by differentiating the right-hand sides.

There is one exception to (14.5). We must not take $n = -1$, as the right-hand side of (14.5) would then become $\frac{1}{0}x^0 + c$, and dividing by zero is forbidden. So (14.5) cannot tell us the integral of x^{-1}, i.e. of $\dfrac{1}{x}$. Luckily, we already know a function whose derivative is $\dfrac{1}{x}$: the natural logarithm; see (12.11). So

$$\int \frac{1}{x} \, dx = \ln |x| + c. \tag{14.6}$$

We must use the absolute value of x inside the logarithm function, as you can only find logarithms of positive numbers.

> **PRACTICE:**
> Find the integrals of $\quad x^7, \quad \dfrac{1}{x^3}, \quad \dfrac{1}{\sqrt{x}}$.
>
> **Answers:** $0.125x^8 + c, \quad -\dfrac{1}{2x^2} + c, \quad 2\sqrt{x} + c.$

14.1.3 The sum rule, and the integrals of polynomial functions

There is a sum rule for integration, analogous to that for differentiation. We differentiate an expression term-by-term, so we integrate term-by-term too.

> **EXAMPLE**
> Find $\displaystyle\int \left(x^4 + \frac{1}{x} \right) dx.$
>
> **Solution:** $\displaystyle\int \left(x^4 + \frac{1}{x} \right) dx = \int x^4 \, dx + \int \frac{1}{x} \, dx = \tfrac{1}{5}x^5 + \ln |x| + c.$

Although we have performed two integrations, which would create two constants of integration, adding them together just gives a single arbitrary constant, so again we just write '$+ c$' at the end. Unless you understand this reasoning, it may seem that the main purpose of the requirement to write '$+ c$' at the end of each integration, is to provide the lecturer with an excuse to deduct one mark if you forget it.

The sum rule for integration, using functions $f(x)$ and $g(x)$, is:

> **Sum rule:**
>
> $$\int (f(x) + g(x)) \, dx = \int f(x) \, dx + \int g(x) \, dx$$
>
> $$\int (f(x) - g(x)) \, dx = \int f(x) \, dx - \int g(x) \, dx. \tag{14.7}$$

Using the coefficient and sum rules, we can now integrate all polynomial functions.

> **EXAMPLE**
> Find $\displaystyle\int (x - 1)(x^2 + 3) \, dx.$
>
> We must first multiply out the polynomial:
>
> $$I = \int (x - 1)(x^2 + 3) \, dx = \int (x^3 - x^2 + 3x - 3) \, dx.$$
>
> Now integrate term-by-term, using (14.5):
>
> $$I = \int x^3 \, dx - \int x^2 \, dx - +3\int x \, dx - 3\int 1 \, dx$$
> $$= \tfrac{1}{4}x^4 - \tfrac{1}{3}x^3 + \tfrac{3}{2}x^2 - 3x + c.$$

Here is an application which we cited in Section 11.1.2:

APPLICATION

The new VW Polo can accelerate from 0 to 60 miles per hour in 16 seconds. Assuming constant acceleration, how far does the car travel during this time?

We know that the speed v at time t is the instantaneous rate of change of distance s with time: $v(t) = \dfrac{ds}{dt}$. Also, the acceleration $a(t)$ is the rate of change of speed with time: $a(t) = \dfrac{dv}{dt}$. At time $t = 0$, we have $s(0) = 0$ (the car starts from this position) and $v(0) = 0$ (the car starts from rest). The driver presses the accelerator pedal fully down throughout the travel, so we can assume the acceleration is constant: $a(t) = A$, say. Measuring time in seconds, we convert 60 mph into miles per second; then $v(16) = \dfrac{60}{3600} = \dfrac{1}{60}$ miles per second.

From the above, $\dfrac{dv}{dt} = A$, a constant, so $v = \int A\, dt = At + c$. We can find c by substituting in the initial values of v and t. Our initial condition is that $v = 0$ when $t = 0$, which means that $c = 0$. So we have $v = At$.

Now we also know that at $t = 16$ seconds, the speed is $v(16) = \dfrac{1}{60}$ miles per second. Inserting in $v = At$, we can calculate A: $\dfrac{1}{60} = A.(16)$, so $A = \dfrac{1}{16 \times 60}$ m s^{-2} (miles per second squared).

Now we find s, from $v(t) = \dfrac{ds}{dt}$. Integrating,

$$s = \int v\, dt = \int At\, dt = A\int t\, dt = A\left(\tfrac{1}{2}t^2\right) + c = \tfrac{1}{2}\cdot\dfrac{1}{16 \times 60}t^2 + c.$$

So $s = \dfrac{1}{32 \times 60}t^2 + c.$

Inserting the initial conditions, namely that $s = 0$ when $t = 0$, we see that

$$0 = \dfrac{1}{32 \times 60}.0^2 + c.$$

So $c = 0$. The constant of integration in this equation is also zero. Thus

$$s(t) = \dfrac{1}{32 \times 60}t^2.$$

We can now find the answer to the problem: the distance travelled during the 16 seconds. Substituting $t = 16$:

$$s(16) = \dfrac{1}{32 \times 60}16^2 = \dfrac{16^2}{32 \times 60} = \dfrac{16}{2 \times 60} = \dfrac{2}{15}\text{ miles,}$$

which is 704 feet, or 213.3 metres.

Finally, some integrals for you to try, to check that you can use the coefficient and sum rules together with (14.5):

PRACTICE:

Find   (i) $\int(3x^2 - 6x + 5)\, dx$,   (ii) $\int(2x - 1)^2\, dx$,   (iii) $\int\left(\dfrac{2}{x^3} - \dfrac{3}{x}\right)dx$.

Answers: (i) $x^3 - 3x^2 + 5x + c$,   (ii) $\dfrac{4}{3}x^3 - 2x^2 + x + c$,   (iii) $-\dfrac{1}{x^2} - 3\ln|x| + c$.

Problems: Now try the end-of-chapter problems 14.1–14.6.

14.1.4 Integrals of some standard functions

Using the summary table of derivatives in Section 12.9, together with 'inverse logic', we can now write down the integrals of some standard functions:

- $\dfrac{d}{dx}(\sin x) = \cos x$, so $\int \cos x \, dx = \sin x + c$;

- $\dfrac{d}{dx}(\cos x) = -\sin x$, so $\int \sin x \, dx = -\cos x + c$;

- $\dfrac{d}{dx}(e^x) = e^x$, so $\int e^x \, dx = e^x + c$.

We can generalize the standard integrals in (14.8) using the rule in Section 11.5 for the derivative of a function of a linear function of x:

$$\text{If } y = f(ax + b), \quad \text{then} \quad \frac{dy}{dx} = a.f'(ax + b). \tag{11.13}$$

For example, this rule tells us that

$$\frac{d}{dx}(e^{ax}) = a.e^{ax},$$

> So $\int ae^{ax} \, dx = e^{ax} + c$, and hence $a\int e^{ax} \, dx = e^{ax} + c$

so

$$\int e^{ax} \, dx = \frac{1}{a} e^{ax} + c.$$

Similarly,

$$\int \sin(ax + b) \, dx = -\frac{1}{a}\cos(ax + b) + c.$$

Also,

$$\frac{d}{dx}(\ln | ax + b |) = a.\frac{1}{ax + b},$$

so

$$\int \frac{1}{ax + b} \, dx = \frac{1}{a}\ln | ax + b | + c.$$

Here is a table of standard integrals which we have already seen, or can deduce in this way:

$f(x)$	$\int f(x) \, dx$
$x^n \quad (n \neq -1)$	$\dfrac{1}{n+1}x^{n+1} + c$
e^{ax}	$\dfrac{1}{a}e^{ax} + c$
$a^x \quad (a > 0)$	$\dfrac{1}{\ln a}a^x + c$
$\sin(ax + b)$	$-\dfrac{1}{a}\cos(ax + b) + c$

$f(x)$	$\int f(x)\,dx$
$\cos(ax+b)$	$\dfrac{1}{a}\sin(ax+b)+c$
$\dfrac{1}{ax+b}$	$\dfrac{1}{a}\ln\lvert ax+b\rvert+c$
$\dfrac{1}{\sqrt{1-x^2}}$	$\sin^{-1}x+c$
$\dfrac{1}{1+x^2}$	$\tan^{-1}x+c$

(14.8)

We still don't know the integrals of $\ln x$, $\tan x$, or the inverse trigonometric functions. We will find these using the techniques of integration set out in the following sections, before seeing some mathematical applications of this theory.

PRACTICE:

Find $\int(4\cos 2x - 3\sin x)\,dx$, $\int\left(6e^{2x} - \dfrac{7}{3x-1}\right)dx$.

Answers: $2\sin 2x + 3\cos x + c$, $3e^{2x} + \dfrac{7}{3}\ln\lvert 3x-1\rvert + c$.

Problems: Now try the end-of-chapter problems 14.7–14.12.

CASE STUDY C16: Integrating the hyperbolic and exponential models of animal speed

We have seen two models for the speed $v(t)$ of an animal accelerating from rest with initial acceleration a_0 and reaching a maximum velocity $v_{\max}$. By integrating such a model we can obtain an expression for the distance $s(t)$ travelled by the animal. The integration will introduce a '+ c' constant of integration; to deal with this we need an initial condition. We'll say that at time $t = 0$ the animal is at a position d metres from the origin, so that $s(0) = d$.

Exponential model:

In this model

$$v(t) = v_{\max}(1 - e^{-kt}) \tag{10.29}$$

where $k = \dfrac{a_0}{v_{\max}}$.

The integration is straightforward, using the standard integrals in (14.8):

$$s(t) = \int v(t)\,dt = v_{\max}\int (1 - e^{-kt})\,dt$$

$$= v_{\max}\left[t - \left(\frac{1}{-k}\right)e^{-kt}\right] + c$$

$$= v_{\max}\left[t + \frac{1}{k}e^{-kt}\right] + c.$$

Use the initial condition to find c:

$$s(0) = d = v_{max}\left[0 + \frac{1}{k}e^0\right] + c$$

$$c = d - \frac{v_{max}}{k}.$$

Substitute this expression for c into our model for $s(t)$:

$$s(t) = v_{max}t + \frac{1}{k}v_{max}e^{-kt} + d - \frac{v_{max}}{k}$$

which we can write as:

$$s(t) = v_{max}t - \frac{v_{max}}{k}\left[1 - e^{-kt}\right] + d.$$

Hyperbolic model:
In this model

$$v(t) = \frac{v_{max}t}{t + b}. \tag{8.5}$$

where $b = \dfrac{v_{max}}{a_0}$. To use the standard integral we need to do a bit of algebra:

$$\frac{t}{t+b} = \frac{(t+b)-b}{t+b} = \frac{(t+b)}{t+b} - \frac{b}{t+b} = 1 - \frac{b}{t+b}.$$

Then integrating (8.5):

$$s(t) = \int v(t)\,dt = v_{max}\int \frac{t}{t+b}\,dt = v_{max}\int\left(1 - \frac{b}{t+b}\right)dt$$

$$= v_{max}\left[t - b\ln|t+b|\right] + c.$$

Again, use the initial condition to find c:

$$s(0) = d = v_{max}[0 - b\ln|0+b|] + c$$

$$c = d + v_{max}b\ln b.$$

Substituting back into our model:

$$s(t) = v_{max}[t - b\ln|t+b|] + v_{max}b\ln b + d$$

$$= v_{max}t - v_{max}b(\ln|t+b| - \ln b) + d$$

which using the laws of logarithms we can write as

$$s(t) = v_{max}t - v_{max}b\ln\left|\frac{t+b}{b}\right| + d$$

and since $b = \dfrac{1}{k}$ we can write this, for comparison with the exponential model, as

$$s(t) = v_{max}t - \frac{v_{max}}{k}\ln|kt + 1| + d.$$

You should try programming these two curves in FNGraph.

14.2 Integration by substitution

The rule from Section 11.5 which we just used was an example of the chain rule (Section 12.2), which is used to differentiate a function of a function of x. The chain rule involves inventing an intermediate function $u(x)$, so that $y = y(u)$, and then $\dfrac{dy}{dx} = \dfrac{dy}{du} \cdot \dfrac{du}{dx}$. In the technique called 'integration by substitution', we also invent an intermediate function $u(x)$. Here's how it works:

EXAMPLE

Find the integral $I = \displaystyle\int \frac{1}{(3x-2)^2}\, dx$.

Solution: The integrand is $\dfrac{1}{(3x-2)^2} = (3x-2)^{-2}$. This is a function of a function, as we learned to recognize in Section 12.2; it can be written as u^{-2}, using the intermediate function $u(x) = 3x - 2$. Differentiating this with respect to x, we get $\dfrac{du}{dx} = 3$.

We now treat $\dfrac{du}{dx}$ as if it were a fraction, and multiply both sides by dx, to get

$$du = 3\, dx.$$

Solve for dx:

$$dx = \frac{1}{3}\, du.$$

Now we can go back to our integral $I = \displaystyle\int \frac{1}{(3x-2)^2}\, dx$, and replace $(3x-2)$ by u, and dx by $\frac{1}{3} du$. This gives us a function of u only, which is to be integrated with respect to u:

$$I = \int \frac{1}{(3x-2)^2}\, dx = \int \frac{1}{u^2}\frac{1}{3}\, du = \frac{1}{3}\int \frac{1}{u^2}\, du = \frac{1}{3}\int u^{-2}\, du \text{ using the coefficient rule.}$$

This is just like having to integrate $\displaystyle\int x^{-2}\, dx$, which we do using (14.5):

$$I = \int \frac{1}{(3x-2)^2}\, dx = \frac{1}{3}\int u^{-2}\, du = \frac{1}{3}(-u^{-1}) + c = -\frac{1}{3u} + c.$$

The final step is to substitute back for u, using $u = 3x - 2$, to get the integral as a function of x:

$$I = -\frac{1}{3(3x-2)} + c.$$

Time and energy permitting, you should check your answer by differentiating it, hopefully obtaining the original function.

Here's another example:

EXAMPLE

Find the integral $I = \int \sin(ax+b)\, dx$, where a, b are constants.
Solution: Use integration by substitution. Let $u = ax + b$.

Differentiate u with respect to x: $\dfrac{du}{dx} = a$.

Solve for dx:

$$du = a\, dx$$
$$dx = \frac{1}{a}\, du$$

Substitute into I, and integrate with respect to u:

$$I = \int \sin(ax + b)\, dx = \int \sin u \left(\frac{1}{a}\, du \right)$$
$$= \frac{1}{a} \int \sin u\, du = \frac{1}{a}\left[-\cos u \right] + c = \frac{-1}{a}\cos u + c.$$

Substitute back for x:

$$I = \frac{-1}{a}\cos(ax + b) + c.$$

In the next example it is not obvious what to choose for the intermediate function $u(x)$.

EXAMPLE

Find the integral $I = \int x e^{-x^2}\, dx$.

Solution: Use integration by substitution. Let $u = -x^2$.

Differentiate with respect to x: $\dfrac{du}{dx} = -2x$,

Then $du = -2x\, dx$, so $x\, dx = -\frac{1}{2}\, du$.

Now I can be rearranged into $I = \int e^{-x^2} x\, dx$, and we can substitute u for $-x^2$, and $-\frac{1}{2}\, du$ for $x\, dx$:

$$I = \int e^{-x^2} x\, dx = \int e^{u} \cdot -\tfrac{1}{2}\, du = -\tfrac{1}{2} \int e^{u}\, du = -\tfrac{1}{2}\left[e^{u} \right] + c.$$

Finally, substituting back, the solution is

$$I = \int x e^{-x^2}\, dx = -\tfrac{1}{2} e^{-x^2} + c.$$

Check: By the chain rule $\dfrac{dI}{dx} = -\tfrac{1}{2} e^{-x^2} \dfrac{d}{dx}(-x^2) = -\tfrac{1}{2} e^{-x^2} \cdot (-2x) = x e^{-x^2}$.

We are beginning to see that integrations that look quite complicated may be integrated relatively easily, while similar, simpler functions may be more difficult or even impossible to integrate. The integral $\int e^{-x^2\, dx}$ looks simpler than the one we have just solved, but in fact it is impossible to integrate! This is very different from the situation with differentiation, where a complicated function can always be differentiated by breaking it down into sub-expressions, and using the product, quotient and other rules systematically.

Note: We have been writing integrals in the form $\int f(x)\, dx$. As you'll see in the questions below, sometimes the 'dx' is written as part of the function, e.g. writing $\int \frac{1}{x}\, dx$ as $\int \frac{dx}{x}$, or $\int \frac{x}{1 + x^2}\, dx$ as $\int \frac{x\, dx}{1 + x^2}$. You may also see $\int dx$, meaning $\int 1\, dx$, which is $x + c$. This habit of mathematicians can seem confusing, but it should always be possible to adjust the expression into the form $\int f(x)\, dx$.

PRACTICE:

Use integration by substitution to find: (i) $\displaystyle\int \frac{dx}{(1 - x)^5}$,   (ii) $\displaystyle\int \frac{x\, dx}{1 + x^2}$.

Answers: (i) $\dfrac{1}{4(1 - x)^4} + c$,   (ii) $\tfrac{1}{2}\ln | 1 + x^2 | + c$.

We'll aim to write integrals in the form $\int f(x)\, dx$ from now on.

Application: temperature-dependent insect development

The obliquebanded leafroller caterpillar is a pest affecting fruit crops such as apples, in North America. Its development rate is dependent on the ambient temperature, and Jones et al. (2005) fitted a least-squares straight line to experimental data. Based on this, we will use the relationship

$$\frac{dL}{dt} = k(\theta - 10) \tag{14.9}$$

where the growth rate $\frac{dL}{dt}$ is the rate of change of the length $L(t)$ of the caterpillar (in mm) at time t (in months after hatching), and θ is the ambient temperature (in degrees C).

We consider a similar insect species in which the eggs hatch in early spring, the larva lives throughout the summer, then pupates over winter before emerging as a moth the following spring. For this species we will use a growth coefficient $k = 0.15$ mm per month per degree C. The larva is 2 mm long when it hatches from the egg. We wish to estimate the length of the adult caterpillar at the end of the summer when it becomes a pupa, using a simple model of annual temperature variation.

For the climate model, we will use a sine function to approximate the temperature variation over 12 months from December. We assume a minimum temperature of 5 °C in December ($t = 0$), rising to an annual maximum temperature of 25 °C in June ($t = 6$). The sine function has a period of 12 months. Using this information to fit the general sine function in Section 9.5:

$$g(t) = A \sin\left(\frac{2\pi}{T}(t - t_0)\right) + d \tag{9.20}$$

produces the function

$$\theta = 10 \sin\left(\frac{\pi}{6}(t - 3)\right) + 15. \tag{14.10}$$

Substituting (14.10) into (14.9) gives $\frac{dL}{dt}$ as a function of time:

$$\frac{dL}{dt} = k\left(10 \sin\left(\frac{\pi}{6}(t - 3)\right) + 5\right). \tag{14.11}$$

We can now find L as a function of t, by integrating (14.11) with respect to t. In integrating the sine function, we use integration by substitution with an intermediate function $u = \frac{\pi}{6}(t - 3)$. Then $\frac{du}{dt} = \frac{\pi}{6}$, so $dt = \frac{6}{\pi} du$.

$$L = k\int\left(10 \sin\left(\frac{\pi}{6}(t - 3)\right) + 5\right) dt$$

$$= k\int 10 \sin\left(\frac{\pi}{6}(t - 3)\right) dt + k\int 5\, dt$$

$$= 10k\int \sin\left(\frac{\pi}{6}(t - 3)\right) dt + 5kt + c$$

$$= 10k\int \sin u\left(\frac{6}{\pi}\right) du + 5kt + c$$

$$= \frac{60k}{\pi}\int \sin u\, du + 5kt + c$$

$$= \frac{60k}{\pi}[-\cos u] + 5kt + c.$$

Substituting back for u, we obtain

$$L = -\frac{60k}{\pi}\cos\left(\frac{\pi}{6}(t-3)\right) + 5kt + c. \tag{14.12}$$

We consider a caterpillar that hatches in March ($t = 3$) and starts the pupal stage in September ($t = 9$) to overwinter. We can find the constant of integration by using the initial condition, that at $t = 3$ months, the larval length is $L(3) = 2$ mm. Substituting into (14.12) and solving for c:

$$2 = -\frac{60k}{\pi}\cos\left(\frac{\pi}{6}(3-3)\right) + 15k + c$$

$$c = 2 + \frac{60k}{\pi} - 15k$$

since $\cos 0 = 1$. Substitute this value of c back into (14.12):

$$L(t) = 2 + \frac{60k}{\pi}\left(1 - \cos\left(\frac{\pi}{6}(t-3)\right)\right) + 5k(t-3). \tag{14.13}$$

Finally, to find the caterpillar length in September, set $t = 9$ in (14.13)

$$L(9) = 2 + \frac{60k}{\pi}\left(1 - \cos\left(\frac{\pi}{6}(9-3)\right)\right) + 5k(9-3)$$

$$= 2 + \frac{60k}{\pi}(1 - \cos\pi) + 30k$$

$$= 2 + \frac{120k}{\pi} + 30k$$

since $\cos\pi = -1$.

Using the growth coefficient $k = 0.15$ mm per month per °C, we find that the adult length is 12.23 mm. FNGRAPH 14.2 has a plot of the function $L(t)$ in (14.13), from $t = 3$ to $t = 9$ months.

Problems: Now try the end-of-chapter problems 14.13–14.18.

14.3 Integration by parts

This technique is the counterpart of the product rule for differentiation, and is used in some (but not all) situations where we need to integrate a product of two functions, for example $\int x\cos x\,dx$. It is derived from the product rule (12.3), which we can write as:

$$\frac{d(uv)}{dx} = u\frac{dv}{dx} + v\frac{du}{dx}.$$

Rearrange this to:

$$u\frac{dv}{dx} = \frac{d(uv)}{dx} - v\frac{du}{dx}.$$

CASE STUDY A15: Solving the differential equation for exponential growth

Recall that in exponential growth, the instantaneous rate of population increase is proportional to the current population size:

$$\frac{dN}{dt} = rN. \qquad (11.2)$$

In Chapter 16 we will show how such differential equations can be solved, using an approach very similar to integration by substitution. Here is a foretaste:

In (11.2), treat $\frac{dN}{dt}$ as a fraction, and multiply both sides by dt:

$$dN = rN \, dt.$$

Divide both sides by N, to collect all terms involving N on the left-hand side:

$$\frac{1}{N} dN = r \, dt.$$

Now integrate both sides:

$$\int \frac{1}{N} dN = \int r \, dt.$$

On the left-hand side we are integrating $\frac{1}{N}$ with respect to N, which from the table in (12.8) gives the natural logarithm of N. On the right-hand side we are integrating the net growth constant r with respect to t, which gives $r.t + c$. We only need to include a constant of integration in one of the integrals:

$$\ln N = rt + c.$$

We don't need to write $|N|$, as the population size is always positive.

To solve for N, take the exponential of each side. This is the inverse function of the logarithm; recall that $e^{\ln a} = a$, as well as the rules of exponents:

$$e^{\ln N} = e^{rt+c}$$

$$N = e^{rt}.e^{c}$$

$$N = Ae^{rt}$$

where A is an arbitrary positive constant (replacing e^c, which will always be a positive number). This is the general equation for exponential growth, which we used back in Chapter 10.

When time $t = 0$, $N(0) = Ae^{r.0} = A.1 = A$, so the coefficient A is the initial population size N_0. Thus we have obtained the equation we saw in Part 9 of this Case Study:

$$N(t) = N_0 e^{rt}. \qquad (10.9)$$

Integrate both sides with respect to x:

$$\int u \frac{dv}{dx} dx = \int \frac{d(uv)}{dx} dx - \int v \frac{du}{dx} dx.$$

Cancel the 'dx's:

$$\int u \, dv = \int d(uv) - \int v \, du.$$

But $\int d(uv)$ means the integral of 1 with respect to uv, which is $uv + c$. We don't need to include the constant of integration, as this will appear from the remaining integrals. The formula for integration by parts is thus:

Integration by parts:

$$\int u\, dv = uv - \int v\, du. \tag{14.14}$$

Here is how it is used to find $\int x \cos x\, dx$:

The first step is to match our problem to the left-hand side of (14.14). We must choose part of the integrand to be u, and the rest (including the dx) will be dv. We will try

Let $u = x$ and $dv = \cos x\, dx$.

Now differentiate the first equation to find du, and integrate the second equation to find v:

$$\frac{du}{dx} = 1 \quad \text{and} \quad \int dv = \int \cos x\, dx$$
$$du = dx \qquad\qquad v = \sin x.$$

Again, there is no need for a '$+ c$' until the last step.

Now substitute for u, du, v, dv in (14.14):

$$I = uv - \int v\, du$$
$$= (x)(\sin x) - \int \sin x\, dx$$
$$= x\sin x - (-\cos x) + c$$
$$= x\sin x + \cos x + c.$$

> Use the product rule of differentiation

Finally, **check** your answer by differentiating it: $\dfrac{dI}{dx} = (x \cos x + \sin x) - \sin x = x\cos x.$

The alternative choice of u and dv would have been:

Let $u = \cos x$ and $dv = x\, dx$.

Then: $du = -\sin x\, dx$ and $v = \frac{1}{2}x^2$.

Inserting into (14.14) would give:

$$I = \tfrac{1}{2}x^2 \cos x - \int \tfrac{1}{2}x^2(-\sin x)dx.$$

This is a dead end – we are faced with an integral that is more difficult to solve than our original one. In using integration by parts, the correct choice of u and dv is crucial. As you need to differentiate u, and integrate dv, you should choose u to be an expression that becomes simpler after differentiation (e.g. a power of x), and dv to be an easily integrable function.

In integrating $\int v\, du$, you may need to use integration by substitution, or integration by parts again:

EXAMPLE

Find $I = \int x^2 \sin x\, dx$.

Solution: Use integration by parts.

Let $u = x^2$ and $dv = \sin x\, dx$.

Then:

$$du = 2x\, dx \quad \text{and} \quad v = \int \sin x\, dx = -\cos x.$$

Substituting in (14.14):

$$I = (x^2)(-\cos x) - \int (-\cos x)2x \, dx$$
$$= -x^2 \cos x + 2 \int x \cos x \, dx$$
$$= -x^2 \cos x + 2 [x \sin x + \cos x] + c \quad\longleftarrow \boxed{\text{Using the result of the previous example!}}$$
$$= (2 - x^2)\cos x + 2x \sin x + c.$$

We can use integration by parts to find $\int \ln x \, dx$. This doesn't look like a product – but there's a trick:

EXAMPLE

Find $I = \int \ln x \, dx$.

Solution: Use integration by parts.
Let $u = \ln x$ and $dv = dx$.
Then:

$$du = \frac{1}{x} dx \quad \text{and} \quad v = \int dx = \int 1 \, dx = x.$$

Substituting in (14.14):

$$I = (\ln x)(x) - \int (x)\frac{1}{x} dx$$
$$= x \ln x - \int 1 \, dx$$
$$= x \ln x - x + c.$$

Check: $\dfrac{dI}{dx} = \left(x.\dfrac{1}{x} + 1.\ln x \right) - 1 = 1 + \ln x - 1 = \ln x.$

We will use integration by parts again in Section 14.5, when we find $\int e^x \sin x \, dx$.

PRACTICE:
Find (i) $\int xe^{-x} \, dx$ (ii) $\int x\ln x \, dx$.
Answers: (i) $-(x + 1)e^{-x} + c$ (ii) $\frac{1}{4}x^2(2 \ln x - 1) + c.$

Note: You'll need to do some algebra, taking out common factors, to get your answers into the forms given above.

Problems: Now try the end-of-chapter problems 14.19–14.22.

. . . .

14.4 Integration by partial fractions

This is a technique we will need when solving the differential equation for logistic growth. It is used when the integrand is a rational function, whose denominator is a factorized polynomial. (This is a mathematician's way of saying that we have to integrate a fraction, where there is a product of two factors on the bottom.)

For example,

$$\text{find } I = \int \frac{9x+1}{(2x+1)(x-3)} \, dx.$$

This is a technique of algebra rather than integration. We aim to write the integrand as the sum of two simpler fractions; each of these can then be integrated using substitution. In the example above, we want to find constants A and B such that

$$\frac{9x+1}{(2x+1)(x-3)} = \frac{A}{2x+1} + \frac{B}{x-3}. \tag{14.15}$$

The above equation is an identity, true for all values of x. We put the right-hand side over a common denominator:

$$\frac{9x+1}{(2x+1)(x-3)} = \frac{A(x-3) + B(2x+1)}{(2x+1)(x-3)}.$$

Now the denominators on both sides are the same, so the numerators must be identical:

$$9x + 1 = A(x-3) + B(2x+1). \tag{14.16}$$

To find A and B there are two methods: the easier one, and the one that mathematicians prefer:

(1) Method of substitution

In the simpler method, we choose values of x to substitute in, which will make either A or B disappear. In this case, first set $x = 3$ in (14.16):

$$9 \times 3 + 1 = A(3-3) + B(2 \times 3 + 1)$$
$$28 = A.0 + 7B = 7B$$
$$B = \frac{28}{7} = 4.$$

To find A, we want the factor $(2x + 1)$ multiplying B to disappear, so we set $x = -\frac{1}{2}$

$$9\left(-\tfrac{1}{2}\right) + 1 = A\left(-\tfrac{1}{2} - 3\right) + B\left(-2 \times \tfrac{1}{2} + 1\right)$$
$$-4.5 + 1 = -3.5A + B.0$$
$$-3.5 = -3.5A$$
$$A = 1.$$

So $A = 1$ and $B = 4$, so that

$$\frac{9x+1}{(2x+1)(x-3)} = \frac{1}{2x+1} + \frac{4}{x-3}.$$

(2) Method of equating coefficients

Multiply out the right-hand side of (14.16), and collect the terms in x together:

$$9x + 1 = Ax - 3A + 2Bx + B$$
$$9x + 1 = (A + 2B)x + (-3A + B).$$

On each side of this equation we have a linear polynomial (i.e. an expression of the form $mx + c$), but as this is an identity (true for all values of x), they must be exactly the same polynomial. That is, the coefficients of x must be the same, and the constant terms must be the same.

Thus we can equate the constant terms:

$$-3A + B = 1$$

and the coefficients of the terms in x:

$$A + 2B = 9.$$

We have two simultaneous equations in two unknowns, which we can solve using the technique in Section 5.7: solving simultaneous linear equations. Multiply the second equation by 3, so that we have

$$-3A + B = 1$$
$$3A + 6B = 27.$$

Now add the two equations together. The terms in A will cancel out:

$$(-3A + 3A) + (B + 6B) = 1 + 27,$$

giving us $7B = 28$, so that $B = 4$. Substituting this into either of the original equations, and solving for A, gives $A = 1$.

This method is more work in this case, but it can be applied in more complicated factorizations where it is difficult to guess the values of x to substitute in method (1).

By either method, we now have a sum of partial fractions, which we can integrate term-by-term:

$$I = \int \frac{9x+1}{(2x+1)(x-3)}\,dx$$
$$= \int \left(\frac{1}{2x+1} + \frac{4}{x-3} \right) dx$$
$$= \int \frac{1}{2x+1}\,dx + \int \frac{4}{x-3}\,dx$$
$$= \int \frac{1}{2x+1}\,dx + 4\int \frac{1}{x-3}\,dx$$
$$= \tfrac{1}{2}\ln|2x+1| + 4\ln|x-3| + c.$$

Using the laws of logarithms (10.13), we could write the solution as

$$I = \ln\left(\sqrt{2x+1}(x-3)^4 \right) + c.$$

PRACTICE:

Find (i) $\int \frac{1}{x^2-1}\,dx$ (ii) $\int \frac{3}{(1+x)(1-2x)}\,dx$.

Answers: (i) $\tfrac{1}{2}\ln\left|\dfrac{x-1}{x+1}\right| + c$ (ii) $\ln\left|\dfrac{1+x}{1-2x}\right| + c$.

Problems: Now try the end-of-chapter problems 14.23–14.26.

14.5 Integrating trigonometric functions

In this section we'll collect results for integrands involving trigonometric functions. These will be needed when you apply calculus to problems with data that is periodic; we have seen in Chapter 9 how such data (e.g. tidal flow or seasonal temperature) fits trigonometric functions. You may want to revise the definitions and identities in Chapter 9, as we'll be using them below.

14.5.1 The general sine and cosine functions

We have already seen an example of integrating a sine function with prescribed period, in (14.10) above. In Section 9.5 we gave the general cosine function with period T, amplitude A, phase t_0, and average value k as:

$$f(t) = A\cos\left(\frac{2\pi}{T}(t - t_0)\right) + k. \tag{9.19}$$

Using integration by substitution, setting $u = \frac{2\pi}{T}(t - t_0)$ so that $\frac{du}{dt} = \frac{2\pi}{T}$, you will find that

$$\int\left(A\cos\left(\frac{2\pi}{T}(t - t_0)\right) + k\right) dt = \frac{TA}{2\pi}\sin\left(\frac{2\pi}{T}(t - t_0)\right) + kt + c. \tag{14.17}$$

Similarly for the general sine function from (9.20):

$$\int\left(A\sin\left(\frac{2\pi}{T}(t - t_0)\right) + k\right) dt = \frac{-TA}{2\pi}\cos\left(\frac{2\pi}{T}(t - t_0)\right) + kt + c. \tag{14.18}$$

14.5.2 The tangent function

Integration by substitution will enable us to find the integral of $\tan x$. Using the definition of the tangent function (9.9), we write

$$I = \int\tan x\, dx = \int\frac{\sin x\, dx}{\cos x}.$$

Choose $u = \cos x$, and differentiate: $\frac{du}{dx} = -\sin x$. Then $du = -\sin x\, dx$.

Notice that the numerator of the integrand is the expression $\sin x\, dx$. So we can substitute $-du$ for this expression, as well as substituting u for $\cos x$ on the denominator:

$$I = \int\frac{\sin x\, dx}{\cos x} = \int\frac{-du}{u} = -\int\frac{1}{u}du = -\ln|u| + c.$$

Finally, substitute back for u and we have the result:

$$\int\tan x\, dx = -\ln|\cos x| + c. \tag{14.19}$$

From the rules of logarithms, this could also be written as $\int\tan x\, dx = \ln|\sec x| + c$.

14.5.3 Powers of sines and cosines

We now think about finding $\int \sin^n x \, dx$ and $\int \cos^n x \, dx$. For this we will need two trigonometric identities from Chapter 9: the basic identity

$$\cos^2 \theta + \sin^2 \theta = 1 \tag{9.18}$$

and the identity found from combining this with the double-angle formulae (9.32):

$$\cos(2A) = \cos^2 A - \sin^2 A = 1 - 2\sin^2 A = 2\cos^2 A - 1. \tag{9.33}$$

These identities hold for all values of the angles θ and A.

From (9.33) we can obtain $\cos^2 x$ in terms of $\cos 2x$, which we already know how to integrate:

$$\cos^2 x = \tfrac{1}{2}(1 + \cos 2x).$$

Then

$$
\begin{aligned}
\int \cos^2 x \, dx &= \int \tfrac{1}{2}(1 + \cos 2x) \, dx \\
&= \tfrac{1}{2} \int (1 + \cos 2x) \, dx \\
&= \tfrac{1}{2} \int 1 \, dx + \tfrac{1}{2} \int \cos 2x \, dx \\
&= \tfrac{1}{2} x + \tfrac{1}{2}\left[\tfrac{1}{2} \sin 2x \right] + c \\
&= \tfrac{1}{2} x + \tfrac{1}{4} \sin 2x + c.
\end{aligned}
\tag{14.20}
$$

How about $I = \int \sin^3 x \, dx$? Here we use a different trick:

$$I = \int \sin^3 x \, dx = \int (\sin^2 x)(\sin x) \, dx = \int (1 - \cos^2 x)(\sin x) \, dx.$$

Now set $u = \cos x$. Then $du = -\sin x \, dx$, and our integral becomes

$$
\begin{aligned}
I &= \int (1 - u^2)(-du) = -\int (1 - u^2) \, du \\
&= -\left[u - \tfrac{1}{3}u^3 \right] + c \\
&= -\cos x + \tfrac{1}{3}\cos^3 x + c.
\end{aligned}
\tag{14.21}
$$

> **PRACTICE:**
> Find (i) $\int \sin^2 x \, dx$ (ii) $\int \cos^3 x \, dx$.
> **Answers:** (i) $\int \sin^2 x \, dx = \tfrac{1}{2}x - \tfrac{1}{4}\sin 2x + c$
> (ii) $\int \cos^3 x \, dx = \sin x - \tfrac{1}{3}\sin^3 x + c.$

In the Problems you can find the integrals of higher powers of sine and cosine. For even powers you need to use the double-angle formulae, and for odd powers you use the trick we've used for $\sin^3 x$.

14.5.4 Integrating $e^x \cos x$

In Section 14.3 we found $\int x \cos x \, dx$ using integration by parts. Can we do the same thing for $I = \int e^x \cos x \, dx$? Let's try integration by parts again:

Let $\quad u = e^x \quad$ and $\quad dv = \cos x \, dx$.

Then:

$$du = e^x \, dx \quad \text{and} \quad v = \int \cos x \, dx = \sin x.$$

Substituting in the integration by parts formula (14.14):

$$I = uv - \int v \, du = (e^x)(\sin x) - \int (\sin x)e^x \, dx$$
$$= e^x \sin x - \int e^x \sin x \, dx.$$

This doesn't look promising – we have obtained another integral $J = \int e^x \sin x \, dx$ which is equally difficult to solve. But all is not lost: try using integration by parts on J:

Let $u = e^x \quad$ and $\quad dv = \sin x \, dx$.

Then:

$$du = e^x \, dx \quad \text{and} \quad v = \int \sin x \, dx = -\cos x.$$

Substituting in the integration by parts formula (14.14):

$$J = uv - \int v \, du = (e^x)(-\cos x) - \int (-\cos x)e^x \, dx$$
$$= -e^x \cos x + \int e^x \cos x \, dx$$
$$= -e^x \cos x + I.$$

So we have two simultaneous equations:

$$I = e^x \sin x - J$$
$$J = -e^x \cos x + I.$$

Substituting the second into the first, and solving for I:

$$I = e^x \sin x - (-e^x\cos x + I) = e^x \sin x + e^x\cos x - I$$
$$2I = e^x \sin x + e^x\cos x$$
$$I = \tfrac{1}{2}e^x(\sin x + \cos x).$$

So

$$\int e^x \cos x \, dx = \tfrac{1}{2} e^x(\sin x + \cos x) + c.$$

Check: Using the product rule to differentiate the right-hand side,

$$\frac{dI}{dx} = \tfrac{1}{2}(e^x(\cos x - \sin x) + e^x(\sin x + \cos x))$$
$$= \tfrac{1}{2}e^x(\cos x - \sin x + \sin x + \cos x)$$
$$= \tfrac{1}{2}e^x(\cos x + \cos x) = e^x \cos x.$$

PRACTICE:

From the simultaneous equations, find $J = \int e^x \sin x \, dx$.

Answer: $\int e^x \sin x \, dx = \tfrac{1}{2}e^x(\sin x - \cos x) + c.$

14.5.5 Integrating inverse trigonometric functions

Recall the definition of $\sin^{-1} x$ from Section 9.6.2:

The **inverse sine** of x (written $\sin^{-1} x$ or arcsin x) is the angle between $-90°$ and $+90°$ whose sine is x, i.e. $\theta = \sin^{-1} x$ means that $\sin \theta = x$.

In Section 12.7 we used implicit differentiation to find its derivative:

$$\text{if} \quad y = \sin^{-1} x, \quad \text{its derivative is} \quad \frac{dy}{dx} = \frac{1}{\sqrt{1-x^2}}. \tag{12.13}$$

To integrate $\sin^{-1} x$, we use the same trick as we used in Section 14.3 to find $\int \ln x \, dx$:

Find $I = \int \sin^{-1} x \, dx$.

Use integration by parts.

Let $u = \sin^{-1} x$ and $dv = dx$.

Then:

$$du = \frac{1}{\sqrt{1-x^2}} \, dx \quad \text{and} \quad v = \int dx = x.$$

Substituting in the integration by parts formula (14.14):

$$I = uv - \int v \, du$$

$$= (\sin^{-1} x)(x) - \int (x) \frac{1}{\sqrt{1-x^2}} \, dx$$

$$= x \sin^{-1} x - \int \frac{x \, dx}{\sqrt{1-x^2}}$$

$$= x \sin^{-1} x + \sqrt{1-x^2} + c.$$

How was that last integration done? This is another example of integration by substitution:

Let $w = 1 - x^2$. Then $dw = -2x \, dx$, so $x \, dx = -\frac{1}{2} \, dw$.

Substituting in:

$$\int \frac{x \, dx}{\sqrt{1-x^2}} = -\frac{1}{2} \int \frac{dw}{\sqrt{w}} = -\frac{1}{2} \int w^{-\frac{1}{2}} \, dw = -\frac{1}{2} \left[\frac{w^{\frac{1}{2}}}{\frac{1}{2}} \right] + c$$

$$= -w^{\frac{1}{2}} + c = -\sqrt{1-x^2} + c.$$

Check:

$$\frac{dI}{dx} = \left(x.\frac{1}{\sqrt{1-x^2}} + 1.\sin^{-1} x \right) + \frac{1}{2}(1-x^2)^{-\frac{1}{2}}.(-2x)$$

$$= \frac{x}{\sqrt{1-x^2}} + \sin^{-1} x - \frac{x}{\sqrt{1-x^2}} = \sin^{-1} x.$$

PRACTICE:

Find $\int \tan^{-1} x \, dx$.

Answer: $\int \tan^{-1} x \, dx = x \tan^{-1} x - \frac{1}{2} \ln(1+x^2) + c.$

14.6 Extension: integration using power series approximations

We said in Section 14.2 that there are some functions which it is impossible to integrate, and we cited $\int e^{-x^2} dx$ as an example. By 'impossible to integrate', I mean that you cannot use the techniques of integration to obtain the integral as a function that, when you differentiate it, gets you back to the integrand e^{-x^2}. There are other ways of approaching the problem, though. One is to use numerical integration; see Section 15.5. Another is to approximate the integrand by a truncated power series, and then to integrate that polynomial. We introduced power series expansions of functions in Section 13.4. Here's an example:

Suppose we want an approximation to $I = \int e^{-x^2} dx$ in the region of the graph around $x = 0$. We will use a Maclaurin series expansion (Section 13.4.3):

$$f(x) = a_0 + a_1 x + a_2 x^2 + a_3 x^3 + \cdots + a_n x^n + \cdots \tag{13.16}$$

where the coefficients are given by

$$a_n = \frac{1}{n!} f^{(n)}(0).$$

Here, $f(x)$ is e^{-x^2}. We need to find the derivatives of e^{-x^2}, which we do using the product rule and chain rule. I have worked out the first six derivatives (please check this!):

$$
\begin{aligned}
f(x) &= e^{-x^2} & f(0) &= 1 \\
f'(x) &= -2x e^{-x^2} & f'(0) &= 0 \\
f''(x) &= -2(1 - 2x^2) e^{-x^2} & f''(0) &= -2 \\
f'''(x) &= 4(3x - 2x^3) e^{-x^2} & f'''(0) &= 0 \\
f^{(4)}(x) &= 4(3 - 12x^2 + 4x^4) e^{-x^2} & f^{(4)}(0) &= 12 \\
f^{(5)}(x) &= 8(-15x + 20x^3 - 4x^5) e^{-x^2} & f^{(5)}(0) &= 0 \\
f^{(6)}(x) &= 8(-15 + 90x^2 - 60x^4 + 8x^6) e^{-x^2} & f^{(6)}(0) &= -120
\end{aligned}
\tag{14.22}
$$

In the right-hand side column, I have evaluated the derivatives at $x = 0$.

If I use these values in (13.16), I obtain the Maclaurin series expansion of e^{-x^2} as:

$$e^{-x^2} = 1 - x^2 + \tfrac{1}{2} x^4 - \tfrac{1}{6} x^6 + \tfrac{1}{24} x^8 - \cdots \tag{14.23}$$

I confess I guessed the fifth term, by spotting a pattern in the preceding ones; can you see it? You need to think about the factorials…

The graph of e^{-x^2} and the approximations given by taking increasing numbers of terms, are sketched in FNGRAPH 14.3 (Figure 14.1). You can see that by taking the first five terms, we have an approximation of degree 8 which closely matches e^{-x^2} for values of x in the interval $-1 < x < 1$. If we truncate the series after five terms, and then integrate this polynomial, we will obtain an approximation to the integral in this region:

$$I = \int e^{-x^2} dx \approx x - \tfrac{1}{3} x^3 + \tfrac{1}{10} x^5 - \tfrac{1}{42} x^7 + \tfrac{1}{216} x^9 \quad \text{for} -1 < x < 1. \tag{14.24}$$

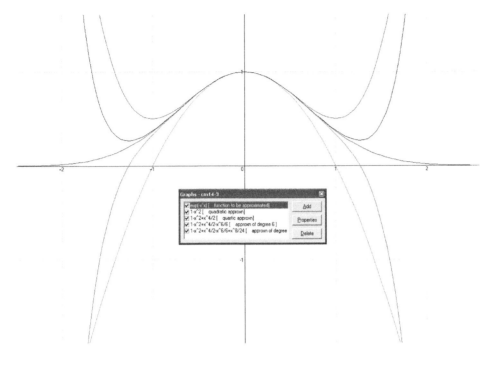

Figure 14.1 Power series approximations to e^{-x^2}

14.7 Summary of standard integrals

With selected results from the previous sections, and some extra generalization, we can now expand the table of standard integrals that we provided in (14.8):

$f(x)$	$\int f(x)\,dx$
$x^n \quad (n \neq -1)$	$\dfrac{1}{n+1}x^{n+1} + c$
$\dfrac{1}{ax+b}$	$\dfrac{1}{a}\ln|ax+b| + c$
e^{ax}	$\dfrac{1}{a}e^{ax} + c$
$a^x \quad (a>0)$	$\dfrac{1}{\ln a}a^x + c$
$\sin(ax+b)$	$-\dfrac{1}{a}\cos(ax+b) + c$
$\cos(ax+b)$	$\dfrac{1}{a}\sin(ax+b) + c$
$\tan ax$	$-\dfrac{1}{a}\ln|\cos ax| + c$
$\cos^2 x$	$\frac{1}{2}x + \frac{1}{4}\sin 2x + c$

$f(x)$	$\int f(x)\,dx$
$\sin^2 x$	$\frac{1}{2}x - \frac{1}{4}\sin 2x + c$
$\sin^{-1} x$	$x\sin^{-1}x + \sqrt{1-x^2} + c$
$\tan^{-1} x$	$x\tan^{-1}x - \frac{1}{2}\ln(1+x^2) + c$
$\dfrac{1}{\sqrt{a^2-x^2}}$	$\sin^{-1}\left(\dfrac{x}{a}\right) + c$
$\dfrac{1}{a^2+x^2}$	$\tan^{-1}\left(\dfrac{x}{a}\right) + c$
$\dfrac{1}{a^2-x^2}$	$\dfrac{1}{2a}\ln\left|\dfrac{a+x}{a-x}\right| + c$
$\ln x$	$x\ln x - x + c$

(14.25)

? PROBLEMS

Find the integrals of the following functions:

14.1: $3x^5 - x^2$

14.2: $(3x^2 + 1)(x - 1)$

14.3: $\dfrac{3}{x^2} - \dfrac{7}{x^4}$

14.4: $\left(x^2 + 1\right)\left(\dfrac{1}{x^2} - 1\right)$

14.5: $3\sqrt{x} - \dfrac{7}{\sqrt{x}}$

14.6: $7x\sqrt{x} - \dfrac{2}{x}$

14.7: $3e^{5x} - 5e^{-3x}$

14.8: $3\sin 5x - 5\cos 3x$

14.9: 3^{4x}

14.10: $\dfrac{5}{2x-3}$

14.11: $\dfrac{x+1}{x-1}$

14.12: $\dfrac{2x+1}{x-1}$

14.13: $\sin x \cos x$

14.14: $\dfrac{\cos x}{1+\sin x}$

14.15: $\sin x \cos^3 x$

14.16: $\dfrac{\ln x}{x}$

14.17: $x\sin(1+x^2)$

14.18: $x\ln(1+x^2)$

14.19: xe^{5x}

14.20: $x\sin x$

14.21: $x^2 \ln x$

14.22: $(\ln x)^2$

14.23: $\dfrac{1}{x^2+3x-4}$

14.24: $\dfrac{x-1}{3x^2-8x-3}$

The definite integral

Until now we have thought of the integral of $f(x)$ as the anti-derivative, i.e. a function that produces $f(x)$ when you differentiate it. In Section 15.1 we'll show that there is a geometric and physical interpretation of the integral. This leads on to the concept of the definite integral, which is evaluated to be a numerical quantity instead of a function. We'll see some practical applications of the definite integral, and we'll conclude by seeing a powerful method for evaluating definite integrals numerically. This relates back to the techniques of interpolation in Chapter 3 and Chapter 7.

. . . .

15.1 The integral as area under the curve

Imagine that we take a function $f(x)$, and sketch the graph of $y = f(x)$. At a particular value of x, the derivative $\dfrac{df}{dx}$ tells us the slope of the tangent to the curve at that point. The integral $\int f(x)dx$ also has a geometrical significance: it tells us the area under the curve up to that point. By 'area under the curve' we mean the area of the region on the graph from the curve down to the x-axis, as indicated by the shaded area in Figure 15.1. I won't give a formal proof of this claim, but here is a quick flavour of it.

15.1.1 The link between the integral and area

In Figure 15.1 the area $A(x)$ is the shaded area below the curve $y = f(x)$ and above the x-axis, as far as the vertical line BC at the value of x as its right-hand boundary. $A(x)$ is a function of x because a change in x moves the right-hand boundary, thus changing the area. Suppose we now increase x by a small amount Δx. Then the right-hand boundary moves to the line DE, and the area A increases by the small area ΔA, which is the area of the vertical strip BCED. We can approximate this strip by cutting off the top horizontally to give the rectangle BCGD. The rectangle has height $f(x)$ and width Δx, so

$$\Delta A \approx f(x).\Delta x.$$

Thus

$$\frac{\Delta A}{\Delta x} \approx f(x).$$

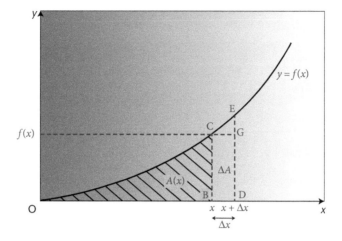

Figure 15.1 The integral as an area under the curve

This approximation becomes closer when we make the strip narrower. In fact, we can take the limit as $\Delta x \to 0$, which using (11.4) gives us the derivative of A with respect to x:

$$\lim_{\Delta x \to 0} \frac{\Delta A}{\Delta x} = \frac{dA}{dx} = f(x).$$

So $\frac{dA}{dx} = f(x)$, which means that $A = \int f(x)\, dx$.

This is not a very precise argument. We haven't properly defined $A(x)$, because we haven't really specified where the left-hand end of the area is. The choice of the starting value for A will determine what the constant of integration will be. The way to get round this problem is to define our area as the difference of two areas that have different right-hand boundaries. We'll illustrate what we mean by returning to our example of the travelling car in Section 11.1, and looking at speed–time graphs.

15.1.2 Speed–time graphs

We use $v(t)$ to denote the speed of the moving car at time t, and we'll use $s(t)$ to mean the distance travelled by the car since time $t = 0$. Suppose we want to work out the distance travelled between times t_1 and t_2. We can draw a speed–time graph, plotting v against t. Figure 15.2 shows three examples of such graphs.

(i) In Figure 15.2(i) the car is travelling with a constant speed V (i.e. zero acceleration). Then we can use the rule from schooldays which we recalled in the introduction to Chapter 14:

$$\text{speed} = \frac{\text{distance}}{\text{time}}.$$

Rearranging, we get:

$$\text{distance} = \text{speed} \times \text{time}.$$

So the distance travelled between times t_1 and t_2 is $V.(t_2 - t_1)$. This is the shaded area A_1 in the graph, the area of the rectangle BCED. But notice that

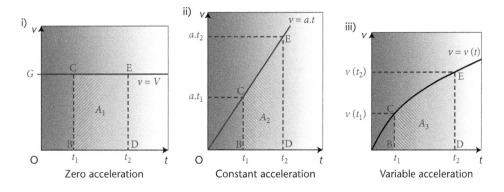

Figure 15.2 Speed–time graphs

$$A_1 = V(t_2 - t_1) = V.t_2 - V.t_1,$$

which is the difference in area between two rectangles: the area of the rectangle OGED, minus the area of the rectangle OGCB.

(ii) In Figure 15.2(ii) the car starts from rest (i.e. $v(0) = 0$) and travels with constant acceleration a. Then as we saw in the example of the VW Polo in Section 14.1.3, its speed at time t is given by $v(t) = a.t$. As $v(t) = \dfrac{ds}{dt}$, we can integrate to find $s(t) = \int a.t\, dt = \dfrac{1}{2}at^2$. The constant of integration is zero, because $s(0) = 0$.

So the distance covered from time $t = 0$ to time t_1 will be $s(t_1) = \frac{1}{2}at_1^2$. If we write this as $\frac{1}{2}(at_1)(t_1)$, you can see that this is the area of the triangle OBC (area of triangle = $^1\!/_2 \times$ base $\times$ height). Similarly, the distance travelled from time $t = 0$ to time t_2 will be $s(t_2) = \frac{1}{2}at_2^2$, the area of the triangle ODE. Subtracting $s(t_1)$ from $s(t_2)$ gives the distance travelled between times t_1 and t_2, which will be the area A_2:

$$A_2 = \tfrac{1}{2}at_2^2 - \tfrac{1}{2}at_1^2.$$

(iii) Finally, consider the general case of travel with variable acceleration $a(t)$, as illustrated in Figure 15.2(iii). Now $s(t) = \int v(t)\, dt$, and the distance travelled between times t_1 and t_2 will be the area A_3:

$$A_3 = s(t_2) - s(t_1).$$

In each of the above examples, the area under the curve between t_1 and t_2 is found by subtracting $s(t_1)$ from $s(t_2)$, where s is the integral of v.

15.1.3 Definition of the definite integral

We can generalize the result from the previous section. We take a continuous function $f(x)$ and write its integral as $F(x)$:

$$F(x) = \int f(x)\, dx.$$

The integrals such as $\int f(x)\, dx$, which we were working out in the previous chapter, producing functions of x with a '$+ c$' at the end, are called **indefinite integrals**.

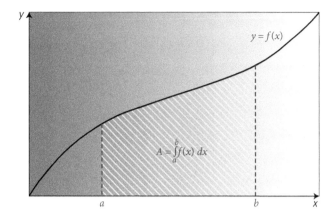

Figure 15.3 The definite integral

Let a and b be two fixed values of x, where $a < b$. Then the area under the curve between $x = a$ and $x = b$, the shaded area A in Figure 15.3, is found by subtracting $F(a)$ from $F(b)$:

$$A = F(b) - F(a).$$

This quantity is called the **definite integral** of $f(x)$. The definite integral is defined by the function $f(x)$ and the values of a and b. The notation used for it, which combines this information, is $\displaystyle\int_a^b f(x)\, dx$:

> If $F(x) = \int f(x)dx$, then the **definite integral** of $f(x)$ from $x = a$ to $x = b$, is
> $$\int_a^b f(x)\, dx = F(b) - F(a).\qquad\qquad(15.1)$$

When working out $F(x)$, we can omit the constant of integration. The values a and b are called the **limits of integration**; a is the lower limit, and b is the upper limit.

EXAMPLE

Find $\displaystyle\int_1^4 3x^2\, dx$.

Solution: Here, $f(x) = 3x^2$. The indefinite integral is $x^3 + c$, so we take $F(x) = x^3$. The area under the curve $y = 3x^2$ between $x = 1$ and $x = 4$ is then given by the definite integral

$$\int_1^4 3x^2\, dx = F(4) - F(1) = 4^3 - 1^3 = 64 - 1 = 63 \text{ square units.}$$

As the definite integral is the difference between two evaluations of the indefinite integral, we are able to forget about the constant of integration. Had we included '$+ c$' in each evaluation, they would have cancelled out, in the same way as the calculation in example (iii) of the previous section has no reference to where the left-hand end of the two areas lies.

To calculate a definite integral $\displaystyle\int_a^b f(x)\, dx$, the procedure is:

1. Work out the indefinite integral $F(x) = \int f(x)\, dx$, omitting the constant of integration.
2. Evaluate $F(x)$ at $x = b$ and $x = a$.
3. Subtract the two evaluations, i.e. evaluate $F(b) - F(a)$.

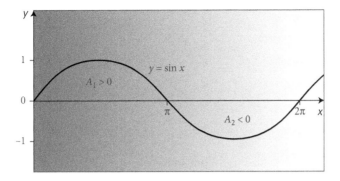

Figure 15.4 Positive and negative areas

The notation used for this procedure is to put the function $F(x)$ within square brackets, with the limits written at top and bottom:

$$\int_a^b f(x)\,dx = [F(x)]_a^b = F(b) - F(a).$$

Here's our example, using this notation:

$$\int_1^4 3x^2\,dx = [x^3]_1^4 = 4^3 - 1^3 = 64 - 1 = 63.$$

Another example:

$$\int_1^2 \frac{1}{x^2}\,dx = \int_1^2 x^{-2}\,dx = [-x^{-1}]_1^2 = \left[-\frac{1}{x}\right]_1^2 = \left(-\frac{1}{2}\right) - \left(-\frac{1}{1}\right) = -\frac{1}{2} + 1 = \frac{1}{2}.$$

And another:

$$\int_0^\pi \sin x\,dx = [-\cos x]_0^\pi = (-\cos\,\pi) - (-\cos\,0) = (1) - (-1) = 2.$$

So the area A_1 enclosed by the sine curve in Figure 15.4, is exactly 2 square units! What would we get if we integrated $\sin x$ over one complete cycle, from 0 to 2π?

$$\int_0^{2\pi} \sin x\,dx = [-\cos x]_0^{2\pi} = (-\cos\,2\pi) - (-\cos\,0) = (-1) - (-1) = 0.$$

Here, the total area has come out as zero, because the area A_2 in the diagram counts as -2 square units because it lies below the x-axis. The positive and negative areas cancel, leaving zero.

It is essential to remember that **areas below the x-axis are counted as negative.** So if you are trying to find an area using the definite integral, you should check that the function does not cross the x-axis anywhere within the interval.

EXAMPLE

Find $I = \int_0^4 (x - 3)(x^2 + 1)\,dx.$

Solution: If we simply want to evaluate the definite integral, and are not concerned about areas, we proceed as in the examples above. We will have to multiply out the function $f(x) = (x - 3)(x^2 + 1)$ and then integrate term-by-term:

$$I = \int_0^4 (x - 3)(x^2 + 1)\, dx$$

$$= \int_0^4 (x^3 - 3x^2 + x - 3)\, dx$$

$$= \left[\frac{1}{4}x^4 - x^3 + \frac{1}{2}x^2 - 3x \right]_0^4$$

$$= (64 - 64 + 8 - 12) - (0) = -4.$$

However, if we are interested in finding out the area enclosed by the curve, the x-axis, the y-axis and the vertical line at $x = 4$, we notice that because of the factor $(x - 3)$, $f(x)$ is negative for $0 < x < 3$, and positive for $3 < x < 4$ (since the other factor $(x^2 + 1)$ is always positive). The graph is sketched in Figure 15.5. So we must find the areas A_1 and A_2 in these two sub-intervals separately:

$$A_1 = \int_0^3 (x - 3)(x^2 + 1)\, dx$$

$$= \int_0^3 (x^3 - 3x^2 + x - 3)\, dx$$

$$= \left[\frac{1}{4}x^4 - x^3 + \frac{1}{2}x^2 - 3x \right]_0^3$$

$$= \left(\frac{81}{4} - 27 + \frac{9}{2} - 9 \right) - (0) = -9.25$$

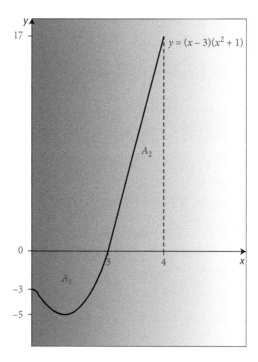

Figure 15.5 The integral $\int_0^4 (x - 3)(x^2 + 1)\, dx$

is the area A_1 below the x-axis, while

$$A_2 = \int_3^4 (x-3)(x^2+1)\,dx$$

$$= \int_3^4 (x^3 - 3x^2 + x - 3)\,dx$$

$$= \left[\frac{1}{4}x^4 - x^3 + \frac{1}{2}x^2 - 3x\right]_3^4$$

$$= (64 - 64 + 8 - 12) - \left(\frac{81}{4} - 27 + \frac{9}{2} - 9\right) = (-4) - (-9.25) = 5.25$$

is the area A_2 above the x-axis. The total area is the sum of the absolute values of A_1 and A_2, namely $9.25 + 5.25 = 14.5$ square units.

If we don't take absolute values, you notice that

$$A_1 + A_2 = \int_0^3 (x-3)(x^2+1)\,dx + \int_3^4 (x-3)(x^2+1)\,dx = -9.25 + 5.25 = -4$$

$$= \int_0^4 (x-3)(x^2+1)\,dx.$$

This illustrates the

Addition of areas rule for definite integrals:

$$\int_a^b f(x)\,dx + \int_b^c f(x)\,dx = \int_a^c f(x)\,dx. \tag{15.2}$$

If we know that the area A_1 lies wholly below the x-axis, then instead of integrating from 0 to 3 and obtaining a negative value which we must change the sign of, we could interchange the limits of integration to obtain a positive result:

$$\int_3^0 (x-3)(x^2+1)\,dx = \left[\frac{1}{4}x^4 - x^3 + \frac{1}{2}x^2 - 3x\right]_3^0$$

$$= (0) - \left(\frac{81}{4} - 27 + \frac{9}{2} - 9\right) = 0 - (-9.25) = 9.25.$$

As $\int_a^b f(x)\,dx = [F(x)]_a^b = F(b) - F(a)$, interchanging the limits of integration gives $\int_b^a f(x)\,dx = [F(x)]_b^a = F(a) - F(b)$. This provides another rule for definite integrals:

Interchange of limits rule:

$$\int_b^a f(x)\,dx = -\int_a^b f(x)\,dx. \tag{15.3}$$

PRACTICE:

Find (i) $\int_2^3 \frac{1}{2x+1}\,dx$ (ii) $\int_{-1}^2 (3e^{2x} - 2x^3)\,dx$.

Answers: (i) $\frac{1}{2}\ln 1.4$ (ii) $\frac{3}{2}(e^4 - e^{-2}) - 7.5$.

Problems: Now try the end-of-chapter problems 15.1–15.4.

. . . .

15.2 The integral as limit of a sum

We could only define the indefinite integral as the anti-derivative, but now we have a concept of the definite integral (as an area) that is independent of the idea of differentiation. In fact, we can use this concept to define the definite integral mathematically as a process involving taking the limit. This is called **Riemann**[1] **integration**.

15.2.1 The Riemann integral

The definition of the gradient $\frac{dy}{dx}$ involves approximating it as the slope $\frac{\Delta y}{\Delta x}$ of a secant line, and then taking the limit of this slope as $\Delta x \to 0$ (see equation 11.7). We do something very similar in defining the area $A = \int_a^b f(x)\,dx$ as shown in Figure 15.3. We split the area into n vertical strips, each of width Δx, as shown in Figure 15.6. Since the width of the whole area is $(b-a)$, the width of each strip is $\Delta x = \frac{1}{n}(b-a)$.

We want each strip to be a rectangle, so we must make the top of the strip horizontal. We take its height to be the value of $f(x)$ at the right-hand end of the strip. The vertical lines are at $x_0, x_1, x_2, \ldots, x_n$, where $x_0 = a$, $x_1 = a + \Delta x$, $x_2 = a + 2.\Delta x$, etc. up to $x_n = a + n.\Delta x = a + (b-a) = b$.

So the area A_i of the ith rectangle is: $A_i =$ height $\times$ width $= f(x_i).\Delta x$.

We can add up the areas of all the rectangles, and this will give us an approximation to the area A under the curve between $x = a$ and $x = b$:

$$A \approx A_1 + A_2 + \cdots + A_n$$
$$= f(x_1).\Delta x + f(x_2).\Delta x + \cdots + f(x_n).\Delta x$$
$$= \sum_{i=1}^n f(x_i).\Delta x. \tag{15.4}$$

We introduced the summation notation in Section 8.7, when we saw the equation of a trend line (linear regression). Here we use it with limits, indicating that within the summation it is the variable i that changes with each term in the sum, going from 1 to n.

[1] Bernhard Riemann, German mathematician, 1826–1866, died of tuberculosis aged 39. The Riemann hypothesis in number theory is the subject of Marcus du Sautoy's book *The Music of the Primes* (du Sautoy 2004).

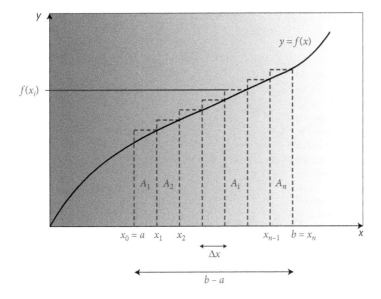

Figure 15.6 The limiting process

The more strips we use, the better this approximation will be. The true area, and hence the definite integral, is the limit of (15.4) as the number of strips n approaches infinity. The mathematical symbol for 'infinity' is ∞, so we can write this limiting process as

$$\int_a^b f(x)\ dx = \lim_{n \to \infty} \left(\sum_{i=1}^n f(x_i).\Delta x \right) \tag{15.5}$$

where $\Delta x = \frac{1}{n}(b - a)$. This is the formal definition of the definite integral $\int_a^b f(x)\ dx$. As the number of strips becomes infinite, so the width of each individual strip tends to zero: $\Delta x \to 0$ as $n > \infty$.

We'll see this technique for approximating the definite integral again in the last section, on numerical integration.

As each strip has the same width, Δx is a common factor of each term in the summation $\sum_{i=1}^n f(x_i).\Delta x$. This means it can be taken outside the summation. For instance, if we had four strips, then

$$
\begin{aligned}
A &\approx A_1 + A_2 + A_3 + A_4 \\
&= f(x_1).\Delta x + f(x_2).\Delta x + f(x_3).\Delta x + f(x_4).\Delta x \\
&= \Delta x \big(f(x_1) + f(x_2) + f(x_3) + f(x_4) \big) \\
&= \Delta x . \sum_{i=1}^n f(x_i).
\end{aligned}
$$

Also, $\Delta x = \frac{1}{n}(b - a)$, so from (15.5)

$$\int_a^b f(x)\,dx = \lim_{n\to\infty}\left(\sum_{i=1}^{n} f(x_i).\Delta x\right)$$

$$= \lim_{n\to\infty}\left(\Delta x.\sum_{i=1}^{n} f(x_i)\right)$$

$$= \lim_{n\to\infty}\left(\frac{1}{n}(b-a).\sum_{i=1}^{n} f(x_i)\right)$$

$$= (b-a).\lim_{n\to\infty}\left(\frac{1}{n}\sum_{i=1}^{n} f(x_i)\right).$$

In the last step, we can take $(b-a)$ outside the limit sign, since a and b are fixed, and independent of n.

In the large brackets, we have the sum of n function values, divided by n. This is how an average value is calculated. By working out this average of f-values using more and more sampling points, in the limit we are obtaining **the average value of $f(x)$ over the interval** $[a,b]$. We will denote this by $\bar{f}$.

$$\int_a^b f(x)\,dx = (b-a).\bar{f}, \quad \text{or} \quad \bar{f} = \frac{1}{b-a}\int_a^b f(x)\,dx. \tag{15.6}$$

What this means graphically is illustrated in Figure 15.7. The area of the rectangle ADEB, of height $\bar{f}$, is the same as the area ACFB under the curve. We have finally obtained a result that has a practical use for bioscientists!

15.2.2 Application: chemotherapy drug delivery

This example uses the theory of drug administration which we discussed in Section 10.5.3; it is adapted from Jones and Sleeman (2003). When planning a regime for administering a chemotherapy drug to a cancer patient, the aim is to inject as much of the drug as possible (to maximize the number of cancer cells killed) within safe limits (as the drug is toxic, and damages blood cell

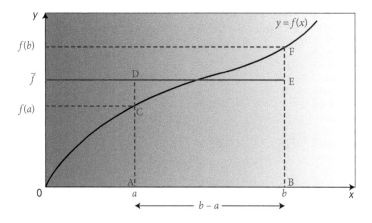

Figure 15.7 The mean value of a function

production). The safety limit is a combination of the drug concentration $C(t)$ in the blood-stream, and how long it remains in the body. Observations may have determined a 'maximum safe drug concentration' C_m; patients should not have a long-term exposure to concentrations higher than C_m. Thus, during a treatment period of length T days, we should ensure that the average bloodstream concentration should not exceed C_m. Moreover, the concentration left in the patient at the end of the treatment period should also not exceed C_m, i.e. $C(T) \leq C_m$.

(i) Suppose that an amount V_0 of the drug is administered at the start of the treatment period, causing an initial concentration C_0 in the patient's blood. We assume that the concentration is proportional to the amount delivered: $C_0 = \alpha V_0$, say. Over time, this concentration will decrease, which we can best model as an exponential decay process: $C(t) = C_0.e^{-kt}$. So it would be possible for the initial concentration to be greater than C_0, provided that, to satisfy the first safety condition,

$$\frac{1}{T}\int_0^T C(t)\, dt \leq C_m. \tag{15.7}$$

The expression on the left-hand side is the average concentration over the period $0 \leq t \leq T$, as in (15.6). The graph of $C(t)$ is in Figure 15.8(i). Assuming we want to inject the maximum safe dose, we can use an equality sign in (15.7) and solve for C_0:

$$\int_0^T C_0 e^{-kt}\, dt = C_m.T$$

$$C_0\left[-\frac{1}{k}e^{-kt}\right]_0^T = C_m.T$$

$$-\frac{C_0}{k}(e^{-kT} - 1) = C_m.T$$

$$C_0 = \left(\frac{kT}{1 - e^{-kT}}\right)C_m.$$

(ii) We might design an improved treatment regime by administering the drug in two doses: an initial amount V_1, causing an initial concentration $C_1 = \alpha V_1$, and a second dose V_2 at a time τ during the treatment period (where $0 < \tau < T$). See Figure 15.8(ii). We now have to consider two separate sub-intervals:

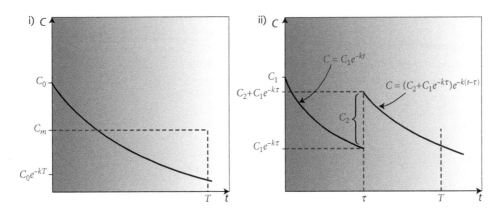

Figure 15.8 Drug delivery curves

- In the interval $0 \leq t \leq \tau$, the first dose concentration decays according to $C(t) = C_1 . e^{-kt}$, as in the previous regime. At time τ the concentration will be $C_1 . e^{-k\tau}$ just before the second dose is injected.

- Now the second dose V_2 is injected; this increases the blood concentration by C_2. The concentration thus shoots up to $C(\tau) = C_2 + C_1 . e^{-k\tau}$. Then, during the remaining sub-interval $\tau \leq t \leq T$, the concentration decays from this initial value. To write $C(t)$ for this interval we need to use an exponential decay function that is shifted horizontally so that it starts when $t = \tau$ (see Section 5.6.1).

The function for $C(t)$ over the whole interval is thus

$$
C(t) = \begin{cases} C_1 . e^{-kt} & \text{when } 0 \leq t < t \\ (C_2 + C_1 e^{-kt}) . e^{-k(t-t)} & \text{when } t \leq t \leq T. \end{cases} \tag{15.8}
$$

To work out the area under the curve, we have to use the addition of areas rule (15.2), and integrate over each sub-interval separately:

$$
\int_0^T C(t) \, dt = \int_0^t C_1 . e^{-kt} \, dt + \int_t^T (C_2 + C_1 . e^{-kt}) . e^{-k(t-t)} dt. \tag{15.9}
$$

We will not take this example further. It is a challenging optimization problem to find values of V_1 and V_2 that maximize the total amount of drug delivered, while satisfying (15.7). We would need to choose the values of V_1 and V_2, as well as the time τ at which the second dose is administered. If this is too close to the end of the treatment period, the additional safety requirement that $C(T) \leq C_m$ might not be satisfied.

The function $C(t)$ in (15.8) has a discontinuity at $t = \tau$, but we were still able to integrate it over the interval $0 \leq t \leq T$ by integrating over each sub-interval and then adding the areas. So definite integration can cope with functions that have one or more discontinuities.

15.2.3 Application: laminar blood flow

When blood flows through a vein or artery, it is not all travelling at the same speed. The blood travelling up the central axis of the artery moves fastest, while the blood in contact with the artery wall is stationary. This is called **laminar flow**; the blood moves like a set of concentric thin cylindrical sheets (laminas) sliding along the axis, with the inner cylinders moving faster than the outer ones, and the outermost cylinder not moving at all.[2] This is illustrated in Figure 15.9(a). We model the artery as a hollow tube or pipe of radius a, so with cross-sectional area $A = \pi a^2$. Let $v(r)$ be the velocity of the blood at radius r from the axis $(0 \leq r \leq a)$.

Fluid mechanics tells us that the velocity–radius graph in Figure 15.9(a) is a quadratic curve:

$$
v(r) = c(a^2 - r^2) \tag{15.10}
$$

---

[2] At higher speeds, laminar flow breaks down and **turbulence** occurs.

where c depends on the pressure gradient and the viscosity of the fluid. We ask: what is the mean (or average) blood velocity? We cannot simply integrate $v(r)$ as this is not a two-dimensional problem, but a three-dimensional one with symmetry around an axis; there is a greater amount of blood in the outer cylinders than in the inner ones. We must adapt the limit-of-a-sum process of Section 15.2.1 to this axisymmetric situation.

Instead of an interval on the x-axis, our domain is now a disk of radius a. We divide it up into n thin concentric rings, each of width Δr, as shown in Figure 15.9(b). Consider the shaded

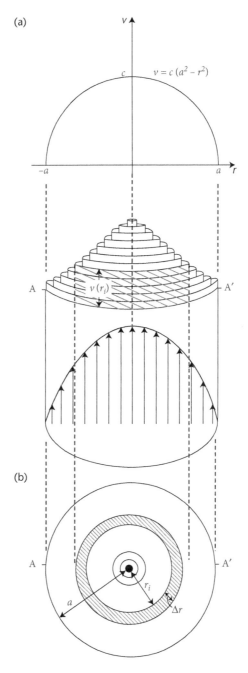

Figure 15.9 Laminar flow through an artery

ring, which is at radius r_i. Its area A_i can be found approximately by: ring area = ring circumference × ring thickness, or:

$$A_i \approx (2\pi r_i).\Delta r.$$

To calculate the mean blood velocity, we need to calculate the amount of blood passing the cross section AA′ in Figure 15.9(a) per unit time. In a single unit of time, the cylinder at radius r_i will slide forward by a distance $v(r_i)$, as shown. The volume of blood is thus

$$V_i = A_i.v(r_i) \approx (2\pi r_i).v(r_i).\Delta r.$$

We can now sum the volumes for all the cylinders, as we did with the areas in Section 15.2.1:

$$\begin{aligned} V &\approx V_1 + V_2 + \cdots + V_n \\ &= 2\pi r_1.v(r_1).\Delta r + 2\pi r_2.v(r_2).\Delta r + \cdots + 2\pi r_n.v(r_n).\Delta r \\ &= \sum_{i=1}^{n} 2\pi r_i.v(r_i).\Delta r \\ &= 2\pi \sum_{i=1}^{n} r_i.v(r_i).\Delta r. \end{aligned}$$

(15.11)

We have taken the common factor of 2π outside the summation sign. We now perform the limiting process, taking an ever-greater number of ever-thinner cylinders, and in the limit approach the true volume V:

$$V = 2\pi \int_0^a r.v(r)\, dr.$$

(15.12)

The limits of integration are from $r = 0$ to $r = a$. Insert the function $v(r)$ from (15.10):

$$\begin{aligned} V &= 2\pi \int_0^a r.c(a^2 - r^2)\, dr \\ &= 2\pi c \int_0^a (a^2 r - r^3)\, dr \\ &= 2\pi c \left[a^2.\frac{1}{2}r^2 - \frac{1}{4}r^4 \right]_0^a \\ &= 2\pi c \left(\frac{1}{2}a^4 - \frac{1}{4}a^4 \right) \\ &= \frac{1}{2}\pi c a^4. \end{aligned}$$

This is the total flow across the artery cross section in unit time. To find the mean flow $\bar{v}$ per cross-sectional unit area we must divide by the area A:

$$\bar{v} = \frac{V}{A} = \frac{\frac{1}{2}\pi c a^4}{\pi a^2} = \frac{1}{2}ca^2.$$

We have observed that the fastest flow occurs down the central axis; here, $v_{max} = v(0) = c(a^2 - 0^2) = ca^2$, so we have shown that $\bar{v} = \frac{1}{2}v_{max}$.

Problems: Now try the end-of-chapter problems 15.5–15.10.

15.3 Using techniques of integration with definite integrals

In the examples of definite integrals given so far, we have not needed to use any of the techniques of integration described in Chapter 14; the integrand $f(x)$ in $A = \int_a^b f(x)\, dx$ was easy to integrate using standard integrals. To remind you, these techniques were:

- Integration by substitution (Section 14.2).
- Integration by parts (Section 14.3).
- Integration by partial fractions (Section 14.4).

If you did need to use one of these techniques in order to evaluate a definite integral, you could work out the indefinite integral on a separate piece of paper, then insert the result into the square brackets, and apply the limits. But it is neater to adapt the technique to handle definite integrals; this can also save some time and effort. We'll show how this is done, for each of the techniques listed above.

15.3.1 Integration by substitution

Recall the steps in finding $I = \int f(x)\, dx$ using substitution:

1. Choose a new intermediate variable $u = u(x)$.

2. Find dx in terms of du.

3. Substitute into I, to convert it to an integral $I = \int g(u)\, du$, which we can integrate to obtain $G(u) + c$.

4. Substitute back for u, to convert this to $F(x) + c$.

Here's an example, finding $I = \int x \sin(1 + x^2)\, dx$

> Rewrite as
> $\int \sin(1 + x^2).\, x\, dx$

1. Let $u = 1 + x^2$.

2. Differentiating, $du = 2x\, dx$, so $x\, dx = \frac{1}{2}\, du$.

3. Then $I = \int \sin(u)\, \frac{1}{2}\, du = \frac{1}{2} \int \sin(u)\, du = -\frac{1}{2} \cos u + c$.

4. Substituting back, $I = -\frac{1}{2} \cos(1 + x^2) + c$.

Now suppose we need to evaluate the definite integral $I = \int_0^2 x \sin(1 + x^2)\, dx$. We choose the new variable $u = 1 + x^2$ as before, and find $x\, dx = \frac{1}{2}\, du$. But we cannot then write $I = \int_0^2 \sin(u)\, \frac{1}{2}\, du$, because the limits of integration are values of x, not of u. We need to convert these to limits of u:

When $x = 0$,   When $x = 2$,

$u = 0^2 + 1 = 1$ $u = 2^2 + 1 = 5$

So our integral becomes

$$I = \int_1^5 \sin u \; \tfrac{1}{2} du = \tfrac{1}{2}[-\cos u]_1^5$$

$$= -\tfrac{1}{2}(\cos 5 - \cos 1) = -\tfrac{1}{2}(0.2837 - 0.5403) = 0.1283$$

(The limits are angles, in radians.)

There was now no need to substitute back to find a function of x.

EXAMPLE

Find $I = \int_0^1 \dfrac{e^x \; dx}{3e^x - 2}$.

Solution: Let $u = 3e^x - 2$. Then $du = 3e^x \; dx$.

Convert the limits to limits of u:

When $x = 0$,   When $x = 1$,

$u = 3e^0 - 2 = 3 - 2 = 1$ $u = 3e^1 - 2 = 3e - 2$

Substitute in:

$$I = \int_1^{3e-2} \frac{\tfrac{1}{3} \; du}{u} = \tfrac{1}{3}[\ln u]_1^{3e-2}$$

$$= \tfrac{1}{3}(\ln \left| 3e - 2 \right| - \ln 1) = \tfrac{1}{3} \ln(3e - 2)$$

$$= \tfrac{1}{3} \ln(6.1548) = 0.6057.$$

15.3.2 Integration by parts

Recall the formula for integration by parts:

$$\int u \; dv = u.v - \int v \; du. \tag{14.14}$$

To find the definite integral $\int_a^b u \; dv$, we apply the limits to the $\int v \; du$ term. Also, the $u.v$ term is the result of an integration, so we must put it in square brackets with the limits. The integration by parts formula for definite integrals is thus:

$$\int_a^b u \; dv = [u.v]_a^b - \int_a^b v \; du. \tag{15.13}$$

EXAMPLE

Find $I = \int_1^2 x^3 \ln x \; dx$.

Solution: Use integration by parts.

Let $u = \ln x$ and $dv = x^3 \; dx$.

Then:

$$du = \frac{1}{x} \; dx \text{ and } v = \int x^3 \; dx = \frac{1}{4} x^4.$$

Substituting in (15.13):

$$I = \left[(\ln x)\left(\frac{1}{4}x^4\right)\right]_1^2 - \int_1^2 \frac{1}{4}x^4 \frac{1}{x}dx$$

$$= \frac{1}{4}\left[x^4 \ln x\right]_1^2 - \frac{1}{4}\int_1^2 x^3 dx$$

$$= \frac{1}{4}\left(16\ln 2 - 0\right) - \frac{1}{4}\left[\frac{1}{4}x^4\right]_1^2$$

$$= 4\ln 2 - \frac{1}{16}(16 - 1) = 4\ln 2 - \frac{15}{16} = 1.8351.$$

15.3.3 Integration by partial fractions

This is straightforward: just apply the limits to each integral:

EXAMPLE

Find $I = \int_3^5 \frac{dx}{x(2x - 1)}$.

Solution: We need to find numbers A, B that satisfy the identity

$$\frac{1}{x(2x - 1)} = \frac{A}{x} + \frac{B}{2x - 1}.$$

Multiplying both sides by $x(2x - 1)$:

$$A(2x - 1) + Bx = 1.$$

To find A, substitute $x = 0$: $-A = 1$,   so $A = -1$.
To find B, substitute $x = \frac{1}{2}$: $\frac{1}{2}B = 1$,   so $B = 2$.
Our integral then becomes

$$I = \int_3^5 \frac{dx}{x(2x - 1)} = \int_3^5 \left(-\frac{1}{x} + \frac{2}{2x - 1}\right)dx$$

$$= -\int_3^5 \frac{1}{x}dx + 2\int_3^5 \frac{1}{2x - 1}dx$$

$$= -[\ln x]_3^5 + 2\left[\frac{1}{2}\ln|2x - 1|\right]_3^5$$

$$= -(\ln 5 - \ln 3) + (\ln 9 - \ln 5) \quad \boxed{\ln 9 = \ln 3^2 = 2\ln 3}$$

$$= -\ln 5 + \ln 3 + \ln 9 - \ln 5$$

$$= 3\ln 3 - 2\ln 5 = \ln 3^3 - \ln 5^2 = \ln\left(\frac{27}{25}\right) = \ln(1.08) = 0.0770.$$

PRACTICE:

Find (i) $\int_3^4 \frac{x\,dx}{25 - x^2}$ (ii) $\int_1^4 \sqrt{x}\ln x\,dx$.

Answers: (i) $\ln\left(\frac{4}{3}\right)$ (ii) $\frac{32}{3}\ln 2 - \frac{28}{9}$.

Problems: Now try the end-of-chapter problems 15.11–15.18.

15.4 **Improper integrals**

In the discussion of chemotherapy drug delivery in Section 15.2, we defined the patient's exposure to the drug over the time period $0 \le t \le T$ as $\int_0^T C(t)\, dt$, where $C(t)$ is the drug concentration in the bloodstream at time t. When the drug is given in a single dose at time $t = 0$, causing an initial concentration C_0, we modelled the subsequent dissipation of the drug over time by an exponential decay function: $C(t) = C_0 e^{-kt}$.

By this model, the drug never disappears completely from the bloodstream. The graph of the function is shown in Figure 15.10. The horizontal axis is an asymptote to the curve. Instead of thinking only about the patient's exposure over the treatment period, we should rather measure the total exposure over the patient's lifetime. That is, we ought to evaluate $\int C(t)\, dt$ over the whole of the positive t-axis. This integral is written as $\int_0^\infty C(t)\, dt$. It represents the whole shaded area under the curve in Figure 15.10. This area can have a finite value, even though it has an infinite length.

To calculate $\int_0^\infty C(t)\, dt$, the trick is to work out $\int_0^T C(t)\, dt$, for a large but finite value T as the upper limit, and then to consider what happens to the integral as $T \to \infty$. As we have seen,

$$\int_0^T C_0\, e^{-kt} dt = C_0 \left[-\frac{1}{k} e^{-kt} \right]_0^T = -\frac{C_0}{k} \left(e^{-kT} - e^0 \right) = \frac{C_0}{k} \left(1 - e^{-kT} \right)$$

$$= \frac{C_0}{k} - \frac{C(T)}{k}.$$

As we can see from the graph in Figure 15.10, as $t \to \infty$ then $C(t)$ approaches the asymptotic t-axis, i.e. $C(t) \to 0$. So as $T \to \infty$ the integral above will approach the limit

$$\frac{C_0}{k} - \frac{0}{k}, \text{ that is } \frac{C_0}{k}. \quad \text{So } \int_0^\infty C_0\, e^{-kt} dt = \frac{C_0}{k}.$$

Thus, there is a finite lifetime exposure from a single dose of the drug, which is directly proportional to the initial concentration, and inversely proportional to the decay rate.

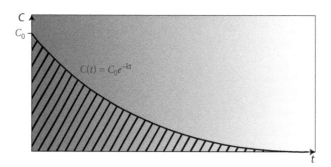

Figure 15.10 Total drug exposure

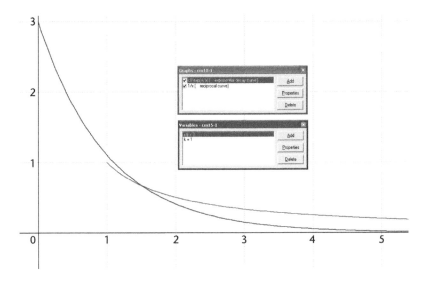

Figure 15.11 Graphs of $y = \dfrac{1}{x}$ and $y = 3e^{-x}$ (FNGRAPH 15.1)

Integrals such as this, where the upper limit is ∞ (and/or the lower limit is $-\infty$) do not fit into the definition in Section 15.2.1 of the definite integral (which uses a finite-length interval and approximates the area by a finite number of strips). They are called **improper integrals**.

Not all improper integrals will have a finite value. Consider the area under the curve $y = \dfrac{1}{x}$, along the positive x-axis starting from $x = 1$. This curve is sketched in FNGRAPH 15.1 (Figure 15.11), alongside the exponential decay curve $y = 3e^{-x}$. Both curves approach the asymptotic x-axis as $x \to \infty$, but $y = \dfrac{1}{x}$ does this more slowly, so the area under the curve will be greater. This area is expressed as the improper integral $\displaystyle\int_{1}^{\infty} \dfrac{1}{x}\, dx$, and we can try to evaluate it by finding the proper integral $\displaystyle\int_{1}^{X} \dfrac{1}{x}\, dx$ for some large positive upper limit $x = X$, and then asking what happens as $X \to \infty$:

$$\int_{1}^{X} \frac{1}{x}\, dx = [\ln x]_{1}^{X} = \ln X - \ln 1 = \ln X.$$

Now as X goes to infinity, the logarithm of X also grows without bound, so the area under the curve is infinite. We could write that

$$\int_{1}^{\infty} \frac{1}{x}\, dx = \infty.$$ We say that the integral **diverges**.

PRACTICE:

Try to evaluate the improper integral $\displaystyle\int_{1}^{\infty} \dfrac{1}{x^2}\, dx$.

Answer: $\displaystyle\int_{1}^{\infty} \dfrac{1}{x^2}\, dx = 1.$

Problems: Now try the end-of-chapter problems 15.19–15.20.

15.5 **Extension: numerical integration**

In Chapter 14 we claimed that there exist functions that cannot be integrated – meaning that the indefinite integral cannot be found. You cannot use any techniques of integration to find a function $F(x)$ so that $\int e^{-x^2}\,dx = F(x) + c$. But the graph of $y = e^{-x^2}$ is perfectly well-behaved, and we can see on it the area under the curve that is expressed by the definite integral $\int_1^2 e^{-x^2}\,dx$, for instance; it is shown in Figure 15.12. We can't find this area by integrating the function and putting in the limits, but we can approximate it by using the approach of Section 15.2, through dividing the area into vertical strips and adding up their areas. Moreover, by taking a sufficiently large number of sufficiently thin strips we can make this numerical approximation as close to the true value of the area as we wish.

There are two standard algorithms (called rules) for performing this numerical integration: a basic one which produces a good approximation, and a slightly more sophisticated version which produces a much better result with virtually the same amount of effort. They are called the trapezium rule and Simpson's rule.

Given a continuous function $f(x)$, and limits a and b, our task is to calculate a numerical value for the definite integral $\int_a^b f(x)\,dx$.

15.5.1 **The trapezium rule**

The starting point for the rule is to choose n, the number of vertical strips. The width of each strip will be $h = \frac{1}{n}(b - a)$. Then, as in Riemann integration, the vertical lines are at x_0, x_1, x_2, ..., x_n, where $x_0 = a$, $x_1 = a + h$, $x_2 = a + 2h$, etc. up to $x_n = a + nh = a + (b - a) = b$. This is shown in Figure 15.13. We will call these values of x the **ordinates**.

To find the heights of the strips, we'll need to evaluate the function at these x-values. We'll write $f_0 = f(x_0)$, $f_1 = f(x_1)$, $f_2 = f(x_2)$, and so on, up to $f_n = f(x_n)$. So the first strip lies between x_0 and x_1, and its height is f_0 on the left-hand side, and f_1 on the right-hand side. The top of the strip is curved, but we'll approximate its area by joining up the points (x_0, f_0) and (x_1, f_1) by a straight line. The resulting shape is called a trapezium, and its area A_1 is calculated as:

area of trapezium = width × (average height).

That is,

$$A_1 = h \times \tfrac{1}{2}(f_0 + f_1).$$

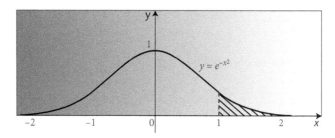

Figure 15.12 Graph of $y = e^{-x^2}$

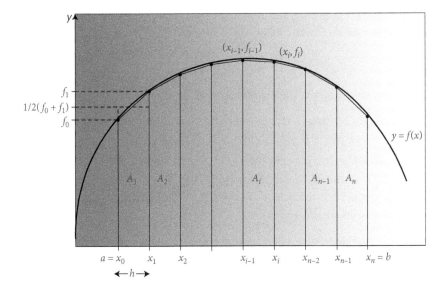

Figure 15.13 The trapezium rule

Similarly, the second strip is approximated by a trapezium, joining up the top corners (x_1, f_1) and (x_2, f_2) by a straight line. This trapezium area is then

$$A_2 = h \times \tfrac{1}{2}(f_1 + f_2).$$

We continue this process up to the final strip: $A_n = h \times \tfrac{1}{2}(f_{n-1} + f_n)$. Adding all the trapezium areas together, gives us the approximation to the area under the curve between $x = a$ and $x = b$:

$$\int_a^b f(x)\,dx \approx A_1 + A_2 + A_3 + \cdots + A_n$$

$$= \tfrac{1}{2}h\big((f_0 + f_1) + (f_1 + f_2) + (f_2 + f_3) + \cdots + (f_{n-1} + f_n)\big)$$

$$= \tfrac{1}{2}h\big(f_0 + 2f_1 + 2f_2 + 2f_3 + \cdots + 2f_{n-1} + f_n\big).$$

The trapezium rule can be summarized as:

Trapezium rule for approximating $\displaystyle\int_a^b f(x)\,dx$:

Choose the number of strips n. Strip width is $h = \tfrac{1}{n}(b - a)$.

Calculate the ordinates $x_0, x_1, x_2, \ldots, x_n$, where $x_0 = a$, $x_1 = a + h$, $x_2 = a + 2h$, etc.

Evaluate the function at these points: $f_0 = f(x_0)$, $f_1 = f(x_1)$, $f_2 = f(x_2)$, and so on. Then

$$\int_a^b f(x)\,dx \approx \tfrac{1}{2}h\big(f_0 + 2f_1 + 2f_2 + 2f_3 + \cdots + 2f_{n-1} + f_n\big). \qquad (15.14)$$

The formula in brackets is a weighted sum of function values $\sum c_i f_i$: each function value f_i is multiplied by the appropriate coefficient c_i (where $c_1 = c_n = 1$ and all the other coefficients c_i

are 2). This weighted sum is finally multiplied by the scaling factor $\frac{1}{2}h$. The calculation is best laid out in a table, as shown in the following example:

EXAMPLE

Estimate $\int_1^5 e^{-x}\ dx$, using the trapezium rule with 8 strips.

Solution: Here, the limits of integration are $a = 1$ and $b = 5$, and the number of strips is $n = 8$. So the strip width is $h = \frac{1}{n}(b - a) = \frac{1}{8}(5 - 1) = 0.5$.

The ordinates are at $x_0 = a = 1.0$, $x_1 = a + h = 1.5$, $x_2 = a + 2h = 2.0$, ..., $x_n = b = 5.0$. We need to evaluate the function at these ordinates, i.e. $f_i = f(x_i) = e^{-x_i}$, then multiply each f_i by the appropriate coefficient c_i, add up the weighted sum and finally multiply by $\frac{1}{2}h$. Here is a table where this is done:

Number of strips, $n = 8$				
Strip width, $h = (b - a)/n = 0.5$				
i	x_i	$f(x_i)$	c_i	$c_i{}^*f(x_i)$
0	1.00	0.3679	1	0.36788
1	1.50	0.2231	2	0.44626
2	2.00	0.1353	2	0.27067
3	2.50	0.0821	2	0.16417
4	3.00	0.0498	2	0.09957
5	3.50	0.0302	2	0.06039
6	4.00	0.0183	2	0.03663
7	4.50	0.0111	2	0.02222
8	5.00	0.0067	1	0.00674
$h/2 =$	0.25		$\Sigma c_i{}^*f(x_i) =$	1.47454
$h/2{}^*\Sigma c_i{}^*f(x_i) =$ **0.3686**				

We have found that $\int_1^5 e^{-x}\ dx \approx 0.3686$.

The table in this example was taken from **SPREADSHEET 15.1**, where the trapezium rule is used with $n = 2, 4, 8,$ and 16 strips. If you want to use these tables to calculate your own definite integrals, you need to change the limits of integration in cells E4 and H4. You also need to program your own function $f(x)$ in cells C11, I11, O11, and U11, then drag the formulas down the columns.

For this problem, we are able to find the true value of the integral:

$$I = \int_1^5 e^{-x}dx = [-e^{-x}]_1^5 = -(e^{-5} - e^{-1}) = \frac{e^4 - 1}{e^5} = \frac{53.598}{148.413} = 0.361141...$$

We can then work out the relative error in I in each instance of the trapezium rule in the spreadsheet, this time defining relative error as

$$\text{relative error} = \frac{\text{exact value} - \text{approximate value}}{\text{exact value}} \times 100\%.$$

The results are:

no of strips	approx value	exact value	relative error
2	0.4742	0.3611	−31.3%
4	0.3907	0.3611	−8.2%
8	0.3686	0.3611	−2.1%
16	0.3630	0.3611	−0.5%

Each time the number of strips doubles, the relative error is reduced by the same factor – in this case a factor of about 4. We say that the trapezium rule **converges linearly** to the exact solution.

PRACTICE

Estimate $\int_0^3 x^3 \, dx$ using the trapezium rule with 6 strips.

Answer: 20.8125.

An alternative way of describing what is happening in the trapezium rule is as follows: we take the function $f(x)$ and evaluate it at the set of equally spaced x-ordinates $x_0, x_1, x_2, ..., x_n$, giving us a set of data-points $(x_0, f_0), (x_1, f_1), (x_2, f_2), ..., (x_n, f_n)$. We then approximate the function by using a linear (straight-line) approximation between each pair of successive data-points. This is called a **piecewise linear approximation**. If we integrate the piecewise linear approximation, we obtain the trapezium rule. We used just such a piecewise linear approximation in Section 3.6.3, to estimate the value of y at an intermediate value of x between x_0 and x_1, using linear interpolation between the data-points (x_0, f_0) and (x_1, f_1). This was illustrated in Figure 3.9 (page 99), where you can now see the trapezium. The only difference is one of notation: we are now using the function $f(x)$ in place of the dependent variable y.

Using function notation, the linear interpolation formula (3.11) becomes

$$\phi(x) = f_0 + s.\Delta f_0 \tag{15.15}$$

where $\Delta f_0 = f_1 - f_0$ is called the first difference of f_0, and the interpolating factor s is related to x by

$$s = \tfrac{1}{h}(x - x_0). \tag{15.16}$$

Remember that $x_1 = x_0 + h$. Then $s = 0$ when $x = x_0$, and $s = 1$ when $x = x_1$. We write the linear function as $\phi(x)$ to distinguish it from the function $f(x)$ which it is approximating.

In the Problems you can check that if you integrate (15.15) with respect to x between the limits x_0 and x_1, you obtain the area of the trapezium, i.e.

$$\int_{x_0}^{x_1} \phi(x) \, dx = A_1 = \tfrac{1}{2}h(f_0 + f_1).$$

15.5.2 Simpson's rule

So, if we take the formula (15.15) for linear interpolation between x_0 and x_1, and integrate it over the interval from x_0 to x_1, we obtain an approximation to the area under the curve between those limits. Adding together these approximations for each strip in Figure 15.13 produces the trapezium rule approximating the definite integral $\int_a^b f(x)\,dx$.

But in Section 7.5 we saw a much better interpolation rule: quadratic interpolation. In this we use (x_0, f_0), (x_1, f_1) plus a third data-point (x_2, f_2), where $x_2 = x_1 + h = x_0 + 2h$. The rule is given by equation (7.40), which is written in function notation as:

$$\phi(x) = f_0 + s.\Delta f_0 + \tfrac{1}{2}s(s-1).\Delta^2 f_0. \tag{15.17}$$

The interpolating factor s is the same as in (15.16), as is the first difference Δf_0. We define another first difference $\Delta f_1 = f_2 - f_1$, and then the second difference of f_0:

$$\Delta^2 f_0 = \Delta f_1 - \Delta f_0 = (f_2 - f_1) - (f_1 - f_0) = f_2 - 2f_1 + f_0. \tag{15.18}$$

Now $\phi(x)$ is a quadratic function whose curve passes through the data-points (x_0, f_0), (x_1, f_1) and (x_2, f_2). If we integrate it over the interval from x_0 to x_2, we will obtain an approximation to the area under the curve $y = f(x)$ between those limits, i.e. the area of the first two strips in Figure 15.13. Here is the integration – you may want to skip this at first reading and go on to see how the resulting equation (15.19) is used to form Simpson's rule:

We want to calculate $A = \int_{x_0}^{x_2} \phi(x)\,dx$, but in (15.17) we have $\phi(x)$ as a function of s, not x. We will therefore use integration by substitution.

We use (15.16) to convert the limits of integration to limits of s:

when $x = x_0$, $\quad s = \tfrac{1}{h}(x_0 - x_0) = 0$

when $x = x_2$, $\quad s = \tfrac{1}{h}(x_2 - x_0) = \tfrac{1}{h}(x_0 + 2h - x_0) = 2$

Also, from (15.16), $s = \tfrac{1}{h}(x - x_0) = \tfrac{1}{h}x - \tfrac{1}{h}x_0$, so differentiating, we find

$ds = \tfrac{1}{h}dx$ since x_0 is constant,

so $dx = h\,ds$.

Thus our integral becomes

$$A = \int_{x_0}^{x_2} \phi(x)\,dx = \int_0^2 \phi\, h\, ds = h\int_0^2 \phi\, ds \text{ since } h \text{ is constant.}$$

Substituting (15.17) and remembering that f_0 and the first and second differences are constants:

$$A = h\int_0^2 \left(f_0 + s.\Delta f_0 + \tfrac{1}{2}(s^2 - s).\Delta^2 f_0 \right) ds$$

$$= h.\left[s.f_0 + \tfrac{1}{2}s^2.\Delta f_0 + \tfrac{1}{2}\left(\tfrac{1}{3}s^3 - \tfrac{1}{2}s^2 \right).\Delta^2 f_0 \right]_0^2$$

$$= h.\left(2f_0 + 2\Delta f_0 + \tfrac{1}{2}\left(\tfrac{8}{3} - 2 \right).\Delta^2 f_0 \right) - h.(0)$$

$$= h.\left(2f_0 + 2\Delta f_0 + \tfrac{1}{3}.\Delta^2 f_0 \right).$$

If we use the definitions $\Delta f_0 = f_1 - f_0$ and (15.18), this reduces to

$$A = \tfrac{1}{3}h.\left(f_0 + 4f_1 + f_2\right). \tag{15.19}$$

This is the quadratic approximation to the area of the first two strips in Figure 15.13, i.e. from $x = x_0$ to $x = x_2$. By the same argument, the area under the next two strips, from $x = x_2$ to $x = x_4$, will be $A = \tfrac{1}{3}h.(f_2 + 4f_3 + f_4)$, and so on. Summing the areas of these double strips together, the area of the definite integral is approximated by

$$\int_a^b f(x)\,dx \approx \tfrac{1}{3}h\left(\left(f_0 + 4f_1 + f_2\right) + \left(f_2 + 4f_3 + f_4\right) + \cdots + \left(f_{n-2} + 4f_{n-1} + f_n\right)\right)$$
$$= \tfrac{1}{3}h\left(f_0 + 4f_1 + 2f_2 + 4f_3 + \cdots + 2f_{n-2} + 4f_{n-1} + f_n\right).$$

As we are using double strips, the total number of strips n must be an even number.

This is Simpson's[3] rule, which we can define by

Simpson's rule for approximating $\int_a^b f(x)\,dx$:

Choose the number of strips n, which must be an even number. The strip width is $h = \tfrac{1}{n}(b - a)$.

Calculate the ordinates $x_0, x_1, x_2, \ldots, x_n$, where $x_0 = a$, $x_1 = a + h$, $x_2 = a + 2h$, etc.

Evaluate the function at these points: $f_0 = f(x_0), f_1 = f(x_1), f_2 = f(x_2)$, and so on. Then

$$\int_a^b f(x)\,dx \approx \tfrac{1}{3}h\left(f_0 + 4f_1 + 2f_2 + 4f_3 + \cdots + 2f_{n-2} + 4f_{n-1} + f_n\right). \tag{15.20}$$

You can see that the work involved in applying Simpson's rule is virtually the same as for the Trapezium rule with the same number of strips. The example from Section 15.5.1 is re-worked with Simpson's rule in **SPREADSHEET 15.2**. Here is the calculation for $n = 8$ strips:

Number of strips, $n = 8$				
Strip width, $h = (b - a)/n = 0.5$				
i	x_i	$f(x_i)$	c_i	$c_i * f(x_i)$
0	1.00	0.3679	1	0.36788
1	1.50	0.2231	4	0.89252
2	2.00	0.1353	2	0.27067
3	2.50	0.0821	4	0.32834
4	3.00	0.0498	2	0.09957
5	3.50	0.0302	4	0.12079
6	4.00	0.0183	2	0.03663

---

[3] Thomas Simpson (1710–1761), English mathematician. There is an interesting biography of him at http://www-groups.dcs.st-and.ac.uk/~history/Biographies/Simpson.html.

i	x_i	$f(x_i)$	c_i	$c_i{}^*f(x_i)$
Number of strips, $n = 8$				
Strip width, $h = (b - a)/n = 0.5$				
7	4.50	0.0111	4	0.04444
8	5.00	0.0067	1	0.00674
$h/3 = 0.1667$			$\Sigma c_i{}^*f(x_i) =$	2.16758
		$h/3{}^*\Sigma c_i{}^*f(x_i) = \mathbf{0.3613}$		

The relative errors from using 2, 4, 8, and 16 strips are now:

no of strips	approx value	exact value	relative error
2	0.3825	0.3611	−5.92%
4	0.3629	0.3611	−0.50%
8	0.3613	0.3611	−0.03%
16	0.3611	0.3611	0.00%

You can see that the approximations converge very rapidly towards the exact solution. Look at the result using Simpson's rule with 4 strips; the Trapezium rule needs to take 16 strips to achieve the same accuracy.

> **PRACTICE**
>
> Estimate $\int_0^3 x^3 \, dx$ using Simpson's rule with 6 strips.
>
> **Answer:** 20.25 (If you check, you'll find that this is the exact solution!)

Problems: Now try the end-of-chapter problems 15.21–15.26.

15.5.3 Using Simpson's rule with data-sets

We have described numerical integration as a way of estimating the definite integral for functions that cannot be integrated analytically. The rules involve first evaluating the function at a set of ordinates, to produce a data-set. A more important practical use of numerical integration, is where we are provided directly with a data-set, from measurement or experiment.

As an example, we have taken river flow data from the UK Centre for Ecology & Hydrology.[4] There are data-sets providing daily flow rates (in m^3 s^{-1}, i.e. cubic metres per second) for around 200 rivers in the United Kingdom. We have chosen the River Spey in Scotland, measured at the Boat o' Brig gauging station since 1952. The water in this river is mainly runoff from the Cairngorm mountains, and flow has recently been affected by the construction of

---

[4] Available at http://www.ceh.ac.uk/data/nrfa/river_flow_data.html

Daily flow rates of River Spey (guaged at Boat o Brig), in cubic metres per second

July	c_i	1956	1966	1976	1986	1996	2006
1	1	27.45	75.56	17.49	27.37	23.76	21.417
2	4	33.68	65.94	19.97	26.55	23.36	20.336
3	2	38.21	56.32	26.81	25.88	22.57	22.227
4	4	34.81	52.92	21.75	25.39	22.15	26.77
5	2	41.89	47.54	18.97	24.86	31.6	22.981
6	4	70.19	43.87	17.34	24.26	36.03	20.735
7	2	48.96	43.87	16.49	23.74	32.44	18.868
8	4	40.19	41.6	15.77	23.69	29.65	18.07
9	2	38.77	37.64	15.28	23.19	29.79	19.154
10	4	36.51	36.22	15.91	23.27	27.36	25.269
11	2	35.09	38.49	16.18	23.07	24.46	21.5
12	4	31.13	35.09	15.52	22.52	22.61	21.249
13	2	27.45	40.19	15.91	22.27	21.95	19.256
14	4	24.71	67.92	16.99	22.08	20.22	18.044
15	2	25.24	54.62	17.88	21.55	18.97	17.091
16	4	23.6	46.98	20.07	21.29	18.39	16.381
17	2	21.96	40.75	20.06	20.91	17.76	15.791
18	4	20.58	35.66	18.29	20.57	17.28	15.496
19	2	20.07	32.55	18.19	20.51	16.92	15.187
20	4	19.39	30.28	20.6	20.31	16.39	14.706
21	2	18.93	28.87	20.68	19.96	16.04	19.183
22	4	19.16	28.02	19.8	20.75	17.44	19.652
23	2	20.07	26.83	18.54	22.79	37.27	15.642
24	4	20.29	27.05	18.61	23.47	50.7	14.606
25	2	19.84	27.05	18.51	22.68	32.56	14.307
26	4	18.48	28.27	17.63	21.62	25.82	14.415
27	2	17.12	26.57	16.74	20.84	22.42	14.616
28	4	17.8	28.58	15.91	22.04	20.65	15.137
29	2	100.5	30.56	15.51	23.28	19.42	15.215
30	4	727.3	127.1	15.71	24.94	18.76	14.466
31	1	291.5	80.09	16.8	63.01	18.23	14.131
$\Sigma c_i * f(x_i) =$		5818.43	4001.35	1625.27	2092.44	2197.57	1638.912
$h/3 =$	28800						
$h/3 * \Sigma c_i * f(x_i) =$		167570784	115238880	46807776	60262272	63290016	47200666
Monthly total: (million m3)		167.5708	115.239	46.808	60.262	63.29	47.201

Figure 15.14 Integrating the flow rates (SPREADSHEET 15.3)

hydroelectric schemes. In SPREADSHEET 15.3 (Figure 15.14) the daily data for the month of July are tabulated for the years 1956, 1966, 1976, 1986, 1996, and 2006. By integrating the flow rate data over the whole month, we can estimate the total flow of water for the month. The spreadsheet is reproduced below.

As July has 31 days, there are 30 strips. The Simpson's rule coefficients c_i are written in column C, and in row 39 the summation is performed using the Excel function SUMPRODUCT. The interval length h is the number of seconds from one day to the next, which is $60 \times 60 \times 24 = 86400$. The result of (15.17) appears in row 42, in units of m³, and this is more conveniently expressed in million cubic metres in row 44. The results indicate a substantial drop in mid-summer water volume since the 1970s.

PRACTICE

The following table shows the daily flow rates for the River Wear at Sunderland Bridge, in August 2008. Use the trapezium rule to estimate the total monthly flow (in millions of cubic metres). Repeat the calculation using Simpson's rule. Perform the calculation by hand, using your calculator, and the check your answers by adapting SPREADSHEET 15.3.

Tip: to perform the calculation by hand, don't multiply every single data value f_i by its coefficient c_i. Instead, first add up all the data values which will be multiplied by c_i, and then multiply the sum. Repeat for each coefficient c_i.

August 2008	flow rate f_i (m^3 s^{-1})	Trapezium Rule		Simpson's Rule	
		c_i	$f_i * c_i$	c_i	$f_i * c_i$
1	13.10				
2	11.10				
3	6.61				
4	5.35				
5	4.42				
6	6.04				
7	7.33				
8	7.98				
9	7.13				
10	10.60				
11	8.13				
12	15.70				
13	17.50				
14	20.20				
15	13.80				
16	11.00				
17	60.20				
18	21.80				
19	66.10				
20	29.00				
21	14.70				
22	10.90				
23	9.07				
24	7.93				
25	6.95				
26	8.05				
27	7.80				
28	6.30				
29	5.37				
30	4.82				
31	4.88				

Answers: (i) 36.363   (ii) 34.424.

? PROBLEMS

15.1–15.4: Evaluate the following definite integrals:

15.1 $\int_1^2 x(x^3 - 4)\, dx$ 15.2 $\int_1^{\frac{\pi}{4}} \tan x\, dx$ 15.3 $\int_{-1}^1 e^{3x}\, dx$ 15.4 $\int_1^4 \frac{1}{\sqrt{x}}\, dx$.

15.5–15.8: Find the mean value of x over the interval of integration, in each of the above problems.

15.9: Using the exponential model for animal motion which we last saw in Case Study C13, namely:

$$v(t) = v_{max}(1 - e^{-kt}) \qquad\qquad (10.29)$$

where $k = \dfrac{a_0}{v_{max}}$ show that the animal's mean speed between time t_0 and time t_1 is given

by: $\bar{v} = v_{max}\left(1 - \dfrac{1}{k.\Delta t}\left(e^{-kt_0} - e^{-kt_1}\right)\right)$

where $\Delta t = t_1 - t_0$.

For the lion accelerating from rest ($v_{max} = 14$, $a_0 = 6.25$), calculate its average speed in the first two seconds of its run, and in the subsequent two seconds (i.e. from $t = 2$ to $t = 4$).

Repeat the calculations for the gazelle ($v_{max} = 25$, $a_0 = 3.75$).

15.10: Repeat problem 15.9 using the exponential model for animal motion, namely:

$$v(t) = \frac{v_{max}.t}{t + b} \qquad\qquad (8.5)$$

where $b = \dfrac{v_{max}}{a_0}$. Show that the animal's mean speed in this model is

$$\bar{v} = v_{max}\left(1 - \frac{b}{\Delta t}\ln\left|\frac{t_1 + b}{t_0 + b}\right|\right).$$

15.11–15.14: Integrate the following functions, using an appropriate technique of integration:

15.11 $\int_0^2 \dfrac{x}{\left(2x^2 + 1\right)^2}\, dx$ 15.12 $\int_1^{\frac{\pi}{2}} \dfrac{\cos x}{2 - \sin x}\, dx$

15.13 $\int_0^1 x\cos\left(\dfrac{\pi}{2}x\right) dx$ 15.14 $\int_1^e x\ln(5x)\, dx$.

15.15–15.18: Find the mean value of x over the interval of integration, in each of the above problems.

15.19: Do you think that the area between the curve $y = xe^{-x}$ and the positive x-axis (see Figure 13.7) is finite or infinite? Find out, by evaluating the improper integral $I_1 = \int_0^\infty xe^{-x}\, dx$.

You may use the fact that $\lim_{x \to \infty} x^n e^{-x} = 0$ for $n = 1,2,3,\ldots$. That is, no matter how large a power n you choose, the function e^{-x} will sooner or later tend to zero faster than x^n tends to infinity.

15.20: Repeat problem 15.19 for the function $y = x^2 e^{-x}$ (also sketched in Figure 13.7). That is, evaluate the improper integral $I_2 = \int_0^\infty x^2 e^{-x}\, dx$.

15.21: Use the trapezium rule to estimate (to 4 decimal places) the value of $I = \int_1^3 \frac{1}{x}\, dx$, taking (i) 4 strips and (ii) 8 strips. Calculate the true value of I, and work out the relative errors in the two estimations.

15.22: Repeat problem 15.21 using Simpson's rule, taking (i) 4 strips and (ii) 8 strips. Compare the accuracy with that from the trapezium rule.

15.23: Modify SPREADSHEET 15.2 to use Simpson's rule with $n = 2,4,8,16$ strips to estimate the integral in problem 15.8: $\int_1^e x \ln(5x)\, dx$. Check the accuracy of each approximation by evaluating the analytic solution of problem 15.8.

15.24: Modify SPREADSHEET 15.2 to use Simpson's rule $n = 2,4,8,16$ strips to estimate the integral $\int_0^4 e^{-x^2}\, dx$.

15.25: Perform the integration of (15.15) for the trapezium rule, to prove that $\int_{x_0}^{x_1} \left(f_0 + s.\Delta f_0 \right) dx = \frac{1}{2} h \left[f_0 + f_1 \right]$. Follow the way we integrated in Simpson's rule.

15.26: In the derivation of Simpson's rule, manipulate $h.\left(2f_0 + 2\Delta f_0 + \frac{1}{3}.\Delta^2 f_0 \right)$ to obtain (15.19): $A = \frac{1}{3} h.\left(f_0 + 4f_1 + f_2 \right)$.

Differential equations I

The study of differential equations is the culmination of the mathematics presented in this book. It is also perhaps the most important mathematical topic for the biological sciences. A **differential equation** is an equation involving a derivative such as $\dfrac{dy}{dx}$; **solving** the differential equation means finding y as a function of x. For instance, I claim that $y = 3x^2 + 5$ is a solution of the differential equation

$$x\frac{d^2 y}{dx^2} - \frac{dy}{dx} = 0. \tag{16.1}$$

You can check this by differentiating the claimed solution:

$$y = 3x^2 + 5, \quad \text{so } \frac{dy}{dx} = 6x, \quad \frac{d^2 y}{dx^2} = 6$$

and inserting these in the left-hand side of (16.1):

l.h.s. $= x(6) - (6x) = 0 =$ r.h.s., so the solution satisfies the differential equation.

The task of finding such solutions will require the basic algebra from Chapters 1–5, the theory of functions in Chapters 6–10, and the differential and integral calculus of the previous five chapters. We start by classifying the different types of differential equation, and then describe methods for solving first-order differential equations. These will include relatively simple but powerful numerical techniques, easily implemented in Excel; these produce the solution as a set of data-points which can be graphed. We will also see these solution techniques applied to mathematical models of biological processes. Before we start, it would be worth looking back to Section 11.1.3, where this topic was introduced with examples of modelling biological processes by differential equations.

16.1 Overview of differential equations

The main classification of differential equations is in terms of their **order**, which we'll define next. This is also a good place to introduce the concept of boundary conditions, which enable us to find the particular solution for our model.

16.1.1 Order of a differential equation

An equation such as

$$y\left(x\frac{dy}{dx} - y\right) = x^2 \tag{16.2}$$

which involves x, y, and the first derivative $\dfrac{dy}{dx}$, is called a **first-order differential equation**. Equations that also involve the second derivative $\dfrac{d^2y}{dx^2}$ are called **second-order differential equations**. Thus, the **order** of a differential equation is the order of the highest-order derivative that appears in it. (The equations we dealt with in Part I, involving x and y but no derivatives, are called **algebraic equations**, so the task of solving a differential equation means finding an algebraic equation relating the variables.)

Any first-order differential equation of practical interest can be manipulated algebraically so that the derivative appears on its own on the left-hand side. Rearranging (16.2) in this way produces

$$\frac{dy}{dx} = \frac{x^2 + y^2}{xy},$$

for example. On the right-hand side we have an expression involving x and y. A general form for the first-order differential equation is thus

$$\frac{dy}{dx} = f(x, y) \tag{16.3}$$

where $f(x, y)$ is a function of the independent variable x and the dependent variable y. We will study methods for solving first-order differential equations in Sections 16.2 and 16.3.

An example of a second-order differential equation is

$$\frac{d^2y}{dx^2} - 2\frac{dy}{dx} - 3y = \cos x. \tag{16.4}$$

The solution of second-order differential equations is a large topic in its own right, which we shall introduce in Section 17.3.1. Luckily, the differential equations that occur in the most important biological models are first-order.

16.1.2 Boundary conditions

In a practical application, a differential equation models a physical or biological process occurring over a certain **domain**. If we are modelling the blood flow around the heart, the differential equation needs to be solved over the three-dimensional volume of the heart, and over the time interval of one heartbeat cycle; in this case the domain is a complex four-dimensional 'hypervolume'. However, we will focus on differential equations where there is only one independent variable x, in which case the domain will be an **interval** on the real line, either finite (e.g. $a \le x \le b$), or infinite (e.g. $x \ge 0$).

If the right-hand side function in (16.3) is a function of x only, then we can solve the equation by simply integrating. For example, to solve the second-order differential equation

$$\frac{d^2y}{dx^2} = 0 \tag{16.5}$$

over the interval $0 \leq x \leq 1$, we would integrate with respect to x once, to obtain

$$\frac{dy}{dx} = c \qquad\qquad (16.6)$$

where c is an arbitrary constant of integration. Integrating both sides with respect to x again, gives us the solution

$$y = cx + k \qquad\qquad (16.7)$$

involving a second constant of integration k. The equation (16.7) gives the **general solution** of (16.5). You should check that the function $y = Ax^2 + B$, where A, B are arbitrary constants, is also a solution of the differential equation (16.1); in fact, this is the general solution of (16.1).

To obtain the solution of a particular problem we need additional information, telling us the value of y and/or of $\frac{dy}{dx}$ for particular values of x. These additional facts are called **boundary conditions**. In the example above, if we were told that:

- $y = 1$ when $x = 0$, and
- $y = 5$ when $x = 1$,

we could insert these two conditions in (16.7), giving us two simultaneous equations to solve for c and k:

$$c.0 + k = 1$$
$$c.1 + k = 5.$$

We find that $k = 1$ and $c = 5 - k = 4$, so inserting these values in the general solution (16.7) gives us the **particular solution**

$$y = 4x + 1.$$

If we are given different boundary conditions, we will obtain a different particular solution. Had we been given the same differential equation (16.5) and the same domain $0 \leq x \leq 1$, but the boundary conditions:

- $y = 2$ when $x = 0$, and
- $\frac{dy}{dx} = 3$ when $x = 0$,

then putting the second boundary condition into (16.6) tells us that $c = 3$, and then using this and the first boundary condition in (16.7) gives $k = 2$, so the particular solution of this problem will be $y = 3x + 2$.

Boundary conditions that are specified at the start of the interval, as in the previous example, are called **initial conditions**. Often, the independent variable is not x, but time t, the domain of solution is $t \geq 0$, and there are initial conditions specified at time $t = 0$. We have used initial conditions in the example of the accelerating VW Polo, in Section 14.1.3.

A complete problem made up of a differential equation plus a set of initial conditions, is called an **Initial Value Problem** or **IVP**.

Problems: Now try the end-of-chapter problems 16.1–16.8.

16.1.3 ODEs and PDEs

So far, we have assumed that the dependent variable y is a function of a single independent variable x, so the derivatives in the differential equation are $\dfrac{dy}{dx}$ and $\dfrac{d^2y}{dx^2}$. But we mentioned that in modelling the blood flow around the heart during one heartbeat cycle, the domain would be a complex shape in four dimensions: the three space dimensions x, y, z, and the time dimension t. The flow v would be a function of all four variables: $v = v(x, y, z, t)$. When the dependent variable is a function of several independent variables, how can we talk about derivatives? Is there a derivative of v with respect to x, another derivative of v with respect to y, and so on? Yes; these are called **partial derivatives**, and an equation involving such derivatives is called a **partial differential equation**, abbreviated to PDE.

As an example of a PDE, the classic diffusion equation is

$$\frac{\partial C}{\partial t} = k\left(\frac{\partial^2 C}{\partial x^2} + \frac{\partial^2 C}{\partial y^2} + \frac{\partial^2 C}{\partial z^2} \right). \tag{16.8}$$

Here, the variable $C(x, y, z, t)$ represents the mass concentration of solute in a solution at the point (x, y, z) in three dimensions, and at time t. The parameter k is the diffusion coefficient, and the strange form of δ indicates partial derivatives of C. I'll say more about partial differentiation in the Extension section at the end of this chapter.

Differential equations where only one independent variable is involved are called **ordinary differential equations**, shortened to ODEs, to distinguish them from PDEs. We will concentrate on solving first-order ODEs. Sections 16.2 and 16.3 will look at analytic methods of solution, while Chapter 17 will describe numerical methods. Then we will consider systems of ODEs (two or more ODEs to be solved simultaneously). We will illustrate these methods through applications and case studies.

16.2 Solution by separation of variables

In this section we show methods for solving a first-order ODE by a technique called separation of variables. We assume that the equation will have been rearranged if necessary into the form (16.3), with the derivative $\dfrac{dy}{dx}$ on the left-hand side, and an algebraic function $f(x, y)$ on the right-hand side. In this method we treat $\dfrac{dy}{dx}$ as if it were a fraction, dy over dx. We already did this trick in integration by substitution (Section 14.2). Recall that to integrate $\displaystyle\int \frac{2x}{x^2 - 3}\,dx$, we choose $u = x^2 - 3$, differentiate to get $\dfrac{du}{dx} = 2x$, and then split this into $du = 2x\,dx$ before substituting.

The details of the method depend on the form of $f(x, y)$, but the aim is always to separate out terms involving the variable y, together with dy, on one side of the equation, and the other terms (involving x and dx) on the other side. Then we can integrate each side separately.

16.2.1 Right-hand side a function of x only

Consider the differential equation

$$\frac{dy}{dx} = p(x) \tag{16.9}$$

where $p(x)$ is a function of x only. We said above that to solve such an ODE you simply integrate the right-hand side. We can write this out more clearly as an algorithm which will be a useful starting-point for the subsequent sections.

Treat $\frac{dy}{dx}$ as if it were a fraction: dy divided by dx, as in integration by substitution. Then multiply both sides of (16.9) by dx:

$$dy = p(x)\, dx.$$

Now place integral signs in front of each side:

$$\int dy = \int p(x)\, dx.$$

The left-hand side looks confusing, but look back to the remark in Section 14.2: this can be written as $\int 1\, dy$, meaning the integral of 1 with respect to y, which is $y + c$. We won't include the '+c' because we will bring in the constant of integration when we integrate the right-hand side. So the left-hand side is simply y:

$$y = \int p(x)\, dx. \tag{16.10}$$

So, by a series of rather dubious steps, we have obtained the correct result: (16.10) follows directly from (16.9) if we interpret the integration as anti-differentiation, as in Chapter 14.

EXAMPLE

Solve $x^2\left(5 + \dfrac{dy}{dx}\right) - 15 = 3\dfrac{dy}{dx} + 2x$, with the boundary condition that $y = 7$ when $x = 2$.

Solution: This looks horrendous, but notice that y only appears in the derivative $\dfrac{dy}{dx}$. First rearrange the equation to isolate $\dfrac{dy}{dx}$ on the left-hand side, so that it will look like (16.9). Here are the algebraic steps:

$$5x^2 + x^2\frac{dy}{dx} - 15 = 3\frac{dy}{dx} + 2x$$

Collect terms in $\frac{dy}{dx}$ on the left-hand side

$$\left(x^2 - 3\right)\frac{dy}{dx} = -5x^2 + 2x + 15$$

$$\frac{dy}{dx} = \frac{-5x^2 + 15 + 2x}{x^2 - 3} = \frac{-5(x^2 - 3) + 2x}{x^2 - 3}$$

$$\frac{dy}{dx} = \frac{2x}{x^2 - 3} - 5.$$

So $p(x) = \dfrac{2x}{x^2 - 3} - 5$, and (16.10) becomes

$$y = \int\left(\frac{2x}{x^2 - 3} - 5\right) dx.$$

Integrating term-by-term:

> Integration by substitution: Let $u = x^2 - 3$. Then $du = 2x\,dx$.

$$y = \int \frac{2x}{x^2 - 3}\,dx - \int 5\,dx$$

$$= \ln\left|x^2 - 3\right| - 5x + c.$$

This is the general solution. Finally, apply the boundary condition to find the constant of integration:

When $x = 2$, $y = 7$. So $7 = \ln(2^2 - 3) - 10 + c.$

As $\ln 1 = 0$, we find that $c = 17$, so the particular solution is

$$y = \ln\left|x^2 - 3\right| - 5x + 17.$$

Problems: Now try the end-of-chapter problems 16.9–16.12.

16.2.2 Right-hand side a function of y only

Consider now the differential equation

$$\frac{dy}{dx} = q(y) \tag{16.11}$$

where $q(y)$ is a function of y only. We can adapt the method of the previous section. Again treat $\frac{dy}{dx}$ as if it were a fraction, and multiply both sides by dx:

$$dy = q(y)\,dx.$$

Now divide both sides by $q(y)$, in order to isolate all the terms involving y on the left-hand side:

$$\frac{dy}{q(y)} = dx.$$

Finally, apply the integral signs:

$$\int \frac{dy}{q(y)} = \int dx.$$

Now it is the right-hand side that is the integral of 1 with respect to x, so we get

$$\int \frac{dy}{q(y)} = x + c. \tag{16.12}$$

Provided we are able to integrate $\frac{1}{q(y)}$ with respect to y (we have already included the constant of integration), we will obtain an algebraic expression involving x and y. It may or may not be possible to manipulate this to express y explicitly as a function of x.

If the right-hand side of (16.11) involves a constant coefficient, this may remain on the right-hand side, as in the following example:

EXAMPLE

Solve $\frac{dy}{dx} = 5y$, given that $y = 100$ when $x = 0$.

Solution: Rearrange the differential equation to

$$\frac{dy}{y} = 5\,dx.$$

Now apply the integral signs, and integrate both sides:

$$\int \frac{dy}{y} = \int 5\,dx$$
$$\ln|y| = 5x + c$$

Finally, solve for y by taking exponents:

$$y = e^{5x+c} = Ae^{5x},$$

where A is a positive constant (since $A = e^c$). Using the initial condition, $y(0) = 100 = Ae^0 = A$, so the particular solution is $y = 100e^{5x}$.

You should recognize this example as the technique we used in Case Study A15, at the end of Section 14.3, where we solved the differential equation of exponential growth

$$\frac{dN}{dt} = rN$$

to obtain

$$N(t) = N_0\,e^{rt}.$$

The value of A is obtained from the initial condition that, when $t = 0$, we have $N(0) = N_0$, the initial population size, so $A = N_0$.

PRACTICE

A biological cell is placed in a tube containing a solution of a certain chemical, at concentration C_1. Over time, the molecules of solute diffuse through the cell membrane into the cell. Let $C(t)$ be the concentration of the solute inside the cell at time t. The rate at which the molecules enter the cell (and hence the rate at which the concentration in the cell increases) is proportional to the concentration difference inside and outside the cell. Assuming that the outside concentration remains constant at C_1, this theory produces the differential equation

$$\frac{dC}{dt} = k(C_1 - C) \tag{16.13}$$

where k is the diffusion coefficient. Use the method of this section to solve this differential equation, obtaining the general solution

$$C = C_1 - Ae^{-kt},$$

where A is a positive constant. Given that there is zero concentration of solute in the cell initially, write down the initial condition and use it to show that the particular solution is

$$C = C_1\left(1 - e^{-kt}\right). \tag{16.14}$$

Check that the intracellular concentration $C(t)$ approaches the outside concentration C_1 as $t \to \infty$.

One reason why differential equations are such a powerful tool for the scientist is that two natural processes in very different fields may be governed by what is essentially the same differential equation. The solution from one process can be used to immediately write down the solution of the other. Newton's law of cooling says that when a hot body cools down, its rate of cooling is proportional to the temperature difference between the body and the outside environment. The differential equation embodying this law is

$$\frac{dT}{dt} = -k(T - \tau) \tag{16.15}$$

where $T(t)$ is the body temperature after time t, τ is the outside temperature, and k is a thermal coefficient. We include the minus sign because $\frac{dT}{dt}$ is negative when the body cools down. This equation is almost identical to (16.13), so we can quickly write down its general solution:

$$T = \tau + Ae^{-kt}. \tag{16.16}$$

The particular solution will be different from (16.14), though, because we have a different initial condition – see problem 16.18.

Application: von Bertalanffy growth of fish

The von Bertalanffy[1] growth model is the most widely used model of how fish grow in length. Its differential equation is

$$\frac{dL}{dt} = K(L_\infty - L) \tag{16.17}$$

where $L(t)$ is the length of the fish at time t after hatching, and K is a growth coefficient (also known as the curvature parameter). L_∞ is the asymptotic length (the length of the mature specimen). Again, this is essentially the same equation as (16.13). Its solution is

$$L(t) = L_\infty \left(1 - e^{-k(t - t_0)}\right). \tag{16.18}$$

The constant t_0 is a shift. When the fish hatches it has a very small positive length, so t_0 is a hypothetical time beforehand when the fish length was zero. Thus $t_0 < 0$. A typical growth curve is shown in Figure 16.1 (after Sparre 1989).

Fish scientists use a graph called a Gulland-Holt plot to estimate the constants L_∞ and K. In Chapter 8 we discussed techniques for fitting curves to data, by transforming the equation into a linear relationship and fitting the best straight line through data-points plotted using the new variables. Suppose the length of a specimen has been measured at regular intervals Δt while it grows. This provides the data-points $(t_0, L_0), (t_1, L_1), (t_2, L_2), \ldots, (t_n, L_n), \ldots$ where $\Delta t = t_{n+1} - t_n$. We work out the average growth rate over each interval: $\frac{\Delta L_n}{\Delta t} = \frac{L_{n+1} - L_n}{\Delta t}$ and plot this against L_n. Since this average growth rate is an approximation to the instantaneous growth rate $\frac{dL}{dt}$, the best-fit straight line:

$$\frac{\Delta L}{\Delta t} = KL_\infty - KL$$

will have slope $-K$ and y-intercept KL_∞.

---

[1] Ludwig von Bertalanffy, 1901–1972, Austrian biologist and systems theorist.

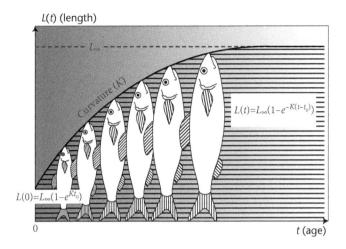

Figure 16.1 von Bertalanffy growth (after Sparre 1989)

Of course, this method will only work to solve differential equations of the form (16.11) if we are able to work out the integral in (16.12). In Case Study B10 we will see the differential equation

$$\frac{dN}{dt} = aN - bN \ln\left(\frac{N}{N_0}\right).$$

This has a function of N only on the right-hand side, but it appears to be quite a daunting task to integrate its reciprocal.

Problems: Now try the end-of-chapter problems 16.13–16.18.

16.2.3 Variables separable

Now suppose that the right-hand side function $f(x, y)$ in (16.3) involves both x and y, but that the variables can be separated out, so that $f(x, y)$ can be written as the product of two functions: one involving only xs and the other involving only ys. Then we can combine the methods of the last two sections to solve the ODE.

That is, $f(x, y)$ can be written as $u(x).v(y)$. Then our ODE (16.3) is

$$\frac{dy}{dx} = u(x).v(y) \tag{16.19}$$

which we rearrange to

$$\frac{dy}{v(y)} = u(x)\,dx.$$

Applying integration to both sides:

$$\int \frac{dy}{v(y)} = \int u(x)\,dx.$$

Performing the integrations (if possible!) will produce an algebraic equation which can hopefully be manipulated to produce y as a function of x.

EXAMPLE

Find the general solution of $\dfrac{dy}{dx} = y^2 \cos x$.

Solution: Rearrange to

$$\frac{dy}{y^2} = \cos x \; dx.$$

Apply the integral signs:

$$\int \frac{dy}{y^2} = \int \cos x \; dx.$$

Perform the integrations:

$$-\frac{1}{y} = \sin x + c.$$

$$\int y^{-2} \, dy = (-1)y^{-1} + c$$

Solve for y:

$$y = -\frac{1}{\sin x + c}.$$

PRACTICE

Find the general solution of $\dfrac{dy}{dx} = xy$.

Answer:

$$y = Ae^{\frac{1}{2}x^2}.$$

Problems: Now try the end-of-chapter problems 16.19–16.22.

CASE STUDY B9: The Gompertz model of tumour growth

In the early stages of a tumour's formation, the cells within it multiply according to the simple exponential growth model:

$$\frac{dN}{dt} = rN \tag{11.2}$$

which we have seen in Case Study A, and which we saw how to solve in the previous section. But doctors have observed in cancer patients that the growth soon becomes slower than predicted by this model. Not all tumour cells are actively dividing, and in large tumours the proportion of active cells is lower than in small ones. One theory is that only the cells in an outer 'crust' of the tumour have access to nutrients, while those in the core die.

The basic model of tumour growth, named after Gompertz,[2] uses (11.2) but the growth coefficient r is no longer constant; it decreases as the tumour grows. Here is one way of deriving the Gompertz equation:

---

[2] The same Benjamin Gompertz whose curve for population mortality appeared in Case Study A11 at the end of Chapter 10. This is another example where a mathematical model proposed for one situation has been used in a totally different application.

We suppose that $N(t)$, the number of cells in the tumour at time t, follows the exponential growth model

$$\frac{dN}{dt} = r(t).N,$$

(16.20)

but that the growth coefficient $r(t)$ follows a model of exponential decay:

$$\frac{dr}{dt} = -br.$$

(16.21)

The solution to (16.21) is $r(t) = r_0\, e^{-bt}$.

Inserting this in (16.20), the differential equation we need to solve is

$$\frac{dN}{dt} = r_0 N e^{-bt}.$$

(16.22)

Here, the right-hand side function is separable, and we can apply the method of separation of variables. Rearrange to

$$\frac{dN}{N} = r_0\, e^{-bt} dt,$$

apply the integral signs:

$$\int \frac{dN}{N} = r_0 \int e^{-bt} dt,$$

and perform the integrations:

$$\ln N = r_0 \left(-\frac{1}{b} e^{-bt} \right) + c.$$

Taking exponentials of both sides:

$$N = A e^{-\frac{r_0}{b} e^{-bt}}$$

(16.23)

where $A = e^c$. Compare this with equation (10.47) in Case Study A11.

Apply the initial condition: $N(0) = N_0$ (the initial population size):

when $t = 0$, $N_0 = A e^{-\frac{r_0}{b}}$, so that $A = N_0\, e^{\frac{r_0}{b}}$.

Inserting this in (16.23) produces the Gompertz growth equation

$$N(t) = N_0\, e^{\frac{r_0}{b}\left(1 - e^{-bt}\right)}.$$

(16.24)

The function is drawn in FNGRAPH 16.1 (Figure 16.2). Notice that the tumour size approaches asymptotically an upper limit N_{max}. If you let $t \to \infty$ in (16.24), you should find that

$$N_{max} = N_0\, e^{\frac{r_0}{b}}.$$

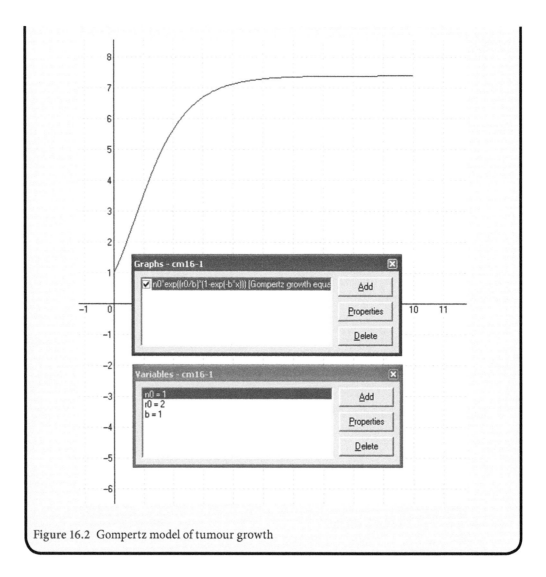

Figure 16.2 Gompertz model of tumour growth

CASE STUDY A16: Solving the ODE for logistic growth

We began our study of population growth models with a discrete-time model of geometric growth, defined by the update equation

$$N_{p+1} = N_p + R.N_p \qquad (1.28)$$

(Part A3, Section 1.7.2). We saw that the solution to this update equation is

$$N_p = (1 + R)^p \, N_0.$$

Later, we introduced a continuous-time model of exponential growth:

$$N(t) = N_0 \, e^{rt} \qquad (10.9)$$

(Part A9, Section 10.2), and showed (in Part A12, Section 11.7) that this function $N(t)$ satisfies

$$\frac{dN}{dt} = rN, \tag{11.19}$$

which we now know to be a differential equation, with initial condition $N(0) = N_0$. In Part A12 we showed the correspondence between the differential equation defining the continuous model and the update equation defining the discrete model. Finally, in Part A15 (Section 14.3) we showed how to solve the differential equation – a technique we have defined properly in this section.

We can apply the same analysis to the discrete and continuous models of logistic growth. The update equation for logistic growth was given in Part A8 in Chapter 7 as

$$N_{p+1} = N_p + r_m.N_p\left(1 - \frac{N_p}{K}\right) \tag{7.33}$$

where K is the carrying capacity. You should use the technique we used in Part A12 (approximating the derivative by a ratio of increments) to show that this update equation corresponds to the differential equation

$$\frac{dN}{dt} = r_m N\left(1 - \frac{N}{K}\right). \tag{16.25}$$

Our challenge is to solve this ODE with initial condition $N(0) = N_0$. The right-hand side is a function of N only, so we rearrange and apply the integral signs:

$$\int \frac{dN}{N\left(1 - \frac{N}{K}\right)} = \int r_m\, dt. \tag{16.26}$$

The integral on the left is not straightforward, but you should recognize the function as the type where we can apply the technique of partial fractions (see Section 14.4). We write

$$\frac{1}{N\left(1 - \frac{N}{K}\right)} = \frac{A}{N} + \frac{B}{1 - \frac{N}{K}}$$

and seek to solve for A and B. Multiplying through by the left-hand side denominator:

$$A\left(1 - \frac{N}{K}\right) + BN = 1.$$

Substitute particular values of N to find A and B. Substituting $N = 0$ we find that $A = 1$, and substituting $N = K$ we find that $B = \frac{1}{K}$. Thus,

$$\int \frac{dN}{N\left(1 - \frac{N}{K}\right)} = \int \left(\frac{1}{N} + \frac{1}{K}\frac{1}{1 - \frac{N}{K}}\right)dN$$

$$= \int \frac{1}{N}dN + \frac{1}{K}\int \frac{1}{1 - \frac{N}{K}}\, dN$$

Integration by substitution: let $u = 1 - \frac{1}{K}N$.

$$= \ln N - \ln \left| 1 - \frac{N}{K} \right| + c$$

$$= \ln \left| \frac{N}{1 - \dfrac{N}{K}} \right| + c.$$

Placing this expression on the left-hand side of (16.26), and integrating the right-hand side (dropping the '+c' from the left-hand side and introducing it on the right, as we have done previously), we can take exponentials and obtain

$$\frac{N}{1 - \dfrac{N}{K}} = Ae^{r_m t}.$$

I will leave it to you to use algebra to rearrange this equation, solving for N as a function of t. You should obtain

$$N(t) = \frac{Ae^{r_m t}}{\left(1 + \dfrac{1}{K} e^{r_m t} \right)}.$$

The final step is to solve for A, using the initial condition $N(0) = N_0$. You should find that

$$A = \frac{N_0}{1 - \dfrac{N_0}{K}},$$

and substituting this back in, with a bit more algebra, produces the final solution of the logistic growth model:

$$N(t) = \frac{K}{1 + \left(\dfrac{K}{N_0} - 1 \right) e^{-r_m t}}.$$

If you look back to Part A10 of this Case Study, in Chapter 10, you will see that this is equation (10.37) which we used to produce the smooth sigmoidal curve of logistic growth (Figure 10.13).

You can check that as $t \to \infty$, the population size approaches the carrying capacity K.

CASE STUDY C17: A harvesting model for fish stocks

Returning to the predator–prey topic of fisheries management, we now introduce the idea of a harvesting model. This is a differential equation modelling a fish stock population $N(t)$, of the form

$$\frac{dN}{dt} = F(N) - H(t, N)$$

where $F(N)$ is a population growth function, usually logistic growth as in (16.25), and $H(t,N)$ is a harvesting function, representing the reduction in population growth caused by continuous removal of fish by fishing fleets (Clark 2005). Here we'll define a simple harvesting function and analyse the long-term effect of the harvesting.

First, it is usual to change the variable so as to eliminate the carrying capacity K from the equation. Let

$$x(t) = \frac{1}{K} N(t), \quad \text{so} \quad N = Kx.$$

Now as N and K both represent numbers of fish, the variable x is dimensionless. Differentiating both sides with respect to time:

$$\frac{dx}{dt} = \frac{1}{K} \frac{dN}{dt}, \quad \text{so} \quad \frac{dN}{dt} = K \frac{dx}{dt}$$

and inserting in the logistic growth equation (16.25):

$$\frac{dN}{dt} = r_m N \left(1 - \frac{N}{K} \right)$$

$$K \frac{dx}{dt} = rKx \left(1 - \frac{1}{K} Kx \right) \quad \boxed{\text{Divide both sides by } K}$$

$$\frac{dx}{dt} = rx(1 - x).$$

This last equation is easier to work with. Our harvesting model will be

$$\frac{dx}{dt} = rx(1 - x) - kx,$$

that is, the catch rate is proportional to the stock level x. This models the situation where fishing fleets leave areas that have low stocks and migrate to exploit fishing zones that are relatively well-populated.

I would urge you to solve this ODE yourself, with initial condition $N(0) = N_0$. Use separation of variables, following the steps in Case Study A16 above. The solution which I reached, once converted back from x to the variable N, is:

$$N(t) = \frac{K \left(1 - \dfrac{k}{r} \right)}{1 + \left(\dfrac{K}{N_0} \left(1 - \dfrac{k}{r} \right) - 1 \right) e^{-(r-k)t}}.$$

Compare this with the analytic solution (10.37) for logistic growth. I have programmed this function into FNGRAPH 16.2. Experiment with different values of k and r. You will find that if $k < r$ (the harvesting rate is less than the growth rate) then $N(t)$ approaches a positive equilibrium value, while if $k > r$ then $N(t)$ tends to zero (the fish stock dies out).

Deduce this behaviour from the analytic solution above, and find an expression for the long-term fish stock level when $k < r$.

Use the fact that as $t \to \infty$, $e^{-at} \to 0$ if a is positive, and e^{-at} grows exponentially if a is negative.

Answer: $N(t) \to \left(1 - \dfrac{k}{r} \right) K$ when $k < r$.

16.2.4 Change of variable

First-order differential equations that do not fall into the above categories, or that produce difficult integrals, can often be solved by using a change of variable. For example, suppose the right-hand side function $f(x, y)$ is a function g of a linear combination of x and y:

$$\frac{dy}{dx} = g(ax + by) \tag{16.27}$$

for some constants a and b. For example, we may have $\frac{dy}{dx} = (2x - 5y)^7$ or $\frac{dy}{dx} = \ln|6x - y|$.
We invent the new variable

$$u = ax + by.$$

Differentiating this with respect to x, using implicit differentiation (Section 12.5), gives

$$\frac{du}{dx} = a + b\frac{dy}{dx},$$

and substituting for $\frac{dy}{dx}$ from (16.27),

$$\frac{du}{dx} = a + b.g(u).$$

This is now a first-order ODE in u and x, which is separable:

$$\int \frac{du}{a + b.g(u)} = \int dx.$$

Once this has been solved by the methods above, substitute back for u to find the solution involving x and y.

> **EXAMPLE**
>
> Solve $\frac{dy}{dx} = (x + y)^2$, with the boundary condition $y(0) = 1$.
>
> **Solution:** Let $u = x + y$. Then $\frac{du}{dx} = \frac{dx}{dx} + \frac{dy}{dx} = 1 + \frac{dy}{dx}$, so our equation becomes
>
> $$\frac{du}{dx} = 1 + u^2.$$
>
> Solving by the method of Section 16.2.2:
>
> $$\int \frac{du}{1 + u^2} = \int dx.$$
>
> The integral on the left-hand side is one of the standard integrals listed at the end of Chapter 14. Using this, we can integrate each side:
>
> $$\tan^{-1} u = x + c,$$
>
> which, using the definition of the inverse tangent, means that
>
> $$u = \tan(x + c).$$
>
> Now substitute back $x + y$ for u, and solve for y:
>
> $$y = \tan(x + c) - x.$$

This is the general solution. Apply the given boundary condition $y = 1$ when $x = 0$:

$$1 = \tan c$$

so $c = \tan^{-1} 1 = \dfrac{\pi}{4}$. Our particular solution is thus

$$y = \tan\left(x + \frac{\pi}{4} \right) - x.$$

A similar approach can be taken when the right-hand side function can be written in terms of a new variable $u = \dfrac{y}{x}$. An ODE of this type, namely:

$$\frac{dy}{dx} = g\left(\frac{y}{x} \right)$$

is called **homogeneous**. In this case we write $y = ux$ before differentiating implicitly with respect to x:

$$\frac{dy}{dx} = u + x\frac{du}{dx}.$$

Substituting in the original ODE

$$\frac{dy}{dx} = g\left(\frac{y}{x} \right) = g(u)$$

produces

$$u + x\frac{du}{dx} = g(u),$$

which is separable:

$$\frac{du}{dx} = \frac{g(u) - u}{x}$$

$$\int \frac{du}{g(u) - u} = \int \frac{dx}{x}.$$

Problems: Now try the end-of-chapter problems 16.23–16.26.

CASE STUDY B10: The Gompertz model revisited

The Gompertz model of tumour growth, as derived in Part B9 of this Case Study, has the advantage that the solution (16.24) produces realistic curves describing tumour growth over time, once suitable values of the parameters are chosen. The disadvantage is that the second differential equation in the model (16.21) has no real justification. There are medical reasons why the growth coefficient will reduce as the tumour size increases, but no reason why this should be a simple exponential decay over time. Researchers therefore looked for a more plausible differential equation that would produce the same or a similar solution.

The modern formulation of the Gompertz model for tumour growth, is the differential equation

$$\frac{dN}{dt} = aN - bN \ln\left(\frac{N}{N_0} \right), \tag{16.28a}$$

which we can rewrite as

$$\frac{1}{N}\frac{dN}{dt} = a - b\ln\left(\frac{N}{N_0}\right).$$

(16.28b)

This resembles an exponential growth model, but with an extra term (a perturbation) on the right-hand side, reducing the growth rate. In the first stage of growth, when N is close to the initial size N_0, this term will be very small (since $\ln 1 = 0$).

We solve this ODE by a change of variable. Let $U = b\ln\left(\frac{N}{N_0}\right) = b(\ln N - \ln N_0)$, so $\frac{dU}{dN} = b.\frac{1}{N} - 0$.

Then by the chain rule $\frac{dU}{dt} = \frac{dU}{dN}.\frac{dN}{dt} = b.\frac{1}{N}.\frac{dN}{dt}$. So (16.28b) becomes

$$\frac{1}{b}\frac{dU}{dt} = a - U.$$

This is much easier to solve. You should now be able to do this, by the method of Section 16.2.2, to obtain

$$U = a - Ae^{-bt}, \text{ where } A > 0.$$

The initial condition $N(0) = N_0$ means that $U = 0$ when $t = 0$. So the constant of integration is $A = a$, giving

$$U = a\left(1 - e^{-bt}\right).$$

Finally, we must substitute back $b\ln\left(\frac{N}{N_0}\right)$ for U:

$$b\ln\left(\frac{N}{N_0}\right) = a\left(1 - e^{-bt}\right)$$

and solve for N:

$$N = N_0 \, e^{\frac{a}{b}\left(1-e^{-bt}\right)}$$

(16.29)

which is identical to the solution (16.24) we obtained in Part B9!

In this new differential equation (16.28a) for Gompertzian growth, the first term on the right-hand side gives the simple exponential growth model, and the second term can be thought of as a perturbation inhibiting growth. The model could be refined by adding further perturbation terms, perhaps representing the inhibition due to cancer therapy.

The 'Gomp-ex' model is a two stage model, in which exponential growth occurs until a critical tumour size N_c is reached ($N_c \approx 10^9$ cells in human tumours), after which growth is Gompertzian. The Gomp-ex differential equation is written as

$$\frac{1}{N}\frac{dN}{dt} = \begin{cases} \lambda & \text{for } N \leq N_c \\ \lambda - \beta\ln\left(\frac{N}{N_c}\right) & \text{for } N \geq N_c. \end{cases}$$

'At present, the Gomp-ex model provides the best available deterministic description of the typical pattern of tumour growth' (Wheldon 1988).

16.3 **Linear first-order ODEs**

An ordinary differential equation is **linear** if it is linear in the variable y and its derivatives. This means that a first-order ODE is linear if it can be written as

$$\frac{dy}{dx} + p(x).y = q(x). \tag{16.30}$$

We will consider the case where the coefficient of y is a constant, i.e.

$$\frac{dy}{dx} + ky = q(x). \tag{16.31}$$

There is a trick to solving such ODEs. Multiply both sides by a factor I defined by $I = e^{kx}$:

$$e^{kx}\frac{dy}{dx} + ke^{kx}y = q(x)e^{kx}.$$

Now consider differentiating the product $(e^{kx}y)$ with respect to x, using implicit differentiation and the product rule:

$$\frac{d}{dx}\left(e^{kx}.y\right) = e^{kx}.\frac{dy}{dx} + \frac{d}{dx}\left(e^{kx}\right).y = e^{kx}.\frac{dy}{dx} + k.e^{kx}.y$$

which is exactly what we have on the left-hand side. So our ODE becomes

$$\frac{d}{dx}\left(e^{kx}.y\right) = e^{kx}.q(x). \tag{16.32}$$

Integrating,

$$e^{kx}.y = \int e^{kx}.q(x)\,dx$$

and solving for y,

$$y = e^{-kx}\int e^{kx}.q(x)\,dx.$$

The factor e^{kx} which we used is called an **integrating factor**. Note that you cannot take the e^{-kx} factor inside the integration sign. You have to work out $\int e^{kx}.q(x)\,dx$ and then multiply this by e^{-kx}.

> **EXAMPLE**
>
> Find the general solution of $\dfrac{dy}{dx} + y = \cos x$.
>
> **Solution:**
> Here, $k = 1$, so multiply both sides by e^x:
>
> $$e^x\frac{dy}{dx} + e^x y = e^x \cos x.$$

The left-hand side is the derivative of $e^x.y$ by the product rule, giving:

$$\frac{d}{dx}\left(e^x y\right) = e^x \cos x.$$

So integrating

$$e^x y = \int e^x \cos x \, dx.$$

We worked out the right-hand side integral in Section 14.5.4:

$$e^x y = \frac{1}{2} e^x (\sin x + \cos x) + c.$$

The general solution for y is then

$$y = \frac{1}{2}(\sin x + \cos x) + ce^{-x}.$$

It is possible to extend this approach to solve the more general linear ODE (16.30); the integrating factor to use is $I = e^{\int p(x)\, dx}$.

16.4 Extension: partial differentiation

Suppose we have a quantity y that depends on two independent variables: a space variable x and a time variable t. We can write this as $y = y(x,t)$. Now there are two ways of thinking about the rate of change of y:

- the rate at which y changes along the x-direction, at a particular fixed instant of time, and
- the rate at which y changes with time, measured at a fixed point x.

If we know y as a function of x and t, we can work out these two derivatives. They are called **partial derivatives**, and are written using the symbol ∂ (a special form of the Greek letter delta) in place of the differential symbol d. Thus:

- The rate of change of y with respect to x, holding t constant, is written $\frac{\partial y}{\partial x}$, and

- the rate of change of y with respect to t, holding x constant, is written $\frac{\partial y}{\partial t}$.

To work out $\frac{\partial y}{\partial x}$, you differentiate y with respect to x, while treating all occurrences of the variable t as if they were constants. For example, if

$$y = e^{3t} \sin x - t^2 \cos x,$$

the partial derivative with respect to x would be

$$\frac{\partial y}{\partial x} = e^{3t} \cos x - t^2(-\sin x) = e^{3t} \cos x + t^2 \sin x.$$

On the other hand, the partial derivative with respect to t (treating x as a constant) would be

$$\frac{\partial y}{\partial t} = 3e^{3t} \sin x - 2t \cos x.$$

Problems: Now try the end-of-chapter problems 16.27–16.31

16.4.1 Reducing a PDE to an ODE

A differential equation that involves partial derivatives is called a **partial differential equa-tion**. This is often shortened to PDE. We gave the example of the three-dimensional diffusion equation (16.8) in the chapter introduction 16.1.3.

Application: laminar flow

A PDE describing a process occurring in two or three dimensions can often be reduced to a one-dimensional differential equation if we consider a simpler situation, where the depend-ent variable remains constant in the other directions. We'll illustrate this by returning to the example of laminar flow along a pipe. In Section 15.2.3 we described laminar flow of blood along an artery, treated as being a pipe of radius a. I claimed that the velocity v of the blood that is at a radius r from the central axis of the pipe is given by

$$v(r) = c\left(a^2 - r^2\right). \tag{15.10}$$

You will probably now realize that this equation must be the solution of a differential equa-tion that models the flow problem. The differential equation from fluid mechanics for this axisymmetric problem can be simplified from the complicated differential equations for fluid flow in three dimensions, known as the Navier–Stokes equations, to

$$\mu \frac{1}{r} \frac{\partial}{\partial r}\left(r \frac{\partial v}{\partial r}\right) = \frac{\partial p}{\partial z}, \tag{16.33}$$

which is called the Poisseuille equation. The parameter μ is the viscosity of the fluid. There are two space dimensions in this axisymmetric model: the radius r and the axial direction z along the pipe. This is shown in Figure 16.3. The fluid velocity at a point (r, z) is $v(r, z)$ and the fluid pressure is $p(r, z)$.

The term $\frac{\partial p}{\partial z}$ on the right-hand side of (16.33) is the rate of change of pressure p as you move along the pipe in the z direction. This is called the pressure gradient. If the pipe has length l and the pressures at either end are p_1 and p_2 pascals respectively, the pressure gradi-ent will be $\frac{\partial p}{\partial z} = \frac{p_1 - p_2}{l}$, a constant. Then there will be a steady flow along the pipe, where the fluid velocity will depend only on the radius r and will be independent of distance along the pipe. So $v = v(r)$ and we can write its derivative as $\frac{dv}{dr}$. Equation (16.33) reduces to an ordinary differential equation:

$$\mu \frac{1}{r} \frac{d}{dr}\left(r \frac{dv}{dr}\right) = \frac{p_1 - p_2}{l}. \tag{16.34}$$

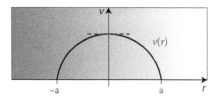

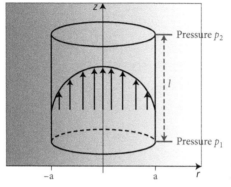

Figure 16.3 Laminar flow through a pipe

The left-hand side looks rather strange, but we can rearrange the equation to:

$$\frac{d}{dr}\left(r\frac{dv}{dr}\right) = \frac{p_1 - p_2}{\mu l}r.$$

Now the equation is saying that the derivative of $\left(r\frac{dv}{dr}\right)$ is the right-hand side function of r. Compare what we had in equation (16.32). We can integrate both sides with respect to r:

$$r\frac{dv}{dr} = \int \frac{p_1 - p_2}{\mu l}r\, dr$$

$$= \left(\frac{p_1 - p_2}{\mu l}\right)\cdot\frac{1}{2}r^2 + A.$$

Divide both sides by r to get just the derivative on the left-hand side:

$$\frac{dv}{dr} = \left(\frac{p_1 - p_2}{\mu l}\right)\cdot\frac{1}{2}r + A.\frac{1}{r}.$$

Now we can integrate again:

$$v = \left(\frac{p_1 - p_2}{2\mu l}\right)\int r\, dr + A\int \frac{1}{r}dr$$

$$= \left(\frac{p_1 - p_2}{2\mu l}\right)\cdot\frac{1}{2}r^2 + A\ln r + B.$$

The general solution is

$$v = cr^2 + A\ln r + B \tag{16.35}$$

where $c = \dfrac{p_1 - p_2}{4\mu l}$.

We need two boundary conditions in order to find the two constants of integration A, B. These are given by physical considerations. First, the fluid flowing down the central axis, where $r = 0$, is travelling fastest, at velocity v_{max}. But if you set $r = 0$ in (16.35) the second term is undefined: $\ln r \to -\infty$ as $r \to 0$. In order to get a finite value for v_{max} this term must disappear, i.e. $A = 0$. The second boundary condition is that the fluid is stationary where it is in contact with the wall of the pipe (this is called the no-slip condition), i.e. $v(a) = 0$. Using this in (16.35) produces $B = -ca^2$, and the particular solution is the equation we used in the last chapter:

$$v(r) = c\left(a^2 - r^2\right). \tag{15.10}$$

16.4.2 Error analysis in several variables

Suppose $f(x)$ is a function of x, and at $x = x_0$, $f(x_0) = f_0$. A small change Δx in x will cause a change Δf in f. In Section 11.7 we related Δf to Δx by

$$\Delta f \approx \frac{df}{dx}.\Delta x. \tag{11.17}$$

We could also obtain this result by truncating the Taylor series (Section 13.4.5) for $f(x)$ around x_0 after two terms, as we did for the Newton–Raphson method (Section 13.5.1):

$$f(x) \approx f(x_0) + f'(x_0).(x - x_0). \tag{13.19}$$

Now what if f is a function of two independent variables x and y? There is a Taylor series for f in two variables around the point (x_0, y_0), the first three terms of which are:

$$f(x,y) \approx f(x_0,y_0) + \frac{\partial f}{\partial x}.(x - x_0) + \frac{\partial f}{\partial y}.(y - y_0). \tag{16.36}$$

The partial derivatives are evaluated at (x_0, y_0).

We want to use this to estimate the size of errors, as in Section 2.6, taking

$$\Delta x = |x - x_0|, \quad \Delta y = |y - y_0|, \quad \Delta f = |f(x,y) - f(x_0,y_0)|,$$

so we take absolute values of the derivatives to avoid the terms partially cancelling:

$$\Delta f \leq \left|\frac{\partial f}{\partial x}\right|.\Delta x + \left|\frac{\partial f}{\partial y}\right|.\Delta y. \tag{16.37}$$

This is the two-variable version of the error inequality (11.18) in Section 11.7.

EXAMPLE

A baked beans tin has a height of 98 mm and a diameter of 74 mm. If the tin manufacturing machine works to an accuracy of ± 2 mm, estimate the possible error in the volume V.

Solution: The volume is given by $V = \pi r^2 h$, and we evaluate V at the point (r_0, h_0), where $h_0 = 98$ and $r_0 = \frac{1}{2} \times 74 = 37$. Then $V_0 = \pi r_0^2 h_0 = 4.2148 \times 10^5$ mm^3.

The partial derivatives of V are

$$\frac{\partial V}{\partial r} = \pi(2r)h = 2\pi rh \quad \text{and} \quad \frac{\partial V}{\partial h} = \pi r^2.(1) = \pi r^2.$$

Evaluating them at (r_0, h_0):

$$\frac{\partial V}{\partial r} = 2.2783 \times 10^4 \quad \text{and} \quad \frac{\partial V}{\partial h} = 4.3008 \times 10^3.$$

Then by (16.37):

$$
\begin{aligned}
\Delta V &\leq \left|\frac{\partial V}{\partial r}\right|.\Delta r + \left|\frac{\partial V}{\partial h}\right|.\Delta h \\
&= \left(2.2783 \times 10^4 \times 2.0\right) + \left(4.3008 \times 10^3 \times 2.0\right) \\
&= (45.566 + 8.6016) \times 10^3 = 54.17 \times 10^3.
\end{aligned}
\tag{16.38}
$$

The error range for V is then $V_0 - \Delta V \leq V \leq V_0 + \Delta V$:

$$3.673 \times 10^5 \leq V \leq 4.756 \times 10^5.$$

In this problem we are also able to express the relative error in V in terms of the relative errors in r and h. Dividing through (16.38) by $V = \pi r^2 h$:

$$
\begin{aligned}
\frac{\Delta V}{V} &\leq \left|\frac{2\pi rh}{\pi r^2 h}\right|.\Delta r + \left|\frac{\pi r^2}{\pi r^2 h}\right|.\Delta h \\
&= \left|\frac{2}{r}\right|.\Delta r + \left|\frac{1}{h}\right|.\Delta h \\
\frac{\Delta V}{V} &\leq 2\left|\frac{\Delta r}{r}\right| + \left|\frac{\Delta h}{h}\right|.
\end{aligned}
$$

In our problem, the relative error in r is $\frac{2}{37} = 0.0541 = 5.41\%$, and the relative error in h is $\frac{2}{98} = 0.0204 = 2.04\%$, so the relative error in V is $2 \times 0.0541 + 0.0204 = 0.1285 = 12.85\%$.

In problem 16.32 you can use this theory to re-solve one of the Revision problems about radiocarbon dating Ötzi the Iceman, from the end of Chapter 10.

16.4.3 Minimization in two variables

To find the turning-points of a function $f(x)$, we look for values of x at which the derivative $\frac{df}{dx} = 0$. To find a local minimum or maximum of a function of two variables $f(x, y)$, we look for points (x, y) at which both partial derivatives $\frac{\partial f}{\partial x}$ and $\frac{\partial f}{\partial y}$ are zero:

Let $f(x, y)$ be a function of two variables. Local maxima and minima of $f(x, y)$ are points (x, y) at which

$$\frac{\partial f}{\partial x} = 0$$

and

$$\frac{\partial f}{\partial y} = 0.$$

(16.39)

As an example, look back to the Extension Section 8.7 where we obtained the formulae (8.33) for calculating the slope m and y-intercept c of the trend line, the best-fit straight line through a set of n data-points. We did this by minimizing $\sum d_i^2$, the sum of the squares of the deviations d_i. The formula we found for $\sum d_i^2$ was

$$\sum d_i^2 = Am^2 - 2Bm + C + 2Dcm - 2Ec + nc^2, \qquad (8.31)$$

where

$$A = \sum x_i^2 \quad B = \sum x_i \cdot y_i \quad C = \sum y_i^2$$
$$D = \sum x_i \quad E = \sum y_i.$$

To obtain the values of m and c that minimize $\sum d_i^2$ we find its partial derivatives with respect to m and c, and set them to zero. This gives us two simultaneous equations to solve. If we write

$$f(m, c) = Am^2 - 2Bm + C + 2Dcm - 2Ec + nc^2$$

you should check that (16.39) becomes

$$\frac{\partial f}{\partial m} = 2Am - 2B + 2Dc = 0$$
$$\frac{\partial f}{\partial c} = 2Dm - 2E + 2nC = 0.$$

Dividing through by 2 in each equation, we obtain the simultaneous equations I quoted in Section 8.7:

$$Am + Dc = B$$
$$Dm + nc = E. \qquad (8.32)$$

PROBLEMS

16.1: Show that $y = Ax^2 + Bx$ is a solution of $\dfrac{d^2y}{dx^2} - \dfrac{2}{x}\dfrac{dy}{dx} + \dfrac{2y}{x^2} = 0$.

16.2: Show that $y = Ae^{3x} + Be^{-3x}$ is a solution of $\dfrac{d^2y}{dx^2} - 9y = 0$.

16.3: Show that $y = (A + Bx)e^x$ is a solution of $\dfrac{d^2y}{dx^2} - 2\dfrac{dy}{dx} + y = 0$.

16.4: Show that $V = \dfrac{A}{r} + B$ is a solution of $\dfrac{d^2V}{dr^2} - \dfrac{2}{r}\dfrac{dV}{dr} = 0$.

16.5–16.8: Rearrange the equations into the form $\dfrac{dy}{dx} = f(x, y)$. Do not try to solve them!

16.5 $y\left(x + y\dfrac{dy}{dx}\right) = 3x$ 16.6 $\dfrac{x + \dfrac{dy}{dx}}{x + y} = 3$

16.7 $\dfrac{x + \dfrac{dy}{dx}}{y + \dfrac{dy}{dx}} = 3x$ 16.8 $\sqrt{1 + \left(\dfrac{dy}{dx}\right)^2} = \dfrac{y}{x}$.

16.9–16.16: Rearrange the following differential equations and find their general solution. Use the given boundary condition to find the particular solution.

16.9 $x\dfrac{dy}{dx} - 3 = 6x^2$, $y = 5$ when $x = 1$.

16.10 $\dfrac{x + 3\dfrac{dy}{dx}}{3 + \dfrac{dy}{dx}} = 5$, $y = 1$ when $x = 0$.

16.11 $\dfrac{e^{3x} + \dfrac{dy}{dx}}{e^x + \dfrac{dy}{dx}} = 3$, $y = -1$ when $x = 0$.

16.12 $\dfrac{dy}{dx}\sec x - x = \tan x$, $y = \pi$ when $x = \dfrac{1}{2}\pi$.

16.13 $\dfrac{dy}{dx} = y + 2$, $y = 5$ when $x = 0$.

16.14 $\dfrac{dy}{dx} = 3y - \dfrac{1}{y}$, $y = 2$ when $x = 0$.

16.15 $\dfrac{dy}{dx} = \tan y$, $y = \dfrac{1}{2}\pi$ when $x = 0$.

16.16 $\dfrac{dy}{dx} = k(y - 1)(y - 2)$, $y = 1.25$ when $x = 0$.

16.17: Find an implicit equation expressing the general solution of

$$\dfrac{dy}{dx} = k\dfrac{y - 1}{y - 2}.$$

16.18: Newton's law of cooling says that the temperature $T(t)$ at time t of a hot object in a cooler environment at constant temperature τ $(T > \tau)$ is governed by the differential equation:

$$\dfrac{dT}{dt} = -k(T - \tau). \tag{16.15}$$

Show that the general solution of this differential equation is:

$$T = \tau + Ae^{-kt}. \tag{16.16}$$

I take a cup of hot coffee at 90 °C into the laboratory which is at a controlled ambient temperature of 20 °C. Find the particular solution for this situation. After 10 minutes have elapsed, the coffee has cooled to 60 °C. Show that the thermal coefficient $k = 0.056$ min^{-1}.

16.19–16.26: Find the general solution of the following ODEs:

16.19 $\dfrac{dy}{dx} = \dfrac{y}{x}$

16.20 $\dfrac{dy}{dx} = \dfrac{y}{x^2 - 1}$

16.21 $\dfrac{dy}{dx} = \cot x \cot y$

16.22 $x^2 \dfrac{dy}{dx} + y = 1$

16.23 $\dfrac{dy}{dx} = \dfrac{y}{x} + 1$

16.24 $\dfrac{dy}{dx} = \dfrac{y(x + y)}{x(y - x)}$

16.25 $\dfrac{dy}{dx} + 5y = e^{-x}$

16.26 $\dfrac{dy}{dx} - y = 3x^2.$

16.27–16.30: Find the partial derivatives of the following functions of two independent variables:

16.27 $f(x, y) = x^5 e^{2y} - y^4 + 3x$

16.28 $f(x, y) = \dfrac{x^2 y}{1 + y}$

16.29 $f(r, t) = \dfrac{rt}{r + t}$

16.30 $f(x, y) = \sin 2x \cos 3y + \cos 2x \sin 3y.$

16.31: Simplify the function in problem 16.30 using a trigonometric identity. Find its partial derivatives and check that, using more trigonometric identities, you can convert them to the derivatives you found in problem 16.30.

16.32: Use the theory of Section 16.4.2 to re-solve revision problem R10.15 at the end of Chapter 10. That is, the radiocarbon age t of a sample is calculated from the equation

$t = \dfrac{-T}{\ln 2} \ln P$ (see problem R10.13)

where T is the half-life of carbon-14, and P is the ratio of ^{14}C: ^{12}C atom proportions in the sample and the atmosphere respectively. In the radiocarbon dating of Ötzi the Iceman, taking $T = 5730$ years and $P = 53\% = 0.53$, we obtain a radiocarbon age of $t = 5249$ years old. Now suppose that the half-life value is $T = 5730 \pm 40$ years and the value of P is correct to $\pm 0.5\%$. Find the partial derivatives $\dfrac{\partial t}{\partial T}$ and $\dfrac{\partial t}{\partial P}$, and hence prove that the error interval for t is 5249 ± 115 years old.

16.33–16.35: Find points (x, y) where the following functions of x and y have local maxima/minima:

16.33 $f(x, y) = 2x^2 - 5xy + 2x + 3y - 6$

16.34 $f(x, y) = x^2 - 6xy + y^2 + 16y + 7$

16.35 $f(x, y) = x^3 + y^2 - 12x - 6y + 7.$

16.36: In Problems 16.27 and 16.28, differentiate $\frac{\partial f}{\partial x}$ with respect to y, and differentiate $\frac{\partial f}{\partial y}$ with respect to x. You should find that the two results are equal:

$$\frac{\partial}{\partial y}\left(\frac{\partial f}{\partial x}\right) = \frac{\partial}{\partial x}\left(\frac{\partial f}{\partial y}\right).$$

This quantity is called the mixed second partial derivative, and written $\frac{\partial^2 f}{\partial x \partial y}$.

Differential equations II

By now, any hope you may have entertained at the start of the previous chapter, that there will be a simple method that can solve all sorts of first-order differential equations, must have disappeared. But there are such 'wonder methods'. The only catch is that they are numerical ones. We have seen numerical methods throughout this book, including:

- linear interpolation (Section 3.6);
- quadratic interpolation (Section 7.5);
- bracketing methods for root-finding and minimization (Section 8.6);
- Newton's method for root-finding and optimization (Section 13.5);
- Numerical integration (Section 15.5).

Numerical methods can be used on problems that are too hard to solve by standard algebra, for example finding roots of complicated equations. The same is true of numerical methods for solving ODEs, and you will see standard numerical methods such as Runge–Kutta used by bioscientists in the research literature to produce numerical and graphical output from their models.

In this chapter we will demonstrate some numerical methods for solving a first-order ODE with an initial condition. We will also look at problems involving two or more simultaneous ODEs, in particular the equations relating the growth of populations of different species in a predator–prey situation.

17.1 Numerical methods for first-order ODEs

In the previous chapter we saw methods for solving ODEs that produce an algebraic equation expressing y as a function of x; these are called **analytic methods**. From the function, one could sketch or plot the curve, and analyse it by the methods of Section 13.1 to extract information about the behaviour of the model. A numerical method, on the other hand, produces a set of data-points (x_i, y_i) that lie on or close to the true solution curve. By fitting a curve through them, we can obtain much the same useful information. All these methods have the same overall structure, which we shall define in describing the simplest method: Euler's method.

17.1.1 Euler's method

The standard problem we shall be solving is:

Differential equation : $\dfrac{dy}{dx} = f(x, y)$ (17.1)

Initial condition: $y(x_0) = y_0$ (i.e. $y = y_0$ when $x = x_0$)

Domain: $x \geq x_0$.

We are given the values of x_0 and y_0, and the function $f(x, y)$. This is called an **Initial Value Problem (IVP)**. The initial condition gives us the first of our data-points, namely (x_0, y_0).

We now choose a value for a small increment in the x-direction, called the **steplength**, and denoted by h. Adding h to x_0 produces the next x-value x_1:

$x_1 = x_0 + h.$ (17.2)

The aim of our numerical method is to estimate the value of y corresponding to this value x_1. If we denote it by y_1, our method will have produced the new data-point (x_1, y_1).

Recall the Newton–Raphson method (Section 13.5.1) for finding the root of an equation $f(x) = 0$. We made a first guess x_0, then used the update equation (13.20) to obtain an improved estimate x_1. By then starting with x_1 and using the update equation again, we obtained a further improved estimate x_2, and so on, producing a sequence of data values. This repetitive process is called **iteration**.

We use the same iterative approach here. We repeat the calculation, starting with (x_1, y_1), to obtain a new y-value y_2 corresponding to the new x-value $x_2 = x_1 + h$. And so on: the iterative algorithm will produce a set of data-points (x_0, y_0), (x_1, y_1), (x_2, y_2), ..., (x_n, y_n).... Figure 17.1 shows the data-points, and the true, analytic solution curve. The hope is, that by using a small enough value of the steplength h, we will be able to produce data-points that are close enough to the analytic curve.

Now you can see why h is called the steplength: we are stepping forward along the interval $x \geq x_0$, and at each step we use the current data-point (x_n, y_n) – the start of the step – to work out the next point (x_{n+1}, y_{n+1}) – the end of the step, and start of the next one. The new x-value

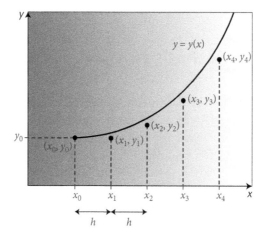

Figure 17.1 An iterative algorithm with steplength h

is given by $x_{n+1} = x_n + h$; the 'meat' of the method is in the way it calculates y_{n+1}, using x_n, y_n and the ODE itself, i.e. the right-hand side function $f(x, y)$ from (17.1).

To make the equations look simpler, we will define the particular method by just saying how to perform the first step, i.e. how to calculate y_1 using x_0, y_0, h, and f.

In Euler's method, we approximate the derivative $\dfrac{dy}{dx}$ as a ratio of increments Δy and Δx, as we saw in Section 11.7:

$$\frac{dy}{dx} \approx \frac{\Delta y}{\Delta x},$$

from which the increment in y is given by

$$\Delta y \approx \frac{dy}{dx}.\Delta x. \tag{11.17}$$

These increments are the changes in x and y when moving from (x_0, y_0) to (x_1, y_1):

$\Delta x = x_1 - x_0 = h$ using (17.2), and

$\Delta y = y_1 - y_0.$

Thus

$$y_1 - y_0 = \frac{dy}{dx}.h$$

or

$$y_1 = y_0 + h.\frac{dy}{dx}.$$

If only we knew the derivative $\dfrac{dy}{dx}$! But we do, from the differential equation (17.1) itself:

$$\frac{dy}{dx} = f(x,y).$$

Our algorithm for calculating y_1 is then:

Euler's method:

Evaluate $f(x, y)$ at the current point (x_0, y_0) and calculate (x_1, y_1) by

$$x_1 = x_0 + h \tag{17.3}$$
$$y_1 = y_0 + h.f(x_0, y_0).$$

One step of Euler's method is interpreted geometrically in Figure 17.2(a). Starting from the point P_0 on the curve at (x_0, y_0), we draw the tangent line. The point on the line where $x = x_1$ gives us the new point P_1 at (x_1, y_1).

Figure 17.2(b) shows how the iteration proceeds, producing the points P_2, P_3, and so on. Even though the points do not lie exactly on the curve, we can still evaluate the function $f(x, y)$ at them, and use this as the slope of the next straight line.

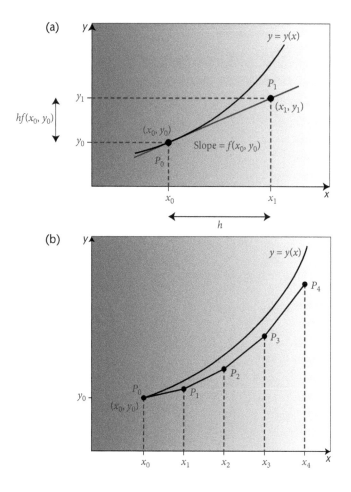

Figure 17.2 Euler's method

Here is a hand-calculation of the first three steps in solving the following IVP:

Differential equation: $\dfrac{dy}{dx} = (x + y)^2$

Initial condition: $y(0) = 1$ (i.e. $y = 1$ when $x = 0$)

Domain: $x \geq 0$.

So in this problem we take $f(x, y) = (x + y)^2$, $x_0 = 0$, $y_0 = 1$.
We choose our steplength to be $h = 0.1$.

Step 1:

Start at $(x_0, y_0) = (0, 1)$

Evaluate $f(x_0, y_0) = f(0,1) = (0 + 1)^2 = 1$

Update x: $x_1 = x_0 + h = 0 + 0.1 = 0.1$

Update y: $y_1 = y_0 + h.f(x_0, y_0) = 1 + 0.1(1) = 1.1$

New data-point is $(0.1, 1.1)$

Step 2:

Start at $(x_0, y_0) = (0.1, 1.1)$

Evaluate $f(x_0, y_0) = f(0.1,1.1) = (0.1 + 1.1)^2 = 1.2^2 = 1.44$

Update x: $x_1 = x_0 + h = 0.1 + 0.1 = 0.2$

Update y: $y_1 = y_0 + h.f(x_0, y_0) = 1.1 + 0.1(1.44) = 1.244$

New data-point is $(0.2, 1.244)$

Step 3:

Start at $(x_0, y_0) = (0.2, 1.244)$

Evaluate $f(x_0, y_0) = f(0.2,1.244) = (0.2 + 1.244)^2 = 1.444^2 = 2.085$

Update x: $x_1 = x_0 + h = 0.2 + 0.1 = 0.3$

Update y: $y_1 = y_0 + h.f(x_0, y_0) = 1.244 + 0.1(2.085) = 1.4525$

New data-point is $(0.3, 1.4525)$

You do not need to do these calculations by hand; they are easy to program into Excel. Look at SPREADSHEET 17.1 (Figure 17.3). The steplength is set to $h = 0.1$ in cell D19. The initial condition is typed in cells B22 ($x_0 = 0$) and D22 ($y = 1$). Then $f(x_0, y_0)$ is calculated in cell F22; the formula in the cell is: =(B22 + D22)^2 for this example. The next y-value is worked out using Euler's formula in cell H22; the formula in the cell is =D22 + $D$19*F22. We use the

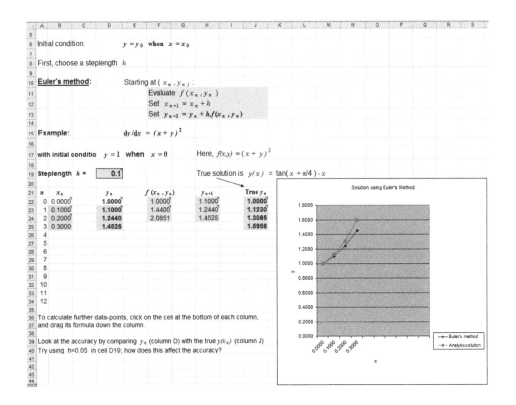

Figure 17.3 Programming Euler's method (SPREADSHEET 17.1)

dollar signs for the steplength cell to ensure this is unchanged when we drag the formulae down the column. This value of y_1 is copied to the row below (cell D23) and the new x_1 value is calculated in cell B23 by adding h to B22. These columns have been copied down to evaluate a further two data-points.

So, starting at $(x_0, y_0) = (0,1)$, we have calculated the data-points (0.1, 1.1), (0.2, 1.244), and (0.3, 1.4525). But how accurate are they? For this problem we do know the analytic solution: we found it using a change of variables to separable form in Section 16.2.4. The solution is

$$y(x) = \tan\left(x + \frac{\pi}{4}\right) - x. \tag{17.4}$$

If we evaluate this solution at $x = 0.3$, we find that $y(0.3) = \tan(1.0854) - 0.3 = 1.5958$. Compare this with the numerical estimate of 1.4525; the relative error in the numerical result is 9.9%. Recall from Section 2.6 that relative error is

$$\textbf{absolute error = true value − approximate value}$$

$$\textbf{relative error} = \frac{\textbf{true value − approximate value}}{\textbf{approximate value}} \tag{2.8}$$

Column J on SPREADSHEET 17.1 evaluates the analytic solution, and the graph of the analytic and numerical results is plotted. This lack of accuracy is quite disappointing, but might have been expected. You can see from Figure 17.2(b) that in cases where the analytic function increases ever more rapidly, the Euler construction will fail to keep up, and will fall increasingly far below it. The analytic solution (17.4) for this example is sketched in FNGRAPH 17.1. It has a vertical asymptote at $x = \frac{\pi}{4}$.

You can calculate further data-points by dragging the cell at the bottom of each column down the column. Do this to find the data-points up to (0.6, 2.9692), which is a long way from the analytic value of 4.7319.

The accuracy of Euler's method can be improved by reducing the steplength, i.e. taking shorter steps between data-points. Change the steplength in cell D19 to $h = 0.05$. Now we need to take 6 steps to reach $x = 0.3$. The numerical estimate at this point is $y = 1.5142$. The relative error is now 5.4%. Thus, by halving the steplength and taking twice as many steps to reach the same x-value, we have halved the relative error in y. A numerical method where there is this linear relationship between the steplength and the accuracy is called a **first-order method**. Note that this has nothing to do with the order of the differential equation. Euler's method is a first-order method.

PRACTICE

Perform a hand-calculation for three steps of Euler's method, with $h = 0.1$, for the differential equation

$$\frac{dy}{dx} = 2xy^2$$

with initial condition $y(0) = 1$.

Also use the method of separation of variables (Section 16.2.3) to find the analytic solution. Modify SPREADSHEET 17.1 to perform Euler's method on this problem.
Answer: see SPREADSHEET 17.2.

For problems where the analytic solution does not grow unboundedly, e.g. logistic growth or exponential decay, Euler's method performs much better. But there is a simple modification to Euler's method that can dramatically improve its accuracy, even in examples such as those above. We will now describe this modified method, and program it in Excel.

17.1.2 Heun's method

Euler's method performs badly when the value of $f(x_0, y_0)$, which is the gradient $\dfrac{dy}{dx}$ at the start of the interval $[x_0, x_1]$, is not a good approximation to the gradient over the whole of that interval. This occurs when the gradient changes rapidly within the interval. The modification to Euler's method, known as the improved Euler, modified Euler, or Heun's[1] method, uses an improved estimate of the gradient. The overall structure is the same as in Euler's method: we are given the differential equation $\dfrac{dy}{dx} = f(x, y)$ and the initial condition $y(x_0) = y_0$. We want to calculate y_1, the y-value at $x_1 = x_0 + h$.

Heun's method is a two-stage method. The first stage is to calculate the Euler estimate of y_1, which we will denote by y_1^e:

$$y_1^e = y_0 + h.f(x_0, y_0).$$

Now evaluate the function f at (x_1, y_1^e). This will be an estimate of the gradient $\dfrac{dy}{dx}$ at $x = x_1$, the right-hand end of the interval $x_0 \leq x \leq x_1$. We now have two gradients: $f(x_0, y_0)$ and $f(x_1, y_1^e)$. We get an improved estimate of the gradient in the interval by taking the average: $\frac{1}{2}\left[f(x_0, y_0) + f\left(x_1, y_1^e\right)\right]$, and we use this new gradient instead of $f(x_0, y_0)$ in the Euler step:

$$y_1 = y_0 + h.\frac{1}{2}\left[f(x_0, y_0) + f\left(x_1, y_1^e\right)\right].$$

The algorithm is illustrated in Figure 17.4. The Euler step takes us from P_0 to P_1^e. The line $P_1^e A$ has slope $f(x_1, y_1^e)$. Moving this line back to pass through P_0 gives the parallel line $P_0 B$. We now draw a third line through P_0 with a slope that is the average of the slopes of the previous two. The point on this line where $x = x_1$ is P_1, whose y-coordinate will be y_1.

Heun's method can be summarized as:

> **Heun's method:**
>
> Evaluate $f(x, y)$ at the current point (x_0, y_0) and calculate (x_1, y_1) by
>
> $$x_1 = x_0 + h$$
> $$y_1^e = y_0 + h.f(x_0, y_0)$$
> $$y_1 = y_0 + \tfrac{1}{2}h\left[f(x_0, y_0) + f\left(x_1, y_1^e\right)\right].$$
>
> (17.5)

Here is a hand-calculation of three steps using Heun's method on the example of the previous section, again with steplength $h = 0.1$:

---

[1] Karl Heun (1859–1929), German mathematician.

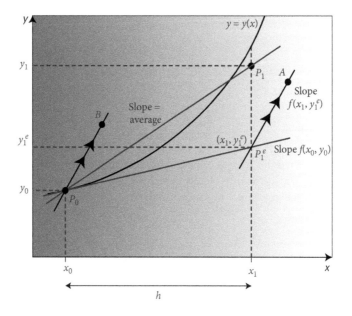

Figure 17.4 Heun's method

Step 1:

Start at $(x_0, y_0) = (0, 1)$

Evaluate $f(x_0, y_0) = f(0,1) = (0 + 1)^2 = 1$

Update x: $x_1 = x_0 + h = 0 + 0.1 = 0.1$

Euler estimate for y: $y_1^e = y_0 + h.f(x_0, y_0) = 1 + 0.1(1) = 1.1$

Evaluate $f(x_1, y_1^e) = f(0.1,1.1) = (1.0 + 1.1)^2 = 1.44$

Modified estimate for y:

$$y_1 = y_0 + \tfrac{1}{2}h\left[f(x_0, y_0) + f\left(x_1, y_1^e\right)\right] = 1 + 0.05[1 + 1.44] = 1.122$$

New data-point is (0.1, 1.122)

Step 2:

Start at $(x_0, y_0) = (0.1, 1.122)$

Evaluate $f(x_0, y_0) = f(0.1,1.122) = (0.1 + 1.122)^2 = 1.4933$

Update x: $x_1 = x_0 + h = 0.1 + 0.1 = 0.2$

Euler estimate for y: $y_1^e = y_0 + h.f(x_0, y_0) = 1.122 + 0.1(1.4933) = 1.2713$

Evaluate $f(x_1, y_1^e) = f(0.2,1.2713) = (0.2 + 1.2713)^2 = 2.1648$

Modified estimate for y:

$$\begin{aligned} y_1 &= y_0 + \tfrac{1}{2}h\left[f(x_0, y_0) + f\left(x_1, y_1^e\right)\right] \\ &= 1.122 + 0.05[1.4933 + 2.1648] = 1.3049 \end{aligned}$$

New data-point is (0.2, 1.3049)

Step 3:

Start at $(x_0, y_0) = (0.2, 1.3049)$

Evaluate $f(x_0, y_0) = f(0.2, 1.3049) = (0.2 + 1.3049)^2 = 2.2647$

Update x: $x_1 = x_0 + h = 0.2 + 0.1 = 0.3$

Euler estimate for y: $y_1^e = y_0 + h.f(x_0, y_0) = 1.3049 + 0.1(2.2647) = 1.5314$

Evaluate $f(x_1, y_1^e) = f(0.3, 1.5314) = (0.3 + 1.5314)^2 = 3.3539$

Modified estimate for y:

$$y_1 = y_0 + \tfrac{1}{2}h\left[f(x_0, y_0) + f\left(x_1, y_1^e\right) \right]$$
$$= 1.3049 + 0.05[2.2647 + 3.3539] = 1.5858$$

New data-point is $(0.3, 1.5858)$

Heun's method for this example is programmed in SPREADSHEET 17.3 (Figure 17.5). There are two extra columns in the table. Compare our y-value of 1.5858 with the analytic solution $y(0.3) = 1.5958$. The relative error (which was 9.9% using Euler's method) is now just 0.6%!

PRACTICE

Repeat the practice question of the previous section, using Heun's method.

Answer: see SPREADSHEET 17.4.

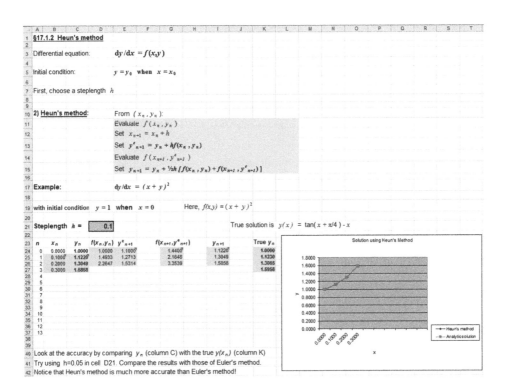

Figure 17.5 Programming Heun's method (SPREADSHEET 17.3)

Heun's method is a second-order method. This means that the relative error is proportional to the square of the steplength, so halving the steplength will improve the accuracy by a factor of 4. The Newton–Raphson method for finding roots was a second-order method, which meant that it converged to the true answer very rapidly. In our example, if you use SPREADSHEET 17.3 to take 6 steps with a steplength of 0.05, you will find the true value of $y(0.3)$, correct to 4 decimal places.

In Heun's method, we have done a little more work at each step (two evaluations of f, rather than one, which is insignificant if we are programming it in Excel!) and obtained a dramatic improvement in accuracy. There is another second-order method called the **midpoint method**; see the Problems. There are many other numerical methods, even more powerful than Heun's method; we will define the most popular of these in the next section.

CASE STUDY C18: Numerical solution of fish harvesting model

In the previous instalment of this Case Study I looked at the analytic solution of a fish stock harvesting model:

$$\frac{dx}{dt} = rx(1 - x) - kx$$

where $x(t)$ is a dimensionless measure of the fish population (we divide the population by the carrying capacity, to eliminate that parameter), r is the logistic growth rate and kx is the harvesting rate (proportional to stock size). A harvesting model that is even simpler to write down – though more difficult to solve analytically – is:

$$\frac{dx}{dt} = rx(1 - x) - k.$$

In this model, fish are harvested at a constant rate, irrespective of the stock size. It models a situation where there is no fisheries management policy, and fishing will continue unchecked until stocks disappear. One drawback of this simple model is that stocks can become negative.

In SPREADSHEET 17.5 I have modified SPREADSHEET 17.3 to solve this model numerically using Heun's method, with initial condition $x(0) = x_0$. Experiment with different values of r and k, and starting-points x_0. For a given value of r, there is a critical harvesting rate k_c: if $k > k_c$ fish stocks drop to zero in a finite time. See if you can deduce k_c as a function of r (the answer is in the next instalment).

If $k < k_c$ the fish stock approaches an equilibrium level over time, irrespective of what starting value you choose (provided it is not too small, making x become negative and the model break down). But there appear to be two equilibrium levels. For example, set $r = 2$ and $k = 0.18$. Then the stock level x tends to 0.9. If you start at a level below 0.1, the stocks go negative. But if you start at exactly $x_0 = 0.1$ the stock stays at that level; it is in equilibrium. However, starting nearby at $x_0 = 0.12$, say, the stocks do not fall slightly towards 0.1 but climb sharply towards 0.9! The equilibrium at $x = 0.1$ is an **unstable equilibrium**. We will explore this important concept in the final chapter.

You should also try modifying SPREADSHEET 17.5 to solve the harvesting model in the first equation in this box, which we solved analytically in the previous instalment. Does this exhibit the same behaviour?

17.1.3 Runge–Kutta method RK4

The main class of numerical methods for solving first-order ODEs is the class of Runge–Kutta[2] methods. The method which is commonly used by scientists, is a fourth-order (and so, very accurate indeed) Runge–Kutta method known as RK4. It involves making four evaluations of $f(x, y)$ at each step. For reference, this is how it is defined:

RK4 method:

Evaluate $f(x, y)$ at the current point (x_0, y_0) and calculate (x_1, y_1) by

$$x_1 = x_0 + h$$

$$k_1 = f(x_0, y_0)$$

$$k_2 = f\left(x_0 + \tfrac{1}{2}h, y_0 + \tfrac{1}{2}hk_1\right) \tag{17.6}$$

$$k_3 = f\left(x_0 + \tfrac{1}{2}h, y_0 + \tfrac{1}{2}hk_2\right)$$

$$k_4 = f(x_0 + h, y_0 + hk_3)$$

$$y_1 = y_0 + \tfrac{1}{6}h\left[k_1 + 2k_2 + 2k_3 + k_4\right].$$

In the problems, you can try this out by programming it in Excel.

Problems: Now try the end-of-chapter problems 17.1–17.7

. . . .

17.2 Systems of first-order ODEs

So far, we have considered differential equations with one dependent variable and one independent variable. But in models of biological and chemical processes there are often more variables than this involved. Where there is more than one independent variable, the mathematical model will involve a partial differential equation (see Section 16.4). Or there could be a single independent variable (usually this is time t), but a number of different dependent variables. The model would involve an ordinary differential equation for each variable. We have already seen this in Case Study B, of angiogenesis in tumours. In this model, which we used to demonstrate algebraic techniques throughout Part I, the three dependent variables were:

- x_0, the density of healthy cells,
- x_1, the density of non-angiogenic cancer cells, and
- y, the density of angiogenic cancer cells in the tumour.

---

[2] Carl Runge (1856–1927), German mathematician and physicist. (He has a crater on the Moon named after him.) Martin Kutta (1867–1944), German mathematician.

Each of these variables changes with time, and in Part B8 of the Case Study, at the end of Section 11.1, we wrote down a system of three first-order ODEs defining the model. This system was of the form

$$\frac{dx_0}{dt} = f_0(x_0, x_1, y)$$
$$\frac{dx_1}{dt} = f_1(x_0, x_1, y)$$
$$\frac{dy}{dt} = f_2(x_0, x_1, y). \tag{17.7}$$

Because in general each equation involves all three dependent variables in the right-hand side function, they cannot be solved separately. It is a **coupled system**.

Sometimes the system can be uncoupled, possibly by using a simplifying assumption. For example, the Goldie–Coldman model of tumour growth (Wheldon 1988) is formulated mathematically in the system

$$\frac{dN}{dt} = \lambda N - \psi \gamma N$$
$$\frac{d\mu}{dt} = \psi \gamma N + \lambda \mu \tag{17.8}$$

where $N(t)$ and $\mu(t)$ are the number of cells in the tumour, and the number of these that have mutated to develop drug resistance. As the first equation involves the dependent variable $N(t)$ only, it can be solved in isolation, and the solution substituted into the second equation, which can then be solved for $\mu(t)$. (The system may be simplified because the mutation coefficient ψ is much smaller than the cell growth coefficient λ. This justifies dropping the second term from the right-hand side of the first ODE, so that the solution is $N(t) = N_0 e^{\lambda t}$.)

To describe analytical methods for solving systems of ODEs of the general form (17.7), we would need concepts from multi-variable mathematics (matrices and determinants). However, it is straightforward to apply numerical methods such as Euler, Heun, and RK4 to such systems. In the next section we will illustrate this, and discuss what information can be extracted from a system of ODEs, by looking at a model of predator–prey dynamics.

17.2.1 Lotka–Volterra models of predator–prey dynamics

Lotka–Volterra[3] models consist of a pair of ODEs describing the variation over time t of an isolated ecosystem consisting of a population of predators $Y(t)$ and a population of prey $X(t)$. This model is a system of the form

$$\frac{dX}{dt} = f(t, X, Y)$$
$$\frac{dY}{dt} = g(t, X, Y). \tag{17.9}$$

[3] Proposed independently in 1925/6 by Alfred Lotka (1880–1949), US physical chemist and statistician, and Vito Volterra (1860–1940), Italian mathematician and physicist. Volterra was forced to resign from his Chair in the University of Turin when he refused to sign an oath of loyalty to the Fascist government of Mussolini.

It is assumed that the prey form an essential part of the diet of the predators, and that all the prey caught is killed and eaten. We will use the example of a population $Y(t)$ of humans (the predators) on an island, and a population $X(t)$ of fish (the prey) in the surrounding sea. We know the initial population sizes X_0, Y_0, i.e. the initial condition is:

When $t = 0$, $X(0) = X_0$ and $Y(0) = Y_0$.

If the islanders are very lazy and/or very incompetent at catching fish, then there are no interactions and the system is an uncoupled one, namely:

$$\frac{dX}{dt} = rX$$
$$\frac{dY}{dt} = -sY$$

where r and s are positive coefficients. That is, the population of fish, in the absence of any predators, grows exponentially, and the human population decays to extinction due to starvation.

Now assume that the human population spends its time fishing. The number of predator–prey interactions (i.e. of humans-catching-fish events) depends on how many humans Y and how many fish X there are. We assume the amount of interaction is proportional to the product $X.Y$. These events will reduce the growth rate of the fish population, and increase the growth rate of the human population. The Lotka–Volterra system becomes

$$\frac{dX}{dt} = rX - aXY$$
$$\frac{dY}{dt} = -sY + bXY$$

(17.10)

involving positive interaction coefficients a, b, which will be much smaller than r, s. We cannot solve this system analytically for $X(t)$, $Y(t)$, but we can deduce significant information about the behaviour of the system.

The first question to ask is: is there an equilibrium state? That is, a state where the populations of humans and fish are in balance, and do not change over time. This would occur when $\frac{dX}{dt} = \frac{dY}{dt} = 0$. If you equate the two right-hand sides to zero, you will find that (apart from the trivial solution $X = Y = 0$) there is such an equilibrium:

$$\overline{X} = \frac{s}{b}, \quad \overline{Y} = \frac{r}{a}.$$

We can also combine the equations to produce an ODE that we can solve. If we take Y to be a function of X, and X to depend on t, the chain rule gives

$$\frac{dY}{dt} = \frac{dY}{dX}.\frac{dX}{dt}.$$

This can be rearranged to

$$\frac{dY}{dX} = \frac{\frac{dY}{dt}}{\frac{dX}{dt}}.$$

(17.11)

Substituting the right-hand sides of (17.10):

$$\frac{dY}{dX} = \frac{-sY + bXY}{rX - aXY} = \frac{Y(-s + bX)}{X(r - aY)}.$$

This is a separable ODE. Using the method of Section 16.2.3,

$$\int \frac{r - aY}{Y} \, dY = \int \frac{bX - s}{X} \, dX$$

$$\int \left(\frac{r}{Y} - a \right) dY = \int \left(b - \frac{s}{X} \right) dX$$

$$r \ln Y - aY = bX - s \ln X + c.$$

Whatever the time t, then, the populations $X(t)$ and $Y(t)$ will satisfy the equation

$$aY + bX - r \ln Y - s \ln X + c = 0 \qquad\qquad (17.12)$$

where c is a constant of integration. This equation describes a curve in the XY-plane. It cannot be rearranged to give Y explicitly as a function of X, so it cannot be sketched using FNGraph, but we could deduce the shape of the curve (for a particular value of c) by evaluating several points on it. To find such a point, choose a value of X and use the Newton–Raphson method to solve for Y. The XY-plane, shown in Figure 17.6(a), is called the **phase-plane**. It contains all possible states of the system: the state with a fish population X and a human population Y is represented by the point (X, Y) on the plane. The initial state is represented by the point (X_0, Y_0).

For a given value of c the equation produces a closed curve (a loop) like a distorted circle enclosing the equilibrium point $\left(\frac{s}{b}, \frac{r}{a} \right)$. Plotting the curves for different values of c produces a set of concentric curves around the equilibrium point, like contour lines around a mountain-top on a map. Typical curves are shown in Figure 17.6(a), from Hoppenstaedt (2006). The values of X_0 and Y_0 determine which of these curves is relevant to our particular problem. Then starting at the point (X_0, Y_0) on the curve that passes through that point, our population state $(X(t), Y(t))$ will move around the loop as time proceeds. This implies that the populations of predator and of prey will vary in cycles, with the same period. Importantly, they will not converge to an equilibrium situation.

But does the population state move clockwise or anticlockwise around the loop? We can deduce this by reasoning about the ecosystem. Consider the loop in Figure 17.6(b), where we start at point A with coordinates (X_0, Y_0). This is in the lower right-hand part of the loop, where X is large and Y is small, meaning that there is a relatively large population of prey, and a relatively small number of predators. There is plenty of food for the predators to eat, so we expect the predator population Y to increase, while the prey will still continue to breed, though more slowly. We thus move anticlockwise around the loop to point B, where the predator population is large enough to halt further growth in prey stocks. Thereafter, the predators continue to feed off the prey stocks and multiply, while those stocks reduce. At point C, the maximum predator population for the available food stocks has been reached. Then, as prey stocks continue to fall, the predator population can no longer be supported, and both populations fall until point D is reached. Now the predator population is so small that prey stocks have a chance to recover. And so we move around to point E, where there is sufficient food to produce an increase in predators again. And so the cycle repeats. The times corresponding to points A – E are marked on the population–time graphs of predators and prey in Figure 17.6(c), also from Hoppenstaedt (2006); we will produce such graphs shortly, in Excel.

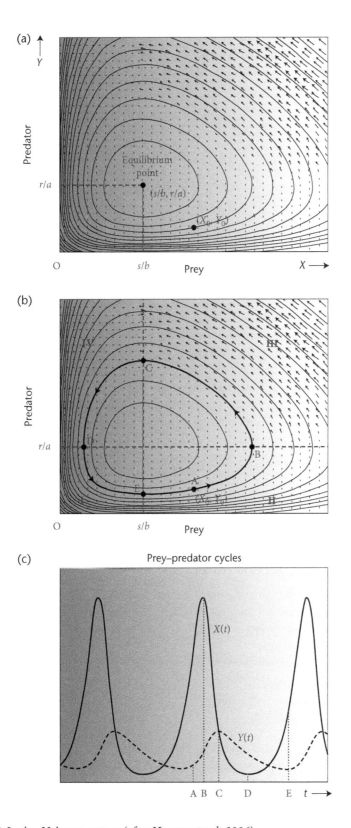

Figure 17.6 A Lotka–Volterra system (after Hoppensteadt 2006)

This descriptive cycle is independent of the particular loop that is followed. In fact, we can divide up the phase plane into four quadrants I, II, III, IV, see Figure 17.6(b) by horizontal and vertical lines through the equilibrium point, and define the behaviour that will occur at any state in each quadrant:

- I: prey increase, predators decrease: $\dfrac{dX}{dt} > 0,\ \dfrac{dY}{dt} < 0;$

- II: prey increase, predators increase: $\dfrac{dX}{dt} > 0,\ \dfrac{dY}{dt} > 0;$

- III: prey decrease, predators increase: $\dfrac{dX}{dt} < 0,\ \dfrac{dY}{dt} > 0;$

- IV: prey decrease, predators decrease: $\dfrac{dX}{dt} < 0,\ \dfrac{dY}{dt} < 0.$

This cyclical behaviour could be disrupted by a sudden external event. Returning to our fish/islanders scenario, the arrival of a fishing fleet from the EU or Japan, which hoovered up a large proportion of the fish stocks and then departed, would push the current population state to the left on the phase plane. The effect of this would be to push the cycle onto a larger or smaller loop, depending on whether we were in quadrants I/IV (Figure 17.7(a)) or II/III (Figure 17.7(b)) at the time. Note that if the curve touches the X or Y axes, one population becomes extinct and the whole ecological relationship is destroyed.

We have deduced a lot about the behaviour over time of the prey and predator populations, without solving the system (17.10). We can now do this, but we need to use numerical methods. It is straightforward to adapt the methods of the previous section to deal with systems of ODEs. For example, the Euler method applied to the general system (17.9) would be:

$$
\begin{aligned}
t_1 &= t_0 + h \\
X_1 &= X_0 + h.f_1(t_0, X_0, Y_0) \\
Y_1 &= Y_0 + h.f_2(t_0, X_0, Y_0)
\end{aligned}
\tag{17.13}
$$

where $h = t_1 - t_0$. To avoid the inaccuracies of Euler's method, we should use Heun's method or RK4, which are just as easy to adapt to systems. In SPREADSHEET 17.6 Heun's method is used to solve the Lotka–Volterra system (17.10) with parameters

$r = 1.0,\quad s = 0.8,\quad a = b = 0.001,$

and initial condition

$X_0 = 1600,\ \ Y_0 = 500.$

A steplength of $h = 0.2$ is used, and the values of X and Y calculated in steps from $t = 0$ to $t = 10.0$. The results are reproduced in Figure 17.8. Charts are drawn to plot $X(t)$ and $Y(t)$ against time (using the data in columns B, C, D), and also a plot of the phase plane using the data in columns C, D. You can see how the predator and prey populations vary in cycles, with the same period but out of phase, and how the phase plane chart corresponds to Figure 17.6(b). Compare this with our simple predator–prey model in Case Study C10 (Section 9.5).

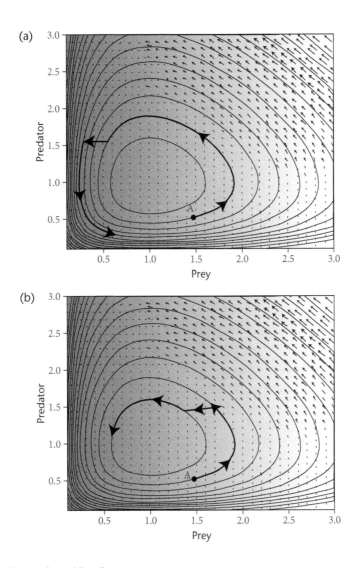

Figure 17.7 Effects of a sudden decrease in prey

You should experiment with this spreadsheet, adjusting the input parameters and the initial point. Try different initial populations, e.g. $X_0 = 400$, $Y_0 = 200$. Also see how the graphs change when you increase the effect of humans-catching-fish interactions by increasing a and b. Note: Excel automatically rescales the graphs to fit the data, so look at the axes to see the maximum and minimum values that the populations reach. Describe in words the effects of the change on the predator population, and on the prey population, and discuss whether these changes seem realistic. To make the effects of changing a parameter more apparent, you can create a slider to adjust the parameter value smoothly; see Problem 6.53 for instructions. You can also try increasing the steplength h. Doubling h means that you see three complete cycles. If you increase h further, round-off errors start to build up and the points in the phase plane no longer lie along a single loop.

The system (17.10) is the simplest model of predator–prey dynamics. Various modifications have been used to improve it, or to reflect the features of particular situations. One

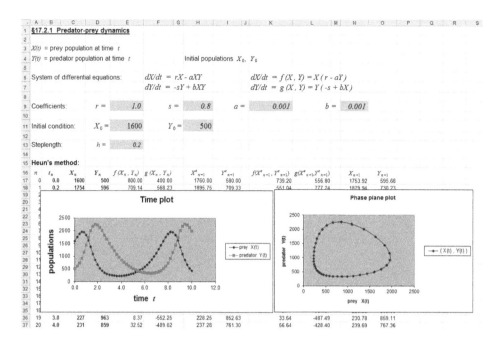

Figure 17.8 Solution of Lotka–Volterra equations (SPREADSHEET 17.6)

improvement to make it more realistic is to use a logistic growth model for the growth of the prey population, in place of exponential growth:

$$
\begin{aligned}
\frac{dX}{dt} &= rX\left(1 - \frac{X}{K}\right) - aXY \\
\frac{dY}{dt} &= -sY + bXY
\end{aligned}
\tag{17.14}
$$

where K is the carrying capacity (see (16.25) in Case Study A16). You can solve this numerically by modifying columns E and K of SPREADSHEET 17.6 (see Problem 17.10).

17.2.2 Kermack–McKendrick[4] model of epidemics

We have already remarked how very different natural processes can be governed by similar differential equations. We will see this if we now look at the basic model for the spread of an epidemic disease through an isolated population.

The population, of size N_0, is divided into three categories:

- $I(t)$ = infected members (those having the disease);
- $S(t)$ = susceptible members (not yet affected);
- $R(t)$ = resistant members (immune to the disease, having already contracted it and recovered).

---

[4] Anderson McKenrick (1876–1943), Scottish physician and epidemiologist. William Kermack (1898–1970) was a Scottish biochemist, who was blinded in a laboratory explosion in 1924.

The disease spreads by infected members coming into contact with susceptible members of the population. We assume that the number of these events is proportional to the numbers of people involved, i.e. $I \times S$, as with the islanders-catching-fish events in the Lotka–Volterra model. The other mechanism is the conversion of infected people to resistant ones, as they recover. This will reduce the infected population and increase the resistant population at the same rate. (If the disease is fatal, then $R(t)$ represents the population who have died, rather than recovered.)

The system of first-order ordinary differential equations is thus

$$\frac{dS}{dt} = \qquad -\beta IS$$
$$\frac{dI}{dt} = -\gamma I + \beta IS \qquad\qquad\qquad (17.15)$$
$$\frac{dR}{dt} = \gamma I$$

which looks very similar to the system (17.10). There are three equations rather than two, but notice that, by adding the three equations, the system implies that

$$\frac{dS}{dt} + \frac{dI}{dt} + \frac{dR}{dt} = 0$$

i.e.

$$\frac{d}{dt}(S + I + R) = 0$$

so it contains the constraint that the total population remains constant:

$$S(t) + I(t) + R(t) = N_0.$$

Therefore, to solve the system we could discard the third equation in (17.15), and solve the remaining pair for $S(t)$ and $I(t)$. These are identical to the Lotka–Volterra equations (17.10) with $a = \beta$, $b = \beta$, $r = 0$, and $s = \gamma$. For each solution point $(S(t), I(t))$ we could evaluate $R(t)$ from

$$R(t) = N_0 - S(t) - I(t).$$

Problems: Now try the end-of-chapter problems 17.8–17.10.

CASE STUDY A17: The peak of an epidemic

It is possible to deduce from the system (17.15) a single ODE in $R(t)$; here's how.
Solve the third equation for I, insert in the first equation and rearrange:

$$\frac{dS}{dt} = -\beta \left(\frac{1}{\gamma} \frac{dR}{dt} \right) S$$

$$\frac{dS}{dt} + k \frac{dR}{dt} S = 0$$

where we will write $k = \dfrac{\beta}{\gamma}$. The left-hand side looks something like the outcome of a product rule. Multiply both sides by the integrating factor e^{kR}:

$$e^{kR}\frac{dS}{dt} + ke^{kR}\frac{dR}{dt}S = 0.$$

Now the left-hand side is the result of differentiating $e^{kR}S$ using the product rule, and the chain rule to differentiate e^{kR}: $\dfrac{d}{dt}\left(e^{kR}\right) = \dfrac{d}{dR}\left(e^{kR}\right).\dfrac{dR}{dt}$. So, as we did in (16.32) in solving linear ODEs in Section 16.3, we can write

$$\frac{d}{dt}\left(e^{kR}S\right) = 0 \text{ so that } e^{kR}S = C$$

where C is a constant. Hence $S(t) = Ce^{-kR(t)}$. We can find C from an initial condition: at the start of the epidemic, no-one has recovered (or died), so $R(0) = 0$. Then $C = S(0) = S_0$, say. So we have

$$S(t) = S_0 e^{-kR(t)}.$$

Now use the fact that $S(t) + I(t) + R(t) = N_0$ to solve for $I(t)$, insert our solution for $S(t)$ and use this in the third equation of (17.15), producing the ODE:

$$\frac{dR}{dt} = \gamma\left(N_0 - R - S_0 e^{-kR}\right).$$

There is a numerical solution in SPREADSHEET 17.7. I will not attempt to solve this analytically, but I can extract some useful information using an idea from Chapter 13. In a fatal epidemic, $\dfrac{dR}{dt}$ represents the death rate from the disease. This will be at a maximum when its derivative is zero, i.e. when $\dfrac{d^2R}{dt^2} = 0$. Differentiating the ODE:

$$\frac{d^2R}{dt^2} = \gamma\left(-\frac{dR}{dt} + S_0 ke^{-kR}\frac{dR}{dt}\right)$$

$$= \gamma\frac{dR}{dt}\left(S_0 ke^{-kR} - 1\right).$$

So $\dfrac{d^2R}{dt^2} = 0$ when $S_0 ke^{-kR} = 1$, i.e. when $k.S(t) = 1$.

But now look at the second equation in (17.15):

$$\frac{dI}{dt} = \gamma I\left(-1 + \frac{\beta}{\gamma}S\right) = \gamma I(kS - 1).$$

So at the time when $\dfrac{dR}{dt}$ is at its maximum, we also have $\dfrac{dI}{dt} = 0$, meaning that $I(t)$, the number of infections, has reached a turning point. This is the **peak** of a fatal epidemic: the time when the death rate is at a maximum and there are the greatest number of infections.

In their 1927 paper, Kermack and McKendrick used their model with appropriate parameters to accurately model the peak of an epidemic of plague which occurred in Bombay in 1906.

17.3 Extension: analytic solutions

The theory of differential equations forms a huge topic in mathematics and engineering, with many books dedicated to the subject or just one particular aspect of it. Further study in this area is not 'core maths' but as a bioscientist you may have to read research papers where such advanced maths has been used. Or you may yourself be part of a team that produces a model that is a second- or higher-order ODE, or a system of ODEs, or a PDE, and need to consult a maths textbook on how to solve it. In this Extension section I'll give a glimpse of the theory in each of these areas.

17.3.1 Solving second-order ODEs

A general second-order linear ODE with constant coefficients, the second-order version of (16.31), would be

$$\frac{d^2y}{dx^2} + b\frac{dy}{dx} + cy = f(x) \tag{17.16}$$

where b, c are constant coefficients. We will consider the case where the right-hand side function $f(x)$ is zero. The first-order version of this is

$$\frac{dy}{dx} - ry = 0,$$

and we saw in Section 16.2.2 that the solution of this is

$$y = Ae^{rx}.$$

We might therefore expect that the solution of

$$\frac{d^2y}{dx^2} + b\frac{dy}{dx} + cy = 0 \tag{17.17}$$

will be an exponential function too. Try setting $y = Ae^{rx}$, where the coefficient r is unknown, find its first and second derivatives and insert in (17.17):

$$Ar^2e^{rx} + Abre^{rx} + Ace^{rx} = 0$$
$$A\left(r^2 + br + c\right)e^{rx} = 0$$
$$r^2 + br + c = 0 \quad \text{since } e^{rx} > 0.$$

The quadratic in r is called the **auxiliary equation** for (17.17). Finding its roots will give us two values of r, and thus two solutions to (17.17). To write the general solution, we take a linear combination of these two solutions.

EXAMPLE

Find the general solution of $\dfrac{d^2y}{dx^2} - 2\dfrac{dy}{dx} - 3y = 0$.

Solution: The auxiliary equation is $r^2 - 2r - 3 = 0$, which factorizes to $(r-3)(r+1) = 0$, so the roots are $r = 3$ and $r = -1$. The general solution is then

$$y = Ae^{3x} + Be^{-x}, \text{ where } A \text{ and } B \text{ are arbitrary constants.}$$

It is in this technique that mathematicians use the notation Dx for $\dfrac{dy}{dx}$, which we listed in the table of notations for the derivative, in Section 12.8. They would solve the above example by writing the ODE as

$$D^2 y - 2Dy - 3y = 0$$
$$\left(D^2 - 2D - 3\right)y = 0$$
$$(D - 3)(D + 1)y = 0.$$

The case where the two roots are equal, needs to be treated differently; we won't go into this. But what if there are no roots to the auxiliary equation, for example with

$$\frac{d^2 y}{dx^2} + y = 0?$$

Rewrite this as

$$\frac{d^2 y}{dx^2} = -y.$$

The solution will be a function whose second derivative is the negative of the function itself. We have seen such functions: sines and cosines.

If $y = \cos x$, then $\dfrac{dy}{dx} = -\sin x$, so that $\dfrac{d^2 y}{dx^2} = -\cos x = -y$.

If we now use time t for the independent variable, and introduce a coefficient ω (Greek letter omega), then the general solution of the differential equation

$$\frac{d^2 y}{dt^2} = -\omega^2 y \tag{17.18}$$

is a cosine function

$$y = A\cos\left(\omega\left(t - t_0\right)\right), \tag{17.19}$$

or alternatively

$$y = A\cos\omega t + B\sin\omega t.$$

This is an important equation describing **simple harmonic motion** (SHM). In SHM, $y(t)$ represents the displacement of an object at time t away from a fixed origin O. As force is proportional to acceleration, equation (17.18) is saying that there is a force pulling the object back towards O, which is proportional to the displacement away from O. This is what happens when a pendulum swings, with gravity pulling the bob back towards the centre-line. Comparing (17.19) with the general cosine function from Section 9.5.1:

$$f(t) = A\cos\left(\frac{2\pi}{T}\left(t - t_0\right)\right) + d, \tag{9.19}$$

we see that $\omega = \dfrac{2\pi}{T}$, where T is the period of one cycle. The parameter ω is the **frequency** of the oscillation. Engineers and scientists will often use the dot notation for the time derivatives in this application (see table in Section 12.8). That is, they write (17.18) as

$$\ddot{y} = -\omega^2 y.$$

A practical situation with harmonic motion is that of a tall tree swaying back and forth in the wind. The equation of motion would be

$$\frac{d^2 x}{dt^2} + \omega^2 x = F(t),$$

where $x(t)$ is the horizontal displacement of the tree-top, and ω is the natural frequency of the tree. The right-hand side function $F(t)$ comes from the varying force of the wind; it's called a **forcing function**. An engineering model based on this equation would be able to predict the relationship between wind force and the stress suffered by the tree roots, and hence the strength of a gale that would likely topple the tree.

In all the above, we have assumed that the coefficients b, c in (17.16) are constant. The more general second-order linear ODE, using x as the dependent variable and t as the independent variable, is

$$\frac{d^2 x}{dt^2} + p(t)\frac{dx}{dt} + q(t)x = f(t). \tag{17.20}$$

In such cases, there is a useful trick: we can transform the equation into a pair of first-order ODEs, such as we have been dealing with earlier in this chapter.

We invent a new variable y, defined by $y = \dfrac{dx}{dt}$. Then if we differentiate both sides with respect to t, we get $\dfrac{dy}{dt} = \dfrac{d^2 x}{dt^2}$. We can substitute for the derivatives of x in (17.20) and rearrange to

$$\frac{dy}{dt} = -p(t)y - q(t)x + f(t).$$

We thus have the first-order system

$$\begin{aligned}
\frac{dx}{dt} &= y \\
\frac{dy}{dt} &= -p(t)y - q(t)x + f(t)
\end{aligned} \tag{17.21}$$

which can hopefully be solved numerically.

Problems: Now try the end-of-chapter problems 17.11–17.20.

17.3.2 Solving first-order systems

So far we have only seen numerical solutions of first-order systems, so we had no idea what an analytic solution of a simple system might look like. From the previous section, you might now guess that it would involve exponential and trigonometric functions – and you would be right. We will not look at methods of solution, but just quote one example which you can check. The general solution of the system

$$\frac{dx}{dt} = -x - y$$
$$\frac{dy}{dt} = x - y$$

(17.22)

with initial condition $x = x_0, y = y_0$ when $t = 0$, is

$$x(t) = e^{-t}\left(x_0 \cos t - y_0 \sin t\right)$$
$$y(t) = e^{-t}\left(y_0 \cos t + x_0 \sin t\right).$$

(17.23)

You should check this by differentiating and inserting into each ODE of the system. Try plotting the functions in FNGraph (the result is in **FNGRAPH 17.2**).

You can also try solving the phase plane equation. As in Section 17.2.1, use

$$\frac{dy}{dx} = \frac{\dfrac{dy}{dt}}{\dfrac{dx}{dt}}$$

(17.11)

to obtain the ODE

$$\frac{dy}{dx} = \frac{x - y}{-x - y}.$$

(17.24)

If we invent the variable $u = \dfrac{y}{x}$, the right-hand side can be written as $\dfrac{1 - u}{-1 - u}$. The ODE is homogeneous, as discussed in Section 16.2.4. Your challenge is to solve it.

Problems: Now try the end-of-chapter problems 17.21–17.22.

17.3.3 Solving partial differential equations

Finally, we want to sketch the solution of a partial differential equation. Recall the example of a PDE quoted in Section 16.1.3:

$$\frac{\partial C}{\partial t} = k\left(\frac{\partial^2 C}{\partial x^2} + \frac{\partial^2 C}{\partial y^2} + \frac{\partial^2 C}{\partial z^2}\right).$$

(16.8)

This is a mathematical model of three-dimensional **diffusion** (also called the heat equation). Diffusion is a basic concept of physics which is a feature of biological and biochemical processes. Recall from Chapter 4 that a **solution** is made up by dissolving a certain amount of chemical (the **solute**) in a fluid medium (the **solvent**). If some of this solution is added to a new quantity of solvent, the solute molecules will gradually disperse throughout the solvent over time; this is the process of diffusion. This occurs when a medicine is swallowed and disperses into the contents of the stomach, or when a solute in the surrounding fluid enters a cell through its plasma membrane.

The diffusion process is measured experimentally by monitoring the mass concentration C of the solute throughout the fluid over time. The dependent variable C is then a function of location (given by the space coordinates (x, y, z) at the sampling point) and the time t when the sample is taken. Thus, C is a function of four independent variables: $C = C(x, y, z, t)$. Equation (16.8) is the mathematical model of three-dimensional diffusion, and k is the diffusion coefficient, which governs how rapidly the diffusion occurs. To simplify matters we may just consider diffusion in one space dimension x (for example, diffusion along a long narrow tube), in which case the equation reduces to

$$\frac{\partial C}{\partial t} = k\frac{\partial^2 C}{\partial x^2}. \tag{17.25}$$

A full solution of (17.25), using appropriate initial and boundary conditions, is beyond our scope, but we will indicate how a solution to the equation itself can be found by the mathematical technique of separation of variables, combined with some educated guesswork. We look for a solution that is a product of a function T of time only, and a function X of space only. That is, we suppose that a solution $C(x, t)$ can be written as

$$C(x,t) = X(x).T(t). \tag{17.26}$$

Recall from Section 16.4 that $\dfrac{\partial C}{\partial t}$, the partial derivative of C with respect to t, is found by differentiating C with respect to t while treating the other variable x as a constant. Thus, $X(x)$ would be treated as a constant coefficient multiplying $T(t)$, and the partial derivative would be

$$\frac{\partial C}{\partial t} = X(x).\frac{dT}{dt}.$$

Similarly, in the partial derivatives with respect to x, we treat $T(t)$ as a constant coefficient, and obtain:

$$\frac{\partial^2 C}{\partial x^2} = \frac{d^2 X}{dx^2}.T(t).$$

I have used the d notation, not ∂, in the derivatives of $X(x)$ and $T(t)$, since they are each functions of only one variable. Inserting these derivatives into (17.25) and rearranging as we did with separation of variables, to get the different variables on different sides of the equation, we arrive at

$$k\frac{1}{X}.\frac{d^2 X}{dx^2} = \frac{1}{T}.\frac{dT}{dt} = -\lambda, \text{ say.}$$

We can write this as a pair of equations:

$$k\frac{1}{X}.\frac{d^2 X}{dx^2} = -\lambda, \quad \text{and} \quad \frac{1}{T}.\frac{dT}{dt} = -\lambda$$

which is an uncoupled system of ODEs in x and t. What is the nature of λ? From the first equation, we see that λ is a function of x only, and not t. But from the second equation we conclude that λ is a function of t only, and not x. As x and t are independent variables, the only way of reconciling these two statements, is if λ is independent of both x and t; that is, λ **is a constant.**

This makes the two ODEs easy to solve.

The solution of $\dfrac{dT}{dt} = -\lambda T$ is $T = Ae^{-\lambda t}$, as we saw in Section 16.2.2.

The first equation is a second-order ODE: $\dfrac{d^2 X}{dx^2} + \dfrac{\lambda}{k} X = 0$ which we can solve by the method of Section 17.3.1. The auxiliary equation $r^2 + \dfrac{\lambda}{k} = 0$ has no roots, and the solution is thus the linear combination

$$ X = B\cos\left(\sqrt{\dfrac{\lambda}{k}}x\right) + C\sin\left(\sqrt{\dfrac{\lambda}{k}}x\right). $$

Inserting these solutions in (17.26), produces our solution for $C(x,t)$:

$$ C(x,t) = e^{-\lambda t}\left(B\cos\left(\sqrt{\dfrac{\lambda}{k}}x\right) + C\sin\left(\sqrt{\dfrac{\lambda}{k}}x\right)\right). \tag{17.27} $$

We don't need to include A, as B and C are already arbitrary constants of integration.

17.3.4 Further reading

Mathematical biology is the study of the application of ordinary and partial differential equations and difference equations (what we've called update equations) in the biological sciences. Mathematical biology textbooks, notably Adler (1997), Britton (2003), and Edelstein-Keshet (2005), pick up from the point at which this book ends. The principal topics include epidemiology, population dynamics, population genetics and evolution, cancer modelling, and molecular and cellular biology. I have introduced some of these topics in Case Studies throughout this book.

These last two chapters have presented the basic theory for the analytical and numerical solution of differential equations, but without much consideration of what these solutions look like or how they behave. In the final, Extension chapter I will introduce the important ideas of equilibrium and stability of solutions. The context will be a relatively new interdisciplinary field which is rapidly gaining importance in biology, as well as being fascinating in its own right: dynamical systems.

 PROBLEMS

17.1: Calculate by hand two steps of Euler's method to solve problem 16.20, namely $\dfrac{dy}{dx} = \dfrac{y}{x^2 - 1}$, starting at $x_0 = 3$, $y_0 = 2$, and with steplength $h = 2.0$.

17.2: Repeat Problem 17.1 but using Heun's method.

17.3: Find the particular solution for Problem 16.20 given the data-point (x_0, y_0). Hence work out the relative errors in the numerical estimates for $y(5.0)$ and $y(7.0)$, in your calculations for Euler's method and Heun's method above.

17.4: Modify SPREADSHEETS 17.1 AND 17.3, to perform Euler's and Heun's methods respectively on the ODE of problem 17.1. For each method, note down the value obtained for $y(7.0)$ when taking $h = 2.0$, $h = 1.0$, and $h = 0.5$, with the appropriate number of steps. Calculate the relative error in each value. You should see that the error in Heun's method is reduced by a factor of about 4 when you halve the steplength.

17.5: There is another modification of Euler's method called the **midpoint method**. In this we make a preliminary Euler step with a steplength of $\frac{1}{2}h$, i.e. to the midpoint of the interval from x_0 to x_1. We then evaluate the gradient $f(x, y)$ at this point, and use that gradient instead of $f(x_0, y_0)$ in the Euler step. The algorithm is thus:

Evaluate $f(x, y)$ at the current point (x_0, y_0) and calculate (x_1, y_1) by

$$x^m = x_0 + \tfrac{1}{2}h$$
$$y_1^m = y_0 + \tfrac{1}{2}h.f(x_0, y_0)$$
$$x_1 = x_0 + h$$
$$y_1 = y_0 + hf\left(x^m, y^m\right).$$

Perform three steps of the midpoint method to solve the example ODE in Section 17.1, namely

$$\frac{dy}{dx} = 2xy^2$$

with initial condition $y(0) = 1$, and taking a steplength $h = 0.1$.

17.6: Modify SPREADSHEET 17.3 (Heun's method) to perform the midpoint method for the example ODE in the previous problem. Hence compile a table showing the relative error in the calculation of $y(0.3)$ with steplengths of 0.1, 0.05, and 0.025. Does it appear from your results that the midpoint method is a second-order method?

17.7: Adapt SPREADSHEET 17.1 to use the RK4 algorithm (17.6) to solve the example problem of Section 17.1.

17.8: Write down an algorithm that uses the Newton–Raphson method to find points (X, Y) on the Lotka–Volterra curve

$$aY + bX - r\ln Y - s\ln X + c = 0 \tag{17.12}$$

using the approach suggested in Section 17.2.1. The value of c should be calculated using the values at the starting-point (X_0, Y_0). Use your spreadsheet, with the values of a, b, r, s, X_0, Y_0 from the example in Section 17.2.1, to calculate the upper and lower points on the phase-plane curve when $X = 500$, 1000 and 1500. What happens to the iteration if you enter $X = 2000$? Explain!

17.9: Write out the algorithm for implementing Heun's method to solve the system of two first-order ODEs (17.9). Check your formulas against those programmed in SPREADSHEET 17.6. Change the starting-point on the spreadsheet to $X_0 = 400$, $Y_0 = 200$, and discuss the effect.

17.10: Modify SPREADSHEET 17.6 to use a logistic growth model for the prey population, i.e. equations (17.14). You will need to add a new input value for the carrying capacity K. Then change the definition of $f(X, Y)$ in cells E17 and K17, and copy these down the columns. Use the initial point $X_0 = 400$, $Y_0 = 200$ as suggested above, with $K = 2000$. When K is large the graphs should appear as before; experiment with the effect of reducing K.

17.11–17.14: Find the general solutions of the following second-order differential equations:

17.11 $\dfrac{d^2 y}{dx^2} - \dfrac{dy}{dx} - 2y = 0$ 17.12 $\dfrac{d^2 y}{dx^2} - 7\dfrac{dy}{dx} + 12y = 0$

17.13 $\dfrac{d^2 y}{dx^2} - 4y = 0$ 17.14 $\dfrac{d^2 y}{dx^2} + 4y = 0.$

17.15–17.18: Find the particular solutions for Problems 17.9–17.12, using the boundary conditions that at $x = 0$ we have $y = 0$ and $\dfrac{dy}{dx} = 1.$

17.19–17.20: Rewrite each of the following second-order ODEs as a pair of first-order ODEs, in the form (17.21):

17.19 $\dfrac{d^2 x}{dt^2} + t\dfrac{dx}{dt} + t^2 x = \sin t$ 17.20 $\dfrac{d^2 x}{dt^2} - t^3 x = \cos t.$

17.21: Check that (17.23) is a solution to the pair of ODEs (17.22).

17.22: Find the solution of (17.24).

Extension: dynamical systems

At the end of the previous chapter we indicated the directions in which the classical theory of differential equations proceeds: essentially, more complicated ODEs and PDEs are used in more complicated mathematical models, and solved by more complicated analytical and numerical techniques to predict more and more complicated behaviour. This branch of applied mathematics has been developing steadily since the nineteenth century (the Navier–Stokes equations describing three-dimensional fluid flow date from 1822).

But in the 1960s and 1970s a radically different mathematics of **dynamical systems** (systems in which the processes are time-dependent) was born. It was discovered that even the simplest nonlinear mathematical models could give rise to incredibly complex behaviour, and it was scientists studying biological and physical processes of Nature who were in the forefront of this revolution in mathematics. Today, many research universities have an interdisciplinary grouping of mathematicians, physicists, biologists, and engineers who meet to discuss the latest findings in and applications of the new science of **complex systems**.

In this final chapter we will pull together mathematics from throughout this book to introduce some basic concepts of complexity theory: equilibria and stability (Section 18.2), bifurcations (Section 18.3), and chaos (Section 18.4). We start with a brief snapshot of the birth of chaos theory. For popular accounts of chaos theory, do get hold of Glieck (1988), and particularly Gribbin (2004), who concentrates on the biological applications, in particular theories of pattern formation and ecological implications.

18.1 The butterfly effect

On a winter day in 1959 Edward Lorenz, a meteorologist, was running his weather model on his new computer (in those days, desk-sized rather than desktop). The program solved numerically a simple system of differential equations modelling the motion of air currents in the atmosphere. In the early days of computers in the 1950s, it was the fashionable belief that as computers grew ever larger and more powerful, they would be able to 'number-crunch' more and more data to produce more and more accurate solutions to more and more difficult questions.[1] Weather forecasting was a prime example: there was confidence that with future generations of super-computers we would be able to accurately predict the weather not just days but weeks and even months ahead.

18.1.1 The birth of a new science

Lorenz wanted to repeat and extend a previous run of his program with the same initial data, but to save time he typed in the output from midway through the previous run. This should have produced identical results to the previous run before continuing to the extended time interval. Perhaps not identical, because the computer stored its data to a precision of 6 decimal places (called **single precision**, see Section 2.1.5), but his program only printed its output correct to 3 decimal places. He typed in the rounded values – we discussed the build-up of rounding errors in Chapter 2 – but he expected that this would make no significant difference to the predictions. In fact, while the results from the two runs did look virtually identical in the early stages, the two graphs soon started to get out-of-phase. Not only that, but the new run began to predict a totally different **pattern** of weather. Lorenz compares this behaviour to the accurate predictions that can be made of future astronomical events, e.g. the precise moment of the next solar eclipse, or of tidal flow (we saw such tidal predictions in Section 9.5.2): 'I used to think of tide forecasts as statements of fact – but of course, you are predicting. Tides are actually just as complicated as the atmosphere. Both have periodic components ... With tides, it's the predictable part that we're interested in, and the unpredictable part is small, unless there's a storm. The average person, seeing that we can predict tides pretty well a few months ahead would say, why can't we do the same thing with the atmosphere, it's just a different fluid system, the laws are about as complicated. But I realized that *any* physical system that behaved non-periodically would be unpredictable.' (Glieck 1988)

The phenomenon that puzzled Lorenz that day has since become popularly known as **The Butterfly Effect** (after Lorenz's 1972 conference paper 'Predictability: does the flap of a butterfly's wings in Brazil set off a tornado in Texas?').

18.1.2 Numerical experiments

Here are the Lorenz equations (Strogatz 1994):

$$\frac{dX}{dt} = \sigma(Y - X)$$
$$\frac{dY}{dt} = rX - Y - XZ \tag{18.1}$$
$$\frac{dZ}{dt} = XY - bZ.$$

This system of three first-order ODEs models the movement of a particle in three-dimensional space; initially its three-dimensional coordinates are (X_0, Y_0, Z_0) and at time t it is at the point $(X(t), Y(t), Z(t))$. They are a huge simplification of the Navier–Stokes equations for three-dimensional fluid flow, and Lorenz used them to predict how convection currents form in the

---

[1] Even, some thought, the Great Question of Life, the Universe and Everything.

atmosphere. There are three physical parameters b, σ, and r, of which the most important is r, called the Rayleigh number.

Like the Lotka–Volterra and Kermack–McKendrick systems we met in Section 17.2, this is a nonlinear system. The nonlinearity comes from the $-XZ$ term in the second equation, and the XY term in the third one. I have programmed this system into SPREADSHEET 18.1, solving it numerically using Heun's method. A run that took Lorenz many hours on his computer, we can now perform instantaneously in Excel (albeit for the movement of a single particle). The initial coordinates are set in cells G11, K11, O11 (see Figure 18.1). The parameters that Lorenz was using that day are $\sigma = 10$, $b = \frac{8}{3}$, $r = 28$; these are set on Row 9.

It is the Rayleigh number r that drives the behaviour. Use the spreadsheet to try the following experiments; the graphs in Figure 18.2 were obtained with a starting-point of (1, 1, 0). (a) If you try $r = 10$ all three coordinates oscillate with rapidly decreasing amplitude and soon settle down to equilibrium values. (b) With $r = 20$ something similar seems to be happening, on a longer timescale. (c) With $r = 23$ the oscillations die down but soon start increasing again. (d) And at $r = 24$ these increasing oscillations suddenly burst into complex, non-periodic behaviour. It is a feature of complex systems that 'past results should not be taken as a guide to future performance', as the investment companies warn us. A system can be meandering along regularly for some time, and then unpredictably swing into wild movement. This is a characteristic also observed in some biological and physical systems. It was previously assumed that such behaviour must be due to a sudden change in the governing equations or boundary conditions. What we are now seeing in SPREADSHEET 18.1 is that such complex behaviour (called **intermittency**) can arise from an apparently simple system.

Try increasing the Rayleigh number further, until you reach Lorenz's value of $r = 28$ shown in Figure 18.1. Stare at the graph while you make a small change to the initial coordinate X_0.

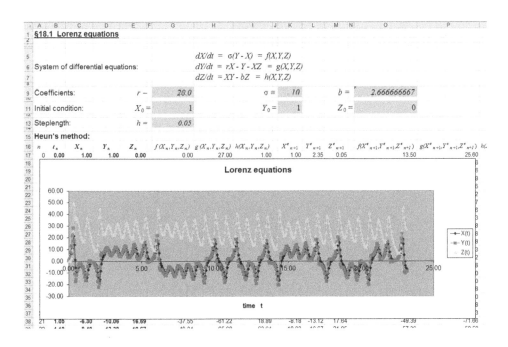

Figure 18.1 Chaotic behaviour of the Lorenz equations (SPREADSHEET 18.1)

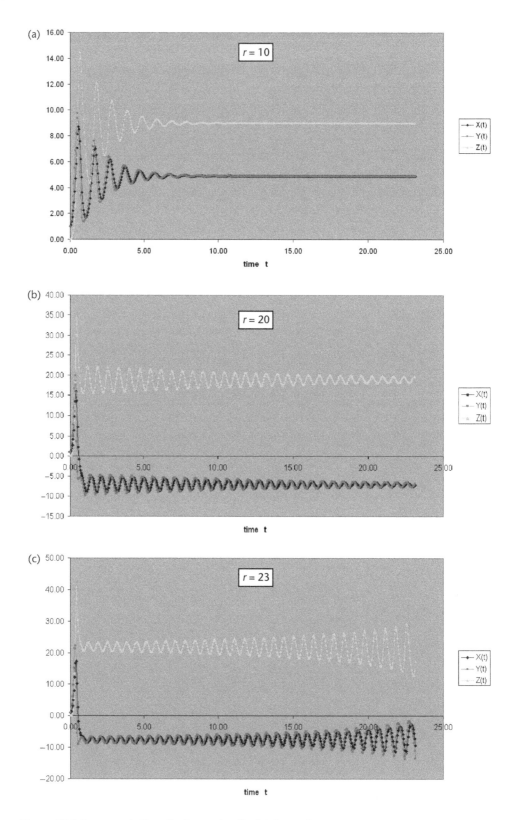

Figure 18.2 Lorenz solutions for increasing Rayleigh numbers

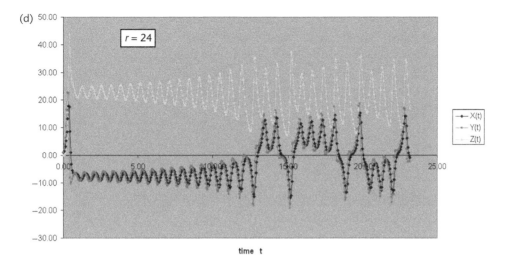

Figure 18.2 (Continued)

Increase it from 1.0 to 1.001 (a 0.1% change). The new graph is in Figure 18.3. While the two graphs look almost identical up until $t = 16.0$, they are thereafter completely different. This is what so surprised Lorenz.

Mathematicians were aware of the danger of large numbers of tiny rounding errors building up in a long sequence of computer calculations. Another source of error in numerical methods is **truncation error**, due to the method approximating the true function by a Taylor series that is truncated after the first two or three terms. You can see the effect of such errors if you increase the steplength h too far. If you used $h = 0.6$ or 0.8 in SPREADSHEET 17.6

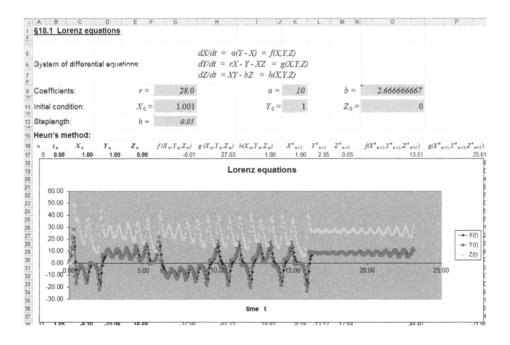

Figure 18.3 Lorenz solutions for $r = 28$, $X0 = 1.001$

solving the Lotka–Volterra system, you saw that the points on the phase plane drift away from the correct loop, but the general behaviour is still correct. More seriously, there is often an upper limit on h beyond which the method breaks down completely. You can see this in SPREADSHEET 18.1 if you increase h from 0.06 to 0.07. But the phenomenon that Lorenz – and we – are observing is quite different. For its explanation we need to study the ideas of stable and unstable equilibria, which I tried to sneak into Case Study C18 in the previous chapter. Have a look back at that, and try out the spreadsheet, before reading on.

. . . .

18.2 Equilibria and stability

In this section we will focus attention on the stability of points of equilibrium in a dynamical system. In the present context, the word 'system' is used to mean a mathematical, physical or biological process that changes with time; it may only involve a single differential equation.

18.2.1 Points of equilibrium for differential equations

We consider a one-dimensional dynamical system, in which a single quantity y varies with time: $y = y(t)$. The system is defined by a first-order differential equation and some initial data. In particular, we consider that class of initial value problem (17.1) in which the rate of change $\dfrac{dy}{dt}$ at any instant depends only on the current value of y and is independent of the time t. We can write this initial-value problem (IVP) as

$$\frac{dy}{dt} = q(y), \quad y(0) = y_0. \tag{18.2}$$

We have seen important practical situations that fall into this category in our Case Study of models of population growth:

- exponential growth: $\dfrac{dy}{dt} = ay$;

- logistic growth: $\dfrac{dy}{dt} = ay\left(1 - \dfrac{y}{K}\right)$.

We saw analytical techniques for solving such IVPs in Section 16.2.2. Even if we can't solve the IVP analytically, we can deduce information about the behaviour of the solution. We did this for the Lotka–Volterra equations (17.10) in Section 17.2.1. The first question we asked was: does the system have any **points of equilibrium**? This is important because if the system is going to settle down to a steady state, that state will be a point of equilibrium – and it's the steady states or long-term behaviour we are usually interested in.

A point of equilibrium is a state in which the system does not change with time. Such a steady-state condition occurs in (18.2) when $\dfrac{dy}{dt} = 0$, i.e. when $q(y) = 0$. Thus,

Equilibrium points of $\dfrac{dy}{dt} = q(y)$:

To find the points of equilibrium of (18.2), set $q(y) = 0$ and solve for y.

EXAMPLE

Find the points of equilibrium of

$$\frac{dy}{dt} = (y-1)(y-2), \quad y(0) = y_0. \tag{18.3}$$

Solution: Here, $q(y) = (y-1)(y-2)$. The solutions of $q(y) = 0$ are $y = 1$ and $y = 2$, so these are the points of equilibrium.

This means that if the initial condition is $y_0 = 1$ then we will have the steady-state solution $y(t) = 1$, and if the initial condition is $y_0 = 2$ then we will have the steady-state solution $y(t) = 2$. But one of these equilibria is stable and the other is unstable. We can see this if we look at the graphs of the solution when the starting-values are close to 1 and 2. First we need to solve (18.2). We have done this in the Problems of Chapter 16: separate the variables and apply integration

$$\int \frac{dy}{(y-1)(y-2)} = \int 1.dt.$$

You need to use partial fractions in the left-hand integral. The solution, when we use the initial condition $y(0) = y_0$ to determine the constant of integration, is

$$y(t) = 1 + \frac{y_0 - 1}{y_0 - 1 - (y_0 - 2)e^t}. \tag{18.4}$$

This is the function sketched in **FNGRAPH 18.1** (see Figure 18.4), for different choices of initial value, namely $y_0 = 0.5, 1.0, 1.5, 1.9, 2.0,$ and 2.1. These can be thought of as perturbations away from the points of equilibrium. You see that when we start at $y_0 = 0.5$ or 1.5, the curves of $y(t)$ approach the equilibrium at $y = 1$. But even if we start at $y_0 = 1.9$, very close to the equilibrium at $y = 2$, the curve still moves to the lower equilibrium point $y = 1$. Even more surprisingly, if we start at $y_0 = 2.1$, the curve moves away from the equilibrium values, but eventually approaches $y = 1$ from below. In fact, the only starting-value that will remain at the equilibrium point $y = 2$ is $y_0 = 2$. If there is even the smallest perturbation from this point, then the equilibrium is lost. This is the characteristic of an **unstable equilibrium point**. The equilibrium point $y = 1$, on the other hand, is a **stable equilibrium**. After a perturbation from this equilibrium, the system returns to it.

A physical example of a system in stable equilibrium is a pendulum hanging vertically down; if the pendulum bob is pulled to one side slightly and released (a perturbation), the pendulum swings back to its original equilibrium state. An example of a system in unstable equilibrium is a pencil balanced vertically on the desk; provided it remains precisely vertical it remains in equilibrium, but the slightest breath of wind and it falls over. There is no point in basing decisions on an unstable equilibrium that will never be attained in practice, so the obvious next question is: how do we tell if an equilibrium point will be stable or unstable?

18.2.2 Stability of equilibria for differential equations

A good example of a dynamical system with a stable equilibrium point is Newton's law of cooling. If an object such as a coffee cup whose temperature is T_0 is brought into a room with

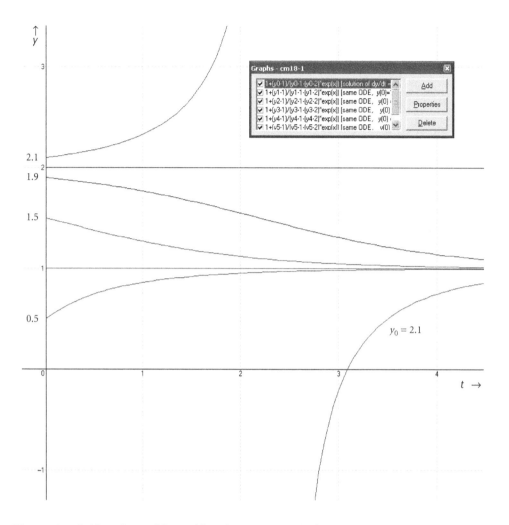

Figure 18.4 Stable and unstable equilibria (FNGRAPH 18.1)

a constant ambient temperature τ, then the IVP describing how the object's temperature $T(t)$ changes with time t, is

$$\frac{dT}{dt} = -k(T - \tau), \quad T(0) = T_0.$$

(18.5)

Here, k is a coefficient of thermal conductivity of the object.

The point of equilibrium is when $T = \tau$, i.e. the object is at the same temperature as the room. This is a stable equilibrium: if the object is initially hotter than room temperature it will cool down, and if it is initially cooler it will warm up. To express this mathematically, note that 'cooling down' means that the temperature T decreases with time, i.e. $\frac{dT}{dt} < 0$ (negative rate of change), and 'warming up' means that the temperature increases with time, i.e. $\frac{dT}{dt} > 0$ (positive rate of change).

Moreover, 'object hotter than room temperature' means $T > \tau$, and 'object cooler than room temperature' means $T < \tau$. So the mathematical expression of our statement defining a stable equilibrium is:

If $T > \tau$ then $\frac{dT}{dt} < 0$, and if $T < \tau$ then $\frac{dT}{dt} > 0$.

You can check that this is the case in (18.4), since $\dfrac{dT}{dt} = -k(T - \tau)$. So, for example,

$$T > \tau \Rightarrow (T - \tau) > 0 \Rightarrow k(T - \tau) > 0 \Rightarrow -k(T - \tau) < 0 \Rightarrow \dfrac{dT}{dt} < 0.$$

In general, suppose that an equilibrium point of (18.1) is at $y = Y$, and it is a stable equilibrium. Then

- if we apply a negative perturbation to y, so that $y < Y$, then $\dfrac{dy}{dt}$ should be positive so that y increases back towards Y, and
- if we apply a positive perturbation so that $y > Y$, then $\dfrac{dy}{dt}$ should be negative so that y decreases back towards Y.

But our differential equation is $\dfrac{dy}{dt} = q(y)$, so our condition for stable equilibrium is:

- if $y < Y$, then $q(y) > 0$ ($q(y)$ is positive when y is less than Y),   and
- if $y > Y$, then $q(y) < 0$ ($q(y)$ is negative when y is greater than Y).

Consider a graph of $q(y)$ plotted against y, as in Figure 18.5. As Y is an equilibrium point we know that $q(Y) = 0$, and for stability we now know that $q(y)$ passes from positive to negative as y moves from left to right along the horizontal axis, in an interval around Y. This means that the function $q(y)$ is **decreasing** over this interval, which means that the derivative $q'(y)$ is negative at $y = Y$ (see Section 13.1.1). Here, $q'(y)$ is the derivative of q with respect to y, which is the slope of the curve in Figure 18.5.

We can test this with Newton's law of cooling. We can write the ODE in (18.5) as $\dfrac{dT}{dt} = q(T)$, where $q(T) = -k(T - \tau)$. Then $q'(T) = \dfrac{dq}{dT} = -k < 0$, so the equilibrium is stable.

We can now expand our rule about the location of equilibrium points, as

Stability of equilibrium points of $\dfrac{dy}{dt} = q(y)$:

To find the points of equilibrium of (18.2), set $q(y) = 0$ and solve for y.

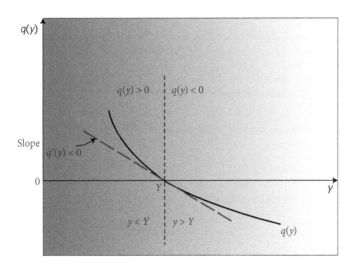

Figure 18.5 $q'(y) = 0$ at a stable equilibrium

An equilibrium point is **stable** if $q'(y) < 0$ at that point, and is unstable if $q'(y) > 0$.
Here, $q'(y)$ means $\dfrac{dq}{dy}$. (18.6)

EXAMPLE

Test the stability of the equilibrium points of

$$\frac{dy}{dt} = (y-1)(y-2), \quad y(0) = y_0.$$

Solution: Here, $q(y) = (y-1)(y-2)$. We have already found that the solutions of $q(y) = 0$ are $y = 1$ and $y = 2$, so these are the points of equilibrium.
Now differentiate $q(y)$. We must first multiply out:

$$q(y) = (y-1)(y-2) = y^2 - 3y + 2, \quad \text{so} \quad q'(y) = 2y - 3.$$

Then at the equilibrium point $y = 1$, $q'(1) = 2(1) - 3 = -1 < 0$, so this point is a stable equilibrium.
But at the equilibrium point $y = 2$, $q'(2) = 2(2) - 3 = 1 > 0$, so this point is an unstable equilibrium.
A sketch of the curves of $q(y)$ and $q'(y)$ is in FNGRAPH 18.2.

Using the same argument, make sure that you can solve the following question, for the ODE that we solved analytically in Case Study A16 in Chapter 16:

PRACTICE

Analyse the equilibrium points of the ordinary differential equations for logistic growth

$$\frac{dN}{dt} = r_m N\left(1 - \frac{N}{K}\right),$$ (16.25)

where $N(t)$ is the population size at time t, the initial population size is $N(0) = N_0$, K is the carrying capacity and r_m is the growth parameter ($r_m > 0$).

Answer: There is an unstable equilibrium point at $N = 0$, and a stable equilibrium point at $N = K$. This means that if you start with a population $N_0 > K$, $N(t)$ will descend towards K as $t \to \infty$, and if $0 < N_0 < K$ $N(t)$ will rise towards K – even if it starts off very close to 0. Only if $N_0 = 0$ will the population be zero.

A differential equation can have many different equilibrium points, and as we adjust the values of the parameters in the equation, these points will move, coalesce, separate, and disappear, so that not just the locations but also the number of stable equilibrium points changes. To see this, consider the harvesting model (Case Study C17):

$$\frac{dx}{dt} = rx(1-x) - k\frac{x^2}{a^2 + x^2}$$

CASE STUDY C19: Analysing the equilibria of the harvesting model

In the previous instalment (Section 17.1) we solved numerically the fish-stock harvesting model

$$\frac{dx}{dt} = rx(1 - x) - k.$$

We can now explain the behaviour we saw in SPREADSHEET 17.5.
Let $q(x) = rx(1 - x) - k$. Then solve $q(x) = 0$ for the points of equilibrium:

$$rx(1 - x) - k = 0$$
$$rx^2 - rx + k = 0.$$

This quadratic will have two roots if the discriminant $b^2 - 4ac$ is positive, i.e. $r^2 - 4rk > 0$ which is true if $k < \frac{1}{4}r$. If this condition is satisfied, the equilibrium points are at

$$x_1 = \tfrac{1}{2}\left(1 - \sqrt{1 - \frac{4k}{r}}\right)$$

$$x_2 = \tfrac{1}{2}\left(1 + \sqrt{1 - \frac{4k}{r}}\right).$$

The graph of $q(x)$ is a parabola that is concave down (Figure 18.6), so the function is increasing at the left-hand root x_1 and decreasing at the right-hand root x_2. This means that x_1 is an unstable equilibrium and x_2 is a stable equilibrium. In the example that I quoted in the previous instalment, taking $r = 2$ and $k = 0.18$, there is an unstable equilibrium at $x_1 = 0.1$ and a stable equilibrium at $x_2 = 0.9$.

If $k > \frac{1}{4}r$ there are no equilibrium points: we found that the fish stock drops to zero in a finite period of time.

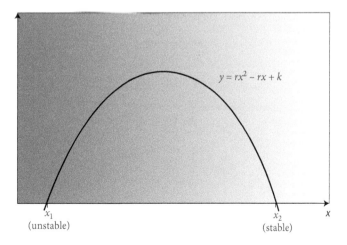

Figure 18.6 Equilibrium points of the harvesting model

in which there is an upper limit of k to the harvesting rate. Setting $\dfrac{dx}{dt} = 0$, we get

$$rx(1 - x) = k\frac{x^2}{a^2 + x^2}$$

so there will be an equilibrium point at $x = 0$ plus further points at any solutions of

$$r(1 - x) = k\frac{x}{a^2 + x^2}. \tag{18.7}$$

Graphically, these will be points where the line $y = r(1 - x)$ cuts the curve $y = k\dfrac{x}{a^2 + x^2}$. You should try sketching the line and curve yourself. They are drawn in SPREADSHEET 18.2 (Figure 18.7). In the spreadsheet, use the horizontal sliders to set values of k and a. Then gradually drag the vertical slider down, reducing the parameter r, and look at the number and locations of any intersection points. For instance, if $k = 1.6$ and $a = 0.13$, there is one intersection until r is about 7.25, then three points until r is down to about 6.0, then one point again. The ODE's behaviour will suddenly change at the states where $r = 7.25$ and $r = 6.0$. At these parameter values, two of the equilibrium points coalesce, or one point splits into two. These parameter states are called **bifurcations**, and are an important property of complex systems (example adapted from Strogatz 1994).

I have used the term 'parameter state' to mean a particular set of values of the parameters k, a, r. I'm treating the growth parameter r as the main parameter driving the equation's behaviour, but this also varies with the choice of k and a. Setting $k = 2.5$ and $a = 0.23$, for instance, there is only one intersection point (so two equilibria, remembering the equilibrium point at $x = 0$) whatever the value of r.

We introduced differential equations as models of population growth in Chapter 16, but throughout Part I of this book we have explored models of population growth defined not by differential equations but by update equations. Can the ideas of equilibrium points and stability be applied to update equations also?

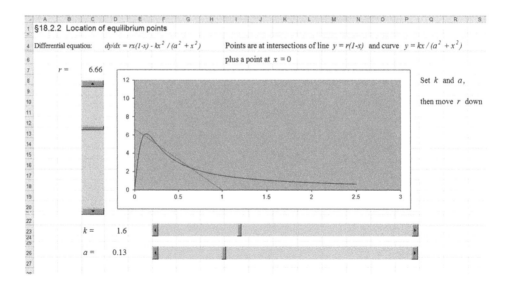

Figure 18.7 Graphical location of equilibrium points (SPREADSHEET 18.2)

18.2.3 Stability of equilibria for update equations

Yes, they can. In Case Study A8 in Section 7.4.2 we saw the update equation model of logistic growth:

$$N_{p+1} = N_p + r_m.N_p\left(1 - \frac{N_p}{K}\right). \tag{7.33}$$

An update equation defines the population N_{p+1} at the $(p+1)$th generation as a function of the previous generation population N_p. Such an update equation has the general form

$$N_{p+1} = f\left(N_p\right). \tag{18.8}$$

In the case of (7.33), the function $f(N)$ is $f(N) = N + r_m.N\left(1 - \frac{N}{K}\right)$.

Whereas the differential equation model (16.25) treats the population $N(t)$ as a variable changing continuously with time (and hence differentiable), the update equation model (7.33) treats the population as jumping from one discrete value N_p to the next N_{p+1}. However, note that the function $f(N)$ is a continuous function of N, and could therefore be differentiated.

In a steady-state situation, the population size remains constant from one generation to the next, i.e. $N_{p+1} = N_p$. This gives us the rule for finding equilibria of update equations:

Equilibrium points of $N_{p+1} = f(N_p)$:
To find the points of equilibrium of (18.8), set $N = f(N)$ and solve for N.

Equilibrium points of update equations are also called **fixed points** of the equation.

EXAMPLE
Find the equilibrium points of (7.33).
Solution: Setting $N_{p+1} = N_p = N$, gives

$$N + r_m.N\left(1 - \frac{N}{K}\right) = N$$

$$r_m.N\left(1 - \frac{N}{K}\right) = 0$$

the solution to which is either $N = 0$ or $\left(1 - \frac{N}{K}\right) = 0$, i.e. $N = K$.

What about stability? Suppose that N^* is a point of equilibrium of (18.8), that is N^* satisfies

$$N^* = f(N^*). \tag{18.9}$$

Now suppose we have a population N_p which is close to N^*. If N^* is a stable equilibrium, successive generations should approach N^*. In particular, the next-generation population N_{p+1} should be closer to N^* than N_p was. We can write this algebraically as

$$\left|N_{p+1} - N^*\right| < \left|N_p - N^*\right|. \tag{18.10}$$

(Remember that the absolute value $|a - b|$ is the distance between points a and b on the real line: Section 1.3.4.) We will use this to derive the criterion for stability, which will be given in the box below.

We know that $N_{p+1} = f(N_p)$, and we can expand $f(N_p)$ using a truncated Taylor series around $N = N^*$, as we did when deriving the Newton–Raphson iterative method: Section 13.5.1, equation (13.19):

$$N_{p+1} = f(N_p) = f(N^*) + f'(N^*).(N_p - N^*)$$
$$= N^* + f'(N^*).(N_p - N^*)$$

since $N^* = f(N^*)$. So

$$N_{p+1} - N^* = f'(N^*).(N_p - N^*),$$

and taking the absolute values of both sides

$$|N_{p+1} - N^*| = |f'(N^*)|.|N_p - N^*|.$$

Compare this with (18.8): $|N_{p+1} - N^*|$ will be smaller than $|N_p - N^*|$ if $|f'(N^*)| < 1$. The rule for stability of equilibria of update equations is:

Stability of equilibrium points of $N_{p+1} = f(N_p)$:

To find the points of equilibrium of (18.8), set $N = f(N)$ and solve for N.

An equilibrium point N^* is **stable** if $|f'(N^*)| < 1$ at that point, and is

unstable if $|f'(N^*)| > 1$. Here, $|f'(N)|$ means $\dfrac{df}{dN}$. (18.11)

This condition can be interpreted geometrically. Taking $f'(N^*)$ as the slope of the tangent to the curve $N_{p+1} = f(N_p)$ on the update graph, the stability condition $|f'(N^*)| < 1$ is saying that there is an acute angle between this tangent and the diagonal line $N_{p+1} = N_p$. Look at Figure 18.8, where the update parabola (7.33) is sketched for two values of the parameter r_m. For the lower curve, the angle θ_1 is less than 90° so the equilibrium point E_1 is stable. For the upper curve, with a greater value of r_m, the angle θ_2 is greater than 90° so the equilibrium point E_2 has become unstable.

EXAMPLE

Analyse the stability of the equilibrium points of (7.33).

Solution: We have seen that the points of equilibrium are $N^* = 0$ and $N^* = K$.

The update function is $f(N) = N + r_m.N\left(1 - \dfrac{N}{K}\right)$. Multiply out:

$$f(N) = N + r_m.N - \dfrac{N^2}{K}$$
$$= (1 + r_m)N - \dfrac{N^2}{K}.$$

Now differentiate with respect to N:

$$f'(N) = (1 + r_m) - \dfrac{2N}{K}.$$

At $N^* = 0$, $f'(0) = (1 + r_m) - \dfrac{2(0)}{K} = 1 + r_m > 1$, so this is an unstable equilibrium.

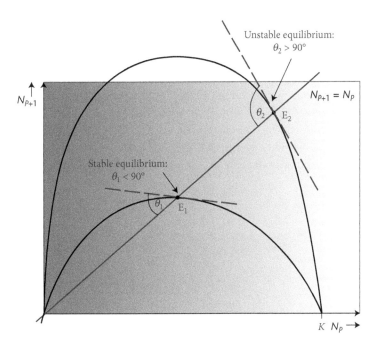

Figure 18.8 Geometric interpretation of the stability criterion

At $N^\star = K$, $f'(K) = (1 + r_m) - \dfrac{2K}{K} = 1 + r_m - 2 = r_m - 1$. So for stability we require $|r_m - 1| < 1$, meaning $-1 < r_m - 1 < 1$, or $0 < r_m < 2$. (We saw the algebra we have used here, in Section 1.8.) We conclude that the equilibrium point at $N^\star = K$ is stable if the growth parameter r_m is less than 2, but unstable when $r_m > 2$.

The rule in (18.11) can be applied to equilibria and stability of any iterative process

$$x_{k+1} = f(x_k), \quad k = 0, 1, 2, \dots$$

starting from x_0 and producing a sequence of iterates $x_1, x_2, x_3, \dots$. An example is the Newton–Raphson method for finding roots, in Section 13.5.1. We have referred to update equations like (18.8) as iterations.

We did experiment with the update equation (7.33), using different values of the growth parameter and carrying capacity, in part 8 of Case Study A. The algorithm was programmed into SPREADSHEET 7.5, and this is reproduced as the first sheet in the Excel file for Chapter 18. Experiment again with different values of r_m, and look at the graph plotting N_p against p. Look also at the 'cobweb graph', which plots the new value N_{p+1} against the previous value N_p; this is a form of phase-plane diagram. The diagonal line has equation $N_{p+1} = N_p$, and the parabolic curve has the equation (7.33). The equilibrium point is where the curve intersects with the line. The cobwebbing process is described in Section 7.4.2.

18.2.4 Numerical experiments with the update equation

Returning to SPREADSHEET 7.5, when $0 < r_m < 2$ the points certainly converge towards K. You can see that with a growth rate $r_m = 0.8$ the data-points (p, N_p) that are plotted, lie on a curve matching the differential equation solution; rapid growth initially, before slowing down and levelling off as the population approaches the maximum carrying capacity $K = 50$. But if you increase the growth rate, some strange behaviour occurs. Try the following experiments, typing the suggested values into cell B6 – and use your own ideas as well.

Figure 18.9 shows what you should see: Figure 18.9(a) shows the smooth convergence to a stable equilibrium for $r_m = 0.8$. For $r_m = 1.4$ you find that the population overshoots slightly above K before settling down. When $r_m = 1.8$ (Figure 18.9(b)) the population oscillates around K although it should converge to K eventually. But with a growth rate of $r_m = 2.2$, (Figure 18.9(c)) the population does not converge to a single level; there seems to be one

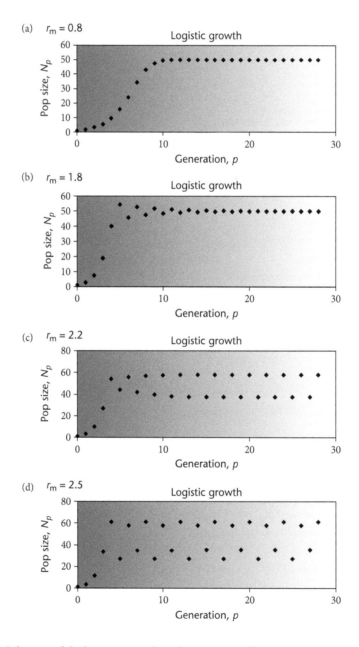

Figure 18.9 Solutions of the logistic growth update equation (from SPREADSHEET 7.3)

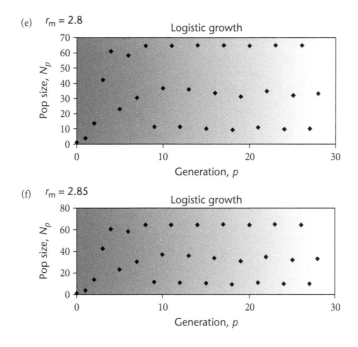

(e) $r_m = 2.8$

(f) $r_m = 2.85$

Figure 18.9 (Continued)

equilibrium pattern for the odd-numbered generations, and another pattern for the even-numbered generations.

For $r_m = 2.5$ (Figure 18.9(d)) each of the two equilibrium patters now has its own oscillation, and for higher values of r_m the iterates shoot around randomly, e.g. Figure 18.9(e), showing the iterates for $r_m = 2.8$. But try $r_m = 2.85$ (Figure 18.9(f))!

We see the same behaviour if we use a different starting-point, or a different carrying capacity. What is going on?

This is what Robert May,[2] a young Australian physicist-turned-biologist, tried to work out in 1971. He wondered if the behaviour he was seeing was related to the chaotic behaviour produced by systems of differential equations, which had been explored by mathematicians and physicists in the wake of Lorenz's published results. We will describe what he found in the next section.

18.3 Bifurcations ...

In designing an experiment you try to reduce the number of variables affecting the result, so May simplified (7.33) so that it involved just one parameter:

$$X_{p+1} = aX_p(1 - X_p) \tag{18.12}$$

where $a = 1 + r_m$.

2 Robert May later became the UK Government's Chief Scientific Advisor, and is now Lord May of Oxford.

Now we have a single parameter a controlling the nonlinear growth. The following discussion shows how the core maths you have learned in this book can be used in an advanced application. This is the sort of thing you will have to do as a bioscientist: apply your mathematics knowledge to new situations.

First, note that the quadratic function $aX(1 - X)$ has roots at $X = 0$ and $X = 1$, with the maximum occurring midway between them, at $X = 0.5$. The function takes the value $\frac{1}{4}a$ at this maximum. So provided we take a in the range $0 < a < 4$, we can be sure that if $0 < X_p < 1$ then $0 < X_{p+1} < 1$ also, since $0 < aX(1 - X) < 1$ for all X in this range.

To find the points of equilibrium, set $X_{p+1} = X_p = X^*$ in (18.12). One solution is $X^* = 0$, the other is

$$X^* = a.X^*(1 - X^*)$$
$$1 = a(1 - X^*) = a - aX^*$$
$$X^* = \frac{a - 1}{a}.$$

To check on the stability of these equilibria, write $f(X) = aX(1 - X)$ and differentiate: $f'(X) = a - 2aX$. Then $f'(0) = a$, and $f'\left(\frac{a-1}{a}\right) = a - 2a.\frac{a-1}{a} = a - 2(a - 1) = 2 - a$. Remember that an equilibrium point X^* is stable if $|f'(X^*)| < 1$. We conclude that:

- the equilibrium point $X^* = 0$ is stable if $0 < a < 1$, and unstable if $a > 1$;
- the equilibrium point $X^* = \frac{a-1}{a}$ is stable if $1 < a < 3$, and unstable if $a > 3$.

This tells us the convergence behaviour of the model, from an initial value between 0 and 1, depending on the value of the parameter a:

- If $0 < a < 1$, the population decays to 0.
- If $1 < a < 3$, the population converges to $\frac{a-1}{a}$.

The simple update equation (18.12) is programmed in SPREADSHEET 18.3 (see Figure 18.10). You can now change the value of the growth parameter a continuously, by dragging the button on the vertical scrollbar, or clicking on the arrows at either end. If you're comparing the graphs with those you obtained from SPREADSHEET 7.3, remember that the two growth parameters are not the same: $a = 1 + r_m$.

Cell D5 shows the value of a, cell B13 works out the value of $X^* = \frac{a-1}{a}$, and cell C13 gives the value of the derivative $f'(X^*) = a - 2aX^*$. Notice that $|f'(X^*)| < 1$ when $1 < a < 3$, and the iteration converges to X^*, as seen in the table of values of X_p and in the top graph. When we plot the cobweb graph, the equilibrium point is where the curve $X_{p+1} = f(X_p)$ intersects the diagonal line $X_{p+1} = X_p$.

If you increase the value of a beyond 3, the value of $f'(X^*)$ drops below -1, and this point of equilibrium becomes unstable. Try increasing a to 3.2. Now there are two equilibrium values. In odd-numbered generations the population is $X_5 = X_7 = X_9 = X_{11} = 0.513$, and in the even-numbered generations the population is $X_6 = X_8 = X_{10} = X_{12} = 0.799$. We will now explain this mathematically. Again, I am presenting the analysis as an example of applying the core mathematics you have learned, in an unfamiliar situation.

An equilibrium value that occurs in every second generation, is one where $X_{p+2} = X_p$. But $X_{p+2} = f(X_{p+1}) = f(f(X_p))$. That is, replace X by $f(X)$ everywhere in the right-hand side of (18.12). We'll write this function as $f^2(X)$. See Section 5.6.3: Composition of functions.

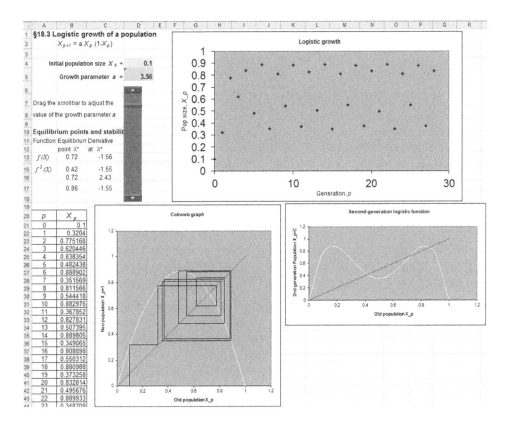

Figure 18.10 A simple dynamical system (SPREADSHEET 18.3)

If we write the second-generation function as $g(X)$ for simplicity, i.e. $X_{p+2} = g(X_p)$, and use x as the variable, then

$$g(x) = f^2(x) = f(f(x)) = a.f(x)(1 - f(x))$$
$$= a.[ax(1 - x)](1 - ax(1 - x)) \quad \text{substituting } f(x) = a.x(1 - x) \text{ twice}$$
$$= a^2 x(1 - x)(1 - ax + ax^2). \tag{18.13}$$

Now the second-generation update function $g(x)$ is a quartic function (polynomial of degree 4). It has factors of x and $(1 - x)$, so there will again be roots at $x = 0$ and $x = 1$. The other factor is the quadratic $(ax^2 - ax + 1)$. Does this have roots? Its discriminant (see equation 7.11) is $D = (-a)^2 - 4.a.1 = a^2 - 4a = a(a - 4)$. But if a lies between 1 and 4, then $(a - 4)$ is negative, so $D < 0$. Thus, the quadratic factor has no roots, and the quartic curve will only cross the x-axis at $x = 0$ and $x = 1$. Figure 18.11 and FNGRAPH 18.3 show the curves $y = f(x)$ and $y = g(x)$, together with the diagonal line $y = x$. The second-generation update graph also appears in SPREADSHEET 18.3 (in the lower right corner of Figure 18.10).

For $a = 3.2$, this curve $y = g(x)$ crosses the diagonal line $y = x$ in three places on the graph. The middle point is the original equilibrium solution for $f(x) : X^* = (a - 1)/a = 2.2/3.2 = 0.6875$. However, this equilibrium is unstable, meaning that if a population approaches it, it gets repelled away again in the next generation. It therefore does not appear in the graph of populations X_p.

The other two points of intersection represent the two second-generation equilibrium values of 0.513 and 0.799 that we observed in the table. These solutions are called **equilibrium solutions of period 2**, because they arise in every second generation. To find these points, we would need to solve the equation $x = g(x)$; the details are in the Problems. One solution is $X^* = \dfrac{a-1}{a}$, the equilibrium point for $f(x)$. The other points will be the roots of a quadratic; these equilibrium points only appear when the original equilibrium becomes unstable. You can see these equilibrium points appear in the update graph for $g(x)$, the lower right-hand graph in the spreadsheet. Once $a > 3$, the diagonal line $y = x$ cuts the quartic curve $y = g(x)$ in three places, corresponding to the three equilibrium points, as seen in Figure 18.11.

The stability of the three equilibrium points is found by testing $|g'(x)|$ at the points. This has been programmed into the cells in rows 15–17. The cells in column B give the equilibrium points, and those in column C evaluate the derivative. You see that while the middle point is unstable, the two new equilibrium points are stable. This explains the observed behaviour of the iteration, with two solutions of period 2. We have seen that the phenomenon of a single equilibrium point splitting into two at $a = 3$, is called a **bifurcation**.

As you increase the value of a further, the two equilibria move further apart. When a reaches about 3.45 (actually $a = 1 + \sqrt{6}$) these points also become unstable, since the values in cells C15 and C17 drop below −1. At this point there is a further bifurcation, into four separate stable equilibria of period 4. For example, if you set $a = 3.50$, the iteration settles down into a cycle: ..., 0.50, 0.87, 0.38, 0.83, 0.50, ...

Try increasing a further. Do you get more bifurcations, into eight equilibria of period 8, then sixteen, and so on?

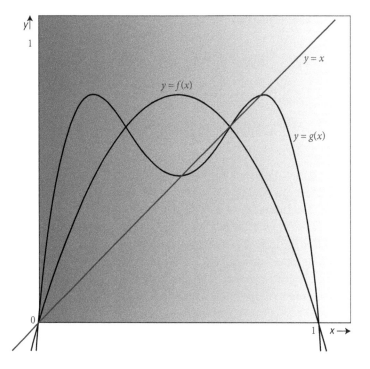

Figure 18.11 Logistic update curve $y = f(x)$ and second-generation curve $y = g(x)$

18.4 ... and Chaos

Robert May and his students spent a long time studying printouts and graphs of the iteration of (18.12) for different values of a, and he eventually published his findings in a review article in *Nature* in 1976, with the title 'Simple mathematical models with very complicated dynamics'. In it, he tried to display this dynamic process in a single graph. On the horizontal axis he plotted the growth parameter a, and on the vertical axis the stable equilibrium points. For a given value of a, we can find the equilibria by running the iteration (18.12) for 50 iterations (until it has settled into an equilibrium cycle) and then plot the next 50 iterations, i.e. plot (a, X_{51}), (a, X_{52}), ..., (a, X_{100}). When $1 < a < 3$ this produces just one point for each value of a, lying on the curve $X = \dfrac{a-1}{a}$.

In SPREADSHEET 18.4 (Figure 18.12) we have produced this plot (called a **bifurcation diagram** or orbit diagram) for values of a going from 2.60 to 3.95 in steps of 0.05. The chart was produced by initially selecting the data in columns B and C to display in an XY scatter chart; all the data-points appear in the same place. Then, with the chart selected, use `Chart > Add data ...` from the menu-bar, and add the data-points from columns D and E as a new series. Continue this process to add each set of iterations as a new series.

You can see the bifurcation at $a = 3.00$. Each arm then bifurcates at around 3.45, as we observed above. Already at $a = 3.55$ there is a further bifurcation, to eight equilibrium solutions. But beyond this point the pattern seems to disappear. We are now seeing **chaos**. This doesn't mean that the equilibria are scattered totally randomly: it looks like there is a deeper structure. In particular, at $a = 3.85$ there is suddenly a 'window' with equilibria of period 3. We saw this behaviour in Figure 18.9(f).

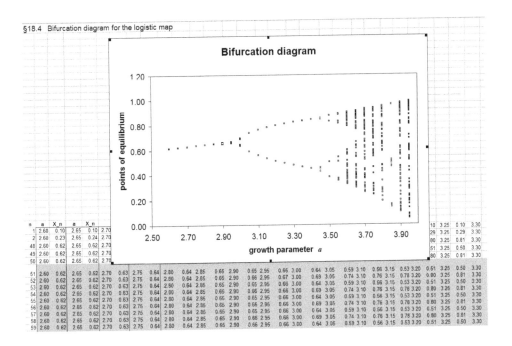

Figure 18.12 Bifurcation diagram in Excel (SPREADSHEET 18.4)

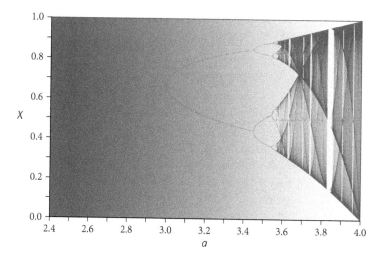

Figure 18.13 Bifurcation diagram for the logistic map

A much more detailed plot of this bifurcation diagram[3] is given in Figure 18.13. This has become a classic image of chaos theory (which you can easily produce for yourself if you know any computer programming[4]).

Now we can see that the 'chaos' (which starts when $a = 3.57$) actually has an amazingly complex structure. An infinitely complex structure, in fact: if you take a small part of the graph, such as the bottom of the 'period 3 window' starting at $a = 3.83$, and magnify it, you will obtain a picture reproducing the whole graph.

Chaos resembles randomness, but there is little that is random about it. The characteristics of chaotic behaviour are:

- an irregular oscillation within certain limits (not increasing unboundedly as with exponential growth);
- completely deterministic (all the values can be deduced from the rule and the initial conditions);
- very highly sensitive to initial conditions (the Butterfly Effect).

Chaotic behaviour is also observed in physical situations, such as turbulent flow. A simple mechanical system that can behave chaotically is a **double pendulum**.[5]

You can also see the Butterfly Effect in SPREADSHEET 18.3. Drag the scrollbar button all the way up, to set $a = 4.0$. Now try changing the value of X_0 from 0.1 to 0.10001, and observe the effect on the points in the upper graph. This tiny change (0.01%) in X_0 causes the later values of the iterates to change to a completely different pattern, as we saw with the Lorenz equations.

In fact, it is a characteristic of chaotic systems that small changes in initial data tend to increase exponentially with time. For the Lorenz equations (18.1) a small change δ_0 in one of the initial coordinates can cause a difference $\delta(t)$ in the position at time t, of size

$$\delta(t) = \delta_0 e^{\lambda t}$$

(18.14)

where $\lambda \approx 0.9$ (Strogatz 1994).

[3] The Wikipedia page: http://en.wikipedia.org/wiki/Logistic_map has animations and links to multimedia resources on chaos.

[4] There's a MATLAB code at http://en.wikipedia.org/wiki/Bifurcation_diagram.

[5] There is an interactive animation of such a pendulum at: http://www.myphysicslab.com/dbl_pendulum.html

You may have been surprised by the emphasis throughout this book on the issues of errors, precision, and accuracy. Here is the last bit of error analysis we will do, quantifying the Butterfly Effect:

Suppose you have a highly-realistic model of the atmosphere, which you program into a supercomputer to make weather predictions. The results will only be as good as the input data, so you obtain measurements of current wind speed and direction etc which are as precise as possible. You need an accuracy of size ε (Greek letter epsilon) in the output in order to make a plausible forecast. This limits how far ahead you can make predictions; you can predict up to t hours ahead, where we insert ε on the left-hand side of (18.14). Solving for t, this 'forecast horizon' is

$$t = \frac{1}{\lambda} \ln \left| \frac{\varepsilon}{\delta_0} \right| \tag{18.15}$$

where δ_0 is the precision of your measurements. For instance, if $\varepsilon = 10^{-3}$ and $\delta_0 = 10^{-6}$, we find that $t = \frac{1}{0.9} \ln \left| \frac{10^{-3}}{10^{-6}} \right| = \frac{3}{0.9} \ln 10 = 7.7$; we can predict up to 8 hours ahead.

Now suppose that over the next decade we hope to improve our meteorological instruments so much that they will produce measurements with a precision that is ten times greater than currently. For how much further ahead will we then be able to make predictions? Using $\delta_0 = 10^{-7}$ we get $t = \frac{1}{0.9} \ln \left| \frac{10^{-3}}{10^{-7}} \right| = \frac{4}{0.9} \ln 10 = 10.2$; we'll be able to predict up to 10 hours ahead.

So a tenfold improvement in precision results in only a 30% extension of the forecast horizon. And this is assuming we have a perfect computer model (no truncation errors) and an infinitely accurate computer (no rounding errors)!

This is why the efforts of Loonquawl and his colleagues to construct the second-greatest computer of all Time to solve the Great Question of Life, the Universe and Everything,[6] were ultimately futile.

18.5 Postscript

The preceding chapters have developed the mathematical theory which is used in all branches of science and technology to model the real world. We started with the real numbers and arithmetic, introduced variables and algebra, defined various types of function, then developed the calculus, ending with differential equations. This roughly parallels the way that 'classical' mathematics has developed historically. In all this there is an underlying assumption of continuity: variables changing continuously with time, or with respect to each other, systems moving smoothly from one equilibrium state to another. This classical mathematics has been remarkably successful in modelling the real world. Differential equations describe how the planets move and apples fall, how water and electricity flow, how bridges and aeroplanes stay up. They also describe many biological processes such as tumour growth. But even these differential equations can give rise to chaotic behaviour when more than two interacting

[6] See the Foreword, page xxi.

elements are involved. Poincaré[7] noted the phenomenon when trying to solve a system describing the movement of three celestial bodies with gravitational interaction, back in 1889. The Lotka–Volterra equations with three competitors can also produce chaotic behaviour.

In this context, mathematicians and physicists often regard difference equations (update equations) as approximations to the true differential equations, just as Euler's method (Section 17.1.1) is an update equation approximating a first-order ODE. But the biosciences are rather different from physics and chemistry in this regard.

A differential equation model of population growth such as (11.2) envisages the population changing continuously and smoothly over time. But there are cases where a discrete-time update equation model is a *more accurate* model of reality: for example, seasonally breeding populations where the generations do not overlap – this includes many economically important crop and orchard pests in temperate climates. Moreover, when mathematical biologists such as Robert May looked at recorded data on species populations over many years, they saw the irregular peaks and troughs that were familiar from chaotic systems. May concluded that 'apparently erratic fluctuations in the census data for an animal population need not necessarily betoken either the vagaries of an unpredictable environment or sampling errors: they may simply derive from a rigidly deterministic population growth relationship' (May 1976).

Epidemics also come and go in a cyclic but unpredictable fashion characteristic of chaos. Moreover, when a vaccination programme is introduced (which would be altering the growth parameter) the short-term effects may be unexpected. Researchers have turned to complexity theory for an explanation.

Current research in the application of chaos theory in biology is largely at the speculative stage: recognizing chaotic behaviour in observed or experimental data, without being able to determine the underlying differential or difference equations and their parameters. One natural phenomenon that is widely accepted to have a Butterfly Effect explanation is the snowflake. It is famously said that every individual snowflake has its own unique shape. How can we explain such an amazing diversity? The physical process of snowflake formation is relatively simple, but if it is a chaotic system, then the extreme sensitivity to initial conditions means that tiny differences in temperature and moisture during formation can produce very different snowflakes.

Parts of the body such as the lungs and the circulatory system have an incredibly complex structure, which medical scientists have compared to the fractal diagrams that come out of chaos theory. We have seen how a simple update equation can produce an infinitely complex structure as in Figure 18.13; can this help to explain the formation of such fine structures in the body? Much progress has already been made in explaining the formation of patterns on animals' skin.

Neuroscience is another area in which chaos theory has been applied. Researchers have interpreted experimental data to conclude that 'signals in the brain are distributed according to chaotic patterns at all levels of its various forms of hierarchy' (Korn and Faure 2003). The

---

[7] Henri Poincaré, 1854–1912, French mathematician and physicist, has been called the Father of Chaos Theory. In 1908 he anticipated the Butterfly Effect: 'If we knew exactly the laws of nature and the situation of the universe at the initial moment, we could predict exactly the situation of that same universe at a succeeding moment. But even if it were the case that the natural laws had no longer any secret for us, we could still only know the initial situation *approximately*. If that enabled us to predict the succeeding situation with *the same approximation*, that is all we require, and we should say that the phenomenon had been predicted, that it is governed by laws. But it is not always so; it may happen that **small differences in the initial conditions produce very great ones in the final phenomena**. A small error in the former will produce an enormous error in the latter. Prediction becomes impossible, and we have the fortuitous phenomenon.' (Poincaré 1914 quoted in Glieck 1988).

brain activity of patients with epileptic seizures is more regular and non-chaotic than in normal functioning. In a similar way, the heartbeats of healthy patients show a chaotic variation in pattern, and this chaos is reduced in cases of chronic heart failure. It seems that we are programmed to live with chaos. For more information and references, see the review article in *Nature* by Coffey (1998), and Gribbin (2004), as well as mathematical biology textbooks (Britton 2003, Edelstein-Keshet 2005). You can find more of the mathematics of chaos theory for update equations in Chapter 10 of Strogatz (1994).

? PROBLEMS

18.1: Find the equilibrium points of the harvesting model

$$\frac{dx}{dt} = rx(1 - x) - kx$$

and determine their stability.

18.2: By using the change of variable $X = \frac{r_m}{(1 + r_m)K} N$, show that the logistic growth equation

$$N_{p+1} = N_p + r_m N_p \left(1 - \frac{1}{K} N_p \right) \tag{7.33}$$

can be rewritten as

$$X_{p+1} = aX_p(1 - X_p) \tag{18.12}$$

where $a = 1 + r_m$.

18.3: Ludwig et al. (1978) proposed a model for the population $N(t)$ of spruce budworm (an insect pest on fir trees in Canada), involving the ODE:

$$\frac{dN}{dt} = RN\left(1 - \frac{N}{K} \right) - \frac{BN^2}{A^2 + N^2}.$$

The second term represents predation by birds. There are four parameters R, K, A, B. This number can be reduced by non-dimensionalizing the equation. We looked at dimensional analysis in Section 2.4. Two of these parameters are dimensionless (in units of 'numbers of insects') and two have dimension T^{-1} (or 'per unit time'). Decide which parameters have dimension T^{-1}. We invent the new variable x as $x = \frac{N}{A}$. We also introduce a new dimensionless time variable τ defined by $\tau = \frac{B}{A}t$. Use the chain rule to relate $\frac{dx}{dt}$ and $\frac{dx}{d\tau}$. Hence convert the ODE to the dimensionless form

$$\frac{dx}{d\tau} = rx\left(1 - \frac{x}{k} \right) - \frac{x^2}{1 + x^2}$$

defining the dimensionless parameters r and k (Strogatz 1994).

18.4: The second-generation update function was given in Section 18.3 as

$$g(x) = a^2 x(1 - x)(1 - ax + ax^2). \tag{18.13}$$

Show that the equation $x = g(x)$ for the fixed points can be rearranged to

$$\left(x - \frac{a-1}{a}\right)\left(ax^2 - (a+1)x + \frac{a+1}{a}\right) = 0.$$

Find the discriminant of the quadratic factor and deduce that the quadratic has two roots when $a > 3$ and no roots when $1 < a < 3$.

18.5: Modify SPREADSHEET 18.3 to use the following update equation in place of the logistic equation (18.12):

$$X_{p+1} = a\sin\left(\pi X_p\right)$$

where the growth parameter a is in the range $0 \le a \le 1$ and the iterates are again in the range $0 \le X \le 1$. Here's how to do it:

- To make the value of a range from 0.0 to 1.0 as you move the slider, change the formula in cell D5 to = (400-D8)/400.

- Program the iteration in cell B22 and drag it down the column. You enter π as PI().

- Delete the graph of the second-generation function. Underneath it you will find columns of x and y values which draw the yellow curve $y = ax(1 - x)$ on the cobweb diagram. Change the formula in cell L24 and drag down, so that the data define points on the curve $y = a \sin(\pi x)$.

- Also sketch the graphs of $f(x) = a \sin(\pi x)$ and the second-generation function $f^2(x)$ in FNGraph.

Now try varying a and explore the update equation's behaviour. You should find that it seems very similar to that of our original equation (18.12). In fact, if you produced the bifurcation diagram for this equation it would look virtually identical to Figure 18.13! A totally different iteration, involving a trigonometric function, produces the same complex structure, down to fine details. Strogatz (1994) has the two diagrams side-by-side on page 371. This amazing phenomenon is called **universality**.

Answers to odd-numbered problems

For problems marked with an asterisk *, there are Excel spreadsheet solutions on the ORC website.

Part I

Chapter 1

1.1 $-2 < -1.5 < \sqrt{2} < 3/4 < 7/8 < \sqrt{3}$

1.3 5/6

1.5 1/21

1.7 2

1.9 80/3

1.11 $\dfrac{y - x}{xy}$

1.13 $-\dfrac{(x - y)^2}{xy}$

1.15 0.0004073...

1.17 1

1.19 1

1.21 0.0130%

1.23

Period	Population	Immigration
1930–1935	3.39%	−85.54%
1935–1940	3.70%	102.86%
1940–1945	0.40%	−46.48%
1945–1950	14.63%	555.26%
1950–1955	8.69%	−4.42%

1.25 0.40% per annum for men, 0.36% per annum for women

1.27 0.39% per annum in shops, −11.61% per annum in pubs

1.29 −14

1.31 $ac - bc - ad + bd$

1.33 $ab - \dfrac{ac}{e} + \dfrac{ad}{e}$

1.35 $\dfrac{1}{2}\sqrt{\left|\left(\dfrac{x+5}{2} - 3\right)^2 - 25\right|} - 6$

1.37 $\sqrt{2\left(2\left(2(x+1)+1\right)+1\right)^2 + 1}$

1.39 3

1.41 70

1.43 -64

1.45 9.22×10^{18} (9 billion billion)

1.47 $\dfrac{b^4 - a^4}{ab^2}$

1.49 $\dfrac{a}{b}$

1.51 $y = -0.5$

1.53 $f \cdot g = 282.36, f = g = 16.8$

1.55 0.5

1.57 $\dfrac{\sqrt{a} + 6\sqrt{b}}{a - 4b}$

1.59 $\dfrac{1}{47}\left(37 - 12\sqrt{2}\right)$

1.61 $-\dfrac{1}{27}\left(44 + 7\sqrt{7}\right)$

1.63 $5 < T < 40$ where T is in Celsius

1.65 $n = 10a + b = 3\left[3a + \dfrac{a+b}{3}\right]$. The trick will only work for 3 and 9.

Chapter 2

2.1 3.12×10^4

2.3 -1.025×10^0

2.5 1.375×10^6

2.7 8.3306×10^8

2.9 4.19958×10^3

2.11 1.764×10^3

2.13 6.71×10^1

2.15 42 kN

2.17 1414.7 kPa or 1.4147 MPa

2.19 $N = 3.25514 \times 10^{79}$ atoms

2.21 350 µl or 3.5×10^{-4} litres

2.23 1.104×10^5 litres

2.25 $V = \dfrac{\pi}{4}\big(0.00216h^3 + 0.2725h^2 + 11.0714h + 146.432\big)\,$mm^3 when h is in cm

2.27 0.625 ms^{-1}

2.29 8.33 ms^{-2}

2.31 (i) $E = mgh$ Dimensions $M \cdot LT^{-2}\,L = ML^2\,T^{-2}$

(ii) $E = \dfrac{1}{2}mv^2$ Dimensions
$M \cdot (LT^{-1})^2 = ML^2\,T^{-2}$

(iii) as in (ii)

2.33 R has dimensions $ML^2\,T^{-2}\,K^{-1}$ (joules per kelvin) or kg m^2 s^{-2} K^{-1}

	2 decimal places	3 sig. figures
2.35	38.24	38.2
2.37	30.26	30.3
2.39	-0.10	-0.103

2.41 $p = 3.142 \pm 0.001$

2.43 $y = 0.9999 \pm 0.0001$

2.45 $v_1 = 10.32 \pm 0.01 \text{ ms}^{-1}$

2.47

	Absolute error	Relative error
Jedward	0.0005 g	0.0001
Britney	0.0005 g	0.02
combined	0.001 g	0.000199

2.49

	Relative error	Absolute error
Jedward	0.0251	6.275 g/l
Britney	0.07	0.0875 g/l
combined	0.0377	4.736 g/l

2.51 $10011 \rightarrow 18$

$111011 \rightarrow 59$

$42 \leftarrow 101010$

Chapter 3

3.1* linear

3.3* approx linear

3.5* approx linear

3.7 (a) 1579 mm (b) 82.35 mm

3.9 (a) 168.7 BPM (b) 225.6 BPM

3.11 15 824 wildebeest, 12 711 zebra, 84 981 gazelles. Each coexistent species will consume the same amount of food per km^2, independent of body weight.

3.13 328.27 ppm

3.15 383.51 ppm

3.17 408.48 ppm

3.19 2005.2, i.e. mid-March 2005

Chapter 4

4.1 17.5 moles

4.3 525.448 grams

4.5 264.97 grams

4.7 10.4 mM

4.9 3.704 M

4.11 20 mM

4.13 8.33 mM

4.15 234.375 mM

4.17 0.524 M

4.19 20 ± 0.5 mM

Chapter 5

5.1 $x = -\dfrac{b + d}{a - c}$

5.3 $x = y + \dfrac{2}{3}$ or $y = 0$

5.5 $x = \dfrac{1 - y^2}{3y - 2}$

5.7 (a) $x + 3$ (b) $6x^2 + x - 2$ (c) $6x - 1$
(d) $6x + 3$ (e) $12x + 1$

5.9 (a) $3x + 2 - \dfrac{1}{x}$ (b) $3 + \dfrac{2}{x}$ (c) $2 + \dfrac{3}{x}$

(d) $\dfrac{1}{3x + 2}$ (e) $6x + 4 + \dfrac{3}{x}$

5.11 (a) 2 (b) 11 (c) -6 (d) $9x + 2$
(e) $(3x + 2)^2$ (f) $3x^2 + 2$ (g) $6t + 5$ (h) $9x + 8$

5.13 (a) 1 (b) 13 (c) 21 (d) $9x^2 + 3x + 1$
(e) $(x^2 + x + 1)^2$ (f) $x^4 + x^2 + 1$ (g) $4t^2 + 6t + 3$
(h) $x^4 + 2x^3 + 2x^2 + 3x + 2$

5.15 $x = 5, y = 7$

5.17 $x = 1.211, y = 3.526$

5.19 $x = 106.8, y = 13.18$

Chapter 6

6.7 slope upwards, y-intercept $y = 2$, x-intercept $x = -0.4$

6.9 slope downwards, y-intercept $y = 3$, x-intercept $x = 3$

6.11 $y = 7x + 3$

6.13 $y = 0.3x + 2.5$

6.15 $y = x + 1$

6.17 $2x - 3y = 8$

6.19 $0\,°F = -17.8\,°C$, $100\,°F = 37.8\,°C$, $-40\,°F = -40\,°C$

6.21 $x + 5y = 36$

6.23 $y = -10x$

6.25 Vertical asymptote: $x = 0$, horizontal asymptote: $y = -5$

6.27 Vertical asymptotes: $x = 0$, $x = 2$, horizontal asymptote: $y = 0$

6.29 Straight line, no asymptotes

6.31 Vertical asymptote: $x = 0$, horizontal asymptote: $y = 1/4$

6.33 Vertical asymptotes: $x = -1/2$, $x = 1/2$, horizontal asymptote: $y = 0$

6.35 Shifted 4 units to left, 1 unit downwards and stretched vertically. Domain: $x \geq -4$

6.37 Shifted 1 unit upwards, and stretched horizontally. Domain: $x \geq 0$

6.39 $y = (x + 3)^2 - 1$

6.41 Shifted 6 units to right. Minimum at $(6, -4)$. Roots when $x = 4$, $x = 8$

6.43 Squashed horizontally. Minimum at $(0, -4)$. Roots when $x = -2/3$, $x = 2/3$

6.45 Shifted 8 units downwards, and flipped about x-axis. Maximum at $(0, -4)$. No roots

6.47 $(3, 5)$

6.49 $(0, -4)$

6.51 $(0, -4)$

Chapter 7

7.5 $f(x) = (x - 6)^2 - 15$

7.7 $(x - 3.5)^2 + 2.75$

7.9 $x = 1$ or $x = 4$

7.11 $x = -3$ (double root)

7.13 $x = -2.0972$ or $x = 1.4305$

7.15 $x = -3.8284$ or $x = 1.8284$

7.17 $y = 5x^2 + 3x - 2$

7.19 $y = -x^2 + 3x - 2$

7.21 $x_0 = 148.10$, $y_0 = -74.05$. He lands 165.6 metres down the slope.

7.23 They meet when $t = 2.158$ seconds; the lion is still accelerating, and catches the gazelle.

7.27* $\dfrac{N_{p+1}}{N_p} \to \dfrac{1 + \sqrt{5}}{2} = \Phi,$

$\dfrac{N_{p+1}}{N_{p-1}} \to \Phi + 1$, $\dfrac{N_{p+1}}{N_{p-2}} \to 2\Phi + 1$

7.29* $N_{12} = 236.27 \pm 8.84$, or $227.4 < N_{12} < 245.1$

7.31 $\dfrac{8}{3} = 2.6667$

7.33 Sum $= 2^{64} - 1 = 1.84467 \times 10^{19}$ grains, which would weigh 9.22×10^{11} tonnes

7.35 (i) $E_2 = 40$, $Y_2 = \$14,000$
   (ii) $E_3 = 57.143$, $Y_3 = \$11,430$
   (iii) $E_1 = 28.57$, maximum net profit is $\$71,142.50$
   (iv) For effort E_2, net profit is $\$6,000$, which is 19.0% lower that at the MEY

7.37 Here, we have assumed that the sequence converges to an equilibrium value. This is not assumed in 7.36

7.39 $y = 327.85$ in 1972 (328.27 by linear interpolation)

7.41 $y = 383.19$ in 2007 (383.51 by linear interpolation)

7.43 $y = 413.07$ in 2020 (408.48 using linear interpolation)

7.45 $P = 8283.5$

Chapter 8

8.1 $a + \dfrac{b}{x}$

8.3 $\dfrac{a}{k}x + \dfrac{b}{k} + \dfrac{c}{kx}$

8.5 (i) no x-intercept (ii) $(0, -1)$ (iii) $y = 0$ (iv) $x = 2/3$

8.7 (i) $(2/3, 0)$ (ii) $(0, 2/3)$ (iii) $y = 3/2$ (iv) $x = 3/2$

8.9 (i) $(-3, 0)$ and $(3, 0)$ (ii) $(0, 1)$ (iii) $y = x$ is a sloping asymptote (iv) $x = 9$

8.11 (i) $(9, 0)$ (ii) $(0, 1)$ (iii) $y = 0$ (iv) $x = -3$ and $x = 3$

8.13 Speeds are equal after 3.40 seconds

8.15* $y = -18.93\dfrac{1}{x} + 8.09$, $R^2 = 0.93$

8.17* $y = \dfrac{1}{-0.02x + 0.39}$, $R^2 = 0.66$

8.19* In 2000, $y = 10{,}285$

8.21 (i) $3.52 < P < 3.90$ kPa
(ii) $37.031 < P < 37.169$ Pa
(iii) $3.705 < P < 3.715$ Pa

8.23* $m = \dfrac{K_m}{v_{\max}}, c = \dfrac{1}{v_{\max}}$,

$v_{\max} = 47.81, K_m = 17.83$

8.25* $m = \dfrac{1}{v_{\max}}, c = \dfrac{K_m}{v_{\max}}$,

$v_{\max} = 5.082, K_m = 1.549$

8.27 Intercepts at $(3/2,0)$ and $(0,3)$
Horizontal asymptote $y = 2/3$
Vertical asymptote $x = 1/3$

8.29 $y = \dfrac{-bx}{ax - 1}$

8.31 $T = 1/9\,(5y - 160)$

8.33* After 4 iterations, $c = 1.1875$ using bisection; and $c = 1.1794$ using regula falsi

8.35 Need 10 steps to achieve precision of ± 0.0005

8.37 $m = \dfrac{\Sigma x_i y_i}{\Sigma x_i^2}$

Chapter 9

9.1 3.927 radians

9.3 −1.257 radians

9.5 36°

9.7 −42.97°

9.9 Amplitude = 3, period = π

9.11 Amplitude = 2, phase $\pi/4$

9.13* $x = 1.8956$

9.15 Tsessebe: $P_1(t) = 150\cos\left(\dfrac{\pi}{2}(t - 5)\right) + 1050$

9.17* Solve
$$150\cos\left(\dfrac{\pi}{2}(t-5)\right) + 70\cos\left(\dfrac{2\pi}{3}(t-1)\right) - 110 = 0.$$
One solution at $t = 4.2$.

9.19* Minimise $P_1(t) + P_2(t)$. One solution at $t = 2.78$.

Chapter 10

10.1 $N_0 = 17$, a = 3

10.3 $N_0 = 52$, a = 7

10.5 0.8^x, 1.25^x

10.7 $P(x) = 2^{x-1}, S(x) = 2^x - 1$

10.9 3^x

10.11 0.3^{2x}

10.13* $y = x + 1$

10.15 9

10.17 3

10.19 $2\ln 2 = 1.3862$

10.21 $2\ln 2 + \ln 3 = 2.4848$

10.23 $n = 21, 22, \ldots, 27$

10.25 $d_n = 2^{-n}d_0$. 11 iterations, 17 iterations

10.27 (i) 2.35 hours, (ii) 3.466 hours, (iii) 6.93 hours

10.29 6.93 hours

10.31 $N(t) = 17e^{1.1t}$, $T_2 = 0.63$ hours

10.33 $N(t) = 52e^{1.95t}$, $T_2 = 0.356$ hours

10.35* $N(t) = 0.0568e^{0.0060t}$ using data 1880 − 1960
$N(1970) = 7825, N(2000) = 9370$
$N(t) = 0.0036e^{0.0074t}$ using data 1780 − 1940
$N(1970) = 8406, N(2000) = 10507$

10.37* $\log V = k\log A + \log b$
$k = 1.2967, b = 40.79$

10.39* Best estimate is $v_{\max} = 45.3$. Then $\eta = 1.6312, K_m = 205.06$

10.43* Using *regula falsi*, $t^* = 4.391 \pm 0.002$

10.45* Taking $K = 16\,500$, the best fit has parameters
$r_m = 0.0098, N_0 = 6.53 \times 10^{-5}$, and the model then predicts $N(1970) = 7690$ and $N(2000) = 8895$

10.49* $k = -\ln|1 - p|$

Revision problems

R10.1 23%

R10.3 $0.989 \times 12 + 0.011 \times 13 = 12.011$

R10.5 1.210×10^{-4} per year, or 3.833×10^{-12} per second

R10.7 1 mg carbon contains 5.014×10^{19} atoms; 1 mg tissue contains 1.146×10^{19} atoms.

R10.9 Decay rate $= kN = (9.630 \times 10^{14}) \times (3.833 \times 10^{-12}) = 3691$ becquerels

R10.11 Taking $\dfrac{P_t}{P_0} = 0.25\%$ gives $t = 49\,500$ years old

R10.13 $t = 5247 \pm 36.6$ years old in 1995

R10.15 $t = 5249 \pm 115$ years old

R10.17 This would mean that Ötzi is 872 years younger than first calculated.

R10.19 I obtained $\tau = 2900$ BC

R10.21 I obtained $f(T) = 1.25T - 2500$
$$i(T) = T - 1950$$

The lines intersect when $T = 2200$ BP, giving $\tau = 250$ BC

R10.23 $h^{-1}(t) = e^{-kt}$

$$g^{-1}(T) = T + 60$$
$$f^{-1}(\tau) = 0.8\tau + 2000$$
$$\tau = 3800 \text{ BC} \rightarrow T = 5040 \text{ BP} \rightarrow t = 5100$$
years old $\rightarrow \dfrac{P_t}{P_0} = 53.95\%$

Part II

Chapter 11

11.1 $\dfrac{\Delta y}{\Delta x} = -0.05, \ -0.055, \ -0.0588, \ -0.0606$

11.3* From calculus, $f'(4) = -0.0625$ for 11.1, and $f'\left(\dfrac{\pi}{4}\right) = 0.7071$ for 11.2.

11.5 $\lim\limits_{\Delta x \to 0}\left(-\dfrac{1}{x(x + \Delta x)}\right) = -\dfrac{1}{x^2}$

11.7 $30x^9 + 25x^4$

11.9 $78x^{12} - 16x^7 - 15x^4$

11.11 $-\dfrac{5}{x^2} + \dfrac{12}{x^3}$

11.13 $\dfrac{1}{3\sqrt[3]{x^2}} - \dfrac{1}{2x\sqrt{x}}$

11.15 $35(5x - 2)^6$

11.17 $-\dfrac{12}{(2x - 1)^3}$

11.19 $15e^{3x} - 4e^{4x} - e^{-x}$

11.21 (a) $\Delta P = -1.855$ MPa

(b) $\Delta P = -185.5$ MPa

11.23 (a) $\Delta Y = 1.31$

(b) $\Delta Y = -2.19$

Chapter 12

12.1 $\dfrac{2}{\sqrt{3 + 4x}}$

12.3 $15(3x - 2)^4 + \dfrac{4}{(2x - 3)^3}$

12.5 $-2x(1 - x) + (1 - x)^2 = (1 - x)(1 - 3x)$

12.7 $(x^2 + 2x)e^x$

12.9 $\dfrac{1}{(1 - x)^2}$

12.11 $\dfrac{2e^x}{\left(1 - e^x\right)^2}$

12.13 $\dfrac{1 - 3x}{2\sqrt{1 - x}}$

12.15 $\dfrac{1}{(1 - x)^2}\sqrt{\dfrac{1 - x}{1 + x}}$

12.17 $x \cos x + \sin x$

12.19 $\dfrac{1}{x}\cos x - \dfrac{1}{x^2}\sin x$

12.21 $\dfrac{\sin x}{\cos^2 x}$

12.23 $\dfrac{dy}{dx} = 2\cos 2x = 2\left(\cos^2 x - \sin^2 x\right)$

12.25 $-\dfrac{x}{y}$

12.27 $\dfrac{2x}{1 + x^2}$

12.29 $-\dfrac{1}{x}$

12.31 $-\dfrac{1}{\sqrt{1 - x^2}}$

12.33 $\dfrac{x}{\sqrt{1 - x^2}} + \sin^{-1} x$

12.35 $\dfrac{dy}{dx} = 2x + 1, \quad \dfrac{d^2y}{dx^2} = 2$

12.37 $y^{(k)} = (x + k)e^x \Rightarrow y^{(k+1)} = (x + k + 1)e^x$

Chapter 13

13.1 Function positive for $-2 < x < 0$ and $x > 2$; Maximum at $\left(-\dfrac{2}{\sqrt{3}}, \dfrac{16}{3\sqrt{3}}\right)$; Point of inflection at $(0, 0)$; Minimum at $\left(\dfrac{2}{\sqrt{3}}, -\dfrac{16}{3\sqrt{3}}\right)$.

13.3 Roots at $x = \pm\sqrt{n\pi}$; Maximum at $\left(-\sqrt{\dfrac{\pi}{2}}, 1\right)$ and $\left(\sqrt{\dfrac{\pi}{2}}, 1\right)$; Minimum at $(0, 0)$.

13.5 $(-4, -27)$

13.7 $(2.171, -1)$

13.9 $a = \dfrac{1}{2}D, \quad A = \dfrac{1}{8}D^2$

13.11 $0.3023 I_0$

13.13 $\ln|1 + x| = x - \dfrac{1}{2}x^2 + \dfrac{1}{3}x^3 - \dfrac{1}{4}x^4 + \cdots$

13.15* $x_4 = 1.246$

13.17* $x_4 = 1.29586$

13.19* $x_4 = 0.669$

Chapter 14

14.1 $\dfrac{1}{2}x^6 - \dfrac{1}{3}x^3 + c$

14.3 $-\dfrac{3}{x} + \dfrac{7}{3x^3} + c$

14.5 $2\sqrt{x}(x - 7) + c$

14.7 $0.6e^{5x} + \dfrac{5}{3}e^{-3x} + c$

14.9 $\dfrac{1}{4\ln 3}.3^{4x} + c$

14.11 $x + 2\ln|x - 1| + c$

14.13 $\dfrac{1}{2}\sin^2 x + c$

14.15 $-\dfrac{1}{4}\cos^4 x + c$

14.17 $-\dfrac{1}{2}\cos(1 + x^2) + c$

14.19 $0.04(5x - 1)e^{5x} + c$

14.21 $\dfrac{1}{9}x^3(3\ln x - 1) + c$

14.23 $0.2\ln\left|\dfrac{x - 1}{x + 4}\right| + c$

Chapter 15

15.1 0.2

15.3 $\dfrac{1}{3}(e^3 - e^{-3}) = 6.6786$

15.5 0.2

15.7 $\dfrac{1}{6}(e^3 - e^{-3}) = 3.3393$

15.9

Average Speed	From $t = 0$ to $t = 2$	From $t = 2$ to $t = 4$
Lion	4.74	10.21
Gazelle	3.40	9.00

15.11 $\dfrac{2}{9}$

15.13 $\dfrac{2}{\pi}\left(1 - \dfrac{2}{\pi}\right) = 0.2313$

15.15 $\dfrac{1}{9}$

15.17 $\dfrac{2}{\pi}\left(1 - \dfrac{2}{\pi}\right) = 0.2313$

15.19 1

15.21 (i) 1.1167 (ii) 1.1032

15.23* $I = 7.2387$

Chapter 16

16.1 $y = 3x^2 - x$

16.3 $y = (1 + 2x)e^x$

16.5 $\dfrac{dy}{dx} = \dfrac{x(3 - y)}{y^2}$

16.7 $\dfrac{dy}{dx} = \dfrac{x(3y - 1)}{1 - 3x}$

16.9 $y = 3x^2 + 3\ln x + 2$

16.11 $y = \dfrac{1}{6}\left(e^{3x} - 9e^x + 3\right)$

16.13 $y = 7e^x - 2$

16.15 $y = \sin^{-1}(e^x)$

16.17 $y - Ae^{y - kx} = 1$

16.19 $y = Ax$

16.21 $\sin x \cos y = A$

16.23 $y = x\ln x + Cx$

16.25 $y = \dfrac{1}{4}e^{-x} + Ce^{-5x}$

16.27 $\dfrac{\partial f}{\partial x} = 5x^4 e^{2y} + 3, \quad \dfrac{\partial f}{\partial y} = 2x^5 e^{2y} - 4y^3$

16.29 $\dfrac{\partial f}{\partial r} = \dfrac{t^2}{(r + t)^2}, \quad \dfrac{\partial f}{\partial t} = \dfrac{r^2}{(r + t)^2}$

16.31 $\dfrac{\partial f}{\partial x} = 2\cos(2x + 3y),$

$\qquad \dfrac{\partial f}{\partial y} = 3\cos(2x + 3y)$

16.33 $(0.6, 0.88)$

16.35 $(-2, 3)$ and $(2, 3)$

Chapter 17

17.1 $y_1 = 2.5, y_2 = 2.7083$

17.3

x	y(true)	y(euler)	error	y(heun)	error
5.0	2.3094	2.5000	7.6%	2.3542	1.9%
7.0	2.4495	2.7083	9.6%	2.5054	2.2%

17.5 $y_1 = 1.121, \quad y_2 = 1.30205, \quad y_3 = 1.5792$

17.7* $y(0.3) = 1.5958$

17.9*

$X_1^e = X_0 + hf(t_0, X_0, Y_0)$

$Y_1^e = Y_0 + hg(t_0, X_0, Y_0)$

$t_1 = t_0 + h$

$X_1 = X_0 + \frac{1}{2}h\left[f(t_0, X_0, Y_0) + f(t_1, X_1^e, Y_1^e)\right]$

$Y_1 = Y_0 + \frac{1}{2}h\left[g(t_0, X_0, Y_0) + g(t_1, X_1^e, Y_1^e)\right]$

17.11 $y = Ae^{2x} + Be^{-x}$

17.13 $y = Ae^{2x} + Be^{-2x}$

17.15 $y = \frac{1}{3}\left(e^{2x} - e^{-x}\right)$

17.17 $y = \frac{1}{4}\left(e^{2x} + e^{-2x}\right)$

17.19 $\dfrac{dx}{dt} = y, \quad \dfrac{dy}{dt} = \sin t - t^2 x - ty$

Chapter 18

18.1 If $k < r$: unstable equilibrium at $x = 0$, and

stable equilibrium at $x = 1 - \dfrac{k}{r} > 0$;

If $k > r$: stable equilibrium at $x = 0$, and unstable

equilibrium at $x = 1 - \dfrac{k}{r} < 0$

18.3 R and B have dimension T^{-1}.

$r = \dfrac{RA}{B}, \quad k = \dfrac{K}{A}, \quad \dfrac{dx}{dt} = \dfrac{B}{A}\dfrac{dx}{d\tau}$

18.5* See solutions spreadsheet

Appendix: The Greek alphabet

The letters in bold type are those most commonly used in mathematics.

Capital letter	Lower-case letter	Greek name	English equivalent
A	**α**	alpha	a
B	**β**	beta	b
Γ	**γ**	gamma	g
Δ	**δ**	delta	d
E	**ε**	epsilon	e
Z	ζ	zeta	z
H	η	eta	h
Θ	**θ**	theta	th
I	ι	iota	i
K	κ	kappa	k
Λ	**λ**	lambda	l
M	**μ**	mu	m
N	ν	nu	n
Ξ	ξ	xi	x
O	o	omicron	o
Π	**π**	pi	p
P	ρ	rho	r
Σ	**σ**	sigma	s
T	**τ**	tau	t
Y	υ	upsilon	u
Φ	**φ**	phi	ph
X	χ	chi	ch
Ψ	**ψ**	psi	ps
Ω	**ω**	omega	o

Common uses

Roots and exponents: α, β, γ, λ, μ
Angles: θ, φ, ψ, ω
Time or temperature variable: τ, θ
Small positive quantity: δ, ε
Small distance in the x-direction: δ^x, Δ^x
Summation notation: Σ
pi = π = 3.14159257 . . .

References

Adams, D. *The Hitchhiker's Guide to the Galaxy*, London: Pan, 2009.

Adler, F.R. *Modeling the Dynamics of Life*, London: Brooks/Cole, 1997.

Alexander, R.M. *Optima for Animals*, London: Edward Arnold, 1982.

Allainé, D., Pontier, D., Gaillard, J.M., Lebreton, J.D., Trouvilliez, J., and Clobert, J. The relationship between fecundity and adult body weight in homeotherms, *Oecologia* 73 (1987): 478–480.

Anderson, L.G. *The Economics of Fisheries Management*, Baltimore MD: Johns Hopkins University Press, 1977.

Atkins, G.L. A simple digital-computer program for estimating the parameters of the Hill equation, *European Journal of Biochemistry* 33 (1973): 175–180.

Audera, C., Patulny, R.V., Sander, B.H., and Douglas, R.M. Mega-dose vitamin C in treatment of the common cold: a randomised controlled trial, *Medical Journal of Australia* 175 (2001): 359–362.

Britton, N.F. *Essential Mathematical Biology*, London: Springer, 2003.

Bryson, B. *A Short History of Nearly Everything*, London: Black Swan, 2004.

Calder, W.A. Ecological consequences of body size, in *Encyclopedia of Life Sciences*, Chichester: Wiley, 2001 (http://www.els.net).

Clark, C.W. *Mathematical Bioeconomics: Optimal Management of Renewable Resources*, 2nd edition, Chichester: Wiley, 2005.

Coffey, D.S. Self-organisation, complexity and chaos: the new biology for medicine, *Nature Medicine* 4 (1998): 882–885.

Davenport, H. *The Higher Arithmetic*, 8th edition, Cambridge: Cambridge University Press, 2008.

Eddison, J.C. *Quantitative Investigations in the Biosciences Using Minitab*, London: Chapman & Hall, 1999.

Edelstein-Keshet, L. *Mathematical Models in Biology*, Philadelphia PA: SIAM, 2005.

Elliott, J.P., McTaggart, C., and Holling, C.S. Prey capture by the African lion, *Canadian Journal of Zoology* 55 (1977): 1811–1828.

Glieck, J. *Chaos*, London: William Heinemann, 1988.

Gribbin, J. *Deep Simplicity*, London: Allen Lane, 2004.

Hagan, D.B. Method for enhancing human skin elasticity by applying octanoyl lactylic acid thereto, Patent 5427772, US Patent & Trademark Office, issued on 27 June 1995.

Hawkins, D. *Biomeasurement*, Oxford: Oxford University Press, 2005.

Hoppensteadt, F. Predator-prey model, *Scholarpedia*, 1(10) (2006): 1563 at http://www.scholarpedia.org/article/Predator-prey_model (accessed 7 July 2010).

Hoppensteadt, F.C. and Peskin, C.S. *Mathematics in Medicine & the Life Sciences*, London: Springer, 1992.

Jones, D.S. and Sleeman, B.D. *Differential Equations and Mathematical Biology*, London: Chapman & Hall, 2003.

Jones, V.P., Doerr, M.D., Brunner, J.F., Baker, C.C., Wilburn, T.D., and Nikolai, G.W. A synthesis of the temperature dependent development rate for the obliquebanded leafroller, *Choristoneura rosaceana*, *Journal of Insect Science* 5 (2005): 24–32.

Korn, H. and Faure, P. Is there chaos in the brain? II. Experimental evidence and related models, *Epilepsia* 44, suppl. 12 (2003): 72–83.

Kutschera, W. Radiocarbon dating of the Iceman Ötzi with accelerator mass spectrometry, in *Nuclear Science: Impact, Applications, Interactions*, NUPECC report, European Science Foundation, 2001.

Ludwig, D., Jones, D.D., and Holling, C.S. Qualitative analysis of insect outbreak systems: the spruce budworm and forest, *Journal of Animal Ecology* 47 (1978): 315–332.

May, R.M. Simple mathematical models with very complicated dynamics, *Nature* 261 (1976): 459–467.

Neal, D. *Introduction to Population Biology*, Cambridge: Cambridge University Press, 2004.

Owen-Smith, N., Mason, D.R., and Ogutu, J.O. Correlates of survival rates for 10 African ungulate populations: density, rainfall and predation, *Journal of Animal Ecology* 74 (2005): 774–788.

Peterson, R.O. *The Wolves of Isle Royale: a Broken Balance*, Ann Arbor MI: University of Michigan Press, 2007.

Poincaré, H. *Science and Method*, London: Thomas Nelson & Sons, 1914.

Prescott, S.A. and Chase, R. Sites of plasticity in the neural circuit mediating tentacle withdrawal in the snail *Helix aspersa*: implications for behavioral change and learning kinetics, *Learning and Memory* 6 (1999): 363–380.

Rothman, T. The short life of Evariste Galois, *Scientific American*, April 1982: 112–120.

Sanjeev, S. and Karpawich, P.P. Superior vena cava and innominate vein dimensions in growing children, *Pediatric Cardiology* 27 (2006): 414–419.

du Sautoy, M. *The Music of the Primes*, London: Harper, 2004.

du Sautoy, M. *Symmetry: a journey into the patterns of nature* (also published as *Finding moonshine*), London: Harper, 2009.

Singh, S. *The Code Book: The Secret History of Codes and Code-breaking*, London: Fourth Estate, 2002.

Sparre, P. Introduction to tropical fish stock assessment part 1. manual. *FAO fisheries technical paper* 306/1, 1989.

Stewart, I. *Why Beauty is Truth: a History of Symmetry*, New York: Basic Books, 2007.

Strogatz, S.H. *Nonlinear Dynamics and Chaos*, Reading MA: Addison-Wesley, 1994.

Traut, T. Enzyme activity: allosteric regulation, in *Encyclopedia of Life Sciences*, Chichester: Wiley, 2007 (http://www.els.net).

Wheldon, T.E. *Mathematical Models in Cancer Research*, Bristol: Hilger, 1988.

Wodarz, D. and Komarova, N.L. *Computational Biology of Cancer*, Singapore: World Scientific, 2005.

Index

#DIV/0! in Excel 29
#NUM! in Excel 32

A

Abel, Niels 332
absolute error 65
absolute value 13, 25, 128
acceleration 51
accuracy 45, 64
acute angles 265
algebraic equations 482
algebraic expression 5
algebraic long division 201
allometry 94, 303
allosteric regulation 235, 305
amplitude 260,272
angiogenesis, see: tumours
angstrom unit 56
anti-derivative 426
arguments of a function 29
arithmetic expression 5
arithmetic operations 9, 143
asymmetric key encryption 44
asymptotes 166, 221, 225
atomic mass 110
average 260
Avogadro number 111
axioms 133

B

Babylonians 331
base 19, 291
becquerel (SI unit), 51
BEDMAS 27
bending moment 407
bifurcation 556
bifurcation diagram 557
binomial theorem 68
birth rate 30, 35
bisection, method of 241, 322
body density 34
Body Fat percentage 34

boundary conditions 482
bracketing methods 239
brackets (use of), 16
Butterfly Effect 537, 559

C

calibration curve, 315
calorie 51
Cardano, Girolamo 331
carrying capacity 207
Cartesian coordinate system 85
cell doubling time 302
cellular learning 309, 359, 372
chaos 557
Chart (in Excel) 86
chemotherapy 460
chess 41, 321, 322
cholesterol level 113, 193, 314
closed interval 33
cobwebbing 102, 208, 551
coefficient 10
coefficient of determination 166,
    229
common logarithms 300
completing the square 186, 253
complex systems 537
concave up/down 177, 390, 395
concentration of a solution 109
constant of integration 428
cosecant 280
cosine 264, 268
cosine function (general) 273
cotangent 280
critical points 393
cube roots, see: roots
curie (unit of radioactivity) 51
curvature 395, 398
curve fitting 227, 302
cycle 259

D

data plots 82

death (or mortality) rate 30, 35
decay constant 307, 313
definite integral 453
    Addition of Areas Rule 457
    as limit of a sum 458
    Interchange of Limits
        Rule 457
degree Celsius 55
degree of a polynomial, see:
    polynomial, degree
degrees (angles in) 262
delta notation 71
denominator 11
density 51
density-dependent growth, see:
    growth models
derivative 340, 344
    function of a function 355,
        367
    higher-order 383
    notation 351
    of exponential functions 357
    of inverse trigonometric
        functions 382
    of linear functions 351
    of logarithmic functions 381
    of polynomial functions 352
    of power functions 347
    of reciprocals 354
    of roots 354
    of trigonometric
        functions 376
    partial 386
differences (in interpolation) 98,
    210
differential calculus 338
differential equations 4, 341
    auxiliary equation 529
    boundary conditions 482
    for exponential growth 363,
        439
    general solution 483
    homogeneous 497

Initial Value Problem 483
integrating factor 499
linear 499, 529
Midpoint Method 535
numerical methods 509
order 482
ordinary differential
    equation 484
partial differential
    equation 484, 532
particular solution 483
solution by change of
    variable 496
solution by separation of
    variables 484, 533
systems 519, 531
differentiation 340
    from first principles 348
    Chain Rule 355, 367
    Coefficient Rule 353
    implicit 379
    partial 500
    Product Rule 370
    Quotient Rule 372
    Sum Rule 353
diffusion 532
Dilution Factor 118
dilutions 116
dimensional analysis 58, 561
dimensionless quantities 59
dimensions 51
discontinuities 366
discriminant 187
distances on the real line 13
Distributive Law 18
division by zero 13
double pendulum 558
double precision 50
double root 392
double-angle formulae 283
drug administration 143, 310,
    460
dynamical systems 537

E

Eadie–Hofstee plot 235, 257
EasyTide 276
endpoints of an interval 33
energy 51
enzyme kinetics 231, 305, 375
epidemics 526
epilepsy 560
equations 15, 26
    linear 134
    of a circle 135, 266
    simultaneous 146, 192, 315
    theory 133

equations of motion 159, 189,
    223, 309, 354, 431
equilibrium point 542
equilibrium solutions of
    period 2 556
equilibrium state 343, 521
errors 64
error analysis 64, 234, 257, 361,
    503, 559
Euler constant 294
Euler's method 510
exponential constant 294
exponential decay 306, 313, 317
exponential growth, see: growth
    models
exponents 19, 68
extrapolation 106, 194, 303

F

factor (real numbers) 10
factorial notation 384
factorization 22, 38, 197
Fibonacci sequence 191, 404
fisheries management 205, 250,
    397, 406,494, 518
fixed points 549
fixed-point notation, 46
flip (transformation), see: linear
    transformations
floating point notation 49, 64
FNGraph 152
force 51
forcing function 531
formulas in Excel 28
fractions 11, 27, 37
fuel consumption 43
functions 125
    arguments 126, 142
    arithmetic operations 143
    average value 460
    composition 145
    constant 155
    continuous 366
    decreasing 392, 398
    differentiable 367
    domain 126
    exponential decay 306
    exponential growth 291
    identity 156
    in Excel 29, 127, 131
    increasing 392, 398
    inverse 237
    inverse trigonometric 280
    limits 166
    linear 159, 164, 390
    logarithmic 297
    logistic 204

negative 391, 398
periodic 259
polynomial 184, 192
polynomial
    approximations 413
positive 391, 398
power 198
power series expansion 417
proportionality 128, 141, 157
quadratic 129, 176, 181, 230,
    390
range 126
rational 221, 224, 230, 231,
    238
reciprocal 220
reciprocal trigonometric 280
roots 131, 161, 392, 398
sawtooth 259
sigmoid 312, 402
sinusoidal 262
smooth 385
tangent 284
turning points 131, 177

G

Galois, Evariste 333
gas constant 78
geometric growth, see: growth
    models
geometric series 198
Golden Ratio, Golden
    Section 215, 248
Golden Section search 249, 322
Goldie–Coldman model 520
Gompertz model of population
    mortality 319
Gompertz model of tumour
    growth 490, 497
goodness of fit 165
gradient 284, 391
gram 50
graphs 226
    in Excel 86
    straight-line 85, 141
group theory 333
growth constant 295, 363
growth models 4
    density-dependent 207
    exponential 291, 303, 341,
        363, 439
    geometric 4, 22, 72, 191, 290
    Gompertz 490, 497
    logistic 204, 312, 341, 492,
        546
    Malthusian 296
    von Bertalanffy 488
Gulland–Holt plot 488

H

habituating sensitisation,
    see: cellular learning
habituation, see: cellular
    learning
half-life 307, 313
Hanes–Woolf plot 235, 257
harvesting model 494, 518, 547
hectare (unit of area), 56
Heun's method 515
Hill coefficient 236
homeopathy 120
hyperbola 221
hypotenuse 264

I

ideal gas law 78
identities, see also: trigonometric
    identities 133
implicit differentiation 379
improper integrals 468
Improved Euler method 515
indefinite integrals 453
inequality 33, 38
infinity 166
initial conditions 341, 343, 483
instantaneous rate of change 338
integer 7
integral 427
    as area under the curve 451
    improper 468
    of inverse trigonometric
        functions 447
    of polynomial functions 430
    of power functions 428
    of standard functions 432
    of trigonometric
        functions 444
    table of standard
        integrals 449
integral calculus 340
integrand 427
integrating factor 499
integration 341
    by partial fractions 441, 467
    by parts 438, 466
    by substitution 435, 465
    Coefficient Rule 428
    definition 427
    numerical 470
    Sum Rule 430
    using power series
        approximations 448
intermittency 539
interpolating factor 98

interpolation 96, 194
    inverse 97, 107
    linear 96
    piecewise linear 100
    quadratic 210
    using Excel 101
interval on the real line 32
inverse problems 362, 426
inverse trigonometric
    functions 282
irrational number 11, 25
isotope 311
iterative methods 239, 419, 509

J

joule (SI unit) 51

K

kelvin (SI unit) 51
Kermack–McKendrick
    model 526
kilogram 50

L

Lagrange, Joseph–Louis 332
laminar flow 462, 501
least-squares 252
limiting process defining the
    derivative 344
limits 166
limits of integration 454
linear approximation of a
    function 414
linear convergence 422
linear regression, see: trend line,
linear relationship 91
linear transformations 159, 169,
    261, 332, 356
Lineweaver–Burk
    transformation 233, 257
litre 55
logarithms 296, 298, 301
logistic function 204
logistic growth, see: growth
    models
log-log plot 304
Lorenz equations 538
Lotka–Volterra models 520
lowest common denominator 12
lowest terms (fractions), 12

M

Maclaurin series 417, 448
Malthusian growth, see: growth
    models

mantissa 46
mass concentration 51, 56
maximum, see: turning-point
May, Robert 553, 560
metre 50
Michaelis–Menten equation, see:
    enzyme kinetics
Midpoint Method 535
midpoint of an interval 33
MindGym exercise 17, 40, 62
minimization, see: optimization
minimum, see: turning-point
Modified Euler method 515
molar absorbance 147
molar mass 111
molarity 110, 112
    calculations 113
    Excel spreadsheets 121
mole (SI unit), 110
multiple root 398

N

nth root, see: roots
nth power, 19
natural logarithms, see:
    logarithms
natural numbers 7
Navier-Stokes equations 501
negation of a real number 9
negative real numbers 8
net growth rate 30
newton (SI unit) 51
Newton's law of motion 52
Newton's law of cooling 488, 543
Newton's method for
    optimization 423
Newton–Raphson method 420
nonlinearity 539
numerator 11
numerical expression 5
    in Excel 28
numerical integration 470

O

objective function 403
Omar Khayyam 331
open interval 33
optimization 246, 403, 423, 504
orbit diagram 557
order 8
    of a differential equation 482
    of operations 18, 27
ordinary differential
    equation 484
Ötzi the Iceman 314, 507
outliers 165

P

parabola 177
partial derivatives 386
partial differential
    equations 484, 532
partial differentiation 500
pascal (SI unit) 51
percentage 14
period 259, 273
phase 262
pi (π), value of 266
Poincaré, Henri 560
point of inflection 347, 396, 398,
    402
Poisseuille equation 501
polynomials 22, 40, 331
    degree 198
    leading term 223
population biology 4
positive real numbers 8
power series 417, 448
precision 45
predator–prey dynamics 5, 274,
    520
pressure 51
principal value 282
product 10, 22
proportion, direct 89, 141
proportion, inverse 93, 141, 228
proportionality 128
protein folding 404
Pythagoras' Theorem 264

Q

quadratic approximation of a
    function 416
quadratic convergence 422

R

radian 265
radioactive decay 307, 313, 317
radioactivity 51
radiocarbon dating 311, 365, 507
rate of change 4, 287, 337
rational 11, 221, 224
Rayleigh number 539
real line 8
real numbers 7, 8
reciprocal 10, 220
regression line, see: trend line
regula falsi 241, 322
related rates 411
relative error 65
Ricker model 397
Riemann integral 458

rollercoaster illustration 286,
    347
roots 23, 37, 161
    double 187
    of a quadratic 187
    of polynomials 331
    root-finding algorithms 241
rounding error 64
rounding numbers 60
r-squared value, see: coefficient
    of determination
rules of algebra 37
Runge–Kutta method 519

S

scientific notation 46, 74
    addition and subtraction 48,
    65
    multiplication and
    division 49, 66
scrollbar (in Excel) 175
secant 280
secant line 287
second derivative 383
semi-log plot 303
semi-width of an interval 33
sensitization, see: cellular
    learning
separation of variables 484, 489
sequences, see also: Fibonacci
    sequences 168
serial dilutions 119
shift (transformation), see: linear
    transformations
SI units 50, 74
significant figures 61, 75
simple harmonic motion 530
simple root 392, 398
Simpson's Rule 473
simultaneous equations, see:
    equations
sine 264, 268
sine function (general), 274
single precision, 50
Siri formula for body
    density 34
slider control (in Excel) 175
slope 91
small changes 361, 503
snowflake 560
specific weight 51
speed 51, 54
square root, see: roots
squash (transformation), see:
    linear transformations
stability of equilibria 545, 549

stable equilibrium 543, 550
statistics 81
steplength 510
stochastic model 317, 327
straight-line graphs, see also:
    graphs 159
stretch (transformation), see:
    linear transformations
surds 25

T

tangent (of an angle) 264
tangent line 286, 394
Taylor series 418
techniques of integration, see:
    integration
temperature 50, 53
terminating decimal 11
tidal data 276
transformation, see: linear
    transformations
Trapezium Rule 470
trend line 88, 107, 165, 229, 251,
    505
Trichotomy of Cardinals 8
trigonometric identities 270,
    272, 283
trigonometry 262
truncation error 541
tumours 4, 6, 342, 519
turbulent flow 559
turning-point 393, 398

U

uncertainty intervals 62
universality 562
Universe 76
unstable equilibrium 518, 543,
    550
update equation 30, 290, 397,
    492, 549
upper bound 71

V

van der Waal's equation 79
variables 5
    dependent and
    independent 82, 126
    physical 45
velocity 51, 54
von Bertalanffy Growth Model,
    see: growth models

Z

Zeno's paradox 217